Metal-Ligand Interactions:
From Atoms, to Clusters, to Surfaces

NATO ASI Series

Advanced Science Institutes Series

A Series presenting the results of activities sponsored by the NATO Science Committee, which aims at the dissemination of advanced scientific and technological knowledge, with a view to strengthening links between scientific communities.

The Series is published by an international board of publishers in conjunction with the NATO Scientific Affairs Division

A **Life Sciences**
B **Physics**

Plenum Publishing Corporation
London and New York

C **Mathematical**
   and Physical Sciences
D **Behavioural and Social Sciences**
E **Applied Sciences**

Kluwer Academic Publishers
Dordrecht, Boston and London

F **Computer and Systems Sciences**
G **Ecological Sciences**
H **Cell Biology**
I **Global Environmental Change**

Springer-Verlag
Berlin, Heidelberg, New York, London,
Paris and Tokyo

NATO-PCO-DATA BASE

The electronic index to the NATO ASI Series provides full bibliographical references (with keywords and/or abstracts) to more than 30000 contributions from international scientists published in all sections of the NATO ASI Series.
Access to the NATO-PCO-DATA BASE is possible in two ways:

– via online FILE 128 (NATO-PCO-DATA BASE) hosted by ESRIN,
Via Galileo Galilei, I-00044 Frascati, Italy.

– via CD-ROM "NATO-PCO-DATA BASE" with user-friendly retrieval software in English, French and German (© WTV GmbH and DATAWARE Technologies Inc. 1989).

The CD-ROM can be ordered through any member of the Board of Publishers or through NATO-PCO, Overijse, Belgium.

Metal-Ligand Interactions: From Atoms, to Clusters, to Surfaces

edited by

Dennis R. Salahub

Département de chimie, Université de Montréal,
Montréal, Québec, Canada

and

Nino Russo

Dipartimento di chimica, Università della Calabria,
Arcavacata di Rende, Cosenza, Italy

Springer-Science+Business Media, B.V.

Proceedings of the NATO Advanced Study Institute on
Metal-Ligand Interactions: From Atoms, To Clusters, To Surfaces
Cetraro, Italy
June 10–21, 1991

Library of Congress Cataloging-in-Publication Data

```
Metal-ligand interactions : from atoms, to clusters, to surfaces /
  edited by Dennis R. Salahub and Nino Russo.
      p.   cm. -- (NATO ASI series. Series C, Mathematical and
  physical sciences ; vol. 378)
    Papers from an institute held in June 1991 near Cetraro, Italy.
    Includes index.
    ISBN 0-7923-1930-3 (HB : alk. paper)
    1. Transition metal complexes--Congresses.  2. Ligands-
  -Congresses.   I. Salahub, Dennis R., 1946-   . II. Russo, Nino.
  III. Series: NATO ASI series.  Series C, Mathematical and physical
  sciences ; no. 378.
  QD474.M45  1992
  546'.6--dc20                                             92-27726
```

ISBN 978-94-010-5254-2   ISBN 978-94-011-2822-3 (eBook)

DOI 10.1007/978-94-011-2822-3

CONTENTS

PREFACE

Understanding the interactions of atoms and molecules with transition metals at various levels of aggregation represents a challenge of the first magnitude. Metal-ligand interactions are at the heart of several disciplines and the points of view of several fields are helping to reveal their nature. Recent progress in many of these fields has been substantial, outstanding in some instances. The inorganic and organometallic chemists have discovered whole new classes of compounds and of reactions; the surface scientists have invented an arsenal of spectroscopic and other probes; catalysis is able to ascertain mechanistic detail at more and more microscopic levels; and the theorists are inventing and improving techniques to treat the energetics and dynamics of more and more complex systems. This multidisciplinarity, as usual, leads to both great opportunities for cross-fertilization and also to great problems in establishing a common language and sharing experience. The NATO ASI on which this book is based was organized to promote interactions across these fields, to expose young researchers at an early stage of their careers to the several conceptual frameworks that currently contribute to the area. We felt that the time was ripe for such interactions; the tools, both theoretical and experimental, are in an advanced stage of development, and fundamental questions remain without answers. The most fundamental of these concerns the nature of the microscopic interactions between metal atoms (clusters, surfaces) and ligands (atoms, molecules, adsorbates, reagents, intermediates, products) and the changes in these interactions during physical and chemical transformations.

Therefore, in June, 1991, about one hundred scientists gathered at the Grand Hotel San Michele, on the Calabrian coast near Cetraro, for a two-week Advanced Study Institute dedicated to the theme of metal-ligand interactions. Leading experts were invited to present series of lectures on *ab initio* theory, on semi-empirical theory, on density functional theory, on experimental studies on complexes and clusters, on surfaces, on catalysis and, most importantly, on the overlaps and potential synergies among these fields. The results of these two weeks, inspired by the Calabrian sunshine and hospitality, are presented here. The lecturers were asked to refine their chapters in light of what they learned at the ASI and also to include overview material. They were also asked to make a special effort to bring out the cross-links between the various communities represented. We believe they have done an exceptional job!

A summer school such as this involves considerable effort on the part of many people. On the Organizing Committee we were joined by Professors Gary Haller and Martin Moskovits who made many valuable suggestions that were incorporated into the proposal. This was received by the NATO Science Committee and we thank NATO and the committee both for funding the ASI and for their helpful interventions. Professor Brian Johnson attended the ASI as the representative of NATO; we could not have chosen a better participant ourselves!

The main lecturers were Jan Andzelm, Alain Dedieu, Tom Ellis, Ha-Jo Freund, Annick Goursot, Mike Hall, Gary Haller, Martin Moskovits, Hiroshi Nakatsuji, Albert Renouprez, Steve Riley, Nino Russo, Dennis Salahub, Bill Trogler, Ulf Wahlgren, Mike Zerner, and Tom Ziegler. Most of the lecturers have generously contributed to this book. We are particularly grateful to Albert Renouprez for preparing his lectures on very short notice when a compatriot failed to show up. To complement the main lectures, late-breaking news talks, poster sessions and a very successful series of Special Research Seminars were organized, to allow more specialized presentations of recent results. In this way, the students received not only the "core" material, but also a snapshot of current activity. We thank all of the "SRS's": Alan Balch, Ivano Bertini, Louis Farrugia, Nick Hadjiliadis, Julius Jellinek, Brian Johnson, Francesco Lelj, Giuliano Longoni, Gianfranco Pacchioni, Theo Theophanides (also a memorable translator at the post-banquet show) and Malgorzata Witko.

Of course, the ASI would not have been possible without the financial aid of NATO and also of the Italian Research Council(CNR). We are also grateful for very important contributions from the Università della Calabria, from the Dipartimento di Chimica, UNICAL, from the Fondazione Catizone, from Convex Italia, Digital Italia, Floating Point Systems, IBM Semea s.r.l, Silicon Graphics and Biosym Technologies. Travel grants for young U.S. scientists were generously provided by the National Science Foundation.

The local arrangements were handled with the cheerful aid of Antonio DaMunno, Mauro Ghedini, Francesco Neve and Marirosa Toscano. The champion chauffeur was Vincenzo Musolino. The Director of the Grand Hotel San Michele, Mr. Giovanni Marra, and his staff bent over backwards to make us comfortable.

One of the best decisions we made in organizing this school was to bring Michelle Piché from Montréal to Cetraro. Her cheerful hard work before, during, and after the ASI is gratefully acknowledged. In addition to her organizational and secretarial duties, Michelle handled the not inconsiderable amount of cold, hard, cash that was needed to reimburse expenses, with only the slightest outward signs of trembling. She has also handled the manuscripts for this book, reformatting most of the chapters, and skillfully catching slips of the authors' word processors. Merci Michelle!!

Dennis R. Salahub
Nino Russo
Arcavacata di Rende, CS, Italy
May, 1992

LIGANDS ON CLUSTERS · ADSORBATES ON SURFACES

M. MOSKOVITS
University of Toronto
Department of Chemistry
80 St. George Street
Toronto, Ontario, M5S 1A1
Canada

ABSTRACT. The similarities and differences between molecules adsorbed on surfaces and ligands bonded to metal clusters are considered. Although important insights obtained in one of these fields can often be applied to the other, there are several properties of surfaces which cannot be understood easily in terms of cluster analogues. Contrarywise, small metal clusters manifest a range of unique properties varying with cluster size that have no easy counterparts in surface science.

1. Introduction

There are three branches of physics and chemistry that seek answers to closely related questions but from, at times, disparate points of view and employing very different language. These are the disciplines of cluster chemistry and surface chemistry. The reason that I refer to them as three rather than two disciplines is that cluster chemistry has taken two paths. The more traditional one seeks to synthesize soluble metal clusters stabilized by ligands. Hundreds of such compounds have been produced over the past two decades largely through the work of Chini [1], Johnson, Stone, Lewis and others [2]. Currently these compounds include such beautiful structures as that of $[Rh_{15}(CO)_{28}C_2]^-$ [3]. Larger cluster compounds are continuously being synthesized so that the field of cluster synthesis is merging seamlessly with that of ligand-stabilized colloid chemistry [4-6] where the goal is to stabilize colloidal metal particles against aggregation and flocculation by attaching ligands, usually long-chain species, to the surface that then keep themselves apart by repulsive steric interactions. Currently, species that are intermediate between true compounds and colloids have been prepared with metal cores that are usually closed shell cub-octahedral crystallites. The number of metal atoms, N_M, in such structures form the series: 13, 55, 147, 309, 561, 923... or in general

$$N_M = 1 + \sum_{N=1}^{M} (10n^2 + 2)$$

D. R. Salahub and N. Russo (eds.), Metal-Ligand Interactions: from Atoms, to Clusters, to Surfaces, 1–15.
© *1992 Kluwer Academic Publishers. Printed in the Netherlands.*

Schmid et al. began this line of research with the report of the compound $Au_{55}[P(C_6H_5)_3]_{12}Cl_6$ [4]. Compounds such as $Pt_{309}(Phen)_{36}O_{30\pm10}$ [5] and the gigantic $Pd_{(570\pm30)}L_{60\pm3}(OAc^-)_{180\pm10}$ [6] (L=phen or bipy) have been reported.

The other branch of cluster chemistry arose through the recent work with metal clusters generated in supersonic nozzle beams. Here the clusters are pre-formed and ligands are allowed to interact with them after the cluster's formation, if at all. Examples of these are the studies of H_2 uptake by iron clusters [7], the large number of reactions carried out on metal clusters by the Exxon group [8] designed to mimic known catalytic processes, and the water and ammonia uptake reactions reported by the Argonne group [9]. The major difference in perspective held by the two communities (those who study clusters that one can put in a bottle, and those who study clusters in beams) comes about from the fact that the former focuses on the synthesis of cluster compounds and the rationalization of their stabilities, and especially the stable metal core geometries and the number and disposition of ligands about them in terms of bonding and electron counting rules familiar to organo-metallic chemists while the other group uses their clusters as microscopic (or not so microscopic) models for adsorption processes at surfaces, or as in the case of the Argonne group, as probes to the geometrical and electronic properties of the cluster itself.

There is, of course, a deeper difference between the two types of experiments that ultimately is related to the temperature of the systems involved. The former group (the organometallists) produce stable compounds whose stability includes the global electronic "needs" of both the metal atoms and the ligands. The latter work with cold beams. They, therefore, form clusters that may not represent the lowest energy structures for any given cluster size. Even when they take pains to ensure that the lowest energy structures are achieved, the ligands are then added after the fact. It is, therefore, not clear that the clusters with their post-attached ligands have sufficient thermal energy to surmount the energy barriers that separate the structure that is formed directly, from the lowest energy form possible with that particular number of metals atoms and ligands. Indeed the work of Riley and coworkers assumes that the cluster geometry is unchanged when ligands are "adsorbed" on their surface so that the ligands become faithful probes of the sites characteristic of the bare metal cluster [9].

Workers in this latter group understand this problem and indeed, dynamical processes and ligand-directed structural modifications of clusters have become an area of focus [10], nevertheless, the fact that one group tries to deal with stable clusters while the other normally deals with entities that are unstable, at the very least unstable towards further aggregation, distinguishes them as separate groups.

This distinction may also determine the fact that many promising cluster compounds generally act as very poor catalysts in reactions that involve its ligands. Several hydrogen-containing metal cluster carbonyls have been synthesized, for example $[H_3Rh_{13}(CO)_{24}]^{-2}$ [11]. None of them have shown good catalytic activity in CO hydrogenation reactions such as Fischer-Tropsch synthesis despite the fact that Ru is a good heterogeneous catalyst in that reaction. The reason for this puzzling observation may lie in the fact that heterogeneous catalysts are prepared by reducing metal from a suitable compound onto an appropriate support (normally an oxide). The structure of the metal clusters so produced are those that minimize the free energy of the bare metal as modified by the interactions with the support. When the synthesis gas interacts with these clusters the metal-ligand structures formed do not normally represent the lowest energy forms of those systems. This is because, in general, there is not sufficient thermal energy available to convert the cluster-adsorbate systems to their most stable forms. (This is not to say that there is no structural modification taking place, only that, in general, the modification can not proceed to its final lowest-energy form). One way of thinking about this process is to consider the difference in the energy between the form that is thermally accessible to the pre-formed cluster after an adsorbate binds to it and that of the system's lowest energy form as strain energy. (The strain induced in the cluster by the adsorbate as the system attempts to restructure towards its preferred structure). This allows primary adsorbates, such as CO, to react more easily with co-adsorbates such as hydrogen. On stable hydrido carbonyl clusters, on the other hand, the system is either already in its lowest energy form or in a low-lying minimum separated from an even lower minimum by large barriers. Either way, reaction, i. e. catalysis, is greatly impeded.

The stability of "bottled" clusters, i. e. the ability to obtain sizable samples in forms that are amenable to routine chemical and structural analysis through, e. g. crystallography, NMR, and other spectroscopic techniques, changes the challenge and the focus of the science in the two groups. The structures of the stable clusters and the disposition of the ligands about them are generally known. Hence, the challenge is to rationalize the structures, to understand their reactivity, and to produce cluster compounds with useful homogeneous catalytic properties. Stability, however, denies one the capability of producing cluster complexes of any desired nuclearity since not all cluster sizes will be stable. In particular it is difficult to produce a series of stable cluster compounds with metal nuclearity increasing continuously by a single metal atom over a large range.

In contrast, the structure of the unstable clusters is generally unknown. Hence structure determination is an important, perhaps **the** important issue here. Similarly, the mode of interaction of adsorbate or ligands with these clusters is unknown and must be determined. These have proved to be difficult problems. By contrast, one has the ability, in this branch of cluster science to generate clusters of continuous nuclearity allowing, thereby, the systematic investigation of physical and chemical properties as a continuous function of cluster size. These include characteristics such as the ionization potential, the electron affinity, the polarizability, the magnetic moment, the rate of hydrogen uptake, etc. eventually determining, in this way, how the analogous properties of the bulk are achieved [12].

The problems and concerns of the third group, surface scientists, are more closely allied with those of the unstable clusters, and if anything, are at an even more primitive level of understanding. Within the context of metal-adsorbate interactions, this group also attempts to determine the structure of the surface and the disposition and orientation of adsorbates upon it. It is also interested in determining what reactions are possible on the surface [13]. All three groups are deeply concerned about detailed reaction mechanisms.

2. Cluster-Surface Similarities

There is a formal similarity between the adsorbates on surfaces and ligands bonded to clusters. These have been wonderfully expressed in the pioneering work of Earl Muetterties [14]. Indeed, the language of organometallic chemistry is normally used to describe surface chemistry, at least in broad terms. For example, the bonding of CO to a metal surface is generally discussed in the σ-donor-π-acceptor language of the Dewar-Chatt-Duncanson model originally formulated for carbonyl bonding.

Familiar species and processes on surfaces have their cluster counterparts. Subsurface and interstitial atoms, well established phenomena in surface science have cluster analogues. For example, (Figure 1) $Fe_4(CO)_{13}C$ and $Ru_5(CO)_{15}C$ [15] have carbon atoms bonded coplanar with the metal "surface" while the molecule $Os_5(CO)_{16}C$ and $Ru_6(CO)_{17}C$ encapsulate "interstitial" carbon atoms. Likewise the reaction pathways on

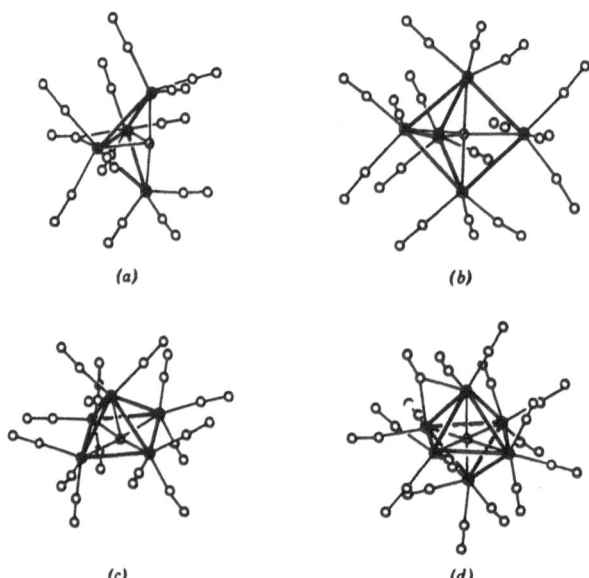

Figure 1. Some examples of carbido clusters of the iron triad. (a) [Fe$_4$(CO)$_{13}$C]; (b) [Os$_5$(CO)$_{16}$C]; (c) [Ru$_5$(CO)$_{15}$C]. (d) [Ru$_5$(CO)$_{17}$C]. Ø = carbido, C atom.

surfaces and on clusters often follow very similar mechanisms. For example the details of the hydrogenation of ethylene on platinum surfaces is very similar to the way in which Wilkinson's catalyst hydrogenates the same alkene [16].

Attempts to rationalize the structures and the number of ligands borne by stable clusters also reveal an incipient bulk metal behavior that is best understood in terms of the condensed matter concepts that are more traditional among surface scientists. So, while the very organometallic 18-electron rule often works for some clusters [15] for others it fails, while an effective atomic number rule that treats all electrons as rather unlocalized, a little like in bulk metals, seems to work. For example the cluster compounds Os$_3$(CO)$_{12}$ and H$_2$Os$_3$(CO)$_{10}$ can be rationalized in terms of the 18-electron rule. (For the former, each metal contributes 8 electrons, 2 electrons each are contributed by the 4 CO's bonded to each metal and 1 electron is contributed by each of the two metal-metal bonds shared by each atom for a total of 18 electrons around each atom). In the second one must postulate a double bond between two of the osmium atoms, in order to satisfy the 18-electron rule for each of the three osmium atoms (Figure 2).

A molecule like Os$_5$(CO)$_{16}$, on the other hand, cannot be understood in terms of the 18-electron rule. However, if one adopts the view that the cluster as a whole rather than each metal atom must attain a closed shell when the total number of valence electrons is a multiple of 18 equal to the number of metal atoms, then the structure can be rationalized. This is an example of the application of the effective electron number rule [17]. Applying it to this particular molecule, 90 electrons are required. Each of the osmium atoms contribute 8 and each of the CO's 2 for a total of 72 electrons. Therefore, there are (90-72)/2=9 metal-metal bonds needed to complete the electron requirements of the cluster

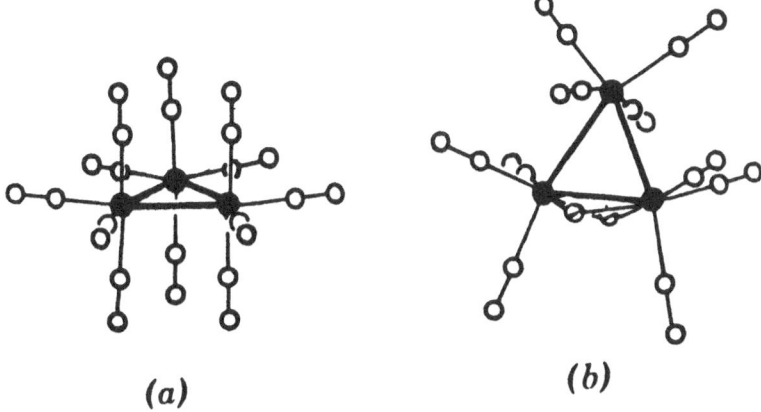

Figure 2. (a) [Os$_3$(CO$_{12}$]; (b) [H$_2$Os$_3$(CO)$_{10}$], Ø = bridging hydride ligand.

cluster complex. This suggests the trigonal bipyramidal structure that is, in fact, adopted.

There is another set of rules that are uniquely applicable to clusters without a straight-forward bulk analogue. This rationalization, based on the rules constructed to understand the structures of the boranes, are often referred to as Wade's rules [18]. They are applied to explain the observation that the connectivity of metal cluster complexes is often too great to be understood in terms of two electrons per bond, requiring instead the participation of multi-center bonding.

The metal atom arrangement in many cluster complexes is the familiar close-packed structure so common in bulk metals. Moreover, the ratio of carbonyl ligands to surface atoms in a series of carbonyl clusters decreases from numbers in the neighborhood of 5 or 6 for mononuclear carbonyls to values near unity, close to the adsorbate/atom ratio often encountered on metal surfaces as the cluster size increases (table 1).

TABLE 1. The CO/metal ratio in a series of transition metal cluster carbonyls

Carbonyl Cluster	CO/metal ratio
Fe(CO)$_5$	5
Rh$_3$(CO)$_{12}$	4
Fe$_3$(CO)$_{13}$C	3.2
[Rh$_6$(CO)$_{15}$C]$^{-2}$	2.5
[Rh$_{10}$(CO)$_{24}$C$_2$]$^{-2}$	2.4
[Rh$_{15}$(CO)$_{28}$C$_2$]$^-$	1.9
[Pt$_{26}$(CO)$_{32}$]$^{-2}$	1.2
metal surface	~1

Not all metal-cluster complexes mimic miniature versions of bulk crystallites covered by adsorbate. A series of Pt clusters with formulas of the form $[Pt_3(CO)_6]_n^{2-}$, for example, take the shape of ladders with triangular platinum rungs, a shape that does not bring to mind any bulk analogue.

3. Cluster - Surface Differences

Clearly there is much to be learned from the analogy between metal-ligand chemistry and the chemical processes that go on at metal surfaces. There are, however, some properties of surfaces, normally physical properties, that impinge of the interpretation of spectroscopic information obtained with the large number of diagnostic probes of the chemistry occurring at surfaces, that do not have direct or facile analogues among the properties of small clusters. Often these physical effects are naively interpreted as indicative of chemical processes such as bond weakening or reaction that are, in fact, not taking place. A few examples of phenomena characteristic of the bulk state follow. Of course as a cluster becomes larger one expects its characteristics to resemble, progressively, those of the bulk surface.

3.1. SURFACE SELECTION RULES

Vibrational spectra are often modified on metal surfaces entirely as a result of the proximity of the surface. The now-famous "surface selection rules" [19] that are, in fact, modifications of the relative intensities of vibrational bands arising from the coherent superposition of the direct and the reflected electromagnetic field at the surface, are examples of surface-specific phenomenon. The most familiar form of these applies to the infrared where the tangential component of the electric field is almost zero, making only dipole-allowed vibrations with non-zero surface-normal components detectable in the spectrum. Analogous rules have been worked out for Raman and other surface spectroscopies [20]. It is important to realize that these surface selection rules are functions of the wavelength characterizing the spectroscopy. The surface-normal component is no longer the dominant one when exciting with visible or UV radiation. Indeed at or near the bulk plasmon resonance the opposite situation should exist -- the tangential components of the transition dipole will become the dominant parameter contributing to the spectrum [20]. This is shown in Figure 3 where the expected absorption spectrum of a molecule physisorbed on silver is shown assuming that the molecule absorbs uniformly throughout the spectral region plotted. Hence all of the spectral features originate from the metal. These features depend on the state of polarization of the light and illustrate the reversal in the magnitudes of the normal and tangential components of the surface field on going from the red towards the bulk plasmon frequency [20].

Performing spectroscopy of adsorbed species as a function of exciting frequency, as is done, for example, in recording the fluorescence excitation spectrum or the excitation-profile of a Raman band may result in a complicated spectrum that may be dominated entirely by the wavelength dependence of the "surface selection rule", or more likely a convolution of the metal's optical characteristics and those of the adsorbate. Failure to take this effect into account may lead to interpretational error. It is also not clear how to correlate this metallic spectral behavior with a molecular electronic band of an analogous metal cluster compound. Is it like the contribution of a charge transfer band to the Raman excitation profile in a molecule? If one opts for a positive reply to this question one has to contend with the fact that the same type of behavior would obtain even if the adsorbate

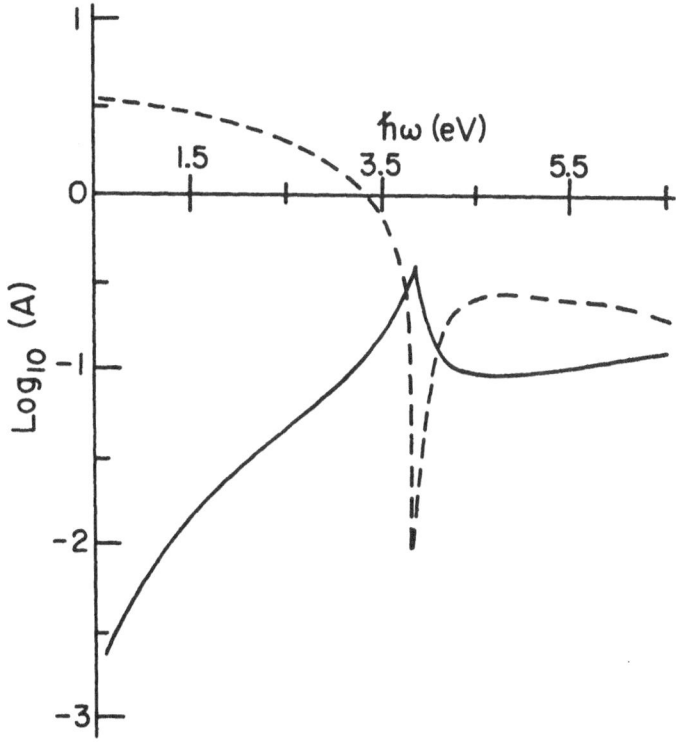

Figure 3. The base ten logarithm of the p-polarized (dashed) and s-polarized (solid) components (in arbitrary units) of the quantity $| \mu.E |^2$ which is proportional to the absorption intensity for a molecule adsorbed on silver as a function of photon energy. Equal x, y, and z components were assumed for the transition dipole μ.

were physisorbed or even merely condensed on the metal surface, a situation without clear-cut parallels in the realm of cluster compounds.

3.2. THROUGH METAL COUPLING

In a normal mode of a molecule, all of the coupled elementary oscillators into which the molecule can be decomposed participate in the vibration. Occasionally, when these elementary oscillators occur either as high frequency or low frequency oscillators the problem can be broken up into an uncoupled high-low frequency problem. This is the idea behind the so-called Cotton-Kraihanzel approximation [21]. The high-low approximation can often be applied, approximately, to molecules such as CO adsorbed on metals. We will use this fact to illustrate the role of the metal in long-range coupling of adsorbed molecules.

In the two fictitious molecules

$$O \equiv C\text{-}M\text{-}M\text{-}C \equiv O \qquad [I]$$

$$\underset{|}{M}\text{-}C \equiv O$$
$$\qquad\qquad\qquad [II]$$
$$M\text{-}C \equiv O$$

the potential energy describing the two coupled CO vibrations, within the terms of the high-low approximation is given by

$$2V = fq_1^2 + fq_2^2 + f_{12}q_1q_2 \quad [1]$$

where q_1 and q_2 are coordinates of the two CO bond stretching motions, f is the primary bond-stretching force constant and f_{12} is the interaction force constant expressing the effect of stretching or compressing one coordinate (bond) on the magnitude of the other.

Solving the vibrational problem with this potential produces the two eigen-frequencies $\nu_{sym} \propto (f + f_{12})^{1/2}$ and $\nu_{asym} \propto (f - f_{12})^{1/2}$. For "molecule" I it is ν_{asym} that is infrared active since its transition dipole is non-zero. The exact opposite situation is true for "molecule" II where it is ν_{sym} that is infrared active. Molecule I is prototypical of the

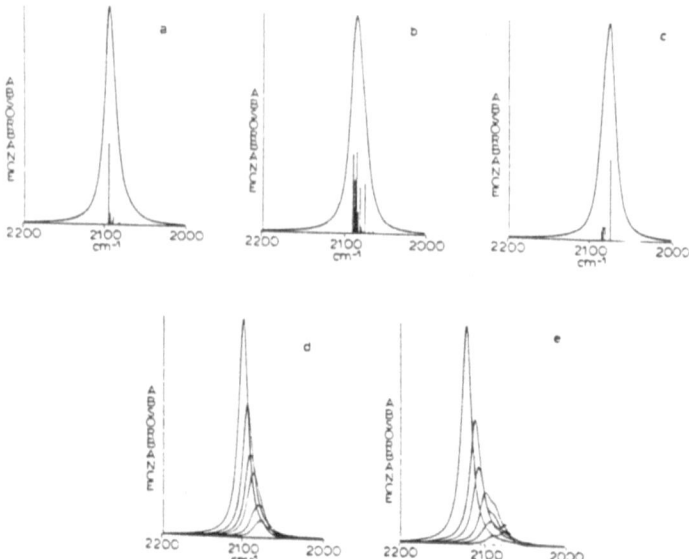

Figure 4. Calculated spectra for ^{12}CO on (100) fcc face: a-c, spectra obtained for coverage values of 0.69, 0.39 and 0.10, respectively. The vertical bars give the frequencies and absorption intensities of individual normal modes making up the "observed" spectrum (solid line). d, a superposition of spectra calculated for coverage values of 0.10, 0.21, 0.39, 0.51, 0.69 and 1.0 scaled according to coverage. Spectra a-d, calculated with $f_{rr} = 0.1$ mdyn/Å. Spectrum e, same as d but with $f_{rr} = 0.2$ mdyn/Å to show more clearly the broadening at intermediate coverages and subsequent narrowing at high coverages of the spectral linewidths.

situation that exists in metal-metal bonded carbonyls such as $Mn_2(CO)_{10}$, while molecule II is reminiscent of bonding of adsorbates on a surface. On a surface one normally has many CO's disposed either randomly or in a two-dimensionally periodic structure with a greater or lesser degree of coupling among them. Equation 1 must be modified in that case to read

$$2V = \sum_{i=1}^{N} f_i q_i^2 + \sum_{i<j}^{N} f_{ij} q_i q_j$$

The result of solving the vibrational problem for this case has been considered by several authors [22]. Briefly, at very low coverage (and in the absence of surface clustering) the spectrum is characteristic of the lone oscillator. The spectrum, therefore, consists of a single line with width reflecting the homogeneous and inhomogeneous properties of the CO-metal interaction [23]. At full coverage the system is capable of sustaining N normal modes for N adsorbed molecules. However only one of them, the one in which all of the molecules oscillate in phase is dipole active (for an infinite surface). In all of the other modes, it can be shown that exactly half of the molecules move in one direction while the other half move in the opposite direction resulting in a net zero transition dipole moment. (For an odd number of molecules there will be a net dipole moment corresponding to one oscillator for any given normal mode. However its contribution will be negligible compared to that of the totally symmetric vibration whose intensity is drawn from all of the coupled oscillators on the surface). For intermediate coverage there are two possible situations. If the adsorbate forms a two-dimensionally periodic structure whose mean unit mesh decreases with increasing coverage then one would always see a single dipole-active mode but one derived from a vibrational frequency incorporating an ever-increasing interaction force constant. Hence the frequency of the band will continue to increase with increasing coverage but with a, more or less, constant band width. In contrast, if at intermediate coverage the adsorbate occupies adsorption sites randomly on the surface, then locally, periodicity is lost and many normal modes of the system become dipole active. This was considered in Ref. 22a. When the band width of each mode is greater than the energy interval between them, as is often the case on surfaces, then the participation of many dipole active normal modes appears as a broadening of the vibrational band. For this case, one sees a simultaneous blue shift in the center frequency of the band together with an initial band broadening, as coverage increases, followed by a re-narrowing when full coverage is approached. This is illustrated in Figure 4.

Another manifestation of this effect is the intensity asymmetry that one observes in the infrared spectrum of adsorbed molecules when adsorbing mixed isotope samples (for example $^{13}CO/^{12}CO$). Depending on the degree of coupling, one sees two bands of sometimes very disparate intensities. If one interprets the two bands (incorrectly) as each belonging to one of the two types of CO on the surface, one would conclude that ^{13}CO binds considerably less strongly to the surface than its more common isotopic counterpart. This would suggest an unusually great isotope effect. Such an interpretation is, in fact, erroneous. The correct explanation is that for a random distribution of the two isotopic forms, formerly forbidden normal modes become allowed. As the interaction force constant is increased the degree of symmetry-breaking is diminished until, for very strong coupling, only a single band is observed although both forms of the molecule are present on the surface [22]. This is illustrated in Figure 5.

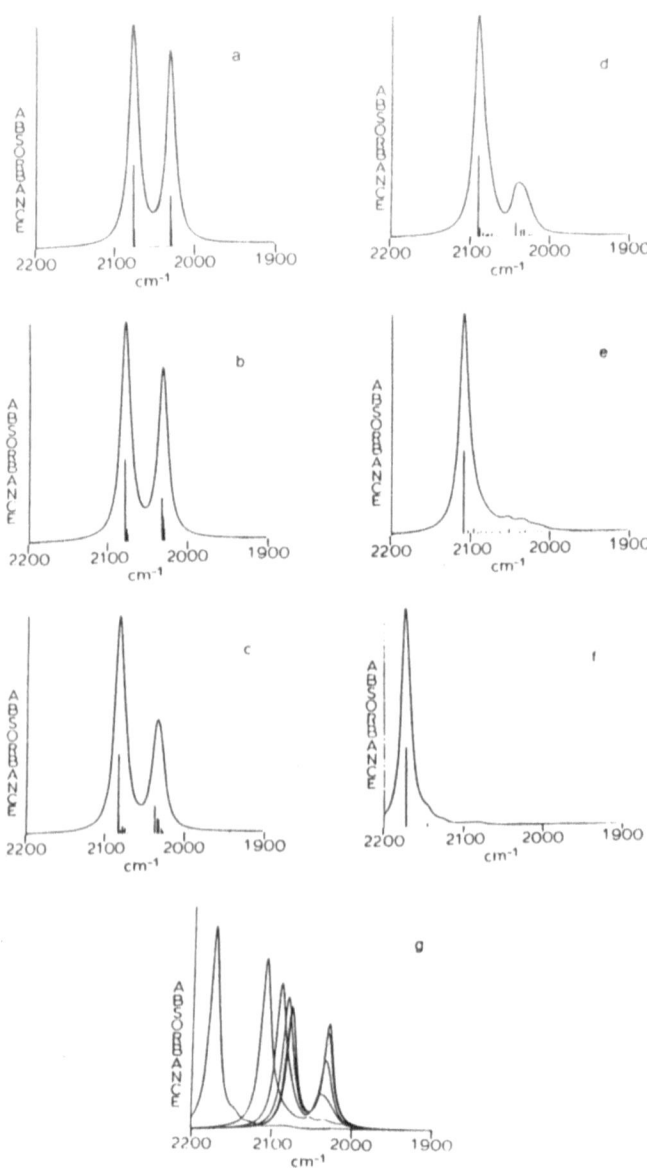

Figure 5. The effect of the value of f_{rr} on the spectrum of an equimolecular mixture of ^{12}CO and ^{13}CO adsorbed on a (100) fcc crystal face at full coverage: a-f, $f_{rr} = 0.01, 0.02, 0.05, 0.1, 0.2, 0.5$, respectively. g, a superposition of the previous spectra. Note that for the two largest values of f_{rr} a spectrum consisting of essentially a single band is expected.

The bottom line is that when one determines quantitatively the magnitude of the interaction force constant that must be assumed to account either for the intensity asymmetry observed in mixed isotope experiments or the observed frequency shifts as a function of coverage one concludes that either the interaction force constant pertinent to molecules adsorbed on some metal surfaces is 5-10 times greater than those operating in metal-metal bonded small metal cluster carbonyls, or that its range is far greater or both [22a], all of which implies that the communication between molecules is far greater when they are adsorbed on a metal surface than when they are bonded as ligands to clusters. This greater difference has been attributed by Scheffler [24] to the attractive role of image charges in the metal. The explanation in terms of image charges simply articulates the role of the metal in other terms. Even so, not all observed features of the vibrational spectra of adsorbed molecules can be accounted for by ascribing the role of the metal entirely to that of a medium for image charges. In particular the broadening of the vibrational bands of adsorbates as a function of coverage is not a feature of Scheffler's treatment.

3.3. SURFACE FIELD EFFECTS

Surface Selection rules are an example of a surface field effect. Other field effects that are unique to surfaces of bulk systems will be listed briefly. (a) Certain surface morphologies such as small particles, periodic grooves and the like can sustain a variety of surface polaritons such as surface plasmon excitations. These can couple to vibrating molecules at the surface, resulting in such effects as the Surface-Enhanced Raman effect as well as other surface-enhanced field effects [25]. (b) Coupling between adsorbate and metal electrons [23] has been invoked to explain the very large infrared linewidths that are normally observed for adsorbed molecules even on single crystal surfaces. It is believed that a large part of the line broadening in these cases is innate and not due to the heterogeneity of the surface. (c) Dipole-allowed electronic transitions in molecules placed near metal surfaces are known to be substantially broadened, or, in the time domain, radiative lifetimes of emissions are dramatically shortened [26]. These effects have been the subject of very many studies. Globally they take place because of coupling between the radiating molecular dipole and various excitations of the metal electrons -- either electron-hole pair production or surface plasmon generation [27]. To manifest these effects the metal system must be characterizable by a fairly high density of states, i. e. by a system containing a large number of metal atoms. (d) Electronic transitions and other excitations occurring in molecules adsorbed on a metal surface will be shifted, sometimes greatly, as a result of "relaxation" processes occurring at the surface [28]. Classically, charged species at the surface will be stabilized by inducing a charge density at the metal surface. This process is normally discussed in terms of the so-called image charge effect. The stabilization can be as large as $-e^2/(4d)$, where d is the charge to surface distance. In real systems the shift is usually less than this but often approaches numbers of this magnitude. Clearly, however, the size of the "relaxation" shift will depend on the precise nature of the excitation process. Intramolecular processes that involve little charge redistribution will not benefit greatly from these effects while in others, such as metal to molecule charge transfer, where the charge redistribution is significant, the energy shift (from the analogous transition in the gas phase) can be large. Hence even physisorption can result in dramatic and non-uniform broadening and shifting of electronic transitions. (e) The presence of a metal surface can result in novel and unique pathways for surface chemistry. The broad electronic bands and low ionization potentials (work functions) can bring about facile electron transfer processes such as the formation of transient, unstable anions resulting from metal to molecule charge transfer that can give rise to unique surface photochemistry [29]. (f) Adsorbates on surfaces can engage in non-specific bonding that does not have analogues in real molecules. The most dramatic is the lateral compression that occurs at high coverage for some adsorbed systems. For example below a coverage of 0.5 the CO's adsorbed on Pd(100) occupy two-fold bridging sites consistent with the infrared spectrum of the system [30]. At 0.5 coverage the system forms a c(4 x 2)R45° system with bridging CO's. Beyond 0.5 coverage no structure commensurate with the underlying metal surface can be constructed that describes the system adequately. In this compressed structure there are CO's on the surface above essentially any point with respect to the center of a metal atom. Uniform bond angles and bond lengths so characteristic of molecules no longer apply here, therefore. Remarkably, this very dramatic change in the character of the bonding is not reflected in the IR spectrum. The increase from 0.5 to 0.53 coverage simply results in a further, slight blue-shift in the CO absorption frequency not very different from what was observed with increasing coverage below the 0.5 coverage point.

Apart from all of these effects it is still not clear how one can make suitable cluster analogues for such effects as face-specific chemistry on single crystals. Is the, often very large, range of binding sites that exist at surfaces analogous to the range of cluster bonding sites? Now that reactivity of clusters has been investigated as a function of a very large range of cluster sizes, the cluster-surface analogy has been turned around to some extent. While we showed above that surfaces can manifest properties that are not normally seen on clusters, clusters have now been found to engage in a range of chemistry of enormous richness and with a size specificity that is unmatched by the face-specificity of most single crystal systems [31]. For example, only two sizes of aluminum clusters are found to dissociate H_2 or D_2: Al_6 and Al_7 [32]. Smaller or larger clusters are essentially inactive in that reaction. Almost equally dramatic is the ability of iron clusters to chemisorb hydrogen [7]. Before settling down to a fairly constant rate of hydrogen chemisorption for clusters with greater than 25 atoms, one encounters iron clusters that are both especially efficient and dismally inefficient at dissociating H_2.

4. CONCLUSION

Surface science and cluster science have a great deal to teach each other. Insights into problems in one field can often result from experiments in the other. Nevertheless, there are unique bulk properties that set in beyond a certain size domain that are characteristic of the bulk. These give rise to the familiar attributes of metal surfaces such as the so-called surface selection rules, through-metal effects, the rapid metal-mediated decay of excited dipoles near surfaces and enhanced field effects resulting from the excitation of polaritons. On the contrary, small metal clusters manifest a range of unique properties of their own, most notably their very size-determined reactivities, that eventually reach a size-independent limit as the cluster size increases. This asymptotic approach to bulk behavior is reached unequally for the variety of physical and chemical properties of the cluster [12].

REFERENCES

1. Chini, P., Longoni, G. and Albano, V.G. (1976), Adv. Organometal. Chem., **14**, 285.

2. Johnson, B. F. G. (ed.), (1980), Transition Metal Clusters, Wiley, New York; Stone, F. G. A. (1981), Acc. Chem. Res., **14**, 318; Johnson, B. F. G. and Lewis, J. (1981), Adv. Inorg. Chem. **24**, 225.

3. Albano, V. G., Sansoni, M., Chini, P., Martinengo, S. and Strumolo, D. (1976), J. Chem. Soc. Dalton, 970.

4. Schmid, G., Pfeil, R., Boese, R., Bandermann, F., Meyer, S., Calis, G. H. M. and van der Velden, J. W. A. (1981), Chem. Ber. **114**, 3634.

5. Schmid, G., Morun, B. and Malm, J.O. (1989), Angew. Chem. Internat. Ed., **28**, 778.

6. Vagaftik, M. N., Zagorodnikov, V. P., Stolyarov, I. P., Moiseev, I. I., Likholobov, V. A., Kochubey, D. I., Chuvilin, A. L., Zaikovsky, V. I., Zamaraev, K. I. and Timofeeva, G. I. (1985), J. Chem. Soc. Chem. Comm., 937.

7. Richtsmeier, S. C., Parks, E. K., Liu, K., Pobo, L. G. and Riley, S. J. (1985), J. Chem. Phys., **82**, 3659; Whetten, R. L., Cox, D. M., Trevor, D. J. and Kaldor, A. (1985), Phys. Rev. Lett., **54**, 1494; Geusic, M. E., Morse, M. D. and Smalley, R. E. (1985), J. Chem. Phys., **82**, 590.

8. Kaldor, A., Cox, D. M. and Zakin, M. R. (1988), Adv. Chem. Phys. 70 pt. **2**, 211; Kaldor, A. and Cox, D. M. (1990), Pure Appl. Chem. **62**, 79.

9. Parks, E.K., Winter, B. J., Klots, T. D. and Riley, S. J. (1991), J. Chem. Phys. (in press); Riley, S. J. (1990), Phase Transitions 24-26, 271; Riley, S.J. (1989), Z. Phys. D: At. 12, 537; Hoffman, III, W. F., Parks, E. K. and Riley, S. J.(1989), J. Chem. Phys., **90**, 1526; Cole, S. K., Liu, K. and Riley, S. J. (1987), NATO ASI Ser., Ser. B 158, 347; Parks, E. K., Nieman, G. C., Pobo, L. G. and Riley, S. J. (1988), J. Chem. Phys., **88**, 6260; Parks, E. K., Weiller, B. H., Bechthold, P. S., Hoffman, W. F., Nieman, G. C. Nieman, Pobo, L. G. and Riley, S. J. (1988), J. Chem. Phys. **88**, 1622; Hoffman, III, W. F., Parks, E. K., Nieman, G. C., Pobo, L. G. and Riley, S. J. (1987), Z. Phys. **D 7**, 83; Parks, E. K., Nieman, G. C., Pobo, L. G. and Riley, S. J. (1987), J. Phys. Chem. **91**, 2671; Parks, E. K., Nieman, G. C. Nieman, Pobo, L. G. and Riley, S. (1987), J. Chem. Phys. **86**, 1066; Liu, K., Parks, E. K., Richtsmeier, S.C., Pobo, L. G. and Riley, S. J. (1985), J. Chem. Phys. **83**, 2882; Parks, E. K., Liu, K., Richtsmeier, S. C., Pobo, L. G. and Riley, S. J. (1985), J. Chem. Phys. **82**, 5470,.

10. Jellinek, J. and Guvenc, Z. B. (1991), Z. Phys. **D 19**, 371; Ragavan, K., Stave, M. S. and DePristo, A. E. (1989), J. Chem. Phys. **91**, 1904.

11. Martinengo, S., Ciani, G., Sironi, A. and Chini, P. (1978), J. Am. Chem. Soc., **100**, 7096.

12. (a) (1990) Proc. Farad. Symp. 25, J. Chem. Soc. Farad. Trans., **86**, No. 13.
    (b) Benedek, G., Martin, T. P. and Pacchioni, G. (eds.), (1987), Elemental and Molecular Clusters, Springer, Berlin.
    (c) Jortner, J., Pullman, A. and Pullman, B. (eds.), (1987), Large Finite Systems, Reidel, Dordrecht.
    (d) Moskovits, M. (ed.), (1986), Metal Clusters , Wiley, N. Y.
    (e) Träger, F., zuPutlitz (eds.), (1986), Metal Clusters, Springer, Berlin.
    (f) (1988), Farad. Disc. Chem. Soc., **86**.
    (g) (1986), Morse, M. D., Chem. Rev., **86**, 1049.
    (h) Maier, J. P. (ed.), (1988), Ion and Cluster Ion Spectroscopy and Structure, Elsevier, Amsterdam.

(i) Scoles, G. (ed.), (1990), The Chemical Physics of Atomic and Molecular Clusters, Proc. Fermi School Phys., Course 107, North-Holland, Amsterdam.

(j) Weltner Jr., W. and Van Zee, R. J. (1984), Ann. Rev. Phys. Chem. **35**, 291.

(k) Jena, P., Rao, B. K., Khanna, S. N. (eds.), (1987), Physics and Chemistry of Small Clusters, Plenum, New York.

(l) Castleman Jr., A. W. and Keesee, R. G. (1986), Ann. Rev. Phys. Chem. **37**, 525; Chem. Rev. **86**, 89.

(m) Kaldor, A., Cox, D. M. and Zakin, M. R. (1988), Adv. Chem. Phys. 70 pt. **2**, 211; Kaldor, A. and Cox, D. M. (1990), Pure Appl. Chem. 62, 79.

(n) Weltner, Jr., W. and Van Zee, R. J. (1990), Chem. Rev. **89**, 1713.

(o) Duncan, M. A. (ed.), (1991), Advances in Metal and Semiconductor Clusters, Vol. 1, Spectroscopy and Dynamics in JAI Press, Greenwich.

(p) Salahub, D.R. (1990), Adv. Chem. Phys. **69**, 447.

13. Lawley, K.P. (ed.), (1989), Adv. Chem. Phys. **76**, Wiley, New York; Albert, M. R. and Yates Jr., J. T. (1987), The Surface Scientist's Guide to Organometallic Chemistry, Amer. Chem. Soc., Washington.

14. Muetterties, E. L. (1978), Angew. Chem. Internat. Ed. 17, 545; (1982), Pure Appl. Chem. **54**, 83.

15. Farrar, D. H. and Goudsmit, R. J. (1986), in Moskovits, M. (ed.), Metal Clusters, Wiley-Interscience, New York.

16. Bochmann, M. (1991), in Twigg, M.V. (ed.), Mechanisms of Inorganic and Organometallic Reactions, Chapt. 14, Plenum, New York.

17. Johnson, B. F. G. and Lewis, J. (1981), Adv. Inorg. Chem. **24**, 225.

18. Wade, K. (1971), J. Chem. Soc. Chem. Comm. 792; (1976), Adv. Inorg. Chem. Radiochem., **18**, 1; Williams, R. E. (1976), ibid. **18**, 67; Mingos, D. M. P. (1977), Adv. Organometal. Chem., **15**, 1.

19. Pearce, H. A. and Sheppard, H. (1976), Surf. Sci., **59**, 205; Dignam, M. J., Moskovits, M. and Stobie, R.W. (1971), Trans. Farad. Soc. **67**, 3306.

20. Moskovits, M. (1982), J. Chem. Phys. **77**, 4408.

21. Cotton, F. A. and Kraihanzel, C. S. (1962), J. Amer. Chem. Soc. **84**, 4432.

22. (a) Moskovits, M. and Hulse, J. E. (1978), Surf. Sci. **78**, 397.
    (b) Hammaker, R. M., Francis, S. A. and Eischens, R. P. (1965), Spectrochim. Acta, **21**, 1295.

23. Metiu, H. (1977), J. Chem. Phys. **68**, 1453; Metiu, H. and Palke, W. E. (1978), J. Chem. Phys. **69**, 2574; Gadzuk, J. W. and Metiu, H. (1982), in Caudano, R, Gilles, J.M. and Lucas, A.A. (eds.), Vibrations on Surfaces, Plenum; Gadzuk, J. W. and Luntz, A. C. (1984), Surf. Sci. **144**, 429; Persson, B. N. (1984), J. Phys. C 17, 4741; Chabal, Y. J. (1988), Surf. Sci. Rep. **8**, 211; Gortel, Z. W., Kreuzer, H. J., Piercy, P. and Teshima, R. (1987), Phys. Rev. B36, 3059.

24. Scheffler, M. (1979), Surf. Sci., **81**, 562; Mahan, G. D. and Lucas, A. A. (1978), J. Chem. Phys. **68**, 1344.

25. Furtak, T. E. and Chang, R. K. (eds.), (1981), Surface Enhanced Raman Scattering, Plenum, New York; Otto, A.(1983), in Cardona, M. (ed.), Light Scattering in Solids IV, Springer, Berlin; Metiu, H. and Das, P. (1984), Ann. Rev. Phys. Chem. **35**, 507; Moskovits, M. (1985), Rev. Mod. Phys., **57**, 783.

26. Chance, R. R., Prock, A. and Silbey, R. (1978), Adv. Chem. Phys., **37**, 1.

27. Weber, W. H. and Eagen, C. F. (1979), Opt. Lett., **4**, 236.

28. (a) Avouris, Ph. and Demuth, J. (1984), Ann. Rev. Phys. Chem. **35**, 49;

(b) Avouris, Ph. and Demuth, J. E. (1985), Surf. Sci., **158**, 21; Avouris, Ph., Bagus, P. S. and Nelin, C. J. (1986), J. Electron and Rel. Phenom., **38**, 269; Avouris, Ph. and Persson, B. N. J. (1984), J. Phys. Chem., **88**, 837.

29. Marsh, E. P., Gilton, T. L., Meier, W., Schneider, M. R. and Cowin, J. P. (1988), Phys. Rev. Lett., **61**, 2725; Gilton, T. L., Dehnbostel, C. P. and Cowin, J. B. (1989), J. Chem. Phys., **91**, 1937; Dixon-Warren, St. J.. Jensen, E. T. and Polanyi, J. C. (1991), Phys. Rev. Lett., **67**, 2395.

30. Bradshaw, A. M.and Hoffman, F. M. (1978), Surf. Sci. **72**, 513.

31. Goodman, D. W. (1986), Ann. Rev. Phys. Chem., **37**, 425; Pritchard, J., Catterick, T. and Gupta, R. K. (1975), Surf. Sci., **53**, 1; Horn, K., Hussain, M. and Pritchard, J. (1977), Surf. Sci., **63**, 244.

32. Cox, D. M., Trevor, D. J., Whetten, R. L., Kaldor, A. (1988), J. Phys. Chem., **92**, 421.

THE CHEMISTRY OF TRANSITION METAL CLUSTERS

S. J. RILEY
Chemistry Division
Argonne National Laboratory
Argonne, Illinois
USA

ABSTRACT. The chemical reactions of isolated clusters of transition metal atoms with small molecules provide information on metal-ligand interactions at the molecular level. This paper shows how the cluster-ligand interaction can be used to estimate cluster ionization potentials and infer something about cluster geometrical structure. Measurements of reaction rate constants demonstrate that large cluster-size dependence of reactivity can be found. Thermodynamic measurements provide information on cluster-ligand binding energies and entropies. Experiments in which ligands decompose on cluster surfaces will be discussed.

1. Introduction

In the past ten years an exciting new area of chemistry has been developed: the chemistry of isolated clusters of transition metal atoms. The technology that has opened up this field is cluster generation by laser vaporization [1]. Using this technique, clusters of even the most refractory materials can be made, and their chemical reactions studied in the gas phase or even in high vacuum. Most of the reactions studied to date have been ones in which intact molecules (or perhaps their decomposition products) bind to the surfaces of the clusters. In this sense, these studies mimic surface science, the reactions being chemisorptions and the bound molecules being adsorbates or ligands. Of course, unlike the usually well-ordered array of metal atoms provided by a single crystal metal surface, clusters offer a variety of metal atom arrangements to the incoming ligand molecule. This variety probably accounts for the wide range of chemical behavior seen in transition metal clusters [2].

Studying the chemistry of metal clusters is relevant to a fundamental understanding of heterogeneous catalysis. Catalytic preparations of finely divided metal particles promote chemical transformations that are essential to the chemical and petroleum industries. The design of catalysts often appears to be a black art, with useful preparations identified by trial-and-error techniques. We hope that the study of the chemical properties of transition metal clusters will reveal the underlying principles governing chemistry on the surfaces of small metal particles, so that more efficient and specific catalysts can be designed.

This paper will review the basic aspects of metal cluster generation, the procedures for studying the gas phase chemical reactions of clusters, and the techniques of cluster detection. Measurements of cluster ionization potentials will be discussed, showing how

D. R. Salahub and N. Russo (eds.), Metal-Ligand Interactions: from Atoms, to Clusters, to Surfaces, 17–36.

the cluster-ligand interaction can be used to estimate ionization potentials. Experiments in which clusters are saturated with ligand molecules will be considered, showing how such studies can often provide information about cluster structure. Examples of large cluster-size dependence of chemical reactivity will be presented. Thermodynamic measurements of cluster-ligand binding energies and entropies will be presented. Finally, processes in which adsorbed molecules decompose on cluster surfaces will be considered.

2. Experimental

The heart of a transition metal cluster chemistry apparatus is the cluster source, in which the beam from a pulsed laser is focussed onto a metal target that is located in a flow channel. Typically, an XeCl excimer laser (308 nm wavelength, 50 to 100 mJ/pulse energy) or the second harmonic of a Nd YAG laser (532 nm, 20 to 40 mJ/pulse) is used to vaporize the metal sample. The plume of metal (~30-50 ng/pulse) produced by the laser pulse is rapidly cooled by inert (usually helium) carrier gas flowing down the channel. Most cluster sources use a pulsed valve to feed the helium into the channel. Since the vaporization laser is pulsed, there is no need to keep the helium flowing between laser pulses, and a pulsed valve greatly reduces the pumping requirements for the vacuum chamber housing the cluster source. However, the use of continuous gas flows allows the accurate and reproducible determination of such reaction conditions as pressure, temperature, and time. Knowledge of these conditions is essential if quantitative measurements of cluster reactivity are to be made [3].

The cooled metal plume, consisting mostly of neutral atoms, is formed in a state of extreme supersaturation, so that cluster growth quickly begins. Under typical conditions of the continuous gas flow source (~20 Torr total gas pressure), the mechanism of cluster growth is usually one of successive addition of atoms to the growing clusters. Cluster growth will be terminated when most of the residual atoms have been lost by diffusion to the flow-tube wall. Reagent molecules are added to the flow-tube reactor (FTR) after cluster growth has stopped. This is important, since if growth continues beyond the point where chemical reactions begin, then any cluster-size-specific chemistry might be lost. Since the chemistry occurs in a bath of inert gas, the reaction exothermicity can usually be dissipated, and the most common type of reaction is a simple association in which the reagent molecule becomes bound to the cluster surface, i.e., becomes a ligand.

At the end of the FTR the clusters and their reaction products exit the channel through a nozzle and are formed into a molecular beam by collimators located in successive vacuum chambers. The purpose of the differential pumping thus provided is to transport the clusters from a region of high (~20 Torr) pressure to one of high vacuum (~10^{-6} Torr) where they can be detected. Prior to detection, however, the clusters must be ionized. Fortunately, transition metal clusters have relatively low ionization potentials, so that they can be readily photoionized using commercially available excimer lasers operating in the ultraviolet. While experiments suggest that metal clusters have very large photoabsorption cross sections in the ultraviolet [4], nonradiative processes such as internal conversion dominate the absorption process. Thus as a rule ionization probabilities are usually in the 0.01 to 0.1 percent range, so that only a few dozen ions might be produced per ionizing laser pulse.

After ionization, the clusters are injected into a mass spectrometer. Since ionization is pulsed (a typical excimer laser pulse width is 10 ns), the most common type of mass spectrometer is a time-of-flight (TOF) unit, in which the ions, accelerated to a common translational energy, enter a field-free drift tube, where they separate in time according to their mass. At the end of the drift tube there is a particle detector, most commonly a microchannel plate electron multiplier. Signal is acquired by a fast transient digitizer or digital oscilloscope, and averaged in a computer. Typically a few thousand laser pulses are sufficient to provide a spectrum of adequate signal-to-noise ratio.

One important consideration in experiments of this sort is to what extent the intensities of ion peaks recorded in a TOF mass spectrum reflect the densities of neutral species that exist at the end of the FTR. There are several processes that might distort this relationship. One is nozzle expansion effects. Studies [5], have shown that if the reagent partial pressure in the FTR exceeds about 20 mTorr, there is the possibility that additional reagent molecules may add to the clusters in the expansion, due to expansion cooling and additional cluster-reagent collisions. The test for such processes is to see if the number of molecules that bind to a given cluster depends on inert gas pressure. If it increases with increasing pressure, than expansion effects may be important. In the free flight from the nozzle to the mass spectrometer there is the possibility that reagent molecules may desorb from the clusters. Again, an inert-gas-pressure dependence of the number of ligand molecules, even at reagent partial pressures below 20 mTorr, would imply that this is occurring. Fortunately, under the typical operating conditions of a continuous-gas-flow source, there is usually sufficient cluster cooling in the expansion that desorption in the beam is not a problem. The relationship between neutral density at the ionization region and ion intensity in the mass spectrum could be distorted by multiphoton or even single-photon ionization processes in which sufficient excess energy is deposited in the clusters to produce ligand desorption. This can be a particular problem when, as will be discussed below, the presence of ligands on cluster surfaces lowers cluster ionization potentials. The test for multiphoton processes is to see if the relative amounts of species in the mass spectrum depends on ionizing laser fluence. To minimize one-photon desorption, a laser whose photon energy is within about 1 eV of the cluster ionization potential should be used for ionization. Finally, even in the absence of ligand desorption there may be coverage-dependent ionization probabilities that distort the ion-neutral relationship. To estimate the magnitude of this effect, the integrated ion signal for a given cluster can be measured as a function of coverage[4]. For most quantitative experiments what is desired is the ratio of one cluster density to that of the cluster with an additional ligand molecule bound to it. As a rule, the ionization probability effect can alter this ratio, as measured by the ion signal ratio, by as much as 20%.

A convenient indication of how a cluster interacts with a particular ligand molecule is provided by the *uptake plot*, a plot of the average number $\bar{m}$ of ligand molecules versus the reagent partial pressure in the FTR [6]. $\bar{m}$ is determined from the distribution of product peak intensities in the mass spectrum from

$$\bar{m} = \sum_m m I_m / \sum_m I_m$$

where I_m is the intensity of the cluster peak having m ligand molecules. An example of an uptake plot is shown in Fig. 1, for the reaction of Fe_{55} with NH_3 [7]. There are several important points about this figure that should be emphasized. First, note the logarithmic scale for NH_3 partial pressure. To fully map out the interaction between a cluster and a ligand molecule can require a five-order-of-magnitude variation in reagent pressure. Also, two sets of data are shown, one in which the reagent reacts with the cluster for a relatively long period of time (~1 ms) and another in which the interaction time is relatively short (~0.2 ms). At the lowest NH_3 pressure, we see a region of behavior in which the NH_3 uptake is fairly rapid, but the extent of ammonia coverage for

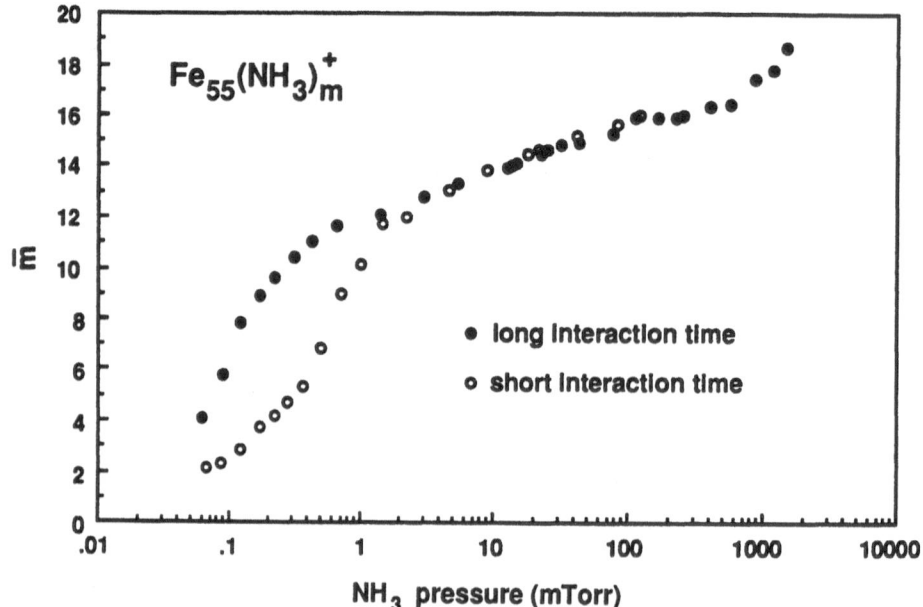

Figure 1. The uptake plot for the reaction of Fe$_{55}$ with NH$_3$. Adapted from Ref. 7.

a given pressure depends quite strongly on interaction time. A reaction whose extent is dependent on time is obviously *kinetically controlled*, i.e., the number of NH$_3$ molecules on the cluster is determined by the total number of cluster-ammonia collisions in the FTR. The study of reactions under these conditions permits the determination of reaction rate constants, as will be discussed below. At about 2 mTorr NH$_3$ pressure, the uptake data for the two interaction times merge, so that the extent of reaction no longer depends on time. A reaction independent of time is obviously an *equilibrium* reaction, and the reaction extent is determined by the thermodynamics of the chemisorption process. Examples of the measurement of adsorption thermodynamics will also be presented below. At still higher pressure the cluster coverage becomes independent of ammonia pressure, indicating that the cluster is *saturated* with the reagent. In this case, we see that Fe$_{55}$ saturates with 16 NH$_3$ molecules. Reagent saturation is an important feature of cluster chemistry, since a determination of the total number of ligand binding sites can provide valuable information about cluster structure, as we will see below. At the highest NH$_3$ pressures in Fig. 1 we see yet another upturn in the cluster coverage. At this point the NH$_3$ pressure is so large that substantial cluster cooling in the nozzle expansion is causing the formation of a second, essentially physisorbed, layer of ammonia on the cluster. As a general rule, for studies of cluster chemistry, the reagent pressure is kept below the range where physisorption sets in.

The reason for the transition from a kinetically controlled reaction to an equilibrium one is that the cluster-ammonia binding energy monotonically decreases with increasing coverage. At low coverages, the binding energy is sufficiently high that the lifetime for desorption is long compared to the residence time of the clusters in the FTR. If an ammonia molecule binds to a cluster, it remains bound, so that the number of molecules depends on the number of collisions (hence the interaction time) and the sticking probability. Eventually, as the cluster becomes more covered, the binding energy decreases to the point where the lifetime for desorption from a thermalized cluster is shorter than the time the cluster spends in the FTR. Molecules will desorb, adsorb, desorb, and so forth, which is essentially a microscopic description of equilibrium. As a rule, we have found that the reactions between ammonia and clusters of iron, nickel, cobalt, and copper show the transition from kinetics to equilibrium as illustrated by the data in Fig. 1.

3. Physical Properties

Substantial effort has been focussed over the years on the determination of cluster physical properties. Among these properties, geometrical and electronic structure should have the most important influence on cluster chemistry. In fact, as we will show here, such physical properties can be probed through chemistry, using cluster-ligand interactions to provide information on cluster ionization potentials and geometrical structure.

3.1. CLUSTER IONIZATION POTENTIALS

The ionization potential (IP) is one cluster physical property that has been studied quite extensively in recent years [8]. Much has been said about the correlation between IPs and chemical properties [9]. In fact, the adsorption of molecules on a cluster's surface can alter the cluster IP. It has been found, for example, that hydrogenation usually increases cluster IPs somewhat [10], and ammoniation lowers IPs considerably [11]. Such effects can provide information on the cluster-ligand interaction. For example, the IP lowering by ammonia is close to linear in coverage, and can be used to provide fairly accurate estimates of cluster IPs and to model the cluster-ligand interaction, as will be demonstrated here.

Fig. 2 shows mass spectra of small ammoniated nickel clusters recorded with photo-ionization by four different excimer lasers. For each laser, there is a minimum number of NH_3 molecules needed to lower the IP below the laser's photon energy, thus allowing ionization. For example, at least six ammonia molecules are needed for Ni_9 and seven for Ni_{10} to lower their IPs below the 3.53 eV of the XeF laser, while only one molecule is needed on each cluster to produce an IP below 5.58 eV. Once a determination is made of the minimum number m_{min} of NH_3 molecules needed to ionize a given cluster with each of the available excimer lasers, then an approximation to the IP vs coverage dependence can be constructed. If, for example, a cluster $M_n(NH_3)_m$ can be ionized by a particular laser, and the $M_n(NH_3)_{m-1}$ cluster cannot, then there is some hypothetical fractional coverage between m-1 and m that would produce an IP exactly equal to that laser's photon energy. Then, on a plot of m_{min} vs photon energy we would place at that photon energy a vertical line extending from m-1 to m. By repeating this process for the other excimer lasers we obtain a plot such as shown in Fig. 3 for Ni_{19}. We now draw the extreme straight lines that simultaneously pass through all vertical brackets, and extrapolate these lines to $m_{min} = 0$. The average of the intercepts we take as the bare cluster IP, and the difference in the intercepts as some measure of the error limits in the IP determination.

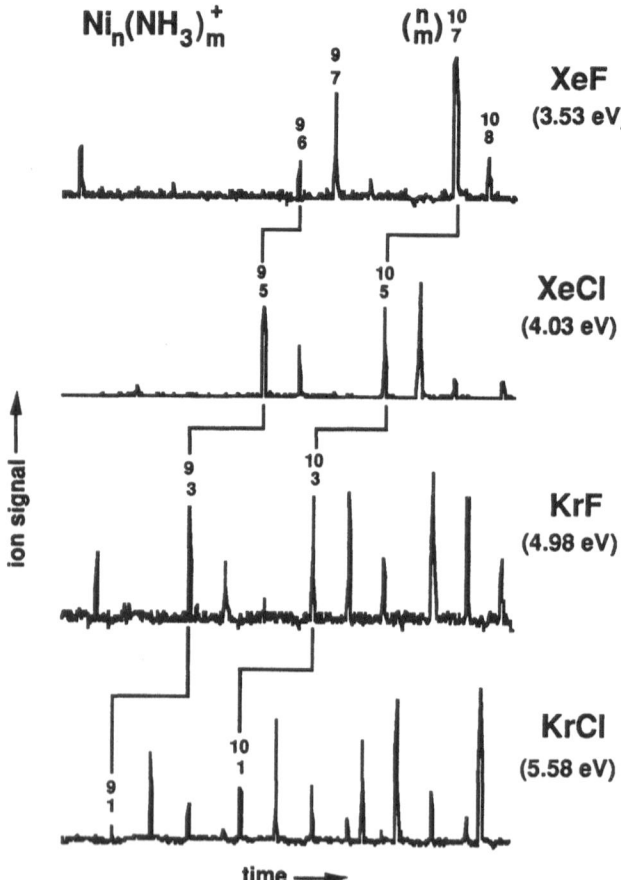

Figure 2. Portions of TOF mass spectra illustrating the determination of the minimum number of NH$_3$ molecules needed to ionize nickel clusters for four different excimer lasers. From Ref. 11.

We find that for virtually all clusters studied it is possible to draw such straight lines through the m_{min} vs photon energy data [11]. This essentially implies that the lowering of cluster IP is a linear function of NH$_3$ coverage. In more recent experiments [12] employing a tunable ultraviolet laser for photoionization in order to measure very accurate cluster IPs, we have shown this relationship to be quite good. For example, Fig. 4 shows a plot of the IP of Ni$_{19}$(NH$_3$)$_m$ vs m measured this way. The straight line is a linear least squares fit and, as can be seen, it falls within the estimated uncertainties of the IP measurements.

This linear dependence of IP on coverage tells us something about the nature of the cluster-ammonia interaction. All evidence from surface science studies suggests that the ammonia binds with the negatively charged nitrogen lone pair electrons directed towards the metal surface [13]. This means that coverage with ammonia will produce a molecular dipole layer that lowers the potential energy of the surface, and in the case of bulk metal surfaces will lower the work function. A simple electrostatic model [14] developed for bulk surfaces can be extended to the case of a cluster by assuming that the cluster is a

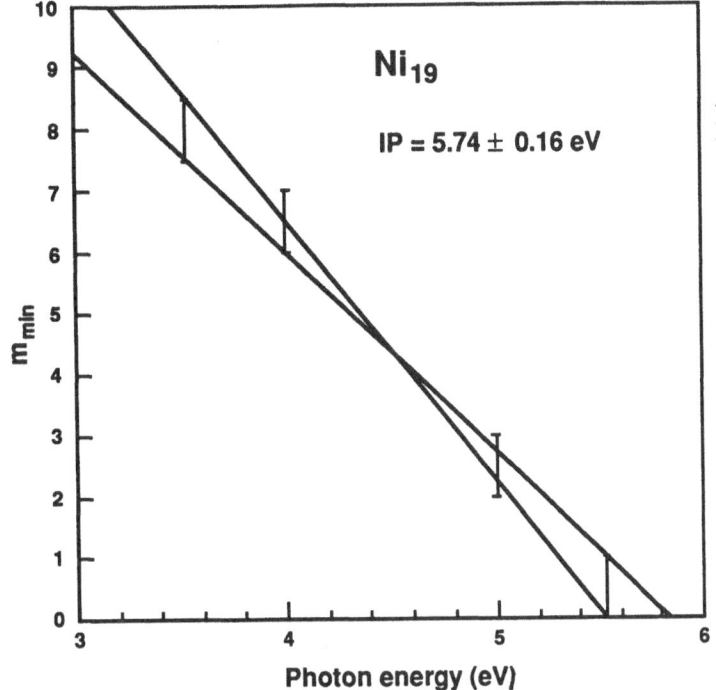

Figure 3. The m_{min} vs photon energy plot for Ni_{19}. The two straight lines are the limiting lines that can be drawn through all of the error bars. Their intercepts with the photon energy axis define the uncertainty of the bare cluster IP.

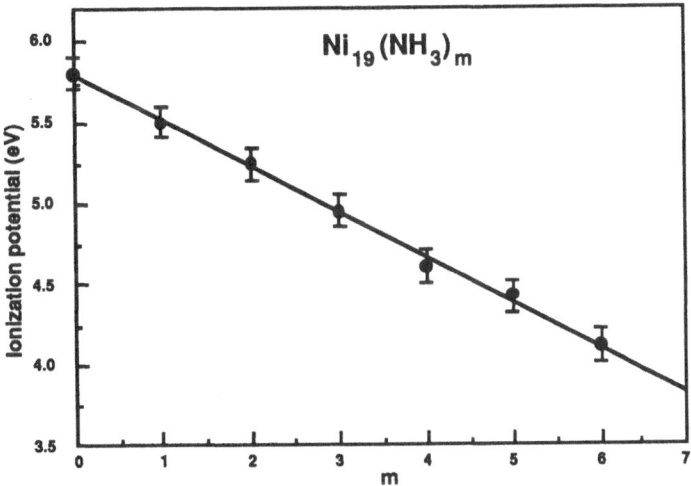

Figure 4. The dependence of the IP of $Ni_{19}(NH_3)_m$, measured by tunable laser photoionization, on m. The straight line is a linear least squares fit to the data points. From Ref. 12.

24

conducting sphere with a dipole on its surface. The result for the IP lowering by m adsorbed species is [11]

$$\Delta IP = \frac{em\mu}{l^2}$$

where e is the magnitude of the electronic charge, μ the dipole moment of an adsorbed molecule, and l the distance from the center of the sphere to the dipole. When this model is applied to IP lowering data for iron, cobalt, and nickel clusters it predicts dipole moments in the 2.0 to 2.5 D range,[11] larger than the 1.47 D of isolated gas phase ammonia [15]. This supports the generally accepted picture of the binding of ammonia to metal surfaces. As the lone pair electrons are drawn into the metal, the charge separation and hence the dipole moment are increased.

As far as the bare cluster IPs are concerned, there is an electrostatic model, the conducting spherical drop model [16], that predicts that the IP of a metal sphere of radius R is given by

$$IP = WF + \frac{\alpha e^2}{R}$$

where WF is the work function of the bulk metal and α is a constant. Different derivations give α as 3/8 [17] or 1/2 [18], and there has been much discussion as to the correct value. The IP data for alkali metal clusters are usually better fit to a value of 3/8 [19]. However, for transition metal clusters the point is somewhat moot. This is illustrated in Fig. 5 for iron cluster IPs measured by tunable laser photoionization [20]. Clearly the model does not predict the observed cluster size dependence of the IPs for either value of α. Of course, a central assumption of the model is that the Fermi level of a cluster is independent of cluster size and equals that of the bulk metal. The measurements on transition metal clusters suggest that in fact the Fermi level must decrease with increasing cluster size, and that more sophisticated theories will be required to explain the size dependence of transition metal cluster IPs.

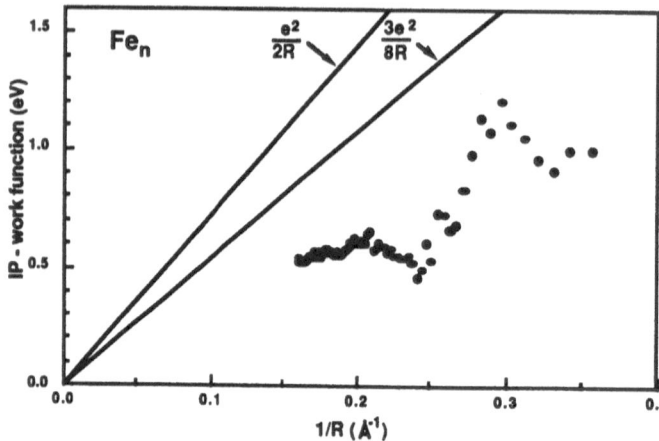

Figure 5. A comparison of the IPs of iron clusters with two forms of the conducting spherical drop model. The bulk work function is taken from the value for polycrystalline iron. Cluster radius is approximated from the structure-less packing model (see Ref. 34). Adapted from Ref. 20.

3.2. CLUSTER STRUCTURE

Perhaps one of the most important physical properties of metal clusters is their geometrical structure, i.e., the way in which the atoms are packed together. Knowledge of structure is essential for theoretical modeling of clusters, and knowing how structure changes with size should provide information on the mechanisms of material growth. But perhaps the most important reason for determining cluster structure is the hope to understand the relationship between structure and chemical reactivity. The most likely reason for strong size dependence of cluster chemical properties is dramatic changes in structure with size. Only with a knowledge of possible structure-reactivity relationships might we be able, for example, to design catalysts having improved properties.

The traditional probe of the structure of isolated clusters is the occurrence of so-called "magic numbers" in cluster mass spectra. These are particular cluster sizes having enhanced intensity in the spectra because they have enhanced stability. For example, alkali metal clusters show magic numbers for sizes corresponding to numbers of electrons (one electron per atom) that fill quantum mechanical levels or shells [21]. Rare gas clusters show a pattern of magic numbers that is consistent with a series of closed shell and subshell clusters having icosahedral packing [22]. Semiconductor clusters have stabilities determined from bonding rules [23]. However, mass spectra of transition metal clusters do not show any magic numbers. This is due to the mechanism of cluster growth in a laser vaporization cluster source. The clusters grow under kinetically controlled, not equilibrium, conditions, so the production of a cluster of a given size does not reflect that cluster's stability, but merely the statistical probability that sufficient collisions occurred to produce that cluster. Thus we need a probe other than traditional magic numbers if we are to determine transition metal cluster structure. This probe can be found in the chemical reactions of these clusters with ammonia and with water.

The essence of this chemical probe of cluster structure is the special nature of the cluster-ammonia and cluster-water interaction. As discussed above, the ammonia binds via donation of the nitrogen lone pair electrons, and the water likewise binds via its lone pair electrons. Both molecules prefer to bind to atop or single atom sites, and the lower the metal-metal coordination of an atom the better a binding site it will be. To probe cluster structure, the ammonia is used to saturate the clusters, and hence to provide a count of the number of atoms on the cluster surface that have lowest metal-metal coordination. The cluster-water bond energy is sufficiently low that the reaction of a cluster with a single water molecule is an equilibrium one under typical FTR conditions. This means that a cluster with a metal atom on its surface with a particularly low metal-metal coordination will show an enhanced propensity to bind water, since the reaction is thermodynamically controlled. We then look for patterns in the cluster-size dependence of the number of ammonia binding sites and the water binding strength that might point to a particular cluster structure.

Such dependences are shown in Fig. 6 for cobalt clusters that have been saturated with hydrogen. The upper panel shows $\bar{m}$ values for ammoniated cobalt clusters $Co_nH_p(NH_3)_m$ in the $n = 82$ to 155 size region. A distinctive pattern of minima can be seen, with $\bar{m}$ repeatedly dipping to a value near 12. Many of the cluster sizes showing minima are the sizes having magic numbers in the mass spectra of rare gas clusters; in other words, they correspond to closed subshells (and in the case of $Co_{147}H_p$ a closed shell) of icosahedra. These closed subshell clusters have maxima in the average number of surface metal atom bonds, and can be generated by successively pulling off adjacent

26

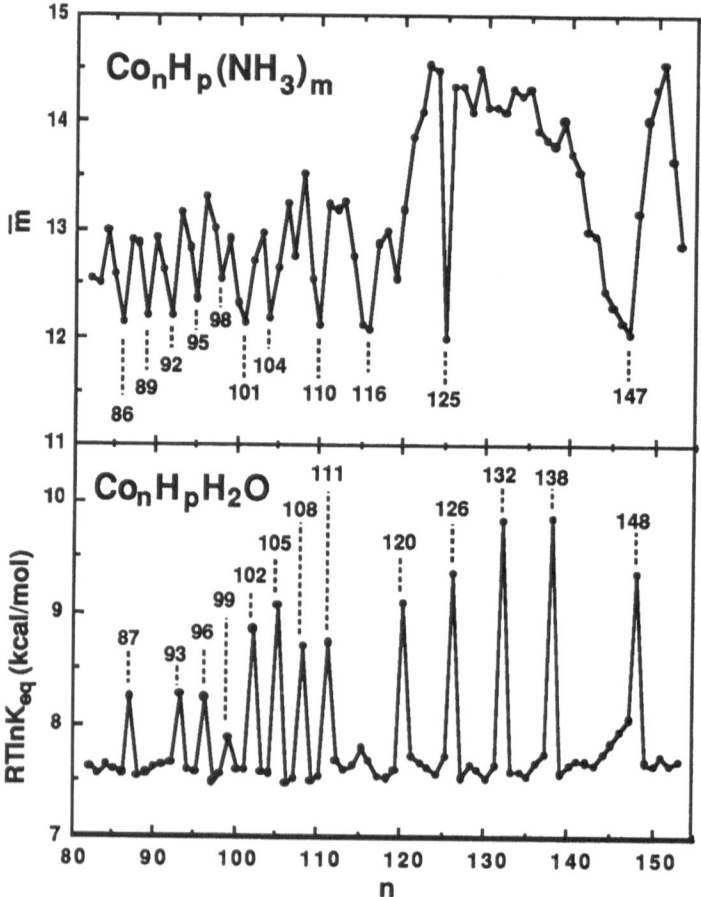

Figure 6. Upper panel: the dependence of $\bar{m}$ on n for $Co_nH_p(NH_3)_m$. Values of n having strong $\bar{m}$ minima are annotated. Lower panel: the dependence of $RTlnK_{eq}$ on n for the reaction of Co_nH_p with H_2O. Cluster sizes having H_2O binding enhancements have their n values annotated.

faces of the 147-atom icosahedron. Calculations on Lennard-Jones clusters show that these closed subshell clusters are more stable than other possible atomic configurations [24]. The closed subshell clusters in most cases have exactly 12 apex metal atoms that have lower metal-metal coordination than the other atoms on the surface, and should serve as primary binding sites for ammonia. $\bar{m}$ minima result for these clusters because a metal atom outside a closed subshell would have even lower metal-metal coordination than the apex atoms, providing a 13th binding site. To form a cluster with one atom less than a closed subshell would most likely require removing one of the apex atoms, since they have minimum metal-metal coordination. This apparently opens up enough room around the apex position to provide two binding sites, so that such a cluster likewise has 13 binding sites. Thus closed (sub)shell clusters have local minima in the number of ammonia binding sites.

While the ammonia binding patterns point to icosahedral structure, it could be argued that saturating a cluster with ammonia might well cause a structural change. This is where the water probe, involving the binding of only a single ligand molecule, comes in. Since the cluster-water reaction is an equilibrium one, the parameter of choice to quantify reactivity is the equilibrium constant K_{eq}. The lower panel of Fig. 6 shows a plot of $RT \ln K_{eq}$ for the reaction between water and hydrogenated cobalt clusters in the same size range as the upper panel. Since $RT \ln K_{eq} = -\Delta G^{\circ} = -\Delta H^{\circ} + T\Delta S^{\circ}$, the vertical axis in the figure is just minus the standard free energy change for the reaction. If we assume that entropy changes are not a strong function of cluster size, then the data in the figure is essentially minus the standard enthalpy of reaction, in other words the cluster-water binding energy. There is clearly a maximization of this energy for certain cluster sizes, and it is apparent that in many cases clusters with maxima have one more metal atom than clusters showing minima in ammonia $\bar{m}$ values. This is consistent with the nature of the water binding discussed above. The extra metal atom on a "closed-subshell-plus-one" cluster will have the lowest metal-metal coordination of all atoms on the cluster, and will provide the strongest binding site for water. Since the rather weak interaction of a single water molecule with a cluster of a hundred or so metal atoms is unlikely to alter the cluster structure, we conclude that nonammoniated hydrogenated cobalt clusters are also icosahedral in structure, i.e., that ammoniation for the most part does not change cluster structure.

Similar studies [25] on nonhydrogenated cobalt clusters show that they likewise tend to adopt icosahedral structure over much of the 50- to 150-atom size range, although the water binding maxima and ammonia binding minima are not as dramatic as for the hydrogenated clusters. Nickel clusters tend to show even more evidence for icosahedral structure [26], while iron clusters show very little [25], so there appears to be a trend across the transition series. This trend is reasonable, given the directional nature of d-orbitals. As these orbitals are filled, the atoms become more spherical, and are more likely to cluster in the configuration adopted by rare gas atoms. Icosahedral packing minimizes surface free energy, and will be preferred when the surface contains a significant fraction of the atoms in the cluster. Of course, eventually for sufficiently large clusters the surface effects diminish, since bulk metals and solid rare gases do not have icosahedral packing.

If the trend across the periodic table continues, we might expect copper clusters to show strong icosahedral features in their binding patterns. In fact they do, having water binding patterns consistent with icosahedral (sub)shells [27]. But in addition, copper clusters show *electronic* shell structure in their reactions with oxygen: clusters with filled electronic shells are unreactive towards O_2 [27]. This electronic behavior is not unexpected. The reaction most likely involves an electron transfer from the cluster to the O_2, so that the higher a cluster's IP, the less energetically favorable this transfer, and the lower its reactivity. Thus the clusters with filled electronic shells, having higher IPs, are less reactive. IP effects should not be as important for the water reaction, since the interaction with water is a local one, dependent on the local arrangement of metal atoms but not particularly sensitive to a cluster-wide property such as IP. These results for copper clusters provide us for the first time with information on both geometrical and electronic structure for metal clusters. This is just the sort of information we will need if we are to develop an understanding of the interplay between the physical properties of metal clusters and the interactions of clusters with ligand molecules.

4. Chemistry of and on Clusters

Studies of metal cluster chemistry encompass several broad categories. Kinetics measurements seek to determine the reactivity of clusters as measured by reaction rate constants. Saturation experiments, as described above, quantify the capacity of clusters to bind particular ligand molecules. Measurements on reactions at equilibrium permit determination of thermodynamic properties. Processes in which ligand molecules decompose on cluster surfaces provide information on metal-promoted breaking and making of chemical bonds. We will consider here representative examples of kinetics, thermodynamics, and ligand decomposition.

4.1. KINETICS

Some of the earliest studies of metal cluster chemistry focussed on the kinetics of their reactions with small molecules. The first system to be quantified was the reaction of iron clusters with hydrogen [3, 28, 29]. Fig. 7 shows a plot of the pseudo-first-order rate constants for the reactions of Fe_{8-29} with an H_2 molecule [29]. It is immediately apparent that cluster reactivity can vary tremendously with cluster size, changing by over two orders of magnitude with the addition of a single metal atom. The chemisorption of hydrogen on iron clusters is dissociative: the hydrogens bind to the clusters as atoms. The rate-limiting step then is most likely the breaking of the H-H bond, and apparently cluster size can have a strong effect on the rate of this process. The early interpretation of this reactivity pattern was based on the observation that in many cases there is an anticorrelation between cluster reactivity and cluster ionization potential [3, 28]. As for the $Cu_n + O_2$ reaction discussed above, there is presumably an electron transfer from the cluster to the molecule, in this case to populate the unoccupied antibonding orbital of the H_2 molecule, weakening the H-H bond and facilitating its breakage. While this model is

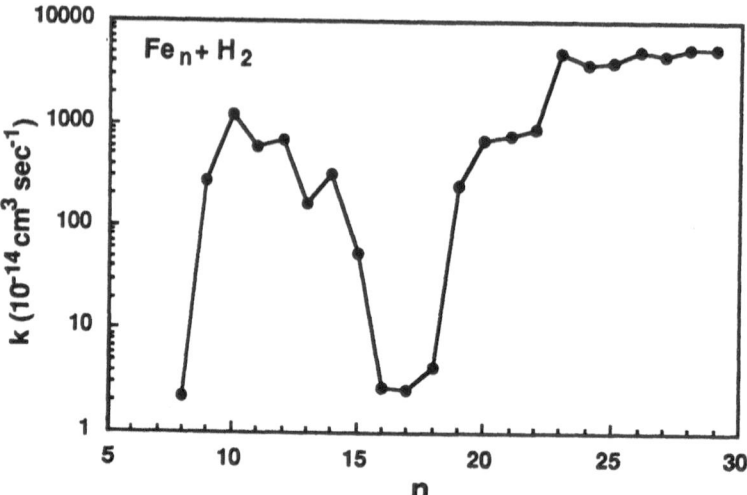

Figure 7. Pseudo-first-order rate constants for the reaction of iron clusters with H_2. Adapted from Ref. 29.

appealing, attempts to quantify the reactivity-IP relationship have not been very successful. More recent IP data[20] show that the anticorrelation with reactivity is not as good as originally thought. Also, thermodynamic data[4] for the equilibrium reaction of iron clusters with D_2O show a reactivity (equilibrium constant) pattern remarkably similar to the hydrogen rate constant data, a trend opposite of that expected if the simple electron transfer model is valid. We now understand that the principal cluster property that is changing with size is structure, and this affects other properties such as reactivity behavior and binding energies (i.e., equilibrium constants). Unfortunately, we do not yet have a clear understanding of the structure of iron clusters in this size range.

The question is often asked: When (at what size) does a cluster begin to behave like bulk metal? The rate constant data in Fig. 7 show a final jump at Fe_{23} to a value that corresponds to a sticking probability of ~5%. (Rate constants have been measured up to Fe_{65}, and there are no further large changes in their values.) 5% is a typical value for the sticking probability of hydrogen on bulk iron [30], so we might say that in this case iron clusters behave like bulk metal abruptly at Fe_{23}. However, other properties, like IPs[11,20] and ligand binding energies [31], have not yet reached their bulk values by a hundred atoms. Clearly the answer to the question above will depend on the specific property.

Kinetics studies have been reported for a series of transition metal cluster reactions with simple molecules [2b] In general, two types of reactivity patterns have been identified: *facile* and *activated*. The iron cluster-hydrogen reaction is characteristic of an activated reaction: relatively small cross sections and strong cluster size dependence. The reaction of hydrogen with nickel clusters, on the other hand, is generally facile: large (nearly unit) sticking probabilities with little or no dependence on cluster size [32]. While kinetics studies *per se* do not really tell us much about actual cluster structure (we need equilibrium processes for that, as described above), they can point to substantial *changes* in structure with size. Once we have a better idea of cluster structure, we should return to the kinetics data to try to arrive at a better understanding of structure-reactivity relationships.

4.2. THERMODYNAMICS

As discussed above, for a cluster-ligand reaction that is at equilibrium, we can measure equilibrium constants K_{eq}. Since $RT\ln K_{eq} = -\Delta H^{\circ} + T\Delta S^{\circ}$, a plot of $R\ln K_{eq}$ vs $1/T$ (a van't Hoff plot) should have a slope of $-\Delta H^{\circ}$ (the binding energy) and an intercept of ΔS°. Fig. 8 shows such a plot for the addition of the 12th NH_3 molecule to Co_{55}. As can be seen, the data are quite linear, supporting the conclusion that at this level of ammoniation such a reaction is an equilibrium one. From the least-squares fit line we derive the ΔH° and ΔS° values given in the figure. Similar measurements on other clusters and for other NH_3 coverages show that indeed the cluster-ammonia binding energy decreases with increasing coverage, but not generally in a linear way [33]. In many cases the nonlinearity can be understood in the context of cluster structure. For example, the binding energies of the 11th and 12th NH_3 molecules to the icosahedral Co_{55} are virtually identical, while the 13th molecule is bound so weakly that it is difficult to measure an equilibrium constant. Also, the binding energy for the 12th molecule on the 54-atom cluster is significantly lower than the 11th, suggesting that the sites created by removing the apex atom are indeed quite close together.

As shown in Fig. 8 the loss in entropy $-\Delta S^{\circ}$ is about 21 e.u. This value is surprisingly smaller than would be expected for the condensation of a gas phase molecule onto a solid surface. Such a process would be dominated by the loss of three translational degrees of

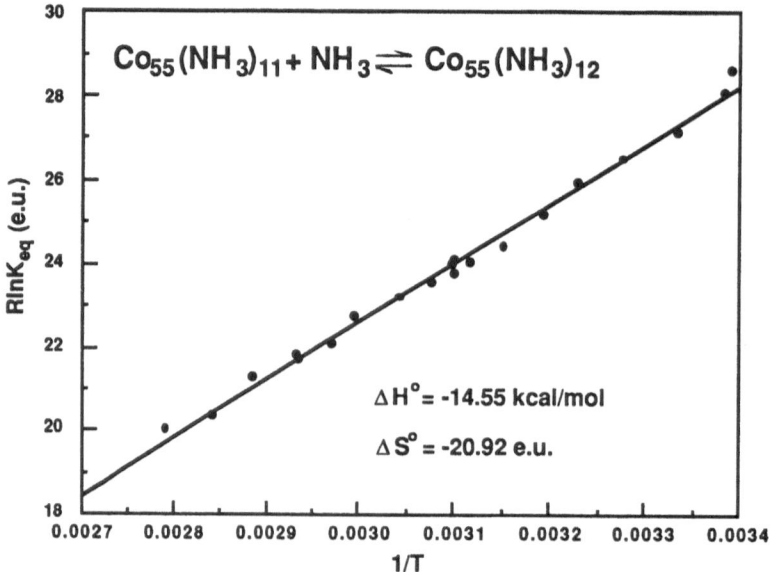

Figure 8. Van't Hoff plot for the equilibrium reaction of $Co_{55}(NH_3)_{11}$ with NH_3. The straight line is a linear least squares fit with the slope (ΔH°) and intercept (ΔS°) values indicated.

freedom, and the entropy loss can be calculated via statistical mechanics to be on the order of 33 e.u. It would seem that ligands on the surfaces of small metal clusters still have substantial freedom of movement, in fact are perhaps almost mobile, so that they retain quite a bit of their gas phase translational degrees of freedom.

4.3. LIGAND DECOMPOSITION

Although strictly speaking the dissociative chemisorption of hydrogen on metal clusters is an example of the decomposition of the ligand molecule, there are other reactions where larger molecules decompose and smaller molecules desorb from the clusters. Such desorption is particularly easy to determine with mass spectrometric detection, since the loss of a different molecule will change the cluster mass in a unique way. We will consider three such reactions: the decomposition of D_2O on iron clusters, the decomposition of NH_3 on nickel clusters, and the decomposition of C_2H_4 on platinum clusters. In each case, the decomposition is accompanied by the loss of H_2 from the clusters. However, as we will see, the detailed mechanisms appear to be quite different.

Fig. 9 shows mass spectra that result from the reaction of iron clusters with D_2O.[4] The upper spectrum was recorded with a D_2O pressure of 20 mTorr and an FTR temperature of 21 ºC and, as indicated, clusters in the Fe_{24-28} size range bind up to three D_2O molecules under these conditions. The lower spectrum shows that when the FTR temperature is raised to about 55 ºC two things happen. First, the distribution of the number of D_2O molecules has shifted to lower values, so that the bare clusters are the most intense peaks in the spectrum. This is exactly what is expected for an exothermic

---

$Fe_n(D_2O)_m^+$

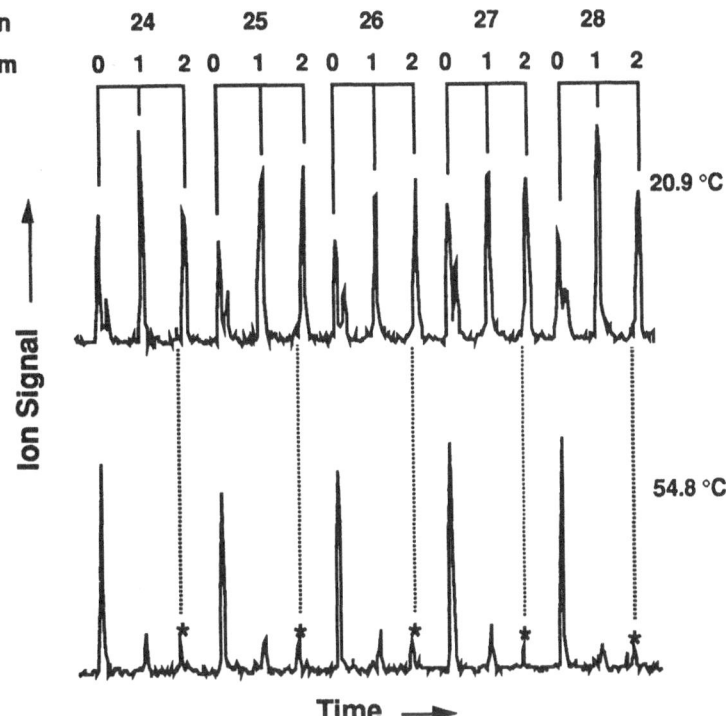

Figure 9. Upper panel: TOF mass spectrum recorded when 20 mTorr D_2O reacts with iron clusters at an FTR temperature of 20.9 °C. The small peaks to the right of the Fe_n peaks are due to $Fe_{n-1}(D_2O)_3$ species. Lower panel: the spectrum recorded when the FTR temperature is raised to 54.8 °C. Adapted from Ref. 4.

reaction that is at equilibrium. Second, in each case the peak corresponding to two D_2O molecules has disappeared, to be replaced with a peak 4 amu to the left, indicating the loss of a D_2 molecule. Notice that a similar peak does not appear for the Fe_nD_2O species.

This implies that it requires two D_2O molecules to produce D_2 loss. Furthermore, we know that iron clusters in this size range can bind many deuterium atoms [34], and bind them quite strongly [31], so we presume that in the desorption process the D atoms never get to the cluster surface. Thus we conclude that the decomposition reaction is a concerted one between two (presumably adjacent) D_2O molecules in which a D atom is contributed from each to form a D_2 molecule that desorbs without interacting directly with the cluster surface.

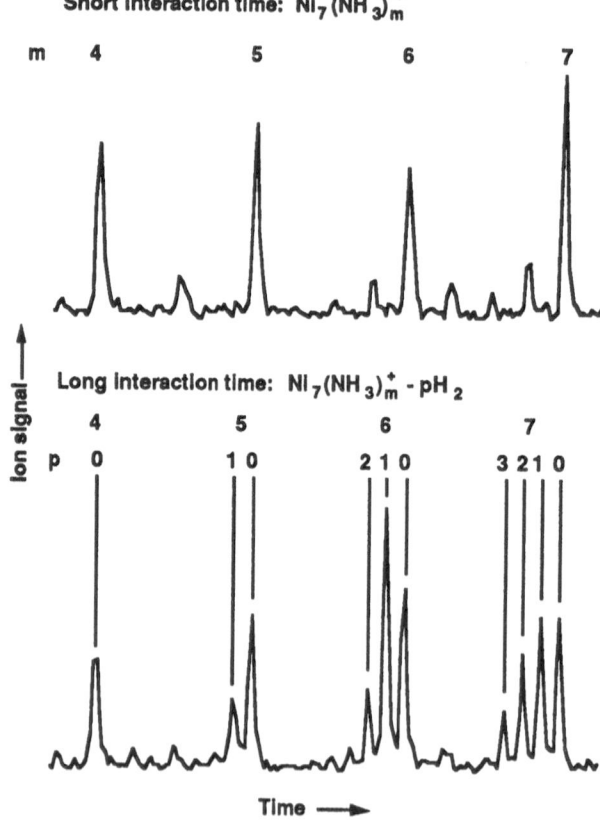

Short interaction time: $Ni_7(NH_3)_m^+$

m   4   5   6   7

Long interaction time: $Ni_7(NH_3)_m^+ - pH_2$

4   5   6   7

p   0   1 0   2 1 0   3 2 1 0

Ion signal ⟶

Time ⟶

Figure 10. Upper panel: TOF mass spectrum recorded when Ni_7 reacts with 220 mTorr of NH_3 for a short interaction time. The principal peaks in the spectrum correspond to $Ni_7(NH_3)_m^+$ with m=4 to 7. Lower panel: the spectrum recorded when Ni_7 reacts with 220 mTorr of NH_3 for a long interaction time. Adapted from Ref. 7.

The upper spectrum of Fig. 10 shows the distribution of reaction products seen for Ni_7 when 220 mTorr of ammonia reacts for a short (~0.2 ms) interaction time at a temperature of 20 °C [7]. Essentially, species with from four to seven NH_3 molecules are seen. The lower spectrum shows the result of allowing the reaction to proceed for a long (~1 ms) time. The additional peaks, separated by 2 amu, indicate the loss of H_2 molecules, with the extent of loss depending on the initial NH_3 coverage. Other than the hydrogen loss, there is approximately the same distribution of ammonias (or, more correctly, nitrogen-containing species) as for the short reaction time, indicating that the initial reaction with ammonia is an equilibrium one. The strong time dependence of the H_2 loss means that some step in the overall reaction sequence that leads to H_2 loss is kinetically limited on the msec timescale.

To probe other aspects of this reaction several additional experiments have been done, including determining temperature and pressure dependence. The most telling experiment, however, is one in which both D_2 and NH_3 are added to the FTR. If the D_2 is reacted under conditions of short cluster-ammonia interaction time, it simply adds to the cluster as though the ammonia were not there. This indicates that under these conditions the NH_3 has not decomposed, since any resulting H atoms would occupy H-atom binding sites and block the binding of some of the D atoms. When D_2 is added to clusters that have interacted with the NH_3 for a long time, a very complex spectrum results, indicative of H/D exchange, implying that N-H bonds have broken and H atoms are bound to the cluster surface. Thus it appears that the rate-limiting step is the breakage of the N-H bond(s), a conclusion that is not surprising in light of many surface science studies of similar processes.

A clue as to the mechanism of the reaction is provided by the systematics seen in the hydrogen loss; up to three H_2 molecules are lost from the cluster having seven NH_3 molecules, two from $Ni_7(NH_3)_6$, one from $Ni_7(NH_3)_5$, and no loss from $Ni_7(NH_3)_4$. If we assume that the final nitrogen-containing species on the cluster is NH, and that it continues to bind to the original NH_3 binding sites (or at least it does not compete with the hydrogen for binding sites), then in each case the largest extent of hydrogen loss leaves eight H atoms bound to the cluster. The number of H binding sites on Ni_7 can be determined in a separate saturation experiment, and is indeed found to be eight. [25]. Thus we summarize the overall reaction sequence as follows. The initial binding of the NH_3 is an equilibrium process. Then, in the rate-limiting step(s), the N-H bonds break, and the H atoms move out onto the cluster surface to find binding sites. Eventually, these sites are filled, and any subsequent N-H bond breakage leads to H atoms that recombine and desorb from the cluster as H_2 molecules. Similar behavior is seen for other small nickel clusters having high ammonia coverage.

The reaction of ethylene with platinum clusters yields product distributions that are in some sense the mirror image of the ammonia-nickel reaction. This is illustrated in Fig. 11, which shows spectra for platinum clusters exposed to 200 mTorr C_2H_4 for short and long interaction times [35]. In this case, the mass spectrometer resolution is insufficient to

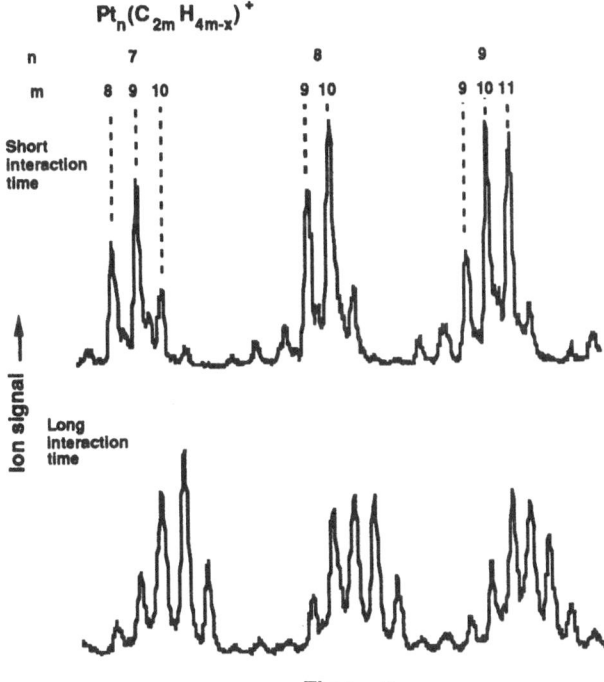

Figure 11. Upper panel: portion of the TOF spectrum recorded when platinum clusters react with C_2H_4 for a short interaction time. Each of the peaks, labeled as to the number of platinum atoms and the number of C_2 units, has a mass corresponding to the loss of significant numbers of H_2 molecules. Lower panel: the spectrum recorded for a long interaction time. Adapted from Ref. 35.

resolve the individual numbers of hydrogen molecules lost, but the masses of the peaks in Fig. 11 do indeed correspond to significant H_2 loss. It is quite clear from the spectra that the distribution of the number of ethylenes (or C_2 units) is in this case quite dependent on reaction time. However, a careful mass analysis indicates that for a given number of C_2 units the extent of H_2 loss is now *independent* of interaction time. These observations suggest a mechanism quite different from that of the ammonia-nickel cluster reaction. It would appear that the initial binding of the ethylene molecules occurs under a rapid equilibrium. Then, in the rate limiting step the ethylenes decompose, with *every* decomposition leading to H_2 loss from the cluster. The resulting carbon-containing species then moves to a binding site different from that initially occupied by the ethylene, allowing an additional molecule to bind to the cluster. Thus the number of C_2 units on the cluster depends on time, since the extent of decomposition depends on time. But for a given number of C_2 units, corresponding to a given number of decompositions, the same number of H_2 molecules will be lost.

These ligand decomposition reactions are prototypes of reactions important in heterogeneous catalysis. Further study of reactions such as these in which chemical bonds are broken and new ones are formed on cluster surfaces will provide us with invaluable insight into the molecular mechanisms of catalysis.

5. Conclusion

This paper has reviewed some of the aspects of the chemistry of isolated transition metal clusters. As was alluded to several times, one important goal of such studies is a determination of the relationships between cluster structure, both geometrical and electronic, and chemical reactivity. We already have some insight into this, in the form of the special binding capabilities of metal atoms at apex sites of icosahedral clusters and lone atoms outside closed (sub)shells of such clusters, and the relative unreactivity of clusters that have filled electronic shells. Future work will be aimed at developing a fuller understanding of the fundamental structure-reactivity relationship for small metal clusters.

ACKNOWLEDGMENTS

We thank Dr. E. K. Parks for many helpful discussions during the preparation of this manuscript. This work was supported by the U. S. Department of Energy, Office of Basic Energy Sciences, Division of Chemical Sciences, under Contract W-31-109-Eng-38.

REFERENCES

1. Deitz, G., Duncan, M. A., Powers, D. E. and Smalley, R. E. (1981), J. Chem. Phys. **74**, 6511.
2. For recent reviews, see (a) Jarrold, M. F. in "Advances in Gas-Phase Photochemistry and Kinetics, Vol. 2: Bimolecular Collisions," edited by M. N. R. Ashford and J. E. Baggot (Royal Society of Chemistry, London, 1989) p. 337; (b) Kaldor, A., Cox, D. M. and Zakin, M. R. (1988), Adv. Chem. Phys. **70**, 211.

3. Richtsmeier, S. C., Parks, E. K., Liu, K., Pobo, L. G. and Riley, S. J. (1985), J. Chem. Phys. **82**, 3659.
4. Weiller, B. H., Bechthold, P. S., Parks, E. K., Pobo, L. G. and Riley, S. J. (1989), J. Chem. Phys. **91**, 4714.
5. Parks, E. K., Nieman, G. C., Pobo, L. G. and Riley, S. J. (1988), J. Chem. Phys. **88**, 6260.
6. Parks, E. K., Nieman, G. C., Pobo, L. G. and Riley, S. J. (1987), J. Chem. Phys. **86**, 1066.
7. Riley, S. J. (1989), Z. Phys. D **12**, 537.
8. Kappes, M. M. (1988), Chem. Rev. **88**, 369.
9. Trevor D. J. and Kaldor, A. (1987), in Suslick, K. S. (ed), ACS Symp., **333**, 43.
10. Zakin, M. R., Cox, D. M., Whetten, R. L., Trevor, D. J. and Kaldor, A. (1987), Chem. Phys. Lett. **135**, 223.
11. Parks, E. K., Klots, T. D. and Riley, S. J. (1990), J. Chem. Phys. **92**, 3813.
12. Knickelbein M. B. and Menezes, W. J. C. (1991), J. Chem. Phys. **94**, 4111.
13. Seabury, C. W., Rhodin, T. N., Purtell, R. J. and Merrill, R. P. (1980), Surf. Sci. **93**, 117.
14. Gasser, R. P. H. (1987), "An Introduction to Chemisorption and Catalysis of Metals", Oxford, New York, p. 161.
15. Marshall M. D. and Muenter, J. S. (1981), J. Mol. Spectrosc. **85**, 322.
16. For a recent discussion, see Schumacher, E. (1988), Chimia **42**, 357.
17. Wood, D. (1981), Phys. Rev. Lett. **46**, 749.
18. Brus, L. E. (1983), J. Chem. Phys. **79**, 5566.
19. Brechignac, C., Cahuzac, Ph., Carlier, F. and Leygnier, J. (1989), Phys. Rev. Lett. **63**, 1368.
20. Yang S. and Knickelbein, M. B. (1990), J. Chem. Phys. **93**, 1533.
21. See, for example, Cohen, M. L., Chou, M. Y., Knight, W. D. and deHeer, W. A. (1987), J. Phys. Chem. **91**, 3141, and references therein.
22. See, for example, Miehle, W., Kandler, O., Leisner, T. and Echt, O. (1989), J. Chem. Phys. **91**, 5940, and references therein.
23. see, for example, Phillips, J. C. (1987), J. Chem. Phys. **87**, 1712.
24. Northby, J. A. (1987), J. Chem. Phys. **87**, 6166.
25. Unpublished results.
26. Parks, E. K., Winter, B. J., Klots, T. D. and Riley, S. J. (1991), J. Chem. Phys. **94**, 1881.
27. Winter, B. J., Parks, E. K. and Riley, S. J. (1991), J. Chem. Phys. **94**, 8618.
28. Whetten, R. L., Cox, D. M., Trevor, D. J. and Kaldor, A. (1985), Phys. Rev. Lett. **54**, 1494.
29. Riley, S. J. and Parks, E. K. (1987) in Jena, P., Rao, B. K. and Khanna S. N. (eds), Plenum, New York, p. 727.
30. Benziger J. and Madix, R. J. (1980), Surf. Sci. **94**, 201.
31. Liu, K., Parks, E. K., Richtsmeier, S. C., Pobo, L. G. and Riley, S. J. (1985), J. Chem. Phys. **83**, 2882.
32. Hoffman III, W. F., Parks, E. K., Nieman, G. C., Pobo, L. G. and Riley, S. J. (1987), Z. Phys. D **7**, 83 (1987).
33. Winter, B. J., Klots, T. D., Parks, E. K. and Riley, S. J. (1991), Z. Phys. D **19**, 381.
34. Parks, E. K., Liu, K., Richtsmeier, S. C., Pobo, L. G. and Riley, S. J. (1985), J. Chem. Phys. **82**, 5470.

35. E. K. Parks and S. J. Riley, in "The Chemical Physics of Atomic and Molecular Clusters," Proc. S. I. F. Course CVII, edited by G. Scoles (North Holland, Amsterdam, 1990) p. 761.

SURFACE SCIENCE STUDIES OF MOLECULAR ADSORBATES ON SOLID SURFACES: A SERIES OF CASE STUDIES

H. KUHLENBECK and H.-J. FREUND
Lehrstuhl für Physikalische Chemie 1
Ruhr-Universität Bochum
Universitätsstrasse 150, 4630 Bochum
Germany

ABSTRACT. One goal of surface science studies is to unravel the changes in electronic and geometric structure a molecule experiences when bound to a substrate surface. In spite of the vast knowledge on molecular adsorbates on solid surfaces there are only very few examples where detailed and rather complete experimental information on the geometric and electronic structure has been collected. One example is the $CO(2x1)p2mg/Ni(110)$ system and we shall discuss this case to exemplify the power of electron spectroscopic studies in this respect.

In light of the question *"How do surface science studies contribute relevant information to mechanistic problems in catalysis?"* we discuss the interaction of molecules with single crystal surfaces of metal and oxide surfaces as monitored with HREELS, TDS, ARUPS, XPS, and NEXAFS.

- We shall discuss the bonding of N_2 to a $Fe(111)$ surface as the prototype system for a small molecule on a metal surface. This system is of some importance with respect to ammonia synthesis.

- In connection with methanol synthesis from CO and H_2 the question of the role of CO_2 in the leading step of the mechanism has turned up. We consider systems where CO_2 is known to chemisorb and discuss possible consequences for the above catalytic reaction.

- As an example for a rather complex reaction we shall discuss results for benzene on $Os(0001)$ where we find interesting intermediate species on the way from molecular adsorption to complete dissociation.

- The overwhelming majority of catalytic reactions takes place on oxide surfaces. Molecular adsorbates on various oxides and their bonding towards the oxide surfaces are discussed and compared with metal surfaces. We consider $NiO(100)$ and $Cr_2O_3(111)$ surfaces interacting with small molecules such as CO, NO, NO_2 and CO_2.

D. R. Salahub and N. Russo (eds.), Metal-Ligand Interactions: from Atoms, to Clusters, to Surfaces, 37–70.

1. Introduction

Among the goals of surface science is the study of the changes in electronic and geometrical structure a molecule experiences when bound to a substrate surface. Quantum mechanically speaking the electronic part of the wavefunction with the wavefunction of the nuclei have to be studied. Of course, a large step towards an understanding of the system is the determination of its symmetry because both the electronic and the nuclear part of the wavefunction transform with respect to the same point or space group. Most of our knowledge of adsorbates stems from electron scattering and/or electron spectroscopy in their variations. Due to the very strong interaction of the electrons used as information carriers with the electrons of the systems, in most cases the geometric (nuclear) structure is indirectly deduced from electronic structure determinations [1,2,3].

This implies that, while we know quite a bit about symmetry properties of adsorbates, i.e. quantities such as orientation of molecular axes or planes, position of mirror and glide planes in two dimensionally ordered systems, our actual knowledge of bond lengths and bond angles is very limited. In the following we use the system $CO(2x1)p2mg/Ni(110)$ as an example [4-11] to demonstrate how electron spectroscopy [6,8,10,11] and electron scattering [9] may be used in this context.

It should be familiar at this point to the reader that CO almost always binds to metal surfaces with its carbon atom oriented towards the surface and the oxygen atom sticking out into the vacuum such that the molecular axis is more or less ($\pm 10°$) oriented perpendicularly with respect to the surface plane [1,2,3]. The so called "light-house" effect in connection with the molecular shape resonance of the oxygen 4σ-lone-pair of CO has been employed to come to this conclusion experimentally [1-3]. We shall come back to this when we discuss the $N_2/Fe(111)$ system [12,13]. The bonding of CO towards metal surfaces may be basically described by the Blyholder model [14], i.e. by the synergetic action of a 5σ-donative and a 2π-backdonative interaction. This leads to a relative shift of the 5σ ionization towards, and often beyond (which means to higher binding energy) the 1π ionization while the relative 4σ ionization energy remains basically unperturbed [1-3]. Our description so far is valid for a single CO molecule interacting with a metal surface. It has to be modified if the adsorbed CO forms an ordered two dimensional array [1-3].

2. How electron spectroscopy is applied to the study of adsorbates

Exposure of a clean Ni(110) surface to CO at T ~ 120 K leads to the formation of a dense CO overlayer with coverage $\theta = 1.0$ which gives rise to a (2x1) LEED pattern [4] (Fig. 1a). This LEED pattern is characterized by spot extinctions along the (110) direction, indicating the existence of a glide plane. Without a detailed I/V analysis it is not possible to tell whether the space group is p1g1 or p2mg. In the former case there is only one glide plane in the (110) direction, in the latter case there is additionally a mirror plane in the (001) direction of the substrate.

Photoemission may be used to decide this problem [10]. Firstly, the symmetry of the wave functions is reflected in the transition matrix elements which determines the differential photoemission current [1-3]:

$$d\sigma/d\Omega \ \alpha \ |<\Psi_f|p\ |\Psi_i>|^2$$

39

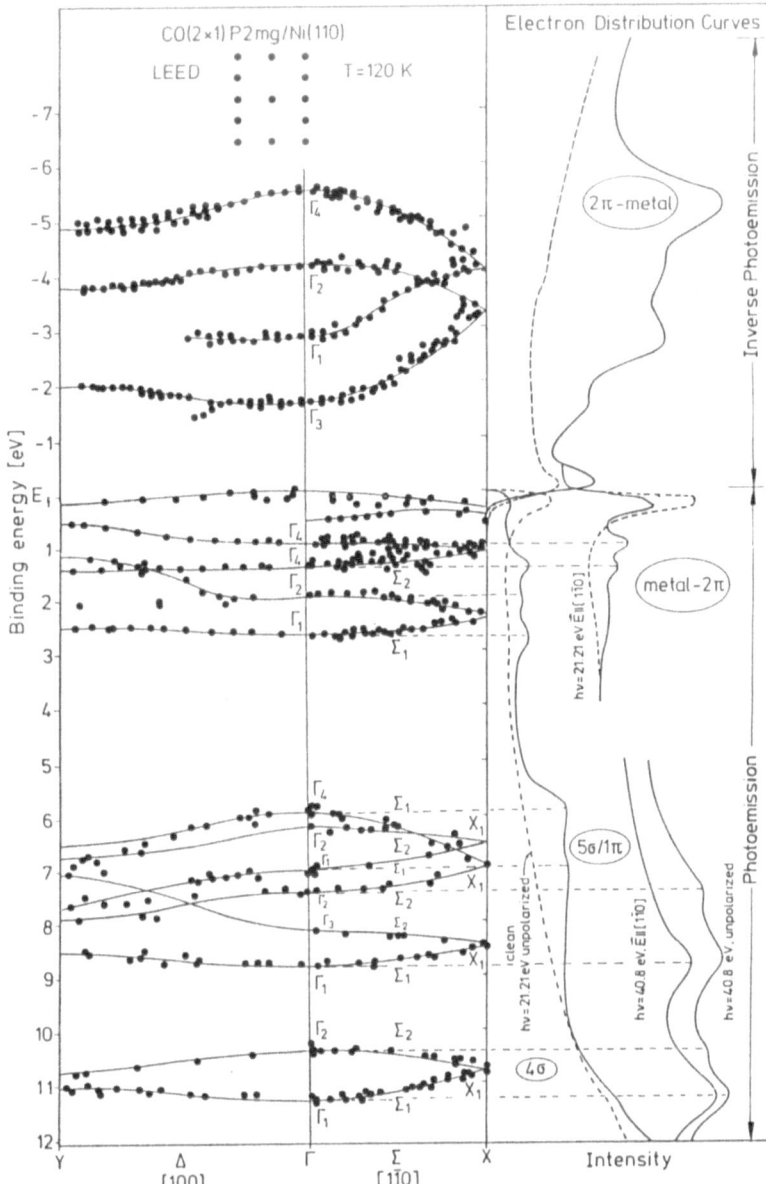

Figure 1. Band structure of the occupied and the unoccupied valence bands of CO(2x1)p2mg/Ni(110). On the right hand side IPE spectra (upper right panel) and photoelectron spectra (lower right panel) of the adsorbate system are shown. The band structure of the unoccupied states and the inverse photoemission spectra have been reproduced from ref. [11].

where Ψ_f and Ψ_i are the final and initial state wave functions involved in the ionization process and p is the momentum operator the direction of which may be varied by changing the polarization of the light. Secondly, the ordering in the adsorbate layer causes a band structure to develop, which reflects the symmetry of the system.

Fig. 1 shows the band structure of the system as determined by measuring the binding energies of the CO induced features as a function of the electron emission angle along two azimuths with respect to the Ni(110) substrate, i.e the (110) and the (001) directions. Together with the experimental data we reproduce the result of a band structure calculation [6]. On the right hand side of the collected dispersion data [6,10,11] we show a set of photoelectron spectra at the Γ-point (normal emission), measured using unpolarized light of a He discharge source. Clearly, the number of outer valence features is larger than four, which would be the maximum number of features for a single molecule within the unit cell, i.e. 5σ, 4σ and two 1π components. This indicates that the unit cell contains more than one molecule. The region of the energetically well separated 4σ emissions points to a splitting into two features consistent with two molecules in the unit cell.

The different dipole selection rules for the p2mg and the p1g1 symmetry groups can be used to decide whether the system posesses p2mg or p1g1 symmetry. According to the p2mg selection rules Γ_4 bands can only be observed with light polarized along the (110) azimuth whereas in p1g1 these states can also be ionized with light polarized perpendicular to the surface. The result of such a test is shown in Fig. 2. For the upper spectrum light polarized predominatly along (110) was used whereas in the lower spectrum the component of the electric field vector along (110) was zero. If the adsorbate symmetry were p1g1 the $1\pi_x{}^+$ state, which transforms according to Γ_4, should also be visible in the lower spectrum. Since this is not the case we conclude that the adsorbate symmetry is most likely p2mg and not p1g1.

Considering the coverage $\theta=1$ and the symmetry of the overlayer, the structure model in Fig. 3 [6] is a reasonable first guess. The CO molecules are canonically bound carbon end down (see above) and tilted along the (001) azimuth in order to avoid the close intermolecular contact along (110) with a separation below 3Å which could occur if the molecules were oriented perpendicularly.

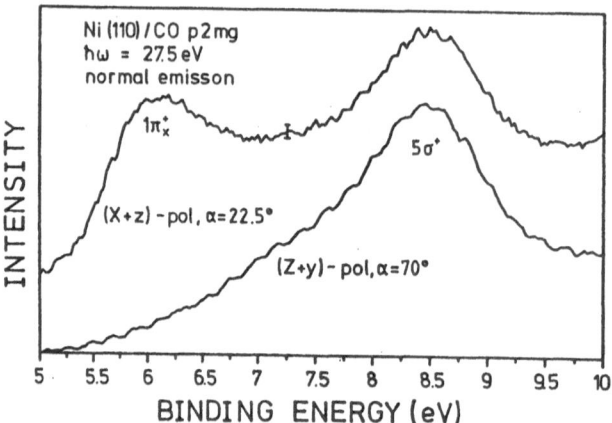

Figure 2. Photoelectron spectra of CO(2x1)p2mg taken with different polarizations of the incident light.

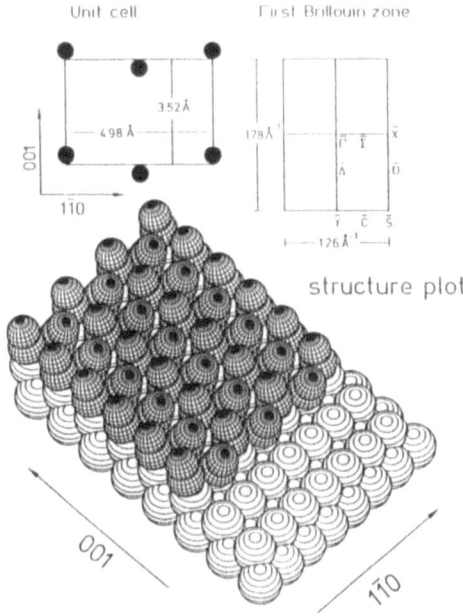

Geometric structure of Ni(110)/CO(2×1)p2mg

Unit cell   First Brillouin zone

structure plot

Figure 3. Upper left panel: Size, points, and lines of high symmetry within the Brillouin zone. Lower panel: Structure plot of the CO(2x1)p2mg adsorbate layer.

The determined band structure in the region of occupied (by angle resolved ultraviolet photoelectron spectroscopy, ARUPS) and unoccupied (by inverse photoelectron emission, IPE) states as shown in Fig. 1 reflects the structure of the adsorbate layer throughout the entire Brillouin zone. Two directions in k-space are plotted, i.e. the (110) direction of the glide plane, and the mirror plane direction (001). The solid lines are the result of a tight-binding calculation and the relatively good agreement between experiment and calculation reflects the fact that the observed band structure is mainly determined by the strong intermolecular repulsion of the CO molecules [6]. As stated above, the presence of two CO molecules per unit cell leads to a doubling of the number of bands at Γ and throughout the Brillouin zone, except for the line X-S, where the glide plane symmetry only allows degenerate bands. The individual band dispersions reflect the different strengths of intermolecular interaction, which the individual CO levels with their different extension of the radial wave functions experience.

Fig. 4 represents schematically two dimensional wave functions consisting of σ-type orbitals (left panel) and of π (π_x, π_y)-type orbitals (two right panels) at the Γ point [6]. The bonding and the antibonding combinations are shown. Consider Fig. 4 (left panel) to represent the 4σ orbital of CO. The splitting between the two components may be taken from experiment to be 0.85 eV. This value should be compared with 1.3 eV for the splitting between the 5σ levels, which hybridize with the $1\pi_x$ component. The increase

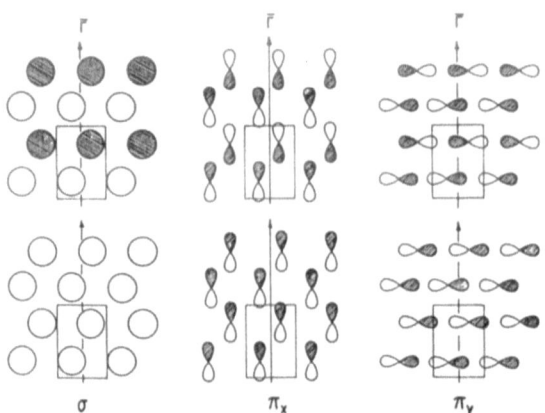

Figure 4. Symmetric and antimetric linear combinations of two dimensional wave functions of σ, π_x, and π_y symmetry at the Γ point of the (2x1) p2mg Brillouin zone.

reflects the larger radial extent of the 5σ-carbon-lone-pair, as compared with the 4σ-oxygen-lone-pair. In addition, however, substrate mediated interaction is important for the 5σ levels, because they establish part of the bonding to the metal. The largest splitting in the occupied bands is observed for the $1\pi_x$ component reaching a value of almost 2.1 eV, while the $1\pi_y$ component splitting (0.8 eV) is relatively small. From Fig. 4 it is obvious that the direct overlap is the origin for this finding. On the other hand the 2π emission observed by IPE [11], and plotted in Fig. 1, exhibits even larger splitting for both, the $2\pi_x$ (3.8 eV) and the $2\pi_y$ (1.2 eV) components. This may be caused again by substrate mediated interaction due to 2π-backdonation, but an additional role may be played by the 2π wavefunctions themselves, which are expected to be considerably more diffuse as compared with the 1π levels.

A direct observation of the substrate mediated interaction is possible by following the CO induced features in the region of the metal band structure, i.e. between the Fermi energy and a binding energy of 3 eV [10]. In this energy range Fig. 1 shows bands which exhibit dispersions of up to 0.5 - 1.0 eV indicating a substantial contribution of this type of interaction. The magnitude of the dispersion may be correlated and explained by the interaction of the metal levels involved in the CO 2π-metal backbonding. We refer to the literature [10] for details.

Even though photoemission [6,10] and inverse photoemission data allow one to draw rather detailed conclusions on the intermolecular interactions and the electronic structure of the CO-metal bond, it is not possible in a straightforward manner to determine the site of adsorption, namely whether the molecule is terminally or bridge bonded to the surface. This may be done considering electron energy loss data recorded with very high resolution by H. Ibach and his group [9]. In the present case a simple decision on the basis of frequencies is not possible because the observed value, close to 1900 cm^{-1}, is on the border between on-top and bridge site frequencies. It is even more complicated in the present case. One spectrum showing all relevant features together with the phonon dispersions is shown in Fig. 5 [9]. The assignment of the modes is discussed in detail by Voigtländer et al [9]. Briefly, for a CO adsorbate system (2000 - 1800 cm^{-1}) we expect four or six normal modes depending on whether the CO molecule is bound on-top or via a bridge site.

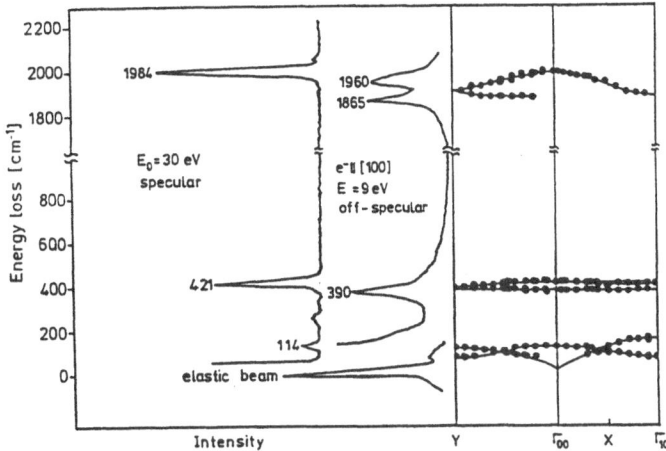

Figure 5. Electron energy loss spectra and phonon dispersion curves for the CO(2x1)p2mg/Ni(110) adsorbate. Data have been taken from ref. [9].

Schematic representations [9] are depicted in Fig. 6. In addition to the CO stretch (υ_2) which are located at different frequencies depending on the site, we expect one degenerate (for on-top site) and two non degenerate (for bridge site) "frustrated" rotations and in the same way "frustrated" translational modes. On the basis of cluster calculations for CO on Ni(100) the vibrational frequencies of the modes have been estimated [15]. Interestingly, the degenerate frustrated rotation for on-top sites should be situated slightly below the metal CO stretch, while for bridge sites the splitting of the degenerate mode is large so that the two components are located well above and below the metal-CO stretch. A measurement of the frustrated rotations of the system would thus allow us to differentiate between on-top and bridge site. However, to apply these considerations to our adsorbate of p2mg symmetry we have to take into account that there are two molecules in the unit cell, which introduces an additional, so called Davydov splitting of the vibrational modes. The splitting documents the fact that even and odd linear combinations of the normal modes of the individual molecules have to be formed and the linear combinations are not necessarily degenerate. The magnitude of the Davydov splitting and thus the observation of additional dispersions depends on the CO-CO intermolecular interaction, which may be classified according to direct through-space-interaction to indirect substrate mediated dipole-dipole interactions, and to interactions arising from the dynamic dipole moments of the CO molecules. Voigtländer et al. [9] have shown that it is mainly dipole-dipole interaction that determines the Davydov splitting although strong direct through-space-overlap causes the adsorbate to exhibit p2mg symmetry in the first place.

The most prominent effect of a Davydov splitting is observed for the CO stretch. The mode at 1984 cm^{-1} (at Γ) is the totally symmetric mode with both CO molecules in the unit cell vibrating in phase. This mode is observed in specular scattering and throughout the entire Brillouin zone. The mode with the CO molecules vibrating out of phase situated at lower frequencies is even with respect to the mirror plane and odd with respect to the glide plane. It is therefore not observed at Γ but along Γ-Y at off-specular scattering along the (001) azimuth. For electron scattering along (110) the even eigenmodes are observed only in the first Brillouin zone, the odd eigenmodes in the second. At the zone boundary

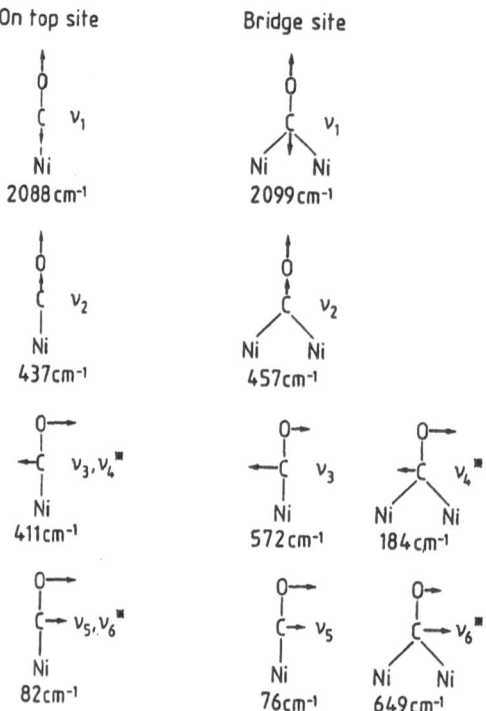

On top site   Bridge site

Figure 6. Schematic diagrams of the normal modes of CO adsorbed on-top sites and on bridge sites, and calculated loss energies for CO(2x1)p2mg/Ni(110). Data have been taken from ref. [9].

both modes are degenerate by symmetry. Therefore the modes appear as one mode without splitting in the dispersion curve from Γ_{00} to Γ_{10}. Also the other modes are expected to be split into even and odd combinations. However, the observed small dispersion, implying a small splitting of the losses at 422 and 385 cm^{-1}, prohibits the observation. The mode split off from the 113-60 cm^{-1} vibration which lies at 60 cm^{-1} at Γ has too low a frequency to be observed.

Summarizing the arguments at this point [9], the modes reproduced in Fig. 6 represent all modes of the system including the additional modes caused by the Davydov splittings. Clearly, the number of modes in the frustrated rotation-translation regime is not consistent with a bridge site. Therefore we must conclude that the molecule is bound on-top of the Ni atoms towards the surface. HREELS (high resolution electron energy loss spectroscopy) provides us thus with a further important structural detail of the adsorbate.

Further information on the CO-Ni bonding may be deduced from the HREELS data: The quantitative theoretical description of the dispersion data is more sensitive to the displacement of the center of gravity of the CO molecules from the ideal on-top position and not so strongly sensitive to the tilt angle of the molecular axis. On the other hand XPD (X-ray photoelectron diffraction) data [8] of the system which are reproduced in Fig. 7 are dominated by the angular information. The XPD data, i.e. the variation of the

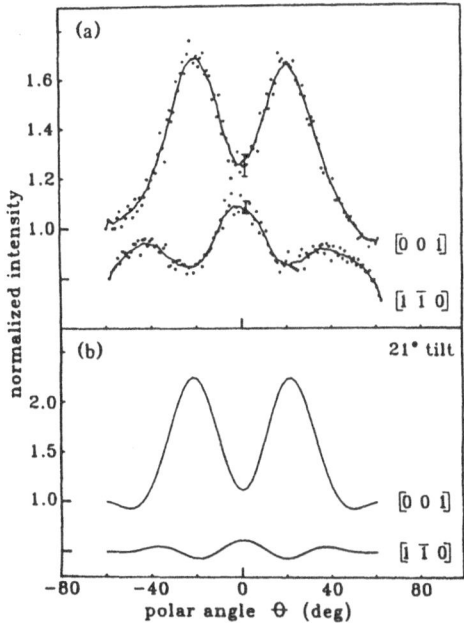

Figure 7. C1s intensity normalized to the O1s intensity as a function of the electron detection angle along two high symmetry azimuths of the CO(2x1)p2mg adsorbate on Ni(110). In the lower panel calculated curves for a CO tilt angle of 21^0 are shown. Data have been taken from ref. [8].

C1s emission intensity as a function of polar angle, indicates a tilt angle of 20^0, very similar to the value deduced from ESDIAD (electron stimulated desorption of ions' angular distribution) [5,7] and ARUPS [6] data. Since the lateral displacement deduced from the vibrational data may be converted to a tilt angle of 20^0 in agreement with XPD we may conclude that the axis of the displaced CO molecule still points towards the center of the Ni atom of the first layer.

In light of the question: "How do surface science studies contribute relevant information to mechanistic problems in catalysis?" we discuss the interaction of molecules with single crystal surfaces of metal and oxide surfaces as monitored with HREELS, TDS, ARUPS, XPS and NEXAFS.

3. Case studies

3.1. CASE 1: N_2/FE(111)

We shall discuss the bonding of N_2 to a Fe(111) surface [13] as the prototype system for a small molecule on a metal surface. This system is of some importance with respect to ammonia synthesis, where it has been shown by Ertl and his group [12] that N_2 dissociation on Fe is the rate limiting step in the formation of NH_3 from N_2 and H_2.

Adsorption of molecular nitrogen and its subsequent dissociation into atomic nitrogen on Fe(111) has been studied in some detail in recent years [12,16-19]. Two weakly chemisorbed molecular N_2 states have been identified mainly via high resolution electron energy loss spectroscopy (HREELS) [20-22], X-ray photoelectron spectroscopy (XPS) [20], and thermal desorption spectroscopy (TDS) [21,22]. In the so called γ-state, with an adsorption enthalpy of ~ 24 kJ/mol [19], the N_2 molecules are terminally bonded to first layer Fe atoms [22]. The slightly more strongly bound α-state, with an adsorption enthalpy of ~ 31 kJ/mol [20], which is the precursor to N_2 dissociation on the surface, has been attributed to N_2 π-bonded to the surface [20]. While the γ-state has recently been found to have a vibrational N-N stretching frequency of 2100 cm^{-1} [20], the α-state exhibits an unusually low stretching frequency of 1490 cm^{-1} [20], which in turn was used to infer together with XPS results the π-bonded nature of the state [20].

The separation of α- and γ-states by an activation barrier allows one to selectively depopulate the γ-states at higher temperature (110 K), while at lower temperature (below 77 K) α- and γ-states are both populated as has recently been clearly demonstrated through HREELS studies [22].

Fig. 8 shows the HREEL spectra at T = 74 K (a) dominated by the γ-state and at T = 110 K (b) dominated by the α-state. After heating the surface to T = 160 K both molecular states are no longer present on the surface [20-22].

Even though the cited HREELS and XPS results suggest that the α-state interacts "side-on" with the metal, the presented HREELS experimental results do not allow conclusions about the geometric structure. Unfortunately, a LEED structure analysis cannot be undertaken due to the lack of any sharp adsorbate induced pattern for molecular N_2 adsorption on Fe(111). Results of theoretical calculations [23] indicate that the γ-state is bonded normal, the α-state parallel to the surface. ARUPS is the method of choice to actually determine the orientation of the molecular axis. The directional properties of the so called shape resonance [24,25] may be used to approach the problem. Fig. 9 shows a schematic representation of the molecular states involved in the excitation processes. The

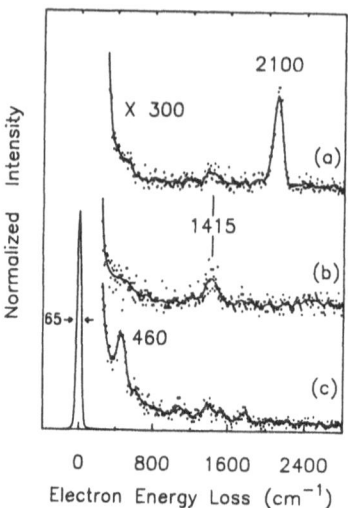

Figure 8. HREELS data for $^{15}N_2$ on Fe(111) taken at different temperatures. a) T = 74 K, b) annealed to T = 110 K, c) annealed to T = 160 K. Data have been taken from Ref. [22].

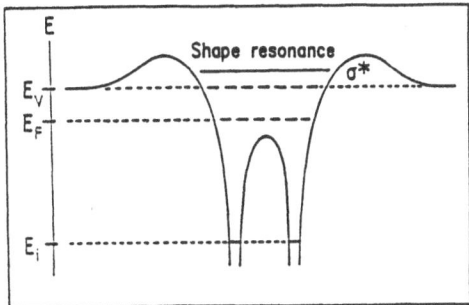

Figure 9. Schematic potential energy diagram for a diatomic molecule like N_2.

molecular potential may be thought of as created by the superposition of the atomic potentials. Due to the centrifugal barrier the molecular potential exhibits potential energy barriers which lead to resonances within the continuum. These resonances may be connected with highly excited unoccupied orbitals. In the case of the diatomics this orbital is the antibonding σ-bond orbital. Since these resonances can be assigned to an irreducible representation of the molecular point group, selection rules are operative in the population of these states. In the present case the σ shape resonance may be only excited with the light polarization vector pointing along the molecular axis if the initial state of the excitation is a σ-state. Consequently, the electron emitted from the resonance by tunneling through the potential high barrier, propagates along the direction of the molecular axis [1-3]. With ARUPS it should be easy to detect this resonance and determine the direction of the axis. Fig. 10 shows a set of angle resolved spectra as a function of photon energy at low temperature (T = 77 K, γ-N_2) and higher temperature (T = 110 K, α-N_2) [13]. Fig. 10a reveals the σ-shape resonance in normal emission for z - polarized light at T ~ 77 K (see inset for the emission geometry of the experiment). Fig. 10b shows a σ-resonance, but only in off-normal emission for σ-polarized light (compare upper and lower panel of the figure). If we choose the same experimental conditions as for Fig. 10a we only see the 1π emission along the surface normal. The comparison clearly reveals the geometrical change between the γ- and the α-state of adsorbed N_2. A detailed comparison of the shifts of the photoemission features with respect to the gas phase indicates that although the orientation of the molecular axis in the α-state is strongly inclined with respect to the surface normal the site appears to be non-symmetric [13]. A possible coordination site is shown schematically in Fig. 11. The tilted N_2 molecule binds five iron atoms, so that the surface can only accommodate a small number of N_2 molecules which explains the observed low coverage (10 - 20% of a monolayer) for the α-state [19].

To understand the bonding of N_2 to the surface it has to be noted that for several chemisorption systems the adsorbate molecule can be considered to be an ion state or an excited state molecule stabilized by the substrate [13].

We invoke this hypothesis in the present study for the α-N_2 species and consider which excited states of molecular N_2 might be reasonable candidates to form a surface complex. The lowest excited state of N_2 is the $A^3\Sigma_u$ state [26], which has a vibrational frequency of 1460 cm^{-1} (N_2 ground state $^1\Sigma_g$: 2360 cm^{-1}) and a bond length of 1.29 Å(N_2 ground state: 1.10 Å) and lies 6.22 eV above the ground state. This state has two attractive features as a candidate for the α-N_2 surface complex: (1) the vibrational frequency is very close to that observed for α-N_2, and (2) the bond length is increased by

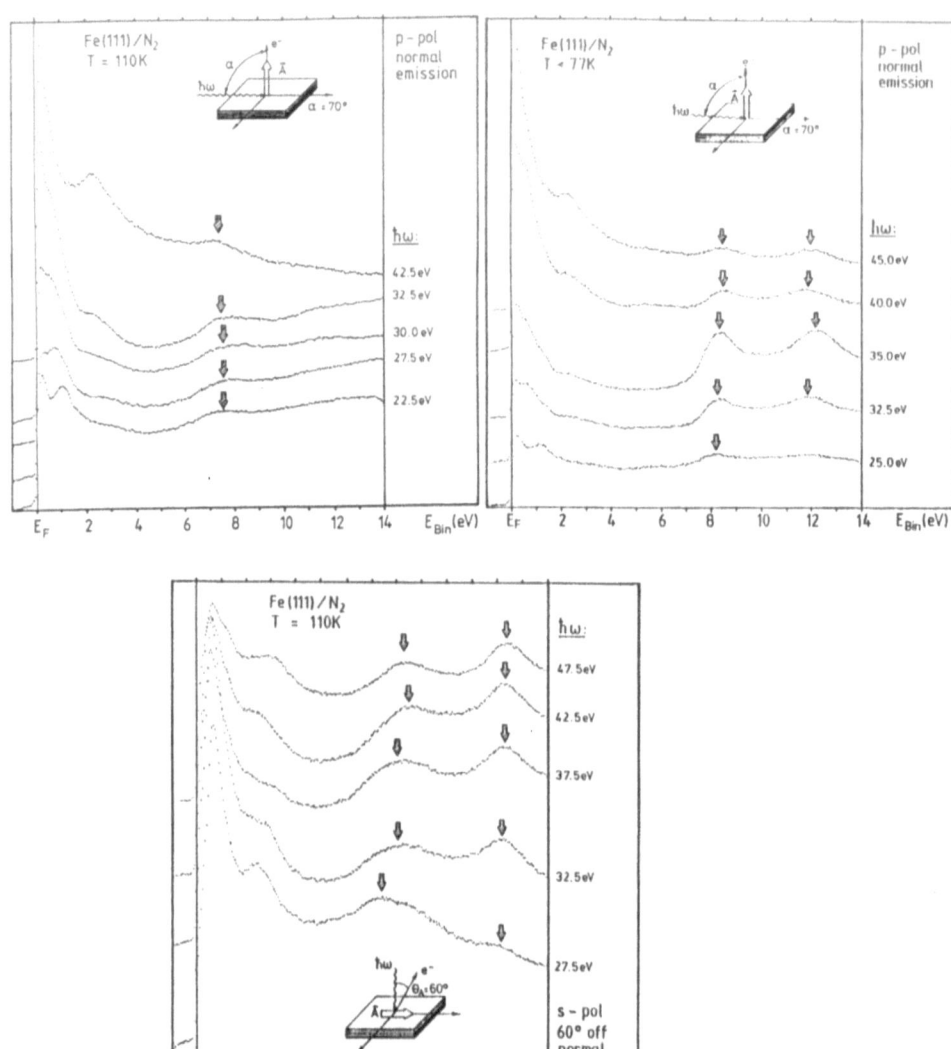

Figure 10a. (upper right panel): Photoelectron spectra in normal emission using p-polarized light as a function of photon energy for the low temperature phase of N_2 on Fe(111). Fig. 10b. (upper left panel): Normal emission photoelectron spectra of the high temperature phase of N_2 on Fe(111) using p-polarized light as a function of photon energy. Fig. 10c. (lower left panel): Normal emission photoelectron spectra of the high temperature phase of N_2 on Fe(111) excited with s-polarized light as a function of photon energy. The electrons have been collected 60° off normal.

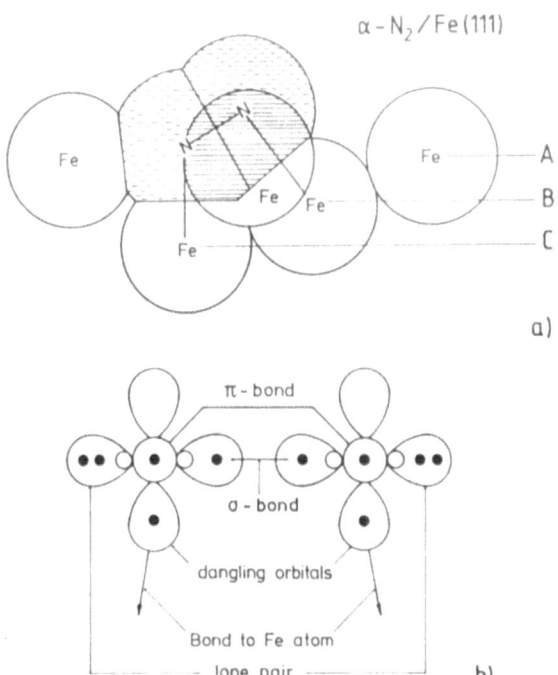

$\alpha - N_2 / Fe\,(111)$

a)

b)

Figure 11. (a) Schematic representation of a possible non symmetric bonding site of α-N_2 on Fe(111). (b) Schematic representation of the $^3\Sigma_u^+$ excited state of N_2 which can form two covalent bonds towards the metal atoms in the geometry shown in (a).

about 0.2 Å from the ground state - an important ingredient for a potential precursor to dissociation. But, the excitation energy of 6.22 eV requires considerable bonding interactions to stabilize this state. However, two covalent nitrogen-iron single bonds can be formed by the N_2 ($^3\Sigma_u$)state; each such bond should contribute a bond energy of about 2.5 eV. This is still not sufficient to counterbalance the 6.22 eV excitation energy and the remaining energy would have to be obtained via several dative bond interactions and electrostatic effects due to polarization. It is not unreasonable, therefore, that the combination of these contributions could yield a species consistent with the known properties of α-N_2. By contrast, we expect the γ-N_2 state to be the ground state of N_2 ($^1\Sigma_g$) weakly bonded to the suface by a single dative interaction.

A schematic representation (in terms of correlated orbitals of σ- and π-symmetry) of this state as it might bond to the surface is given in Fig. 11 [13]. We note that the state corresponds to a broken "π-bond" with the two resulting dangling orbitals available for forming covalent bonds to the surface. In addition there is the potential for forming several dative bonds in this unique geometrical arrangement. Hence, it appears that the proposed site geometry does offer the considerable bonding interactions which are necessary to stabilize the $^3\Sigma_u$ state. It will require considerable further experimental and theoretical work to test the validity of our proposed bonding model for α-N_2, however we feel it may be valuable as a starting point for discussion to have a definite microscopic model in mind.

3.2. CASE 2: $CO_2/Ni(110)$

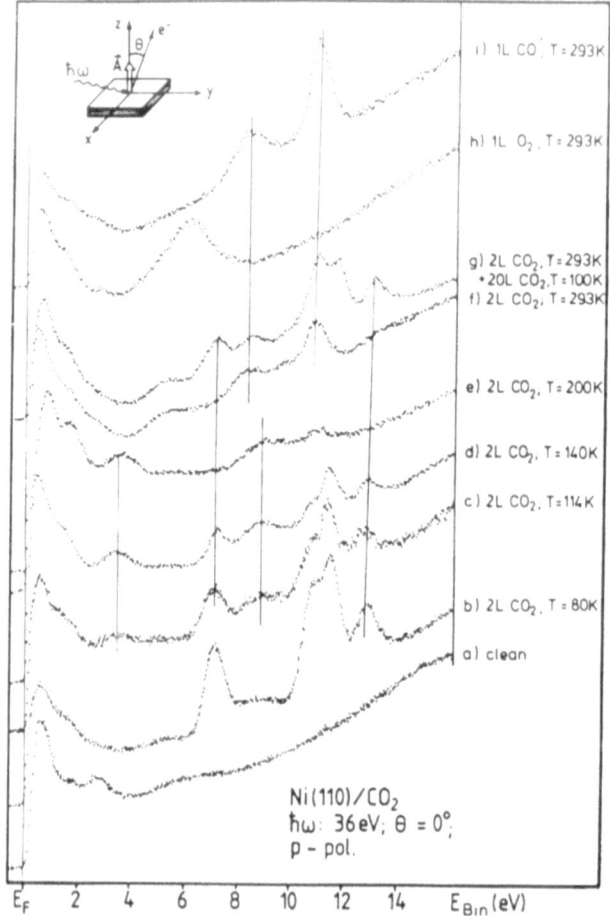

Figure 12. Photoelectron spectra of a $CO_2/Ni(110)$ adsorbate taken at various temperatures in comparison with a clean (a), an oxygen covered (h) and a CO covered surface (i).

In connection with methanol synthesis from CO and H_2 the question of the role of CO_2 in the leading step of the mechanism has turned up [27]. Also, in connection with the elementary steps in the Fischer-Tropsch reaction, CO_2 adsorption may play an important role [28].

Several groups have studied the adsorption of CO_2 on metal surfaces [29-33], and for a long period of time it was assumed that CO_2 does not chemisorb on metal surfaces without dissociation [34,35]. We have investigated, as an example, interaction of CO_2 with a Ni(110) surface, applying a variety of methods [29,30,33,36]. A series of normal emission ARUP spectra of the system $CO_2/Ni(110)$ [29] in Fig. 12 indicates the usefulness of ARUPS to identify reaction intermediates in favourable cases. At low temperature CO_2 adsorption leads to mixed chemisorbed/physisorbed layers (spectrum b), and the physisorbed species desorbs selectively by elevating the temperature (spectra b)-e)). Around 200 K a spectrum of the pure chemisorbed species is found which shows three features. One additional feature around 5 eV is forbidden in normal emission, indicating C_{2v} symmetry of the adsorption site. Comparison with results of cluster calculations [37,38] shows that this is a bent anionic CO_2^--species. Whether the CO_2^--species is carbon or oxygen bound to the surface cannot be decided on the basis of the ARUPS results alone. However, we shall discuss HREELS results showing that the $CO_2^{\delta-}$-species is bound through the oxygen atoms to the surface. Above 200 K (spectrum f)) this species dissociates into CO and O, both adsorbed on the surface, as is clear from a comparison with the spectra of pure CO and O adsorbates (spectra h)-i)). It was concluded from this study that $CO_2^{\delta-}$ is an intrinsic precursor for CO_2 dissociation [29].

Fig. 13 shows a set of HREELS spectra [29] after dosing a Ni(110) surface at low temperature (T~170 K) with CO_2 and heating it up to above 200 K. At low temperature the strong band at 670 cm^{-1} indicates the presence of undistorted CO_2 and has been assigned to the CO_2 bending mode. The loss feature at 2350 cm^{-1} is due to the asymmetric stretch of linear CO_2 and exhibits small but appreciable intensity.

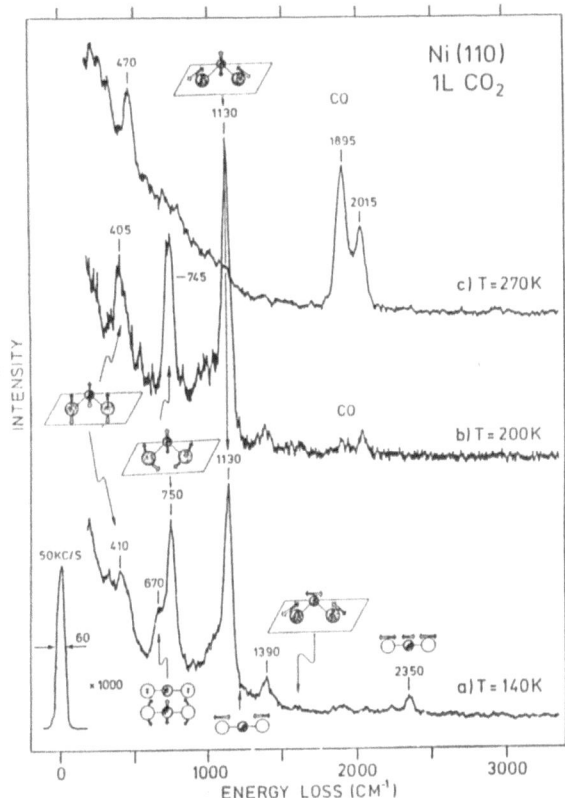

Figure 13. Vibrational electron energy loss spectra of a Ni(110) surface exposed to 1 L CO_2 at 140 K as a function of the annealing temperature.

The dynamic dipoles of the two mentioned CO_2 vibrations behave differently, i.e. the stretching mode is polarized along the molecular axis, the bending mode perpendicularly with respect to the molecular axis. Therefore, if the molecular axis were oriented parallel to the metal surface the stretching mode should be completely screened, if the molecular axis were oriented perpendicular to the metal surface the bending mode should be screened. Both modes are active, however, with the bending mode strongly dominating, which indicates that CO_2 is oriented not quite but almost parallel to the surface.

As has been discussed above in addition to linear CO_2 there is a bent $CO_2^{\delta-}$ species present on the surface which leads to losses at 272 cm^{-1}, at 1103 cm^{-1}, and at 403 cm^{-1}. These losses have been assigned to the bending mode, the symmetric stretch and the molecule-substrate stretching mode of $CO_2^{\delta-}$, respectively. The loss feature at 1352 cm^{-1} has been observed in earlier work [29] and has been attributed to the presence of carbonate. We shall provide conclusive evidence [36] that it is due to the formate symmetric stretch.

If the adsorbate is heated from 90 K to a temperature close to 200 K we observe characteristic changes of the loss intensities.

Firstly, between 120 K and 170 K the features due to linear CO_2 disappear indicating the reaction of this surface species. Note, that upon elevating the temperature the relative intensities of CO_2 bending- and stretching modes change, i.e. the stretching vibration gains intensity. This may indicate a slightly different average adsorbate geometry of linear CO_2 with its axis assuming a smaller inclination with respect to the surface normal at higher surface temperature.

Secondly, while the CO_2 features disappear the $CO_2^{\delta-}$-losses gain intensity suggestive of the earlier finding CO_2 is partly transformed into bent $CO_2^{\delta-}$. As discussed in detail by Bartos et al. [29] before, the asymmetric stretch of the $CO_2^{\delta-}$ is missing on Ni(110) due to the C_{2v} symmetry of the adsorbed species.

Above 200 K the $CO_2^{\delta-}$-signal has disappeared and the typical CO + O spectrum is observed indicating similar to ARUPS the dissociation of CO_2 to CO + O. Such processes have been found on several clean and modified (by alkali metal) surfaces [39-46]. In particular alkali appears to promote the process. We have indications that the geometry of the chemisorbed $CO_2^{\delta-}$ in the latter case is different from the one on pure Ni(110). One indication is the frequency of the molecule-metal vibration. While we observe a band at 405 cm^{-1} for Ni (110) typical for oxygen-metal vibration we find a band at 282 cm^{-1} compatible with a carbon coordination to the metal [39-41]. We may understand this behaviour on the basis of some simple consideration: the bonding on the unmodified surface is dominated by the greater stability of the oxygen-metal dative bond as compared with the covalent C-Ni bond involving a carbon 2p and a Ni-4s electron. However, if we modify the surface with alkali, a strong surface dipole is created which influences the metal-$CO_2^{\delta-}$-dipole in such a way that the molecule in its trying to maximize the compensation of the alkali-metal dipole has to reorient with respect to the surface.

Obviously, the existence of $CO_2^{\delta-}$ on the surface opens up the possibility to try to react this intermediate with other species on the surface. In addition, the reactivity may depend on the $CO_2^{\delta-}$-metal-bond, e.g. it may be different for a clean and a modified surface. In fact, $CO_2^{\delta-}$ on Ni(110) readily reacts with coadsorbed hydrogen as is shown in Fig. 14. At low temperature the situation is very similar to the one described above but starting at a surface temperature of T = 125 K. Above temperatures of T = 200 K the spectrum is very simple. It is characterized by three losses (excluding the CO loss at 1880 cm^{-1}) at 403 cm^{-1}, 727 cm^{-1} and 1353 cm^{-1}, and a very weak loss at 2904 cm^{-1}. The latter one becomes clearly visible at off specular scattering angles (see Fig. 14). Clearly this is the

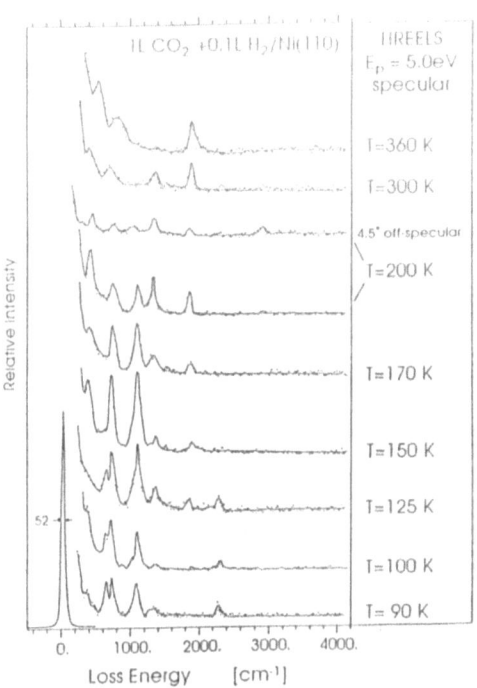

Figure 14. Series of HREEL spectra of 1 L CO$_2$ + 0.1 L H$_2$/Ni(110) as a function of temperature.

loss spectrum of adsorbed formate as reported by Richardson and coworkers [47], who also observed the unusual high intensity of the C-H stretch under off specular conditions. (Note that we are able to reproduce their results by adsorption of formic acid and subsequent heating to room temperatures). The formate species formed on the surface is stable up to temperatures between 300 K and 340 K. The only species stable on the surface close to 360 K appears to be adsorbed CO.

The above analysis of the reaction is also corroborated by the set of XP-spectra taken with synchrotron radiation shown in Fig. 15. The spectrum at the top reflects the relatively high concentration of physisorbed CO$_2$ (binding energy 291.2 eV) with respect to CO$_2^{\delta-}$ (at 286.6 eV). The binding energies are very similar to those reported by Illing et al. [30]. At T = 120 K physisorbed CO$_2$ has been widely attenuated, but CO$_2^{\delta-}$ remains on the surface. At the same time a small CO induced feature grows in at 285.6 eV. At 170 K a new feature has developed at a binding energy of 287.0 eV and the CO induced peak has grown. A surface temperature of 320 K leaves CO as the only remaining surface species. The feature at 287.0 eV observed between 170 K and 300 K must be connected with the formate species. As is shown in the inset of Fig. 15 (top spectrum) a peak due to formate at 287.0 eV is observed if adsorbed formic acid is thermally treated. According to the work of Madix et al. [48] and of Illing [49] we know that the additional feature at 288.5 eV is due to formic acid solvating the formate formed in the monolayer. The lower trace in the inset of Fig. 15 is taken after CO$_2$/H$_2$ adsorption, heating to T = 180 K and

54

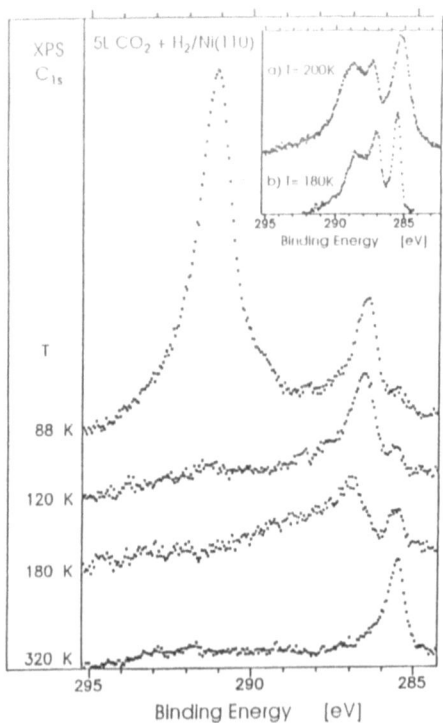

Figure 15. Series of high resolution C1s XP spectra of 5 L CO_2 + H_2/Ni(110) as a function of temperature. The inset compares formate species formed a) (in the top spectrum) out of formic acid by heating to 200 K with formic acid as solvent, and b) after reaction of CO_2 and H_2 at 180 K solvated with readsorbed CO_2.

allowing CO_2 to readsorb at low temperature. As for the formic acid, the CO_2 solvating the formate leads to an additional feature at higher binding energy, a trace of which is also seen in the spectrum shown in Fig. 15. We have shown before [29] that without presence of hydrogen only $CO_2^{\delta-}$ forms on the surface which above 200 K dissociates into adsorbed CO and oxygen. Only with the presence of hydrogen (or deuterium) do we observe formate formation [36].

However, the tendency for formate formation is a rather complex function of the hydrogen coverage. Clearly, if we preadsorb hydrogen (or deuterium) at room temperature to saturation coverage (1 L), CO_2 adsorption and even CO_2 physisorption at low temperature is suppressed. Note, that at saturation coverage, the Ni(110) surface undergoes a (2x1) reconstruction, which was monitored using LEED [50]. We have to go to low H_2 doses (0.2 L) before we see CO_2 physisorption. $CO_2^{\delta-}$ formation appears to be suppressed by higher coverages of hydrogen but favoured by low hydrogen precoverages. It is not clear at present what causes the effect but we have noted earlier [9] that $CO_2^{\delta-}$ formation appears to be correlated with the substrate workfunction for typical δ-type substrates. Hydrogen is known to increase the workfunction by about 0.5 eV (from 4.5 eV to 5.0 eV) for saturation coverage [50-54]. It is not unreasonable to assume that it is the increase in workfunction that quenches $CO_2^{\delta-}$ formation and its further reaction.

3.3. CASE 3: $C_6H_6/Os(0001)$

As an example for a rather complex reaction we discuss the reaction of benzene on Os(0001) [55]. Fig. 16 shows a set of complicated TD spectra obtained after dosing the Os(0001) surface at low temperature (T ~ 200 K) and detecting the desorbing hydrogen. In the temperature range 200 K < T < 290 K a chemisorbed ordered ($\sqrt{7}$ x $\sqrt{7}$)R19.1° monolayer, the low temperature phase, exists as monitored by LEED and ARUPS experiments [18]. Very small amounts of molecular C_6H_6 desorption are observed up to a temperature of 320 K. Increasing the temperature to T > 290 K leads to dissociative chemisorption. Fig. 16 shows a series of H_2 TD spectra from C_6H_6 + Os(0001) with variation of the initial coverage and a constant heating rate β (β = dT/dt = 3.5 K/s). Five desorption states are detected in the range of 300 K < T < 830 K for high coverages. The first H_2 desorption peak α is observed at 326 K followed by a second more intense feature β at 372 K. The combined areas of the structures α and β take about 1/3 of the whole area covered by the spectrum at saturation coverage. This indicates the loss of two H atoms per benzene molecule at 400 K. Further heating leads subsequently to three desorption peaks, γ, δ and ε at 460, 675 and 750 K, respectively. H_2 desorption ends at 830 K and an ordered (9 x 9) graphitic structure is formed as indicated by LEED and AES.

The number of dehydrogenation steps is dependent on the coverage. Lower coverages lead to a broad dehydrogenation feature in the TD spectrum with only two peaks whereas high coverages cause five clearly distinguishable dehydrogenation steps.

No shift of the α-peak maxima at various exposures is observed. This indicates a first order desorption reaction, perhaps related to C-H bond breaking. In contrast to this finding the β-peak maximum shifts towards lower temperatures with increasing coverages. A superposition of a low coverage effect and a high coverage effect has to be considered which may be responsible for the observed shift. In the temperature range 290 K < T < 340 K the observed ($\sqrt{7}$ x $\sqrt{7}$)R19.1° benzene structure develops broad LEED spots with increasing background intensity and vanishes subsequently.

ARUPS data were recorded in the temperature range of 200 K < T < 1000 K [55]. Fig. 17 shows a series of normal emission spectra at different temperatures and saturation coverage with a photon energy of hν = 21.21 eV and the angle of incidence α = 72.5°, i.e. dominant z-polarization [56]. The experiments were performed by heating the sample up to the temperature T noted on the left hand side of the diagram and then cooling it down to ~ 200 K. This procedure was applicable because the observed transition processes turned out to be irreversible.

The lowest spectrum in Fig. 17 is typical for the low temperature phase observed after adsorption of benzene at temperatures below 290 K. Structures close to the Fermi edge up to 4 eV binding energy can be attributed to emission from the osmium 5d bands. Adsorbate induced features are observed in the binding energy range 4-14 eV, with a prominent peak at 11 eV, a double peak around 8 eV and weaker structures around 6 and 4 eV. After heating the sample, the shapes of the spectra vary significantly indicating several adsorption phases. The first variation is observed at T ~ 330 K, when hydrogen desorption starts (cf. Figs. 16 and 17). At T = 340 K desorption of the α-peak is complete and the photoemission spectrum changes. The former intense peak at 11 eV

56

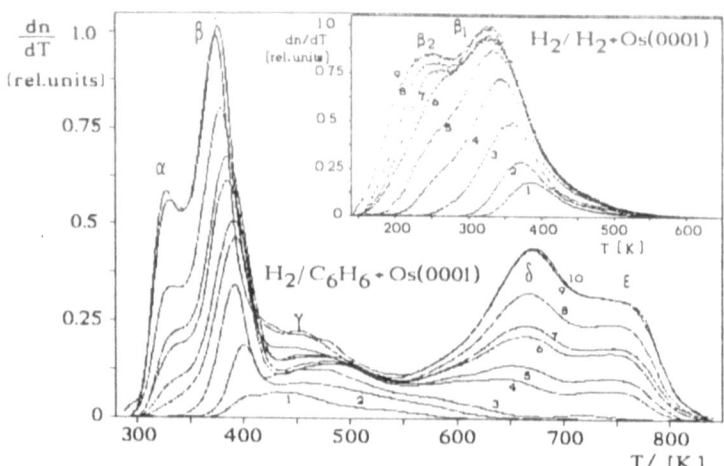

Figure 16. Hydrogen desorption spectra after adsorption of benzene on Os(0001) at T = 180 K for various exposures. The benzene dose is given as the pressure in the gas inlet system (Pa) multiplied by the dosing time (sec). (1) 0.065, (2) 0.13, (3) 0.26, (4) 0.52 (5) 0.63, (6) 0.78, (7) 0.94, (8) 1.25, (9) 3.12, (10) 12.6 Pa sec. The heating rate was 3.5 K/sec. In the inset hydrogen desorption spectra obtained after adsorption of H_2 at T = 180 K are shown. Hydrogen doses: (1) 1.58, (2) 3.15, (3) 6.3, (4) 12.6, (5) 25.2, (6) 50.4, (7) 196.0, (8) 211.0, (9) 660.0 Pa sec. The heating rate was 3.5 K/sec.

weakens and eventually disappears, and the spectrum is now dominated by a peak at 8.5 eV. The work function shows no significant variation during this phase transition. Further heating of the sample to 380 K leads to further desorption of hydrogen (β-peak in Fig. 16) and to the formation of another adsorbate phase. The absence of the peak at 11 eV and the appearance of a new feature at 10.5 eV (Fig. 17) are characteristic for this phase. The work function increases from 3.7 to 3.92 eV. Heating to temperatures above 380 K leads to less structured spectra. Finally the clean surface was obtained by flashing the sample up to very high temperatures above 2000 K for a longer time. The work function of the clean surface was determined to be 5.6 eV and could also be used as a monitor of the cleanliness of the Os(0001) surface. In order to characterize the different adsorption phases in more detail, angle resolved photoemission spectra are presented as a function of the electron emission angle θ.

We use the $2a_{1g}$ emission of the benzene (D-band), i.e. the totally symmetric C-H-bonding orbital as the level to apply the oriented free molecule approach (OFM) for a determination of the orientation of the molecular plane. Fig. 18 shows the polar angle variations of the intensity of the D-band for three different temperatures indicating a geometrical distortion as a function of temperature. The temperatures have been chosen to represent the α, β and γ states as marked in the TD spectra (Fig. 17). Together with the HREELS studies, discussed in the following, we are able to propose a sequence of reactions and interpret the changes as a function of temperature in the ARUP spectra.

Fig. 19a collects a series of HREEL spectra of benzene on Os(0001) as a function of temperature [57].

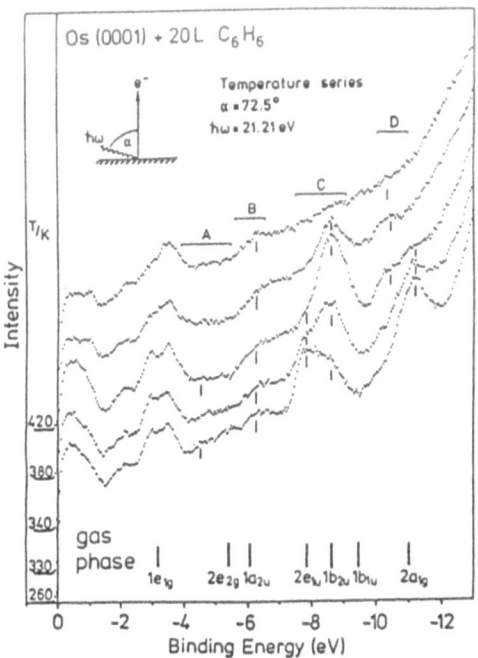

Figure 17. He1 photoelectron spectra of a saturated benzene adlayer on Os(0001) at selected temperatures. Adsorbate levels are labeled A-D.

Spectrum (a) shows the data of a benzene adsorbate phase at T = 273 K . This adsorbate gives rise to a ($\sqrt{7}$x$\sqrt{7}$)R19.1° LEED pattern [55]. The HREEL spectrum in specular scattering is dominated by one transition which can be assigned by comparison with similar spectra on other benzene adsorbate systems [58]. A comparison of specular and 5° off-specular spectra of adsorbed benzene is given in Fig. 19b. This comparison basically reveals the fact that dipole selection rules determine the dominant C_6H_6 vibration at 810 cm^{-1}. The strong band at 810 cm^{-1} in specular scattering geometry has to be assigned to a C-H wagging(γ) mode. These modes lead to strong dynamic dipoles perpendicular to the metal surface, if the molecular plane is oriented parallel to the Os surface. Other vibrational modes show negligible intensities in the specular scattering direction (Fig. 19a), but gain intensity off specular (Fig. 19b). All studies presented so far conclude from such behaviour that chemisorbed benzene lies flat on metal surfaces at room temperature. The small peak at 1885 cm^{-1} is probably due to residual CO. We only note that at even lower temperature (T = 80 K) we find the growth of benzene multilayers, as indicated in the EEL spectra (not shown) by additional lines similar to those reported by Jakob and Menzel for C_6H_6/Ru(0001) [59].

Upon elevating the temperature to T = 325 K (spectrum (b)), which is equivalent to the temperature where the system looses the first hydrogen atom [55], the wagging mode shifts to lower values, namely 760 cm^{-1}. Simultaneously, a sharp band at 400 cm^{-1} in the C-C wagging region gains intensity, while the region of C-H vibrations above 3000 cm^{-1} only shows a slight intensity, considerably more though than for the flat-lying benzene

58

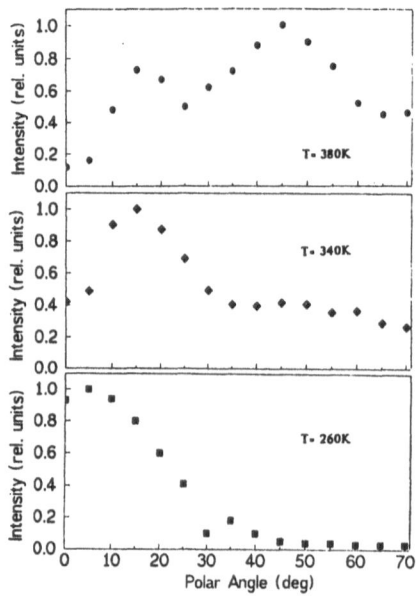

Figure 18. Polar plot of the integrated intensities of the D band ion states (see Fig. 17) of benzene on Os(0001) at different temperatures.

adsorbate. However, C-H bending modes at 1110 and 1390 cm^{-1} are detectable. Further increase of the temperature to a value (T = 382 K), where the second hydrogen atom is lost, leads to several further changes in the EEL spectrum (spectrum (c) in Fig. 19a): firstly, the intensity of the elastic peak decreases by orders of magnitude, indicating a pronounced disordering of the adsorbate layer; secondly, several modes in the region of C-C frame vibrations between 465 and 700 cm^{-1}, as well as in the region of C-H bending modes between 980 and 1400 cm^{-1} grow in. Most pronounced, however, is the considerable intensity in the region of the C-H stretching modes above 3000 cm^{-1}. Note that the observed vibrational frequencies are only slightly shifted with respect to those observed for the benzene moiety. For a definite assignment it would be desirable to compare the observed frequencies with those of the benzyne-Os$_3$ cluster compounds for which X-ray structure determinations are reported showing an inclined C$_6$H$_4$ ring with respect to the Os cluster plane [60,61]. Unfortunately, the vibrational spectra of these compounds in the hydrocarbon region have not been reported. However, the vibrational spectrum of matrix-isolated benzyne has been determined and assigned on the basis of a normal mode analysis [62-65].

A comparison with the benzene spectrum shows that except for one vibration, i.e. the υ_{C-C} mode at 2082 cm^{-1}, the benzyne vibrational energies are in the neighbourhood of those of C$_6$H$_6$. For the adsorbed benzyne we find indications of losses at 2000 and 2565 cm^{-1}, which may be due to the ring vibration. The other observed vibrational losses are in the range of those for benzene. The band at 2000 cm^{-1} might also be due to co-adsorbed CO. On the other hand the changes of the loss intensities in going from benzene to adsorbed benzyne are most characteristic and consistent with a pronounced geometrical

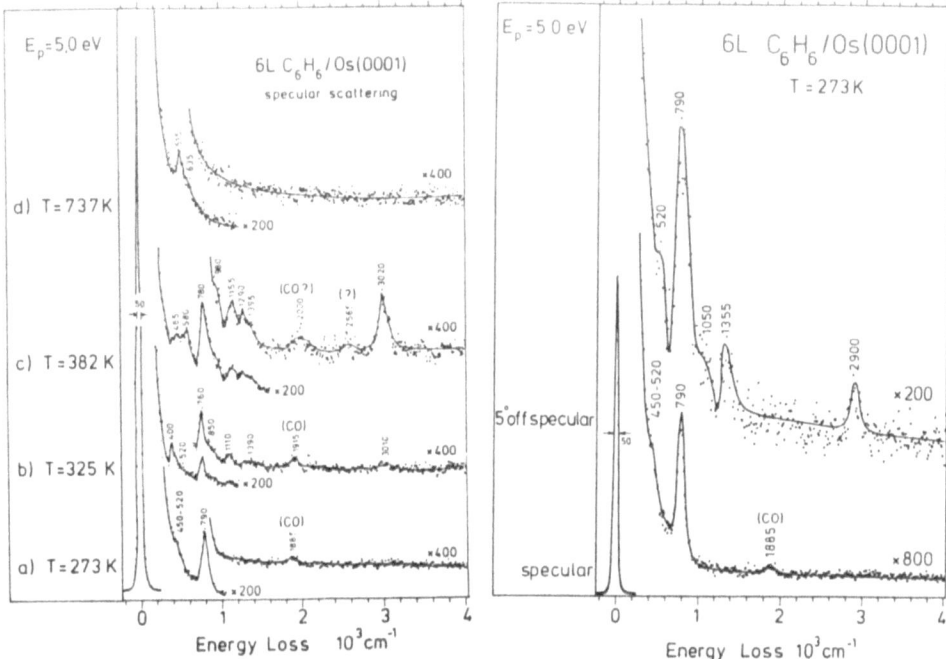

Figure 19. HREEL spectra of benzene on Os(0001). Left panel: HREEL spectra taken for various annealing temperatures. Right panel: HREEL spectra taken at room temperature. A spectrum taken in specular direction is compared with a spectrum taken at 5° off specular.

change of the adsorbed species as shall be alluded to in the following. The simplest way to explain the observed changes of intensities is to consider a molecular plane which is no longer parallel to the surface plane but rather tilted to a certain angle. It is very hard to estimate the tilt angle quantitatively. Jakob and Menzel [59] qualitatively estimate the tilt of the molecular plane by forming the ratio (R_i) of the vibrational loss intensities of the C-H stretch modes above 3000 cm^{-1} with respect to the C-H wagging mode (γ_{C-H}) below 1000 cm^{-1} under specular scattering conditions. These authors find R_i values of 2×10^{-2} for flat lying C_6H_6 and 12×10^{-2} for C_6H_6, where the molecular plane is inclined in a physisorbed C_6H_6 layer. We find values of 1×10^{-2} for adsorbed C_6H_6, 4×10^{-2} for the phenyl phase, and 30×10^{-2} for adsorbed C_6H_4. Since we know that the C-H-framework in phenyl and benzyne stays coplanar within a few degrees in the cluster compound - as it is in gaseous benzene - we may draw the conclusion from the change of R_i as a function of temperature that the tilt angle with respect to the surface normal decreases from 90° in the C_6H_6 adsorbate to a value comparable to or even larger than C_6H_6 physisorbed on Ru(0001). A reasonable value for the physisorbed phase is 45°, so that we can expect a tilt angle in the neighbourhood of this value for the benzyne species on Os(0001). Our photoemission study (see above) is compatible with these findings. Also, as a result of the photoemission study we concluded that in the phenyl adsorbate the tilt angle of the molecular plane increases only slightly with respect to the benzene adsorbate. The

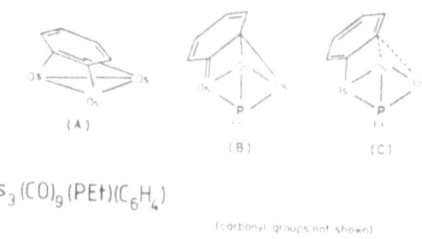

$Os_3(CO)_9(PEt)(C_6H_4)$

(carbonyl groups not shown)

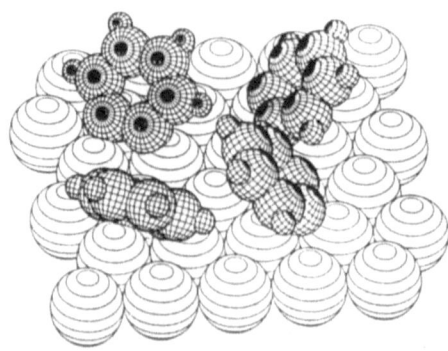

Figure 20. Top: structure formula for possible adsorbate-substrate bondings of C_6H_4 on an Os_3 cluster reported in ref. [60]. Bottom: schematic (not site specific) 3D plot of the proposed ortho C_6H_4 type intermediates at the Os(0001) surface at T ~ 380 K for different domains. The molecules are inclined by 45°.

HREELS data corroborate these findings as well. Upon heating the surface above 400 K the spectra continuously change and, at temperatures above T = 500 K, the vibrational wagging modes, characteristic of the existence of a six-membered ring, have disappeared. This is compatible with the conclusions based on the TDS data [55] that above this temperature the ring structure breaks up. We have not studied this latter temperature region in detail with HREELS and leave this for future studies.

The HREELS data provide further experimental evidence to support a reaction channel of adsorbed benzene proposed above, namely a successive loss of hydrogen accompanied by the formation of at least two intermediate species which we believe to be a phenyl and a benzyne species, before the six-membered ring structure starts to break up, and finally leads to the formation of a carbon overlayer on the Os(0001) surface. To our knowledge there is only one other study reported in the literature where the authors find experimental indications of an adsorbed benzene moiety on a solid surface. Liu and Friend [66] have published XPS and NEXAFS (near edge X-ray absorption fine structure spectroscopy) data for the system C_6H_6/Mo(110). By comparison with reference data (gained via decomposition of C_6H_5SH on Mo(110)) [66] they deduce the presence of a benzyne species on the surface. Evidence from vibrational spectroscopy has so far not been reported. The present study represents the first experimental indication via HREELS for the existence of a benzyne precursor for benzene dissociation.

3.4. CASE 4: ADSORPTION ON OXIDES

Metal oxides, and transition-metal oxides in particular, are in use as catalysts in industrial processes. This is certainly one of the reasons why the study of adsorption and reaction on oxide surfaces has been pioneered rather early in the fifties and sixties. With the advent of surface science the interest has shifted towards clean metal surfaces and the study of metal oxides has been abandoned to some extent. During the last decade or so, however, the interest in oxide surfaces has been revitalized and some clean single-crystal surfaces have been studied by applying surface-science methodology. Henrich [67] has recently published an excellent review of this field. For certain oxides, i.e., semiconducting oxides such as ZnO, a great deal of information already exists even for molecular adsorbates on these surfaces. Much of this literature has been collected in a review by Heiland and Lüth [68]. It appears, though, that ZnO is a singular case. One issue has been that many oxides exhibits only limited conductivity which in turn limits the applicability of electron-spectroscopic techniques which play a central role in the characterization of clean surfaces and of molecular adsorbates on these surfaces. Some of the latter difficulties may be circumvented by looking at thin oxide films grown on metallic substrates.

As an example we briefly discuss the situation of NO on $NiO(100)$ [69].

We have investigated the adsorption of NO on a thin $NiO(100)$ film of several layers thickness grown on top of a $Ni(100)$ surface in comparison with data of an *in vacuo* cleaved $NiO(100)$ single crystal. The layer exhibits a high defect density. We demonstrate via application of several surface-sensitive electron-spectroscopic techniques (XPS, ARUPS, NEXAFS, HREELS) that the occupied (ARUPS) and unoccupied (NEXAFS) electronic states are similar to those of a bulk $NiO(100)$ sample. In spite of its limited thickness, the band structure of the thin film exhibits band dispersion perpendicular to the surface that are compatible with those of bulk $NiO(100)$. It is shown that the electronic structure of the oxygen sublattice can be described in a band-structure picture while for the Ni sublattice electron localization effects lead to a breakdown of the band-structure picture.

NO on NiO desorbs at 220 K. Fig. 21 shows the thermal desorption spectra of NO from a bulk $Ni(100)$ surface in comparison with desorption from the oxide layer [69]. The desorption temperatures for both systems are only marginally different. If we consider the different heating rates and identical, commonly used frequency factors we calculate on the basis of the Readhead formula [P.A. Readhead, Vacuum 12, 203 (1962)] almost identical desorption energies for both cases, i.e., 0.52 eV. This means that NO is weakly chemisorbed on a NiO surface. It is quite surprising that the desorption energies on both surfaces are the same because the defect densities are different by orders of magnitude as judged from the LEED spots [70], and one is tempted to expect a strong influence of the defects on the desorption temperature. However, even though we do not know the exact nature of the defects the similarity of the desorption temperatures indicates that the defects are not the sites of NO adsorption on the oxide film. The NO coverage is close to 0.2 relative to the number of Ni surface atoms as determined by XPS. HREELS reveals that there is only one species on the surface documented by the observation of only one bond-stretching frequency.

Fig. 22 shows some HREEL spectra of NO on a $NiO(100)$ film at different temperatures. Upon exposure to NO at low temperature, we observe in addition to the very strong NiO surface phonons one peak at the high-frequency side of the third multiple-phonon-loss. This peak vanishes at about 200 K surface temperature in agreement with the thermal desorption data which showed a peak temperature only a little

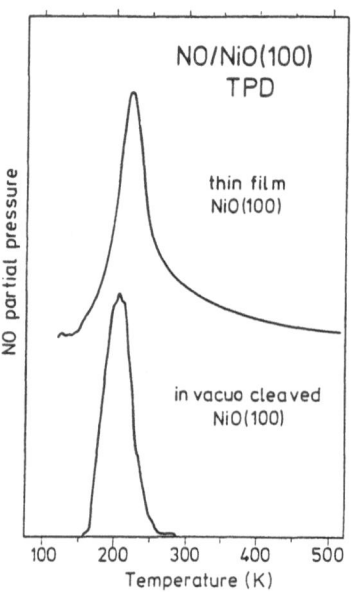

Figure 21. Comparison of TD spectra for NO adsorbed on a NiO(100) single crystal surface cleaved in vacuo with a TD spectrum for NO adsorbed on NiO(100) epitaxially grown on Ni(100).

above T = 200 K. We assign this peak to the N=O bond-stretching vibration of NO adsorbed on top of Ni sites in the NiO layer. This assignment is based on a detailed HREELS study of NO-O-coadsorption on Ni(100) [71]. We have plotted HREELS spectra of NO on Ni(100) and NO + O on Ni(100) for comparison in Fig. 22. Both spectra are rather complex, and a detailed discussion shows that the spectra are caused by the superposition of a set of different species [71]. The important aspect for the present purpose is the appearance of a single peak at 1800 cm^{-1} for adsorption near coadsorbed oxygen. This peak has been assigned to NO adsorbed on top of Ni atoms with a bent Ni-NO bond. The bending of the axis in the coadsorbate is also indicated by the appearance of a bending vibration at 640 cm^{-1}, typical for a strongly bound system [72].

We have transferred this assignment to the oxide surface although we do not observe a bending mode. We cannot exclude at present that such a bending vibration is situated near the position of the NiO phonon loss but this would imply that the force constant of the bending mode on the oxide surface is similar to the adsorbate on the metal surface. However, we know that the molecule-substrate bonding is much weaker on the oxide surface as compared with the metal surface, so that we expect a reduced bending force constant. This would shift the bending mode to lower frequencies which might render the bending mode unobservable under the present conditions. Clearly, an independent experimental clue as to the geometry of the molecular axis is highly desirable. We have therefore performed NEXAFS investigations on the NO/NiO(100) adsorbate [69]. NEXAFS data on the system and a comparison with previous data on the system NO/Ni(100) indicate that the molecular axis of adsorbed NO is tilted by an angle of approximately 45° relative to the surface normal. The N1s XP spectra of the weakly

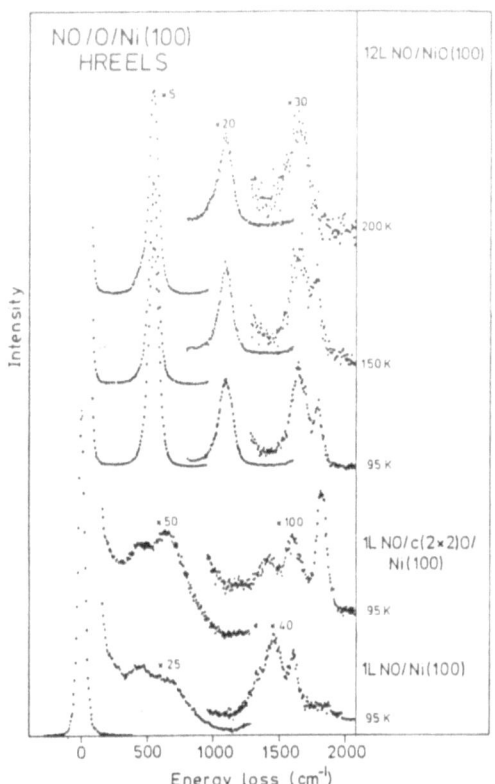

Figure 22. HREEL spectra for NO on NiO(100) in comparison with HREEL spectra of NO on a clean and on an oxygen covered Ni(100) surface.

chemisorbed species show giant satellites similar to the previously observed cases for weak chemisorption on metal surfaces. This is the first observation of an intense satellite structure for an adsorbate on an insulator surface, which shows that there must be sufficient screening channels even on an insulating surface. A theoretical assignment of the peaks is discussed. We compare the spectroscopic properties of the NO species on the thin-film oxide surface, which is likely to contain a certain number of defects, with NO adsorbed on a basically defect-free bulk oxide surface by TD and XP spectra. TD and XP spectra of the bulk system are basically identical as compared with the oxide film, indicating that the majority of species adsorbed on the films is not adsorbed on defects but rather on regular NiO sites. Results of *ab-initio*-oxide-cluster-calculations are used to explain the bonding geometry of NO on regular NiO sites [69].

As a second example we summarize some of the results we have collected for a more reactive oxide surface, namely the $Cr_2O_3(111)$ surface [73].

We have chosen Cr_2O_3 as a substrate for our investigations since it is well known that the catalytic activity of this oxide for various reactions, like for instance polymerization of olefins [74], hydrogenation of alkenes and dehydrogenation of alkanes [75], reduction of NO and decomposition of N_2O_4 [76] is rather high.

For CO we observe a well ordered $(\sqrt{3}x\sqrt{3})R30^\circ$ superstructure in the LEED pattern. The ARUPS and NEXAFS data obtained for this adsorbate are compatible with an adsorption geometry where the CO molecules lie flat on the surface interacting with the oxide via the σ valence states and the π systems. Fig. 23 shows a schematic representation of the CO-adsorbate system as it is deduced from the experimentally available information. The CO molecules interact with two Cr-ions on the surface. The CO $1\pi_z$ electrons interact repulsively with the oxygen layer underneath. The combined interactions lead to interestingly different ionization potentials of CO on an oxide surface as compared with metal surfaces. This is summarized in Fig. 24. Worth noting is the comparison between Cr_2O_3 and Cr metal because in both cases the molecule is known to ly flat on the surface. Clearly, the interaction with the oxide is much different from the CO-metal interaction.

For CO_2 not only adsorption on the surface but also reaction with the surface is observed. At low temperatures CO_2 is partly adsorbed in a molecular physisorbed state and partly it reacts with the oxide surface to form a surface carbonate. After slight annealing only carbonate is found on the surface. It appears that adsorption, respectively reaction of CO and CO_2 only takes place on those parts of the surface that expose chromium atoms to the respective adsorbate since the surface reactivity is strongly suppressed by preadsorption of oxygen. The adsorption of oxygen leads to a surface which exposes far fewer chromium atoms. A clean, freshly flashed oxide surface exhibits chromium atoms that are in a different oxidation state than those in the bulk as judged from EELS. These atoms are most likely responsible for the high reactivity of the $Cr_2O_3(111)$ surface.

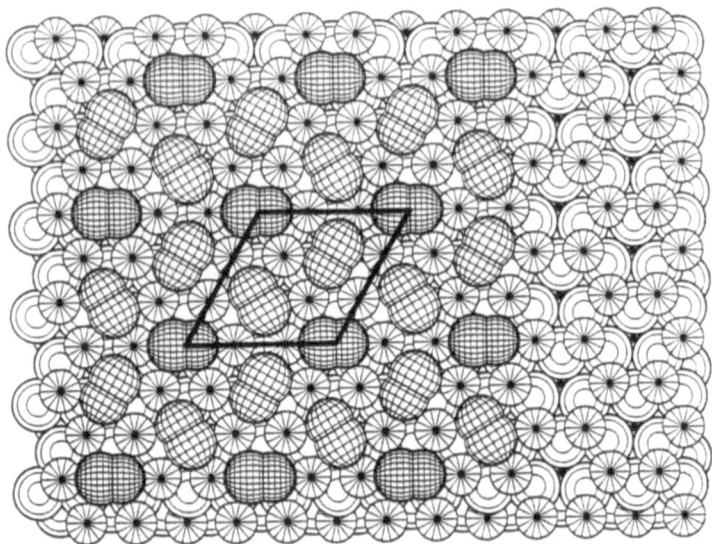

Figure 23. Structure model for $CO(\sqrt{3}x\sqrt{3})R30^\circ/Cr_2O_3(111)$. The unit cell of the adlayer is indicated.

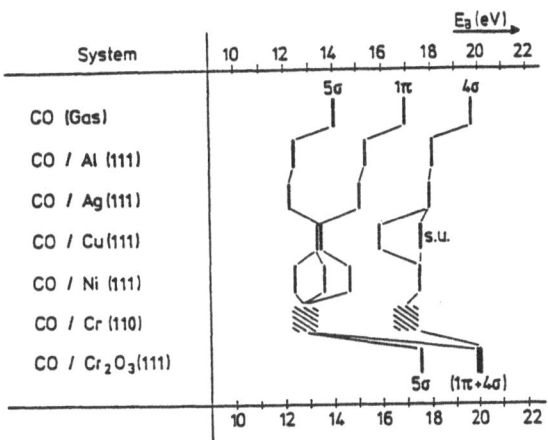

Figure 24. Binding energies of the valence levels of CO adsorbed on different hexagonal or quasihexagonal surfaces with respect to the vacuum level. Since we do not know the work function change for the adsorption of CO on Cr(110) we have hatched the binding energy region in which we expect the ionizations to occur. The data have been taken from refs. [77-82].

Fig. 25 shows a set of off-specular HREEL spectra for $^{12}CO_2$ and $^{13}CO_2$ on $Cr_2O_3(111)$ taken at two different temperatures. The strongest losses in these spectra, i.e. those at 410 cm^{-1}, 700 cm^{-1}, 1420 cm^{-1} and 2100 cm^{-1} are due to the optical phonons of the oxide film. The adsorption of CO_2 and thus the formation of carbonate shifts these losses by some cm^{-1}, indicating the strong interaction of the carbonate with the oxide surface. At T = 100 K the asymmetric stretching vibrations of linear, undistorted CO_2 at 2360 cm^{-1} and 2290 cm^{-1}, respectively, are clearly visible. Upon annealing to 220 K these losses strongly decrease in intensity. Some residual intensity is due to readsoption of CO_2 from the residual gas since the sample was only shortly annealed to T = 220 K. The measurements were performed when the crystal was cooled down to 100 K.

In the spectra of the cold layer some weak structure is visible to the left and the right of the phonon loss at 1420 cm^{-1}. Upon annealing these get stronger so that they are most likely due to carbonate. We identify losses at (1630 ± 20) cm^{-1} [(1600 ± 20) cm^{-1}], (1285 ± 20) cm^{-1} [(1260 ± 20) cm^{-1}], (1035 ± 20) cm^{-1} [(1015 ± 15) cm^{-1}] and (920 ± 20) cm^{-1} [(900 ± 20) cm^{-1}]. The numbers in brackets are the loss energies for $^{13}CO_3^{2-}$. These loss energies are graphically represented in Fig. 26 in comparison with data from the literature for differently coordinated carbonate species. The different surface coordinations of carbonate are schematically shown in Fig. 26. From the data shown in Fig. 26 it is clear that the carbonate species on $Cr_2O_3(111)$ is most likely doubly coordinated to the oxide, i.e. it is bidentate.

4. Synopsis

The system CO (2x1) p2mg/Ni(110) is used to demonstrate the possibilities of electron spectroscopies. Photoelectron and inverse photoelectron spectroscopy are used to probe

66

the electronic structure of the system. The strong intermolecular interactions lead to pronounced E vs. k dispersions which may be monitored through angle dependent measurements. Inelastic electron scattering may be used to deduce the site of adsorption, dispersion measurements infer the lateral displacements and from X-ray photoelectron diffraction results the tilt angle of the molecular axis.

We have used various examples to illustrate how surface science techniques may be applied to identify metal-molecule interactions and how these interactions change in the course of the increase of surface temperature.

Finally we show that these techniques may be applied for semi-conductive and even insulating oxide surfaces if the adsorption is studied on oxide films grown on metal substrates. With non electron spectroscopic techniques such as TDS and high energy electron spectroscopic techniques such as XPS comparison to adsorbation on bulk single crystal is possible. Through this approach we learn about the influence of defects on oxide surfaces onto the adsorption behaviour.

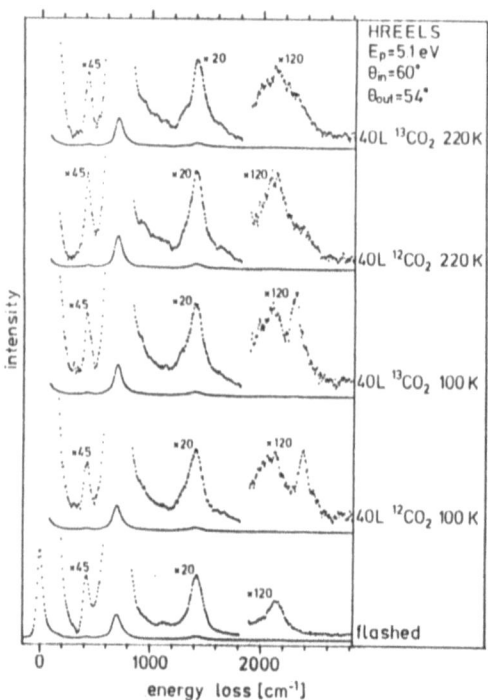

Figure 25. HREEL spectra for $^{12}CO_2$ and $^{13}CO_2$ on $Cr_2O_3(111)$ taken at two different temperatures.

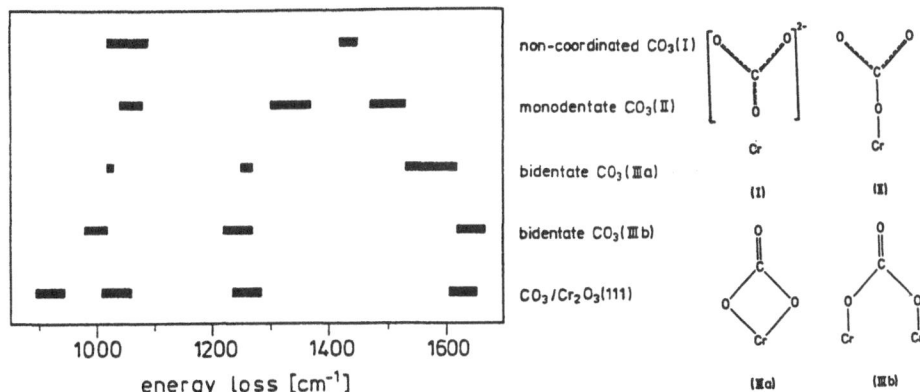

Figure 26. Left panel: Comparison of the vibrational loss energies of carbonate on $Cr_2O_3(111)$ obtained after exposing the surface to CO_2 at 100 K and subsequent annealing to 220 K with data from the literature [83] for different carbonate species on Cr_2O_3. Right panel: Schematic diagram showing different surface coordinations of carbonate.

ACKOWLEDGEMENTS

We are grateful to "Deutsche Forschungsgemeinschaft", "Ministerium für Wissenschaft und Forschung des Landes NRW", "Bundesministerium für Forschung und Technologie", and the "Fonds der Chemischen Industrie" for financial support.

Without the collaboration of many scientists the work would not have been possible. Special thanks are due to Dr. M. Neumann, Osnabrück for a long standing, fruitful collaboration.

We thank Dr. H. Hamann for helping us to prepare the manuscript.

REFERENCES

1. Plummer, E.W., Eberhardt, W. (1982), Adv. Chem. Phys. **49**, 533
2. Richardson, N.V., Bradshaw, A.M. (1982), in Brundle, C.R. and Baker, A.D. (eds.), Electron Spectroscopy, Vol. 4, Academic Press, New York.
3. Freund, H.-J., Neumann, M., (1988), Appl. Phys. **A47**, 3.
4. Behm, R.J., Ertl, G. and Penka, V. (1980), Surf. Sci. **160**, 387.
5. Riedel, W. and Menzel, D. (1985), Surf. Sci. **163**, 39.
6. Kuhlenbeck, H., Neumann, M. and Freund, H.-J. (1986), Surf. Sci. **173**, 194.
7. Alvey, M.D., Dressler, M.J. and Yates Jr., J.T. (1986), Surf. Sci. **165**, 447.
8. Wesner, D.A., Coenen, F.P. and Bonzel, H.P. (1988), Phys. Rev. Lett. **60**, 1045.
9. Voigtländer, B., Bruckmann, D., Lehwald, S. and Ibach, H. (1990), Surf. Sci. **225**, 151.
10. Kuhlenbeck, H, Saalfeld, H.B., Buskotte, U., Neumann, M., Freund, H.-J. and Plummer, E.W. (1989), Phys. Rev. **B39**, 3475.

11. Memmel, N., Rangelov, G., Bertel, E., Dose, V., Kometcr, K. and Rösch, N. (1989), Phys. Rev. Lett. 63, 1884.
12. Ertl, G. (1983), J. Vac. Sci. Techn. **A1**, 1247.
13. Freund, H.-J., Bartos, B., Messmer, R.P., Grunze, M., Kuhlenbeck, H. and Neumann, M. (1987), Surf. Sci. **185**, 187.
14. Blyholder, G. (1968), J. Phys. Chem. **68**, 2772; (1974), J. Vac. Sci. Techn. **11**, 865.
15. Richardson, N.V. and Bradshaw, A.M. (1979), Surf. Sci. **88**, 255.
16. Bozso, F., Ertl, G., Grunze, M. and Weiss, M. (1977), J. Catal. **49**, 18; (1977), **50**, 519.
17. Spencer, N.D., Schoonmaker, R.C. and Somorjai, G.A. (1982), J. Catal. **74**, 129.
18. Ertl, G. (1982), in Critical Review in Solid State and Material Science, CRC Boca Raton, Florida, p. 349.
19. Grunze, M., Golze, M., Fuhler, J., Neumann, M. and Schwarz, E. in Proc. 8th Intern. Congr. on Catalysis, West-Berlin p. IV-13
20. Grunze, M., Golze, M., Hirschwald, W., Freund, H.-J., Pulm, H., Seip, U., Tsai, M.C., Ertl, G. and Küppers, J. (1984), Phys. Rev. Lett. **53**, 850.
21. Tsai, M.C., Seip, U., Bassignana, I.C., Küppers, J. and Ertl, G. (1985), Surf. Sci. **155**, 387.
22. a) Whitman, L.J., Bartosch, C.E., Ho, W., Strasser, G. and Grunze, M. (1986), Phys. Rev. Lett. **56**, 1984.
    b) Whitman, L.J., Bartosch, C.E. and Ho, W. (1986), J. Chem. Phys. **85**, 3788.
23. D. Tomanek, K.H. Bennemann, (1985) Phys. Rev. **B31**, 2488.
24. Davenport, J W. (1976), Thesis, University of Pennsylvania, Phys. Rev. Lett. **36**, 945.
25. Plummer, E.W., Gustafsson, T., Gudat, W. and Eastman, D.E. (1977), Phys. Rev. **A15**, 2339.
26. Huber, K.H. and Herzberg, G. (1979), Constants of Diatomic Molecules, Vol. 4, van Nostrand, Princeton, NJ.
27. Copperthwaite, R.G., Davies, P.R., Morris, M.A., Roberts, M.W. and Ryder, R.A. (1988), Catal. Lett. **1**, 11.
28. Behner, H., Spiess, W., Wedler, G. and Borgmann, D. (1986), Surf. Sci. **175**, 27.
29. Bartos, B., Freund, H.-J., Kuhlenbeck, H., Neumann, M., .Lindner, H and Müller, K. (1987), Surf. Sci. **179**, 59.
30. Illing, G., Heskett, D., Plummer, E.W., Freund, H.-J., Somers, J., Lindner, Th., Bradshaw, A.M., Buskotte, U., Neumann, M., Starke, U. Heinz, K., De Andres, P.L., Saldin, D. and Pendry, J.B. (1988), Surf. Sci. **201**, 1.
31. Asscher, M., Kao, C.-T. and Somorjai, G.A. (1988), J. Phys. Chem. **92**, 2711.
32. Reled, H. and Asscher, M.(1987), Surf. Sci. **183**, 201.
33. Freund, H.-J., Behner, H., Bartos, B., Wedler, G., Kuhlenbeck, H. and Neumann, M. (1987), Surf. Sci. **180**, 550.
34. Weinberg, W.H. (1983), Surf. Sci. **128**, L 224.
35. Dubois, L.H. and Somorjai, G.A. (1983), Surf. Sci. **128**, L 231.
36. Wambach, J., Illing, G. and Freund, H.-J., Chem. Phys. Lett. accepted
37. Freund, H.-J. and Messmer, R.P. (1986), Surf. Sci. **172**, 1.
38. Messmer, R.P., Freund, H.-J. (1988), in Ayers, W.M. (ed.), Catalytic Activation of Carbon Dioxide, ACS Symposium Series 363, Washington, DC, p.16.

39. Ehrlich, D., Wohlrab, S., Wambach, J., Kuhlenbeck, H. and Freund, H.-J. (1990), Vacuum **41**, 157.
40. Wohlrab, S., Ehrlich, D., Wambach, J., Kuhlenbeck, H. and Freund, H.-J. (1989), Surf. Sci. **220**, 243.
41. Wambach, J., Odörfer, G., Freund, H.-J., Kuhlenbeck, H. and Neumann, M. (1989), Surf. Sci. **209**, 159.
42. Solymosi, F. and Berko, A. (1986), J. Catal. **101**, 458.
43. Berko, A. and Solymosi, F. (1986), Surf. Sci. **171**, L 498.
44. Kiss, J., Revesz, K. and Solymosi, F. (1988), Surf. Sci. **207**, 36.
45. Paul, J. (1989), Surf. Sci. **224**, 348.
46. Rodriguez, J.A., Clendening, W.D. and Campbell, C.T. (1989), J. Phys. Chem. **93**, 5238.
47. Jones, T.S., Ashton, M.R. and Richardson, N.V. Richardson (1989), J. Chem. Phys. **90**, 7564.
48. Bowker, M. and Madix, R.J. (1981), Surf. Sci. **102**, 542.
49. Illing, G. (1991), Thesis, University of Bochum.
50. Christmann, K., Chehab, F., Penka, V. and Ertl, G. (1985), Surf. Sci. **152**, 356.
51. Voigtländer, B., Lehwald, S. and Ibach, H. (1989), Surf. Sci. **208**, 113.
52. Penka, V., Christmann, K, and Ertl, G. (1984), Surf. Sci. **136**, 307.
53. Rieder, K.H. and Stocker, W. (1985), Surf. Sci. **164**, 55.
54. Jackmann, T.E., Griffiths, K., Unertl, W.N., Davies, J.A., Gurtier, K.H., Harrington, D.A. and Nortin, P.R. (1987), Surf. Sci. **179**, 297.
55. Graen, H.H., Neuber, M., Neumann, M., Illing, G., Freund, H.-J. and Netzer, F.P. (1989), Surf. Sci. **223**, 33.
56. Netzer, F.P., Graen, H.H., Kuhlenbeck, H. and Neumann, M. (1987), Chem. Phys. Lett. **133**, 49.
57. Graen, H.H., Neumann, M., Wambach, J. and Freund, H.-J. (1990), Chem. Phys. Lett. **165**, 137.
58. Graen, H.H. (1991), Thesis, University of Osnabrück and references therein
59. Jakob, P. and Menzel, D. (1988), Surf. Sci. **201**, 503.
60. Brown, S.C., Evans, J. and Smart, L.E. (1980), J. Chem. Soc. Chem. Commun. 1021.
61. Gallop, M.A., Johnson, B.F.G., Lewis, J., Mc Camley, A. and Perutz, R.N. (1988), J. Chem. Soc. Chem. Comm. 1071.
62. Chapman, O.L., Mattes, K., McIntosh, C.L., Pacansky, J., Calder, G.V.and Orr, G. (1973), J. Am. Chem. Soc. **95**, 6134.
63. Dunkin, I.R. and MacDonald, J.G. (1979), J. Am. Chem. Soc. Commun., 722.
64. Nam, H.-H. and Leroi, G.E. (1985), Spectrochim. Acta **41A**, 67.
65. Nam, H.-H. and Leroi, G.E. (1987), J. Mol. Struct. **157**, 301.
66. Liu, A.C. and Friend, C.M. (1988), J. Chem. Phys. **89**, 4396.
67. Henrich, V.E. (1985), Rep. Progr. Phys. **48**, 1481.
68. Heiland, G. and Lüth, H. (1982) in King, D.A. and Woodruff, D.P. (eds.), The Chemical Physics of Solid Surfaces and Heterogeneous Catalysis, Vol. 3, Woodruff, Elsevier, N.Y.
69. Kuhlenbeck, H., Odörfer, G., Jaeger, R., Illing, G., Menges, M., Mull, Th., H.-J. Freund, Pöhlchen, M., Staemmler, V., Witzel, S., Scharfschwerdt, C., Wennemann, K., Liedtke, T., Neumann, M. (1991), Phys. Rev. **B43**, 1969.
70. Bäumer, M., Cappus, D., Kuhlenbeck, H., Freund, H.-J., Wilhelmi, G., Brodde, A. and Neddermeyer, H., Surf. Sci. in press.

71. Odörfer, G., Jaeger, R., Illing, G., Kuhlenbeck, H., Freund, H.-J. (1990), Surf. Sci. **233**, 44.
72. Jones, L.H., Ryan, R.R. and Asprey, L.B. (1981), J. Chem. Phys. **49**, 581.
73. Xu, C., Dillmann, B., Habel, M., Adam, B., Kuhlenbeck, H., Freund, H.-J., Ditzinger, U.A., Neddermeyer, H., Neuber, M. and Neumann, M., Surf. Sci. submitted
74. Yermakov Yu. and Zakharov,V. (1975), Adv. in Catal. **24**, 173.
75. Connor W. C. and Kokes, R. J. (1969), J. Phys. Chem. **73**, 2436.
76. Kung, H. H. (1989),Transition Metal Oxides: Surface Chemistry and Catalysis, Elsevier Publishing Company, Amsterdam.
77. Turner, D.W., Baker, C., Baker, A.D. and Brundle, C.R. (1970), Molecular Photoelectron Spectroscopy, Wiley-Interscience, London New York.
78. Jacobi, C., Astaldi, C , Geng, P. and Bertolo, M. (1989), Surf. Sci. **223**, 569.
79. Schmeisser, D., Greuter, F., Plummer, E.W. and Freund, H.-J. (1985), Phys. Rev. Lett. **54**, 2095.
80. Freund, H.-J., Eberhardt, W., Heskett, D. and Plummer, E.W. (1983), Phys. Rev. Lett. **50**, 768.
81. Schneider, C., Steinrück, H.-P., Heimann, P., Pache, T., Glanz, M., Eberle, K., Umbach, E.and Menzel, D. (1987), BESSY Annual Report .
82. Shinn, N.D. (1984), J. Vac. Sci. Technol. **A4**, 1351.
83. Davydov, A.A. (1990),in John Wiley & Sons (eds.), Infrared Spectroscopy of Adsorbed Species on the Surface of Transition Metal Oxides, Chichester, England.

CONCEPTS IN HETEROGENEOUS CATALYSIS

G. L. HALLER and R. S. WEBER
Department of Chemical Engineering
Yale University
New Haven, CT 06520
USA

ABSTRACT. We review some general concepts associated with chemisorption and active sites as they apply to heterogeneous catalysis. This is followed by some specific concepts that affect reactivity. These include solid acidity, bifunctionality, structure sensitivity, metal-oxide support interaction and stereochemistry.

1. Introduction

A catalyst is a substance that increases the rate of a chemical reaction but is not consumed or permanently altered by the reaction [1]. Increasing the rate of desired chemical reactions will obviously have commercial applications. It was recently found that more than 60% of the products and 90% of the processes used by the chemical industry rely on catalysts and the reliance is believed to be even greater in the petroleum refining industry [2]. Industrially catalyzed reactions are very much dominated by heterogeneous catalysts, catalysts that are in a phase different from that of the reactants, most commonly a solid catalyst acting on fluid (gas or liquid) reactants. These catalysts are almost always solids and usually metals, e.g., Pt, or inorganic oxides, e.g., alumina. They are often composite, e.g., very small particles (of order of 1 nm) of Pt on a high area (of order of 100 m^2/g) alumina. The mechanism of catalyzed reactions on small metal particles involves intermediates which are chemisorbed molecules on the surface of the small metal particles. While this chemistry is complicated by the size (number of atoms involved) and lacks the symmetry of an inorganic metal complex, the fundamental metal-ligand interactions are essentially the same. One can find many commercial catalytic reactions that involve metal-ligand interactions: from atoms (such as homogeneous catalysts, e.g., $[Rh(CO)_2I_2]^-$), to clusters (such as the 1 nm particles in a $Pt/g-Al_2O_3-Cl$ reforming catalyst), to surfaces (such as the Pt/Rh gauzes used in the air oxidation of NH_3).

Heterogeneous catalysts have several advantages over other kinds of catalysts which are usually classified as either enzyme or homogeneous. First among these is the cost of separation of the reactants from the catalyst, which can be minimal for heterogeneous catalysts since the fluid (gas or liquid) passes over the solid catalyst, reacts, and is displaced by incoming reactant so that reaction and separation are one and the same step. The cost of separation is just the pumping cost to move the fluid through the bed (which is small compared to other kinds of separation, e.g., distillation) and, in any case,

D. R. Salahub and N. Russo (eds.), Metal-Ligand Interactions: from Atoms, to Clusters, to Surfaces, 71–100.
© 1992 *Kluwer Academic Publishers. Printed in the Netherlands.*

necessary for the reaction. A second advantage is that packed beds lend themselves to continuous reactors that can be designed to process huge flows. Because the heterogeneous catalyst is usually a refractory oxide or is supported on one, it can be expected to be quite thermally stable allowing the reactor to be operated over a wide temperature range to take advantage of high rates and/or high equilibrium conversions for endothermic reactions.

A homogeneous catalyst will be uniformly distributed in the reactant phase. There are many examples of gas phase catalyzed reactions which are nearly uniform on a macroscale (meters to hundreds of meters) in stratospheric chemistry, e.g., the catalytic destruction of ozone by decomposition products of chlorofluorocarbons. As an example of a commercial homogeneously catalyzed reaction, we can consider the Monsanto process for methanol carbonylation to acetic acid [3]. Acetic acid is produced continuously by reacting methanol and CO in a liquid phase homogeneous catalytic reactor using $[Rh(CO)_2I_2]^-$ as a catalyst. In the Monsanto process acetic acid is separated from the catalyst by distillation.

Enzymes are biological catalysts and might also be considered homogeneous since they are often uniformly distributed at least on the cellular level. However, enzyme size, particularly if associated with a cell (on the order of microns), can put them somewhere between homogeneous and heterogeneous catalysts. Also the fact that they are biologically active proteins (amino acid polymers) which are sensitive to pH, temperature, solvent, etc., as is true of almost all biological material, distinguishes them from most heterogeneous catalysts, but less so from homogeneous catalysts which are often metal organic complexes and which might have similar thermal stability to that of enzymes. For example, the digestive enzyme, chymotrypsin, catalyzes the break-down of ingested protein molecules at a rate that is about one billion times faster than the spontaneous rate [4]. While chymotrypsin is not as selective as some enzymes (an area where most heterogeneous and inorganic homogeneous catalysts cannot compete), it certainly can make the necessary distinction between proteins that are part of the reactor walls and those that are reactants. An important commercial application of enzymes is as additives to synthetic detergents [5]. This additive is protease or another hydrolase (an enzyme that catalyzes hydrolyses of various kinds) which breaks down a variety of fabric stains and aids in the solubilization of soil.

Heterogeneous catalysts will be our main topic of discussion because they are the kinds of catalysts that completely dominate in applications in the chemical and petroleum industry. The largest monetary use of heterogeneous catalysts is for the application to pollutant control of automobile exhaust [6]. These catalysts are a combination of Pt, Pd and Rh supported on a ceramic matrix (and often have other important components, e.g., ceria, BaO, etc.). These catalysts can effectively convert a wide variety of unburned hydrocarbons and CO to CO_2 and H_2O while also converting NO to molecular N_2. The largest use of catalysts based on either processed capacity or mass of catalyst used is catalytic cracking. We will return to this below when we discuss the concept of solid acidity. The concept of solid acidity will be one which is particular to solid heterogeneous catalysis (although soluble heteropolyanions have similar acid properties and mineral acids are homogeneous catalysts to which we make the analogy). Thus, most of the concepts we discuss either have generality across all kinds of catalysts, e.g., active sites, or some reasonable counterpart in homogeneous and/or enzyme catalysts.

2. Some Fundamental Concepts

2.1. ACTIVE SITES AND HOW WE COUNT THEM

There is no single moiety on the surface of a heterogeneous catalyst that can be called the active site at steady state, all of the intermediates that participate in the main, repeating loop of a catalytic cycle are converted at the same rate and therefore any of them could be used to represent the locus of catalytic activity. Taylor first noted in 1925 [7] that it is logical to define the set of active sites *only* under reaction conditions since the nature and number of the reaction intermediates are dependent upon reaction conditions. However, in the current state of the art, we are grateful for ways to count surface species even far from reaction conditions and even if the species we count are only distantly related to those that evolve into active sites. Therefore we resort to techniques like chemisorption and various spectroscopies that have known deficiencies (Table 1) but which have proved to be useful in the systematic ranking of the activity of catalysts and in comparing results obtained in different laboratories.

TABLE 1. Methods of Characterizing Catalytic Surface

Technique	Information it yields	Deficiencies
Physisorption (e.g. BET method)	Total accessible surface area, pore sizes and pore size distribution	Relies upon empirical correlations; does not differentiate between support and catalyst; is not applicable under reaction conditions
Chemisorption (e.g. hydrogen uptake)	Titer of classes of surface sites	Stoichiometry of adsorption can be variable; does not always differentiate chemically distinguishable sites; requires auxiliary spectroscopies or isotopic labeling to probe surface sites during reaction.
Electron microscopy	Local structure and composition	Difficult to employ under reaction conditions; requires high contrast, thin samples; has a limited field of view; beam damage.
Scattering techniques (e.g. neutron scattering, electron scattering)	Micro- to meso-range structures	Requires special facilities; not surface-sensitive; yields one-dimensional (radial) information.
Diffraction techniques (e.g. x-ray diffraction)	Local structure and composition	Requires long range order in the samples; not surface sensitive.
Absorption spectroscopies (e.g. infrared, nuclear magnetic resonance, x-ray absorption spectroscopy	Vibrational, electronic, magnetic, molecular structures	Bulk-averaging if they can be used under reaction conditions; require calibration using the spectra of species of known structure; may be applicable over a narrow range of elements or compounds.

TABLE 2. Areal Densities of Sites on Catalytic Materials

Material	Surface species	Areal density
Transition metal (e.g. Pt)	metal atom	10^{14} to 10^{15} sites/cm^2
Main group oxide (e.g. SiO$_2$)	Si, OH	O $3\text{-}5 \times 10^{15}$ sites/cm^2 $0\text{-}5 \times 10^{14}$ sites/cm^2
Defect sites at 1000 K	$\Delta H_f = 80$ kJ/mol	10^{10} to 10^{11} sites/cm^2

2.1.1. *Physisorption and Chemisorption.* The areal density of atoms at the surface of solids varies between 10^{14} to 10^{15} atoms per square centimeter for materials encountered in heterogeneous catalysis (Table 2). To enhance the accessibility of these surface sites catalysts are prepared as highly porous materials or as very small particles distributed on the surface of a high surface area support. This means that techniques for measuring the number and type of surface sites of practical catalysts should afford high sensitivity and be capable of probing internal regions of the sample. The BET method [8], which employs physisorption of condensable gases to measure total surface area, was the first development to satisfy those criteria. It continues to be a mainstay for the characterization of supports and unsupported catalysts even though the detailed interpretation of BET data can be problematic when the surfaces being probed have pores of sizes comparable to the adsorbate [8].

Soon after the development of the BET method, selective chemisorption of reactants was introduced to better distinguish the fraction of the surface which is active from that portion of the surface which acts as an inert support or diluent. In the case of pure or supported metals each of the surface metal atoms can be a site of adsorption, particularly for small adsorbates like H atoms or CO molecules whose projection on the surface is smaller than the diameter of the surface atoms. For larger adsorbates and for materials with a less homogeneous surface, alloys, oxides, composites, larger moieties of the surface will comprise the adsorption sites and therefore the areal density of adsorption sites will be smaller than the areal density of surface atoms.

Since adsorption of at least one reaction intermediate is prerequisite to the use of the surface as a catalyst, the number of adsorption sites for a reactant or a product sets an upper bound on the fraction of the surface which can be involved in the catalytic activity of a solid, provided that the surface does not develop new adsorption sites under the conditions of the reaction. A necessary but not sufficient test that a particular adsorbate counts active sites is to normalize the catalytic reaction rate by the amount adsorbed. The calculated rate per site, called the turnover frequency [9], will be invariant with the preparation of the sample if the number of adsorbate molecules is proportional to the number of any of the active sites. For example, when the rates of hydrogenation are normalized by the amount of hydrogen taken up by a metallic catalyst, the turnover frequency under specified conditions remains constant over a considerable range of samples [9]. A contrary example can be found in the first use of selective chemisorption [10], where the chemisorption of CO was employed to count the number of Fe atoms on the surface of an ammonia synthesis catalyst that contained about 10wt% alumina and other promoters. We now know that neither CO uptake nor H$_2$ uptake counts sites correctly but that N$_2$ adsorption at high temperature may, see Figure 1, [11]. Since the adsorption of N$_2$ is the rate limiting step for this reaction, it is likely that the nitrogen uptake succeeds in normalizing the reaction rate because the adsorption sites are indeed active sites in the sense that at least a constant fraction of them participate in the main catalytic cycle for producing NH_3 from N_2 and H_2.

The mixed case can also be found: surface sites counted by hydrogen chemisorption of Pt catalysts each having a different average size of the metal particles produce a nearly constant turnover frequency for the H_2O_2 reaction under lean conditions (H_2/O_2 ratio is low) but not when the reaction is run in the rich regime (see reference [9], chapter 5). Hanson and Boudart [12] attribute this result to corrosive chemisorption, a reconstruction of the Pt surface when exposed to excess oxygen. Their idea is that the reconstruction causes the surface to become much more uniform so that a census of all accessible Pt atoms, even measured under reducing conditions, can provide the proper basis for normalizing the reaction. Under rich conditions, where the reaction is structure sensitive, hydrogen chemisorption counts surface sites that are active as well as those that are not active (or better, do not become active).

2.1.2. *Spectroscopic characterization of adsorbates.* In principle, the distribution of adsorption sites on a surface can be determined through spectroscopic characterization of the adsorbed species in order to resolve the types and number of adsorbed species associated with nonuniformities in the surface and with interactions among the adsorbates themselves. For example, temperature programmed desorption (TPD), which measures the quantity of material that desorbs into vacuum or a stream of carrier gas as the temperature of the sample is increased, is often employed as a way to characterize the strength of interaction between adsorbates and the surface [13]. The qualitative idea, that more strongly bonded species should desorb at higher temperatures is plausible, but, as has been pointed out by a number of recent studies, quantitative interpretation of TPD curves in terms of enthalpies of desorption is rarely possible [14, 15, 16]. Nonetheless, TPD can be used to separate chemisorbed species into different classes, the population of which can be compared to the measured reaction rates.

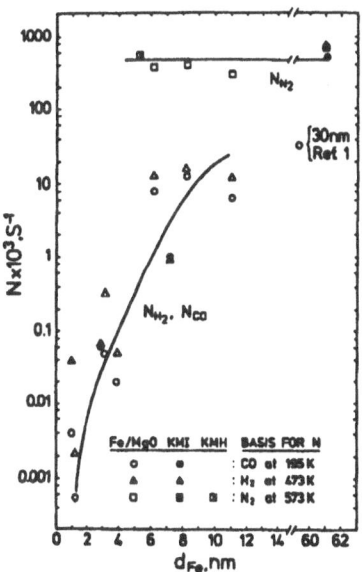

Figure 1. Comparison of turnover frequencies for the synthesis of ammonia as normalized by the uptake of hydrogen, carbon monoxide and dinitrogen. Only the latter gives a constant value of the turnover frequency as the mean size of the iron particles is varied. After reference [11].

Spectroscopies that look at the adsorbed species can reveal even more subtle details about the nature of the adsorbate site. For example, Brønsted and Lewis sites on the surfaces of acidic catalysts may be distinguished through the infrared absorption features of adsorbed pyridine: the Brønsted sites produce pyridinium ions, the Lewis sites produce unprotonated adducts (Figure 2). In the case of silica-aluminates and zeolites it is found that the number of Brønsted sites correlates best with the rates measured for catalytic cracking of hydrocarbons [17]. The accepted inference is that the Brønsted sites are predominately responsible for the cracking.

Dynamic chemisorption (to be distinguished from the dynamics *of* chemisorption, see Section 2.2) is the name given to the class of techniques in which the inventory of species chemisorbed on a surface of a reacting catalyst is probed by changing the isotopic composition of the feed gas. Both the amount and reactivity of different species can be determined from the response in composition in the reactor effluent and in the composition of the surface [18, 19]. The techniques can satisfy the Taylor ideal of studying the surface under reaction conditions, provided of course that the transient is performed without changes in the reactant concentrations and provided that there is no isotope effect on either the kinetics or the thermodynamics of the reaction processes. The latter will have practical consequences when hydrogen isotopes comprise the transient. Two examples suffice to illustrate the techniques. The first comes from the study of ammonia synthesis, the process which has prompted the largest number of seminal ideas in the study of catalysis. By means of isotopic labeling of the nitrogen atoms in the reactant and product molecules, Horiuti has shown that the reactions subsequent to the adsorption of N_2 are rapid (in virtual equilibrium) and that therefore the dissociative adsorption of N_2 constitutes the rate determining step for the reaction sequence [20]. In more recent work, Efstathiou and Bennett [21] employed ^{13}C labeling of CO and spikes of hydrogen to probe the identity and amount of species present during the CO hydrogenation reactions catalyzed by supported Rh catalysts. They found that the

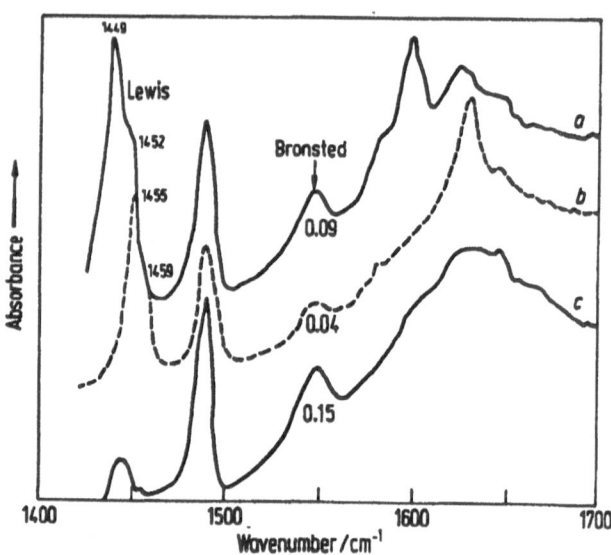

Figure 2. Infrared spectra of pyridine adsorbed on silica-aluminas showing both protonated pyridinium species and the Lewis acid adduct. After reference [17].

catalysts contain a significant pool of some partially hydrogenated carbon that they call an active surface carbon.

2.1.3. *Spectroscopic characterization of catalysts*. Of course spectroscopies can also be used more directly to measure the electronic and molecular structures of catalytically important surfaces. The difficulty only is in finding a technique that is both surface-sensitive and is employable in conditions that approximate those found in a catalytic reactor (to better ensure that the measured surface structures are related as closely as possible to those that participate in the catalytic reaction). Surface sensitivity does not restrict the choice of a spectroscopy when the catalyst is completely exposed, e.g. nanometer sized particles of Pt dispersed on an alumina support, since the form of the sample then permits the use of even a bulk averaging spectroscopy. The list of candidate spectroscopies is considerably longer than that presented in Table 3 but these are frequently encountered in the current literature. They share the use of photons or electrons energetic enough to traverse the windows of the sample cell, the reaction mixture, and the bulk of the sample, while their intrinsic time scales span the range from milliseconds (the diffusive time constant of whole molecules as measured by t_1 in NMR) to femtoseconds (corresponding to the lifetime of electronic transition that contributes to the width of peaks in x-ray absorption spectra). The table does not list the surface science techniques like x-ray photoelectron spectroscopy that may also be useful in the characterization of heterogeneous catalysts if appropriate precautions are taken in interpreting data acquired under conditions so far removed from those of reaction.

For typical heterogeneous catalysts, the composition and chemical state of the surface can be probed with x-ray absorption spectroscopy, ultraviolet-visible spectroscopy, Raman spectroscopy, nuclear magnetic resonance spectroscopy, and Mössbauer effect spectroscopy. They are listed in order of decreasing generality of applicability: all elements can be characterized by their electronic transitions; fewer elements have an adequate abundance of high gyromagnetic isotopes to be useful for NMR; and only a handful of elements possess the necessary conjunction of nuclei required in Mössbauer spectroscopy.

TABLE 3. A short list of spectroscopies employed to characterize the surfaces of heterogeneous catalysts.

Method	Information
Nuclear Magnetic Resonance Spectroscopy	Identity of species, their mobility, and bonding to neighbors
Electron Spin Resonance Spectroscopy	Location and localization of unpaired electrons
Infrared spectroscopy	Vibrational frequencies and local symmetry
UV-Visible spectroscopy	Energies of valence electronic states
X-ray absorption spectroscopy	Energies and symmetries of empty valence electronic states
Extended x-ray absorption fine structure spectroscopy	Radial distribution function of neighbors of the absorbing atom
Mössbauer effect spectroscopy	Symmetry, rigidity, and magnetic coupling of the target atom

The techniques that have clear advantages for determining the molecular structure of catalyst surfaces include transmission electron microscopy (TEM), extended x-ray absorption fine structure spectroscopy (EXAFS), and scanning tunneling microscopy (STM). EXAFS can yield information about the number, type, and symmetry of the neighbors of an atom in the sample. It is well suited to study samples under reaction conditions [22]. It suffers, however, from a still uncertain reliability for the characterization of complex samples, requiring careful calibration and an extended set of reference spectra. Moreover, because it is a bulk-averaging technique, it averages the geometry around *all* of the atoms whose spectrum is being measured. TEM is rarely used *in situ* but there are some noteworthy exceptions [23]. It is more frequently employed to visualize the microscopic morphologies present in samples before or after reaction. STM is the most recent visualization technique to be applied to catalyst surfaces and there as yet few examples to prove it merits. Still, the pictures that result are so appealing [24] that it will undoubtedly be pursued vigorously over the next few years.

A clever, recent example of the use of TEM from Datye's laboratory characterizes a catalyst sample by means of a tangential view across nonporous microspheres of silica in order to follow changes in the local structure of overlayers of titania used as a catalyst in the dehydration of propanol [25]. Such mixtures of metal oxides are difficult to characterize by ordinary electron microscopy or by x-ray absorption spectroscopy because of the lack of contrast between the phases. Moreover, this oxide-supported oxide lacks sufficient long range order to permit examination by x-ray diffraction. The results from the TEM study, presented schematically in Figure 3, show clearly that the initial monolayer dispersion of the titania, which were stable to high temperature in dry air or vacuum, agglomerated into microcrystals of anatase upon exposure to a reaction stream. The transition from monolayer to agglomerated crystals could also be followed by the characteristic vibrational bands in the Raman spectra of the samples, reinforcing other results from Wachs laboratory that indicate the utility of Raman spectroscopy for probing the molecular structures of oxide catalysts [26].

EXAFS data have been used recently [27] to show that the stoichiometry of hydrogen uptake correlates with the average metal-metal coordination in the metal particles in a sample (Figure 4). The results suggest that hydrogen uptake with an assumed stoichiometry of, say, H/M=1 should not be used as a quantitative measure of particle size

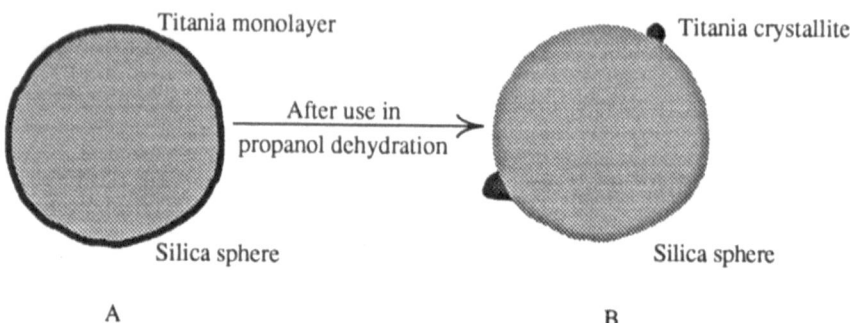

Figure 3. Schematic of the TEM of titania overlayers on silica spheres. A) as prepared. B) following use in the dehydration of propanol. After reference [25].

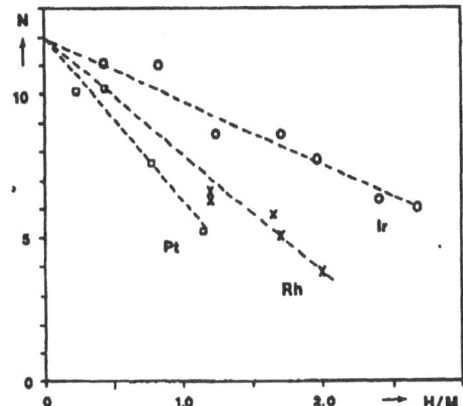

Figure 4. Coordination number of metal atoms in supported metal catalysts and their average uptake of hydrogen. After reference [27].

nor, since samples always contain a typically unknown distribution of particle sizes, as a way to count surface sites. One inference to be drawn from this work is that instances where hydrogen uptake measurements do succeed in providing a basis on which to normalize reaction rates must involve a certain degree of coincidence among the distributions of sites and the distributions of activity in a catalyst sample.

To date, scanning tunneling microscopy and its variants have afforded data that corroborates but does not yet extend our knowledge of the structure of catalyst surfaces. One application in which it may prove to be useful is in the estimation of the local density of filled and empty electronic states of regions of a surface. These quantities can be determined by supplementing the usual topographic scans with measurements of the electron current that flows between the probe tip and the surface as a function of applied voltage at each raster point. At a minimum, identification of the local density of states curve will aid in fingerprinting the local composition of the surface. It may ultimately provide the means to understand how the molecular and electronic structure of a region affects its reactivity.

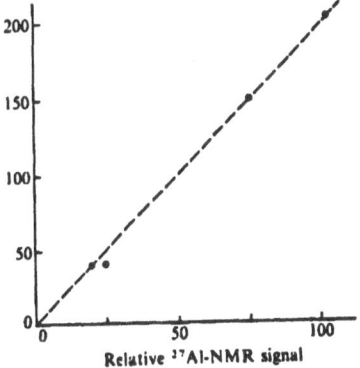

Figure 5. Relative rates of the cracking of n-hexane over ZSM-5 catalysts prepared with varying amounts of framework aluminum. After reference [28].

The use of NMR spectroscopy to characterize the aluminum in H-ZSM-5 [28] comprises one of the cleanest examples we have of the use of a spectroscopy to count the surface species that are associated with catalytic activity. The result is straightforward (Figure 5): the rate of cracking reactions is linearly proportional to the quantity of alumina in tetrahedral sites of the zeolite framework.

A cogent illustration of the use of some of the other spectroscopic techniques and the power of combining them comes from the many studies of catalysts that effect the removal of sulfur from heterocyclic compounds found in petroleum. The optimal catalyst for this process provides sites for the regioselective hydrogenolysis of carbon-sulfur bonds. The materials that have been found to work best contain molybdenum and cobalt sulfides dispersed on the surface of alumina. In the case of catalysts containing just molybdenum sulfide, data from electron microscopy, oxygen chemisorption and other techniques suggests that the hydrodesulfurization activity is associated with sites on the edge planes of MoS_2 crystallites (29). For example, by use of a special optical absorption spectroscopy Roxlo, *et al.* (30) were able to probe the valence electronic states of MoS_2 powders. Their results indicate that the hydrodesulfurization activity of the powders correlates very well with the absorption cross-section in the energy range that can be identified with defect sites on the edges of MoS_2 platelets (Figure 6).

In the case of mixed metal sulfides, the key observation is that the activity of a series of catalysts reaches a maximum at intermediate concentrations of cobalt, a pattern of promotion called synergy. Mössbauer spectroscopy of the cobalt, an experiment that involves the preparation of radioactive samples, shows that hydrodesulfurization activity parallels the formation of a particular cobalt species (Figure 7) that does not resemble any of the known, bulk cobalt sulfides [31]. The same result appears in spectra of supported and unsupported mixtures of sulfided molybdenum and cobalt and has led the Topsøe group to propose a species they call CoMoS for the site of the hydrodesulfurization activity. They have also proposed that the CoMoS species exists on the edges of MoS_2 plates, a result reinforced by EXAFS data on highly dispersed samples that are consistent with the decoration by the added Co of rafts of MoS_2 [32].

Despite the consistency of the spectroscopic data in favor of the involvement in the synergy of a single Mo-Co-S moiety, an alternate explanation has been proposed and should be considered, namely, that it could also arise from the diffusive transport of

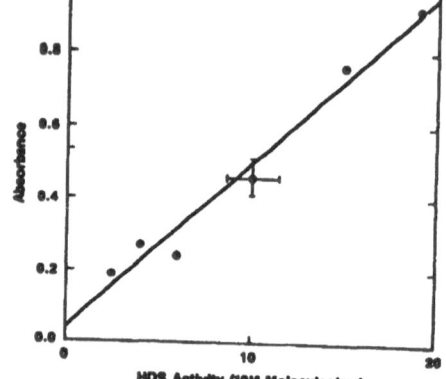

Figure 6. Correlation between optical absorbance and activity for hydrodesulfurization of MoS_2 powders prepared so as to expose varying ratios of edge/basal planes. After reference [30].

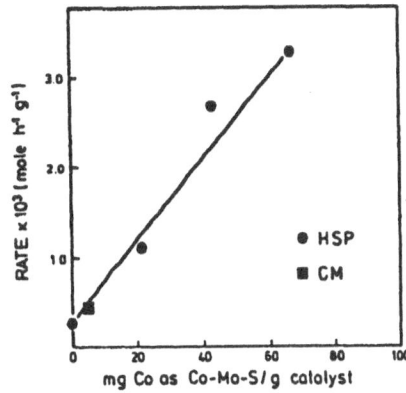

Figure 7. Correlation between the presence of CoMoS phase (determined from Mössbauer spectroscopy) and the activity of hydroprocessing catalysts. After reference [31].

surface species between two or more distinct surface species [33]. Indeed, it would be surprising if such controversies did not arise when the discussion concerns the existence of something as ephemeral as an active site.

2.2. DYNAMICS OF CHEMISORPTION: DIRECT VERSUS PRECURSOR ADSORPTION

In order for a molecule to react with an active site (see preceding section), it must find its way to the surface, chemisorb, react and then reverse the first two processes. It can often happen that the finding its way in part of this, i.e., diffusive transport into small pores of order 1 to 10 nm in diameter, can be the slow or rate determining step [34]. However, molecular transport is not conceptually unique to catalysis; what happens in the next step, chemisorption, is. While we have motivated this discussion by the industrial importance of heterogeneous catalysis, we will now immediately turn to rather esoteric experiments involving molecular-beams and single crystals to get at the heart of the concept of chemisorption [35].

Rettner describes three different dynamical mechanisms for chemisorption, two involving precursors and one being direct. The simplest of these, is the direct mechanism by which whatever barrier exists is surmounted by the kinetic energy with which the molecule approaches the surface. Weinberg has defined a direct reaction as one which occurs at the surface in a single collision from the gas phase with a time scale less than 10^{-12} s [36]. Such a mechanism may not often play a significant role in catalyzed reactions since so few molecules in a Boltzmann distribution may have sufficiently high energy [37]. Such a mechanism will result in the probability of the adsorption being independent of surface temperature in the molecular beam experiment where the energy of the incoming molecule is manipulated in the beam source. However, this will not be the case in the so called bulb experiment where both the surface and the gas temperature would be varied together. In a more complex version of the direct mechanism, other kinds of energy, e.g., vibrational, rotational, etc., may be useful for surmounting the chemisorption barrier. In still another version of this direct mechanism, the vibrational energy of the surface can be transferred to the molecule during collision resulting in a thermally (surface) assisted direct mechanism. Such a mechanism will be surface temperature dependent, but with a much weaker dependence than in the precursor mechanisms to be discussed below. Rettner argues that this is the kind of mechanism that applied for CH_4 chemisorption on the (111) surface of Pt [35]. There is ample evidence that CH_4 chemisorption is generally activated and that a high component of normal momentum will overcome this barrier on many metals [38, 39, 40, 41]. In general, any chemisorption that does not involve some precursor state on the surface before the final chemisorbed state is entered, is a direct mechanism.

We can divide precursor states into intrinsic and extrinsic depending on whether the interaction is with the surface (intrinsic) or with some previously adsorbed species (extrinsic). Intrinsic precursors can be further sub-divided into species that interact with the surface by van der Waal forces (physical adsorption) or chemisorption. (In principle, such a sub-division also exists for extrinsic precursor states but no example of an extrinsic precursor involving chemical bonding comes to mind.) In a study of N_2 adsorption on W(100) by Rettner et al. [42] it was observed that the dissociative chemisorption probability decreased with increasing surface temperature at a fixed incident translational energy and decreased with increasing translational energy at a fixed surface temperature. For incident energies greater than 46 kJ mol^{-1}, the dissociative chemisorption probability was independent of surface temperature and increased as the incident translational energy was increased, evidence that a second, direct and activated, chemisorption channel dominated at higher incident energies. The precursor must have been intrinsic because it mediated the adsorption even near zero coverage. For low beam energies, where the precursor-mediated adsorption dominated, the chemisorption probability was nearly independent of coverage, consistent with a mobile precursor. The independence of the chemisorption probability on surface coverage has been found in several other studies of N_2 chemisorption on W (100) [43, 44, 45].

The kind of behavior outlined above for N_2 on W (100) might be considered somewhat classical in the sense that we can consider the precursor state to be in pseudo-equilibrium with the gas phase, so that the concentration of precursor molecules is determined by equating the net rate of adsorption into the precursor state to the rate of adsorption into the chemisorbed state. The sticking coefficient then becomes:

$$s_O = \alpha k_r / (k_r + k_d) \tag{1}$$

where α is the trapping probability (a function of the angle of incidence, incident energy and surface temperature) and k_r and k_d are the rate coefficients for reaction (chemisorption) and desorption, respectively, out of the precursor state. Both the trapping probability and rate coefficient for chemisorption may be functions of surface coverage.

The chemisorption of N_2 on Fe (111) is conceptually very important, not only because it illustrates a third kind of chemisorption mechanism, through sequential precursors, but also because this step is the rate determining step in a well known commercial process, the synthesis of NH_3 from its elements [9]. The observation has been made that N_2 chemisorption on Fe (111) is strongly dependent on the translational energy [46] but also depends on the surface temperature and that this dependence is similar to that observed for adsorption of ambient N_2 from a Boltzmann distribution [47]. In this case, the first precursor is a molecular N_2 trapped in a classical van der Waal's well, but there is a barrier over which this species moves to form a chemisorbed molecular N_2 and, in turn, the dissociative chemisorbed state is entered through this chemisorbed molecular state [35]. Rettner and Mullins have proposed that a similar sequential set of physical and chemisorbed molecular precursors are involved in the dissociative chemisorption of O_2 on Pt (111) [48]. It should be emphasized that the concept of sequential precursors is not in itself new since this idea had already found its way into a text book some thirty years ago

[49]. What is novel here is that modern molecular beam experiments have been able to establish at least two examples where this is the preferred mechanism of dissociation of a diatomic molecule and that in both cases there exist large scale industrial processes that involve these steps, e.g., the synthesis of NH_3 on Fe and the oxidation of NH_3 on Pt. Weinberg [36] believes that it should be intuitively obvious that trapping-mediated dissociative chemisorption (chemisorption through a precursor) is far more probable than direct dissociative chemisorption because such a mechanism requires the molecule to have both sufficient energy to scatter inelastically from the repulsive part of the potential surface and redirect sufficient energy along the reaction coordinate to result in chemisorption. On the other hand a species that becomes trapped on the surface will have numerous opportunities to sample and surmount this barrier. Several recent investigations of alkanes on metals provide evidence for the dominance of the trapping-mediated dissociative chemisorption [40, 50, 51, 52, 53]. As a specific example, we may consider the chemisorption of C_2H_6 on Ir(110)(1×2) reconstructed surface. Figure 8 shows the appropriate one-dimensional potential energy diagram for this reaction and, from the molecular beam investigation of both the desorption and dissociation of physically adsorbed ethane on the Ir(110)(1×2), the rate constants in eq. (1) are deduced to be $k_r = 3\times10^{10}\exp(-5.5/RT)$ and $k_d =10^{13}\exp(-7.7/RT)$, where E_r and E_d are given in Kcal mol^{-1} [50].

3. Some Concepts used to Affect Reactivity

3.1. SOLID ACIDITY

The use of acids (or bases) to catalyze chemical reactions is perhaps among the oldest commercial applications of catalysts. One example which comes immediately to mind is the acid (or base) catalyzed hydrolysis of animal esters to produce soap, a base catalyzed reaction that reaches back into antiquity. While homogeneous acid catalyzed reactions

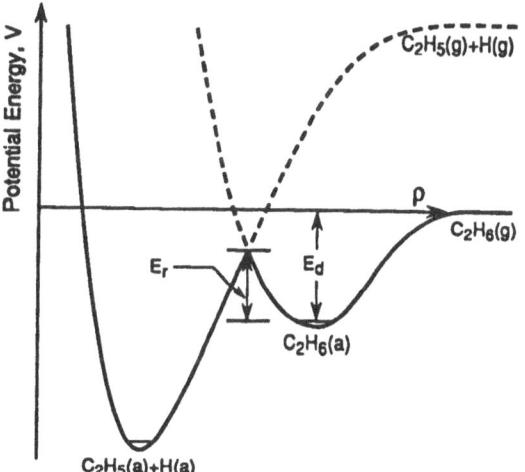

Figure 8. One-dimensional potential energy diagram (along the reaction coordinate) proposed for dissociative chemisorption of C_2H_6 on Ir(110)(1×2). After reference [36].

continue to be used for some large scale reactions such as alkylation, solid acids are catalysts of choice for both direct economical reasons (low separation costs) and indirect economical reasons such as environmental concerns that arise with the disposal of the used catalysts. The largest use of solid acid catalysts (both based on feed-stock processed and mass of catalyst used) is in catalytic cracking, the conversion of long chain hydrocarbons into smaller molecules in the gasoline range. The original catalysts were natural silica-alumina clays which were quickly replaced by synthetic silica-aluminas. A very major improvement in this class of catalysts resulted with the introduction of zeolites, crystalline silica-aluminas [2]. The origin of their acidity can be understood in terms of the fundamental structure of silica and the stable oxidation state of silicon and aluminum. Silicon cations of essentially all forms (including amorphous gels) impose a local tetrahedral structure on the surrounding oxygen anions. Simple substitution of Al^{3+} cations for Si^{4+} in a crystalline (or amorphous structure) thus produces a charge imbalance which must be balanced by a cation. If this is a proton, the structure will have Brønsted acidity.

As described above, the origin of the Brønsted acidity of amorphous and crystalline silica-alumina have the same origin and it might thus be expected that they would have the same acidity, but there is persuasive evidence that this is not the case, and that some zeolites (such as the faujasite structure shown in Figure 9) are stronger acids than amorphous silica-alumina. Just why this is, is a matter of some controversy. We will describe a couple of contributing factors which are direct consequences of the crystalline structure and which may turn out to be determining factors. To make this discussion specific, we will pose it in terms of the properties of L-zeolite which is of research interest to one of the authors.

The discovery of the non-acidic reforming catalyst, Pt/L-zeolite [54], has stimulated a significant body of literature because of its commercial potential for high selectivity for aromatic production and its conceptual uniqueness, see ref.[55] and references cited therein. The conceptual uniqueness referred to here is not based on the monofunctional (metal only) mechanism by which Pt particles catalyze normal alkane to aromatic conversion. This mechanism had been established by the late F. Gault and coworkers for Pt/SiO$_2$ [56]. On the contrary, it is the property which provides for the high aromatic selectivity (perhaps by the same mechanism as on Pt/SiO$_2$) of small Pt particles in the L-zeolite channels, relative to the same size particle on another support, which is unique. From the beginning, the scientific discussion has been couched in terms of either geometric effects of L-zeolite structure [57, 58] or electronic effects associated with the basicity of L-zeolite [59, 60].

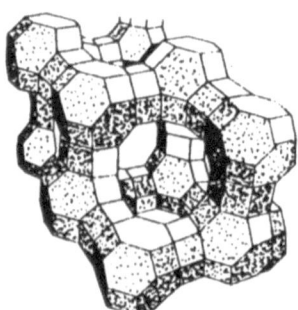

Figure 9. Schematic of the cage structures found in faujasite type zeolites.

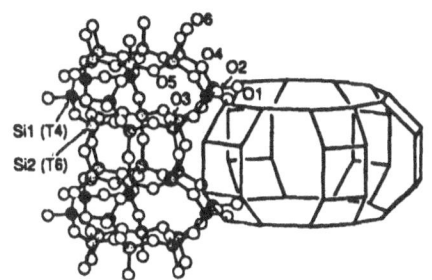

Figure 10. The framework structure of L-zeolite. Two cancrinite cages joined through a hexagonal prism are shown in ball and stick representation and one lobe of the channel is illustrated by straight lines connecting adjacent T (Si or Al) sites. After ref. [61].

The structure of L-zeolite results in parallel channels that undulate with a period of 0.75 nm. The channel diameter has a minimum of about 0.71 nm and a maximum of 1.26 nm (0.92 nm if the channel cations are taken into consideration and the counter ion is K$^+$) [61]. This forms lobed, 12-ring channels which are the host positions for metal particles

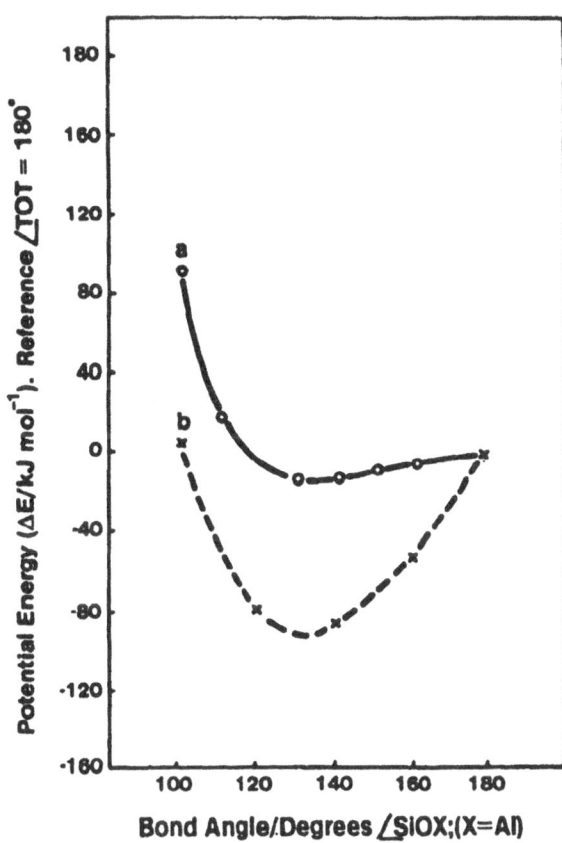

Figure 11. Calculated potential energy as a function of bond angle for SiO$^-$Al (a) and SiOHAl (b). After reference [64].

and passages into which hydrocarbon molecules can move and interact with the metal particles. The walls that form a single channel lobe are created by two cancrinite cages joined through a hexagonal prism, see Figure 10. While these cancrinite cages and the associated hexagonal prism are too small to accommodate a metal particle (and were they occupied by single metal atoms, these would be inaccessible to molecules moving through the channels), they do represent the thermodynamically most stable positions for multiply charged cations. Only the cations along the channel are exchanged at room temperature, but when the zeolite is calcined multiply charged cations will move from the channel to the locked positions inside the cancrinite cages and hexagonal prisms [62].

L-zeolite is inherently a basic zeolite and this can probably be attributed to the size of the T-O-T angles (where T is either Si or Al in a tetrahedral site) which is nearly identical at the two possible Al sites in the L-zeolite structure. These angles average to 142.0° and 142.9° [63] while zeolites that form strong acid sites tend to have some sites with greater bond angles [64]. If one of the T's is Al and the charge is balanced by a proton, SiOHAl, the acidity will be affected by the bond angle. That is, the energy difference between the deprotonated SiO⁻Al and protonated SiOHAl decreases (resulting in increased acidity) as the bond angle approaches 180° and this energy difference is greatest (the acidity is minimum) when the angle is about 130°, see Figure 11. Another factor may be the rigidity of the L-zeolite structure. When the SiOHAl bridge is deprotonated, the angle between SiO⁻Al increases and the SiO and AlO bond lengths decrease [65], a distortion that can be accommodated by more flexible structures, e.g., that of the faujasites. It has been suggested by Rabo and Gajda [64] that one can view the zeolite lattice as if it were a solvent media, i.e., had some average electronegativity, polarizability, etc. Indeed, the O 1s XPS binding energy decreases in the order $H^+ > Li^+ > Na^+ > K^+ > Cs^+$, as might be expected from electronegativity considerations, and suggests that the ionicity of the Ocation bond is $H^+ < Li^+ < Na^+ < K^+ < Cs^+$ [66]. This might provide an explanation for the fact that cation exchange can apparently affect the Pt particles even when there is no direct contact between the cation and the particle.

The general mechanism for acid catalyzed reaction envisions the addition of a proton to the molecule, the rearrangement of the protonated species, and the removal of the proton. For example, we can write for the isomerization of cyclopropane ($c\text{-}C_3H_6$) to propene ($n\text{-}C_3H_6$) on solid acid HA:

$$c\text{-}C_3H_6 + HA \rightarrow c\text{-}C_3H_7^+ + A^-$$
$$c\text{-}C_3H_7^+ + A^- \rightarrow n\text{-}C_3H_7^+ + A^-$$
$$n\text{-}C_3H_7^+ + A^- \rightarrow n\text{-}C_3H_6 + HA$$

This is a quite general mechanism and can be applied to many molecules. When applied to a paraffin, one arrives at a non-classical penta-coordinated carboniun ion [67]. This leads to what is called the monomolecular cracking mechanism [68]. As noted by Haag et al., one kind of evidence for this mechanism is that $C_6H_{15}^+$ formed from n-hexane on zeolite catalysts, e.g., HZSM-5, produces the same distribution of products as when this same ion is produced in the gas phase by chemical ionization by CH_5^+ in an ion cyclotron resonance mass spectrometer. The conditions which favor this mechanism are high temperature and low paraffin partial pressure and/or low conversion. The latter assures low olefin concentrations (as products). When significant olefin concentrations exist in the feed or are formed by the monomolecular cracking mechanism, these are more easily protonated than the paraffin and lead to the classical (bimolecular) mechanism involving a carbenium ion chain reaction which for n-hexane may be written:

$$C\text{-}C\text{-}C\text{-}C\text{-}C\text{-}C + C\text{-}C^+\text{-}C \rightarrow C\text{-}C\text{-}C\text{-}C\text{-}C^+\text{-}C + C\text{-}C\text{-}C \qquad (1)$$

$$C\text{-}C\text{-}C\text{-}C\text{-}C^+\text{-}C \rightarrow \ldots \rightarrow C\text{-}\overset{\overset{\displaystyle C}{|}}{C}\text{-}C\text{-}C\text{-}C^+\text{-}C \xrightarrow{\beta\text{-scission}} C\text{-}C^+\text{-}C + C\text{=}C\text{-}C \quad (2)$$

This is referred to as a bimolecular mechanism because the step (2) involving rearrangement of the carbenium ion and β-scission occurs very rapidly and the rate determining step is the hydrogen transfer reaction of step (1). The conditions where the bimolecular reaction path predominates are low temperature and high olefin concentration. In practical cracking conditions, both mechanism may contribute to varying extents depending on the composition of the feed and the reaction conditions. As discussed by Haag et al., the overall rate (which is a sum of the two mechanisms can be formulated such that in two limiting cases they produce apparent first order kinetics. However, the apparent first order rate constant represents the rate limiting formation of penta-coordinated carbonium ion (in the limit of high temperature and low olefin concentration) and formation of tri-coordinated carbenium ion (in the limit of low temperature and large olefin concentration). If only reactions of paraffins is being considered, the low (high) olefin concentration is equivalent to low (high) partial pressure and conversion.

3.2. BIFUNCTIONAL CATALYSIS AND TRANSPORT

Many heterogeneous catalytic reaction mechanisms involve more than one kind of function and it is possible that a given site may serve more than one of these functions. However, we reserve the classification of bifunctionality to catalysts where not only is the chemical functionality different, but the sites that serve the different functions are physically separated. This then requires a transport mechanism between the two kinds of sites. It is likely that the most widespread example of this kind of behavior is one in which some site on the active phase or support getters a reactant which is then transported by surface diffusion to the active site where reaction occurs. This is a surmise which we can only document in a few isolated cases at the moment. There is the example of oxygen adsorption on the basal plane of graphite followed by transport to a step site where gasification occurs (with or without catalysis) [69]. Also, we can cite the recent description of alumina gettering CO in the vicinity of a Pd particle followed by surface diffusion of the CO to the metal particle where it is oxidized [70]. While such processes, involving adsorption on a large number of sites of one kind (high area portion of the catalyst) and transport to another site where chemistry occurs (small area portion of the catalyst), may be suspected to be prevalent, the experimental evidence in given systems is not overwhelming.

A more chemically interesting bifunctionality is one in which catalytic chemistry occurs on two different sites. Classically, the textbook example is always that of the dehydrogenation function of a metal coupled with the acid function of a support as occurs in conventional reforming [71]. The first commercial catalyst of this kind involved Pt/Al_2O_3 where the acidity of the alumina support is enhanced by chlorine. Figure 12 shows a schematic of the mechanism involved for conversion of hexane into isohexane or methylcyclopentane into benzene. Since methylcyclopentane is another C_6 isomer of hexene, the combined reactions can be viewed as a mechanism for converting hexane into benzene. Note that the function involved in each step is shown either as the metal function (Pt over the arrows) or the acid function (A over the arrows). A particularly interesting

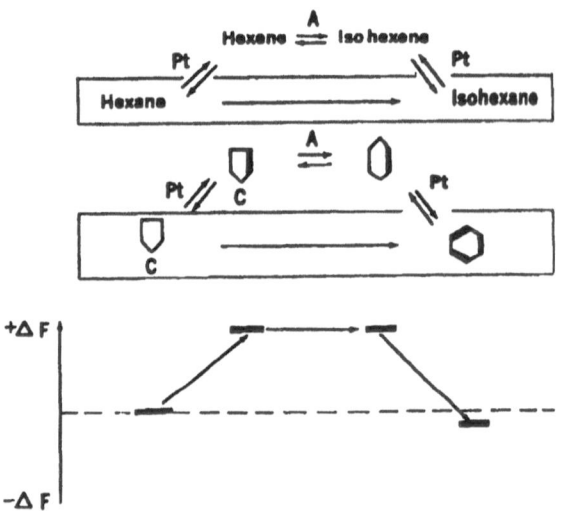

Figure 12. Polystep reactions in the catalytic transformation of hydrocarbon structures, as practiced by the petroleum industry. Paraffin isomerization (above) and aromatization of alkylcyclpentanes are carried out over catalysts that contain both acidic sites (A) and platinum (Pt). The free-energy changes in the reaction sequence are qualitatively similar, and involve a high-energy intermediate. After [34].

aspect, also illustrated in Figure 12, is that the acid function only operates on the intermediate generated by the metal function, e.g., hexene formed by hexane dehydrogenation. These intermediates all have a Gibbs free energy higher than the reactant (although the overall reaction has a negative Gibbs free energy, see bottom panel of Figure 12). Because the formation of the intermediate is thermodynamically unfavorable, it will be of very low concentration, and in the example we are discussing, the mode of transport between the sites is desorption and diffusion through the gas phase. The equilibrium concentration and its rate of diffusion place limits on how far the sites can be separated. Weisz [34] used well known chemical engineering principles (Thiele effectiveness factor analysis [72]) to construct what he calls an intimacy criterion for geometrically separate catalytic regions. In Figure 13 is shown an early demonstration of the determination of critical separation of the catalytic functions. For this particular example, the isomerization of n-heptane, (where the intermediate is n-heptene), this turns out to be about 100 μm and was both predicted by a diffusion analysis (calculation of the

Figure 13. An early demonstration of the application of the intimacy criterion for the polystep reaction of n-heptane isomerization over a mixture of highly porous platinum- and acid-bearing articles. Approach to theoretical (equilibrium) conversion (ordinate) depends critically on the component particle size expressed by their radius R. After [34].

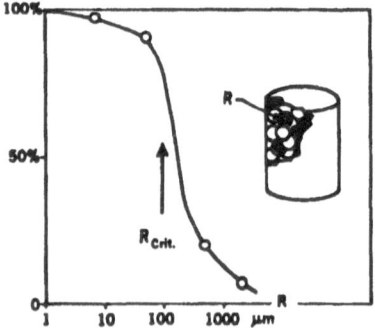

intimacy criterion) and empirical observation of the approach to isomerization equilibrium as the particle size of the metal and acid catalysts were varied.

There is beginning to be substantial evidence for another class of bifunctional reactions which can be distinguished from the first class above (where the support adsorbs reactant directly from the gas phase, but does not cause any reaction) and the second class (where the support adsorbs an intermediate from the gas phase and does cause a reaction) by spillover from one kind of site to another [73, 74, 75]. The reaction we are discussing is the hydrogenation of CO where there is overwhelming evidence for a large support effect for poor catalytic metals such as Pt [76]. It is also clear that in several systems, e.g., Pt [75] and Pd [74], a methoxy intermediate formed on or near the metal moves out onto the support surface. When the rate of hydrogenation of the methoxy species is compared to the rate of hydrogenation of chemisorbed CO in a temperature programmed reaction experiment, it is observed that the methoxy species hydrogenates at a lower temperature which suggests the mechanism involving the formation of this species, migration (which occurs at a temperature below that of its hydrogenation on Pd) and its hydrogenation may be kinetically faster than the direct hydrogenation of the CO on the metal particle. Unfortunately, this has not been established to be the case under steady-state reaction conditions nor is there any adequate explanation for why this should be faster t han the direct hydrogenation. This is particularly perplexing since the initial CO hydrogenation steps which lead to gas phase CH_4 or the formation and spillover of the CH_3-O- species would appear to be common to both routes to methane. This is a system which very much needs to be investigated with transient techniques to see if the support stabilized methoxy intermediate is, in fact, kinetically significant and the reaction bifunctional as we imply.

3.3. STRUCTURE SENSITIVITY

The concept of structure sensitivity for surface reaction arises quite naturally from an understanding of reactivity in chemistry. Structural isomers of inorganic intermediates,

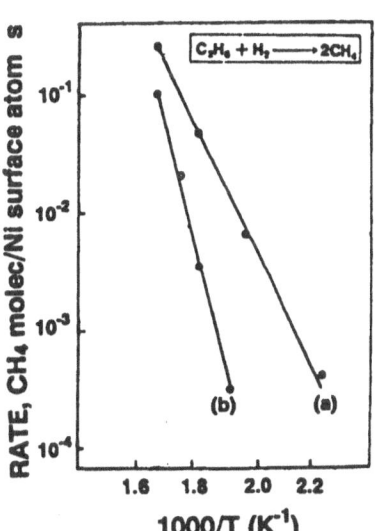

Figure 14. Methane production from ethane hydrogenolysis over (a) Ni (100) and (b) Ni (111) at a total pressure of 100 Torr and $P_{H_2}/P_{ethane} = 100$. After reference [77].

e.g. O-N-O-O (involved in the photo-oxidation of nitrogen oxides) and its isomer where all of the oxygens are equivalent are expected to have a different reactivity. Likewise, any student of elementary organic chemistry would not be surprised that *n*-pentane, isopentane and neopentane have a different reactivity. Thus, it would be natural to expect that the reactivity of surfaces with different structure will have different reactivity. A very nice exhibit of this principle can be found in the comparison of the catalytic activity of single crystal Ni exposing the (100) and (111) planes, where the latter is found to be substantially more active for ethane hydrogenolysis than the former, see Figure 14 [77]. In fact, this concept goes back t o much earlier in catalytic history see Chapter 5 of ref. [9]. As originally enunciated for the case of heterogeneous catalysis, the idea was that different particle sizes (particularly in the range from single atoms up to about 10 nm particles) would have different surface structures and thus if the reaction rate were normalized to the number of exposed surface atoms, those reaction which were structure sensitive would be particle size dependent. The surprising observation is not that there are structure sensitive reactions, but that so many seem not to be structure sensitive, see Table 5.2 from ref. [9]. Thus, we say that structure insensitive reactions will be independent of particle size and plane exposed on single crystals, while structure sensitive reactions will be dependent on both when the rate is normalized to the number of exposed atoms.

One of the most reproduced figures in all of heterogeneous catalysis is the left-hand panel of Figure 15 which shows the relative change in ethane hydrogenolysis and cyclohexane dehydrogenation as a function of percentage Cu in a supported NiCu catalyst. Whatever is happening when the Cu, essentially an inactive metal for these reactions, is added to the Ni, the active metal, it is clearly affecting the two reactions in a different way. In Sinfelt's original paper [79], the rates were normalized to H_2 chemisorption and it was suggested that the structure sensitive reaction required a particular number of Ni atoms to constitute a site and that the site might be disrupted by the substitution of or covering of a Ni atom by Cu. Subsequently, several laboratories have demonstrated that there exists dissociation of hydrogen on the active group VIII metal and spillover onto the group Ib metal, see [80] for a nice example using proton NMR on $RuCu/SiO_2$. This may change the interpretation of the original Sinfelt rates in detail, but will not change the qualitative conclusion that ethane hydrogenolysis is very structure sensitive (to the structure of the surface plane, the particle size, the chemisorption of an inactive metal, chemisorption of a poison, etc.) and cyclohexane dehydrogenation is not. The original purpose of Figure 15 was to draw an analogy between adding an inactive metal (Cu) to the active Ni surface and, the similar behavior of Rh/TiO_2 at different reduction temperatures, adding inactive TiO_x to active Rh (see section 3.4) leading to the conclusion that some species from the support had poisoned the surface of Rh. However, we now have more direct evident that TiO_x decorates the surface of Rh, see Figure 20, so we can now turn the argument around. That is, there is no evidence for spillover of H_2 from Rh to TiO_x and since the Ni-Cu (where spillover exists) and the Rh-TiO_x (where it does not) are quite parallel, we determine that the conclusion of relative structure sensitivity arrived at by Sinfelt [79] is not much compromised by spillover of H_2 from Ni to Cu.

Because our main focus, as outlined in the introduction, is fundamental concepts that have implication for practical catalysis, it is important to tie the concept of structure sensitivity so obvious from the model system, see Figure 14, back to supported metals. This was, in fact, done by Goodman where he showed that Ni/SiO_2 behaves more like Ni(100) than like Ni(111) [77]. However, we have an even nicer example in the recent work of Logan et al. [81]. They were able to draw a very complete parallel between the activity and selectivity of Rh (111) [and Rh(100)] single crystal[s] and Rh/SiO_2 for the

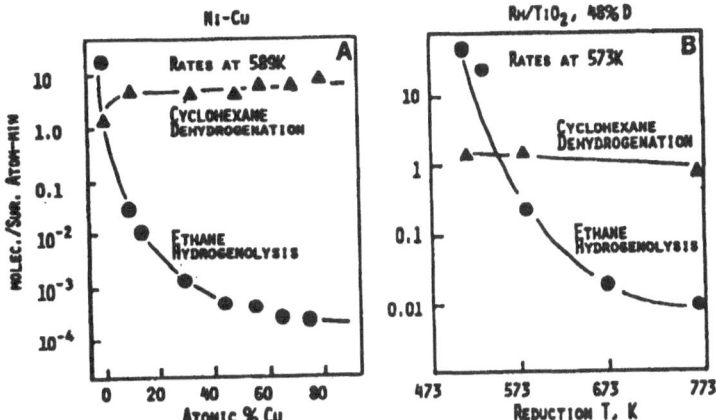

Figure 15. (left panel) Ethane hydrogenolysis and cyclohexane dehydrogenation on Ni-Cu catalysts as a function of Cu content and (right panel) 2 wt% Rh/TiO$_2$ catalysts (H/Rh=0.48) as a function of reduction temperature. After reference [78].

hydrogenolysis of n-pentane. Well annealed (in H$_2$) Rh/SiO$_2$ had an activity quite the same as Rh(111), see Figure 16. When either the single crystal surfaces or the supported small particles were roughened by oxidation followed by low temperature reduction, the activity and selectivity for multiple bond hydrogenolysis increased on both surfaces. Moreover, while the Rh (100) surface is more active than the Rh(111) surface in the ordered (annealed) state, both have the same higher activity after oxidative roughening and this activity and selectivity is comparable to that of the roughened small particles supported on silica, see Figure 17.

Figure 16. Specific reaction rate for n-pentane hydrogenolysis on annealed and preoxidized 2 wt% Rh/SiO$_2$ compared with that on Rh(111) as a function of temperature. After reference [81].

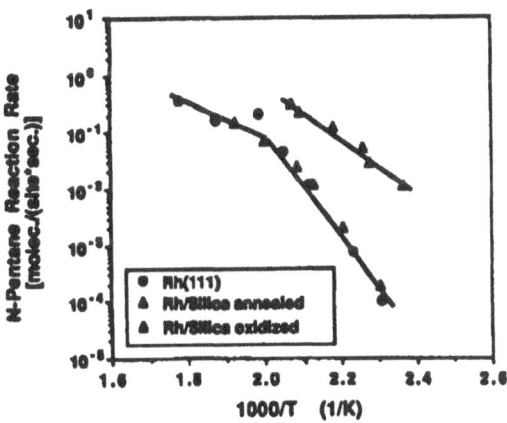

92

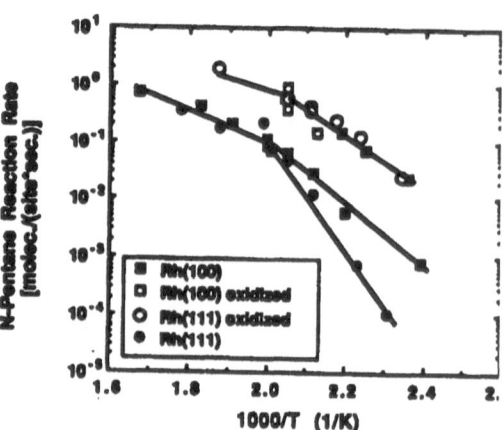

Figure 17. Reactivity of Rh(111) and Rh(100) single crystals for *n*-pentane hydrogenolysis in the annealed state and after preoxidation in 70 Torr at 773 K. After reference [81].

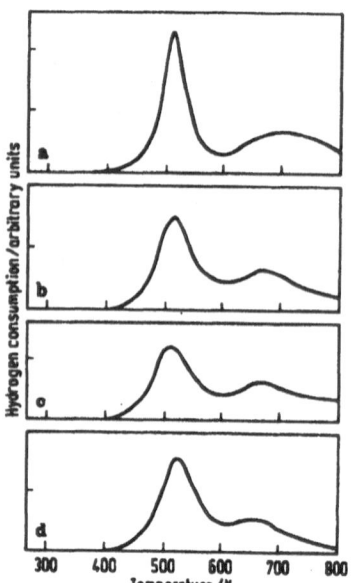

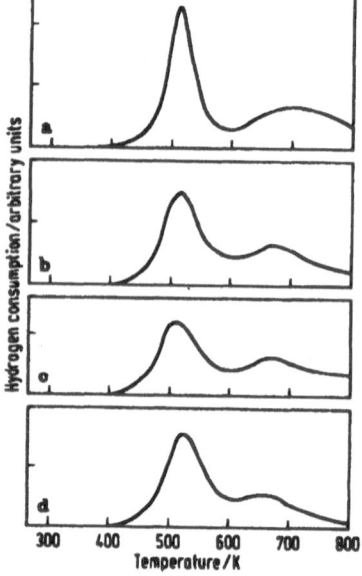

Figure 18. Temperature programmed reduction of H_2PtCl_6 on SiO_2 (Aerosil) after 2 hour treatment in 1% O_2 in He a: 373K; b: 473K; c: 573K; d: 773K. The TPR peak at $T_{max} < 400K$ originates from the reduction of $PtCl_2$. After [82].

Figure 19. Temperature programmed reduction of H_2PtCl_6 on Al_2O_3 (Aerosil) after 2 hour treatment in 1% O_2 in He a: 373K; b: 473K; c: 573K; d: 773K. After [82]

3.4. METAL-OXIDE SUPPORT INTERACTIONS

Whether we have been premature in suggesting that Pt/Al_2O_3 [75], Pt/L-zeolite [73], Pd/Al_2O_3 [74] or Ni/Al_2O_3 [83] are bifunctional CO hydrogenation catalysts, they are clear examples of metal-oxide support interactions of some kind. We are still largely ignorant of the detailed bonding that prevails between a small cluster of metal atoms and an oxide surface, but there is ample empirical evidence that some kind of interaction does occur. On the most fundamental level, this interaction can lead to different degrees of stability of the metal cluster when the same cluster is compared on different supports or clusters of different metals on the same support. As an example of the latter, one can compare Pd to Pt, prepared by parallel methods, on many supports where it is usually observed that Pd percentages exposed are uniformly lower. As an example of the former, we may compare the stability of Pt on SiO_2 and Al_2O_3 where it is generally agreed that Pt is more stable on the Al_2O_3 support. It may be that the underlying chemistry has its origins, in part, in the inorganic chemistry that precedes the metal particle. This is illustrated in Figures 18 and 19 where the temperature programmed reductions of H_2PtCl_6 impregnated SiO_2 and Al_2O_3, calcined at different temperatures are compared [82]. From other experiments, it is known that the calcination leads to increasing number and size of $PtCl_2$ crystals on SiO_2, but the Pt^{4+} state is preserved on Al_2O_3. Is there some residual intermediate, e.g., retained traces of Cl^-, that then lead to the greater stability of the Pt/Al_2O_3 or is the interaction of the Pt cluster inherently greater with Al_2O_3 than with SiO_2? The latter would likely find more support among the experts, but not without significant dissent.

While metal-support interaction of some sort prevails in all systems, rather extreme behavior is observed when reducible oxides are used as support for group VIII metals [84]. Among the reducible oxides, TiO_2 has received the majority of the experimental attention. There is general agreement now that the loss of chemisorption capacity that accompanies high temperature reduction of M/TiO_2 result from the covering or decoration of the surface the metal particle with a suboxide, TiO_x. We found extensive circumstantial evidence for this picture for the dispersed system Rh/TiO_2, see Figure 15 [78], and there was presented direct physical evidence for the parallel model system of Rh films (particles) on single crystal TiO_2 (110) [85]. The latter experiments, using Auger surface analysis to follow temporal sputtering of the Rh films, with and without reduction, demonstrated that the Rh particles were decorated with TiO_x after reduction. The summary of this experiment is shown in Figure 20. As is the case for the dispersed system this interaction can be reversed by a high temperature oxidation followed by a mild reduction. Typically, for dispersed systems, the oxidation is at 673K or above and the reduction is at 473K or below.

In the same manner that we observed (above) that the stability of metal particle dispersion would be specific to the particular metal-oxide combination, so we might suspect a similar specific chemistry for reducible oxide supports. Indeed, when Rh is compared to Pt supported on TiO_2, several differences are observed [84]. Pt enters into a strong interaction at lower temperature than Rh, undergoes a greater degree of extensive change of particle morphology and requires a higher temperature of oxidation to reverse the interaction. Comparing a single metal, Rh, on different reducible oxides, TiO_2 and V_2O_3, there is again the expected evidence for specific chemistry. The vanadia support is more sensitive to preparation variables (because it has more available oxidation states, is more soluble in aqueous solutions, etc.), results in more extensive coverage of the metal particles, enters into an interaction at lower temperature and requires a higher oxidation temperature to reverse than does TiO_2 [84].

94

We have already described the behavior of Pt/L-zeolite as a selective aromatization catalyst, see Section 3.1. Assuming that the conclusion of Mielczarski, et al. [55] is sustained by further experiments, the Pt/L-zeolite system must be considered a most interesting case of metal-oxide interaction. Their conclusion was simply that L-zeolite stabilized very small Pt clusters (several atoms) which remain stable at temperatures up to 773 K under reaction conditions. This implies that the very high selectivity for aromatization is an interesting property of small Pt clusters. The unique selectivity is not observed on other supports because such clusters cannot be formed or maintained. This has recently been confirmed by Iglesia and Baumgarten [86] who find that Pt/L-zeolite and Pt/SiO$_2$ can have comparable initial activities and selectivities but neither is sustained on Pt/SiO$_2$. In this case, they attribute the loss of activity and selectivity of Pt/SiO$_2$ primarily to coke deposition.

3.5. STEREOCHEMISTRY

The sizes and shapes of the fluid phase molecules and the morphology of the surface constrain both the direction and the orientation of approach to and regress from the surface of a heterogeneous catalyst. Therefore there should be ample opportunity in heterogeneous catalysis to employ stereochemistry to direct and to elucidate the course of a catalyzed process. However, side reactions and unexpected flexibility within the adsorbed intermediates (associated with the high temperatures of many processes) complicate the picture and preclude a simple application of stereochemical principles.

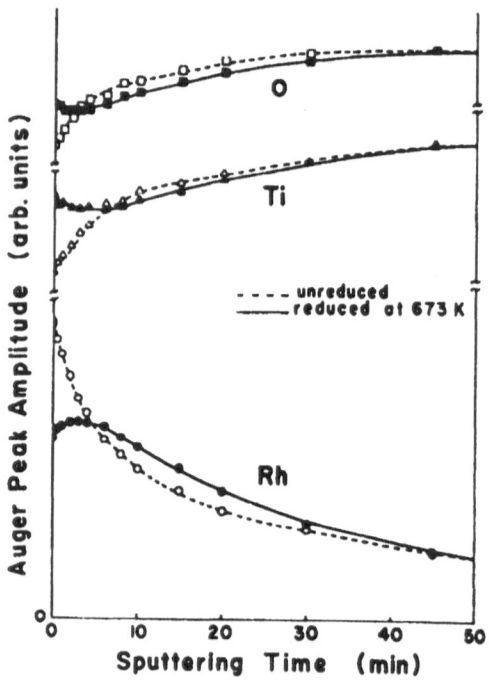

Figure 20. Auger sputter profiles (Auger amplitude vs. ion-bombardment time) for unreduced (open points, dashed curves) and reduced (solid points, solid curves). Rh/(single crystal) TiO$_2$ model catalysts. After reference [85].

3.5.1. *Size and shape specificity*. Stereochemical effects can be very large when the active site is located deep within the narrow channels of a zeolite. An excellent example [87] is the faster cracking of straight chain hydrocarbons than of branched chain hydrocarbons over narrow pore zeolites (Figure 21) The large pore zeolite, Y (see Figure 9), and amorphous silica alumina catalyze the cracking reactions in the reverse order: 3-methylpentane is about twice as reactive as *n*-hexane when there are no steric constraints, as would be expected from the relative stabilities of the tertiary and secondary carbenium ions that form. However, in erionite and in ZSM-5, the straight chain hydrocarbons react faster, consistent with their greater mobility a nd capacity in the narrow pores. The

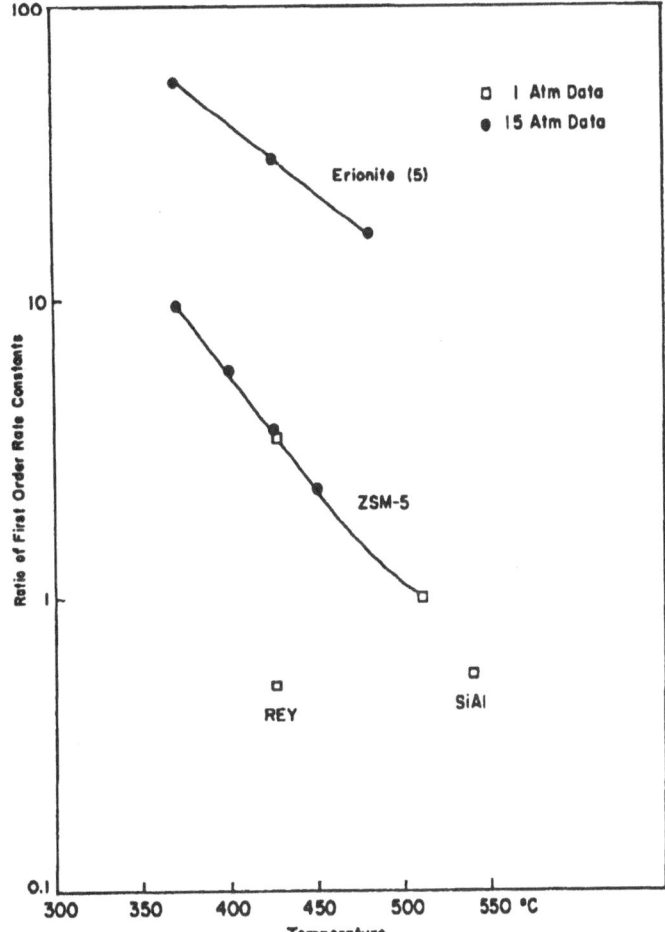

Figure 21. Demonstration of the effect of internal pore structure on the catalytic activity of silica-aluminas for paraffin cracking reactions. Erionite and ZSM-5 have pores that are nearly equal in size to the reactants; the other zeolite and amorphous silica-alumina have larger pores. After reference [87].

96

selectivity decreases with increasing temperature, in part because the higher temperature permits faster diffusion of the branched hydrocarbons to the interior of the erionite and ZSM-5.

3.5.2. *Regioselectivity.* Once a molecule encounters the catalyst site, the mode of binding may direct the reaction to a preferred product. This apparently occurs in the hydrogenation of olefins where, in the absence of a promoter (actually a poison), the predominant reaction pathway involves *cis-* addition of two hydrogen atoms [88]. The most plausible explanation invokes a sequential addition of hydrogen to an initially dicoordinated olefin, the so-called Horiuti-Polanyi mechanism [9, 89].

3.5.3. *Enantioselectivity.* The most subtle effect that can be ascribed to stereochemical constraints occurs when the selectivity involves the handedness of the reactant or product molecules. The production of L-dopa, a drug used in the treatment of Parkinsons disease, has as a key step the enantioselective hydrogenation of a prochiral olefin [90]. The asymmetric ligands attached to the soluble rhodium complex confer chirality on the catalyst. Although it may seem natural to ascribe the selectivity to an enhanced stability of the diastereomer that leads to the preferred product, there is at least one example where it can be shown that the predominant product comes from the less stable adduct [91]. The explanation must involve a lower activation energy for the less stable intermediate (Figure 22).

There are examples of enantioselective conversions catalyzed by heterogeneous catalysts [92, 93], all of which involve either the adsorption of a chiral compound on the surface of the catalyst or an attempt to use a chiral material as a support (e.g. silk α-quartz). However, none of these examples produce as high enantiomeric excesses as can be gained from homogeneous catalysts or enzymes where the selectivity can be nearly 100%.

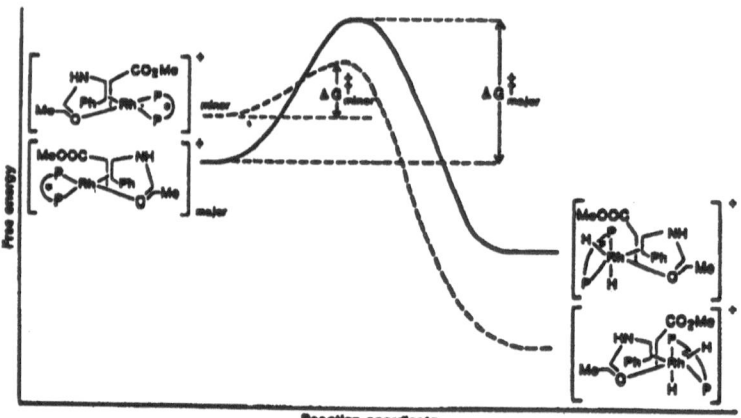

Figure 22. Reaction profiles consistent with the observation that the less predominant diastereomer participates in the main reaction channel. After reference [91].

4. Summary

We have attempted in this short overview to touch upon the chemical concepts that are required to understand many current investigations in heterogeneous catalysis. The processes of adsorption and desorption, and the ideas of an active center and metal support interactions find ready parallels in organic and inorganic chemistry (c.f. ligation, dissociation, chain reactions, and ligand effects). Even some of the more complex phenomena, like bifunctionality and structure sensitivity can be paired with examples drawn from the organometallic chemistry of the transition metals. Indeed, the various spectroscopic techniques employed to confirm the presence of particular intermediates or reaction channels almost always rely on experiences with isolable complexes and compounds for their definitive calibration. However, please do not be lured into believing that the entirety of heterogeneous catalysis can be subsumed into the current framework of inorganic and organic chemistry: the reaction conditions employed in heterogeneous catalysis are often much more extreme than those accessed in more conventional chemical experiments and therefore the species that participate may not have direct analogs in the chemistry based on isolable complexes and compounds. Moreover, we have not dealtwith physico-chemical processes like energy transport, which may be critical to the understanding of highly exothermic reactions, or mass transport, which is certainly involved in reactions where one component like hydrogen or oxygen can diffusive rapidly through the bulk., Nor have we dealt with the thermodynamics of surfaces, particularly the nature and concentration of defect centers. We close by noting that the study of heterogeneous catalysis and heterogeneous catalysts provides a route towards enhancing the efficiency of industrial processes, a way to link fundamental, molecular beam research to more conventional experiments and a fertile source of intriguing phenomena.

REFERENCES

1. Haller, G. L. and Resasco, D. E. (1992), in Trigg,G. L. (ed.), Encyclopedia of Applied Physics, Berlin, vol. 3, pp. 67.
2. Bell, A. T. et al., (1992), Catalysis Looks to the Future, National Academy Press, Washington, D. C.
3. Eby, R. T. and Singleton, T. C. (1983), in Leach, B. E. (ed.), Applied Industrial Catalysis, Academic Press, New York, vol. 1, pp. 275.
4. Dressler, D. and Potter, H. (1991), Discovering Enzymes, Scientific American Library, New York.
5. Bailey, J. E. and Ollis, D. F. (1986), Biochemical Engineering Fundamentals, McGraw-Hill Chemical Engineering Series, McGraw-Hill, Inc., New York.
6. Heinemann, H. (1981), in Anderson, J.R. and Boudard, M. (eds.), Catalysis: Science and Technology, Springer-Verlag, Berlin, vol. 1,.
7. Taylor, H. S. (1925), Proc. Royal. Soc., London, **A108**, 105.
8. Gregg, S. J. and Sing, K. S. W., (1982), Adsorption, Surface Area and Porosity, Academic Press, New York,.
9. Boudart, M. and Djéga-Mariadassou, G. (1984), Kinetics of Heterogeneous Catalytic Reactions, Princeton University Press, Princeton, NJ.
10. Brunauer, S. and Emmett, P. H. (1940), J. Am. Chem. Soc., **62**, 1732.
11. Topsøe, H. , Topsøe, N. , Bohlbro, N. , Dumesic, J. A. (1981), in Seiyama, T. and Tanabe, K. (eds.), Proc. 7th Intern. Congr. Catal., Kodansha, Tokyo, p. 247.

12. Hanson, F. V. and Boudart, M. (1978), J. Catal. **53**, 56.
13. Gasser, R. P. H. (1985), An Introduction to Chemisorption and Catalysis by Metals, Clarendon Press, Oxford.
14. Gorte, R. (1982), J. Catal. **75**, 164.
15. Rieck, J. S. and Bell, A. T. (1948), J. Catal. **85**, 143.
16. Ionnides, T. and Verykios, X. E. (1989), J. Catal. **120**, 157.
17. Peri, J. (1971), Catal. Rev.-Sci. Tech., **5**, 171.
18. Tamaru, (1978), Dynamic Heterogeneous Catalysis, Academic Press, New York.
19. Happel, J. (1986), Isotopic Assessment of Heterogeneous Catalysis, Academic Press, New York.
20. Horiuti, J. and Nakamura, T. (1967), Adv. Catal., **17**, 1.
21. Efstathiou, A. M., and Bennett, C. O. (1989), J. Catal., **120**, 118.
22. Teo, B. K. (1986), EXAFS: Basic Principles and Data Analysis, Springer Verlag, Berlin.
23. Baker, R. T. K. and Harris, P. S. (1978), in Walker, P.L. and Thrower P.A. (eds.), Chemistry and Physics of Carbon, Marcell Dekker, New York, vol. 14,.
24. Watson, B. A., Barteau, M. A., Haggerty, L., Lenhoff, A. M. and Weber, R. S. (1992), Langmuir in press.
25. Srinivasan, S.et al. (1991), J. Catal., **131**, 260.
26. Wachs, I. E. (1990), Chem. Eng. Sci., **45**, 2561.
27. Kip, B. J., Duivenvoorden, F. B. M., Koningsberger, D. C. and Prins, R., J. (1987), Catal. **131**, 260.
28. Haag, W. O., Lago, R. M. and Weisz, P. B. (1984), Nature, **309**, 589.
29. Tauseter, S. J., Pecoraro, T. A. Pecoraro, Chianelli, R. R. (1980), J. Catal., **100**, 176.
30. Roxlo, C. B., Daage, M., Ruppert, A. F. and Chianelli, R. R. (1986), J. Catal. **100**, 176.
31. Candia, R., Clausen, B. S. and , Topsøe, H. (1982), J. Catal., **77**, 564.
32. Boudart, M., Sanchez-Arrieta, J. and Dalla Betta, R. A. (1983), J. Am. Chem. Soc., **105**, 6501.
33. Delmon, B. (1979), in Barry, H.F. and Mitchell, P.C.H. (eds.), Proceedings of the Climax Third International Conference on Chemistry and Uses of Molybdenum, Climax Molybdenum Co., Ann Arbor, Michigan, USA.
34. Weisz, P. B. (1973), Science, **179**, 433..
35. Rettner, C. T. (1992), in Dwyer, D.J. and Hoffmann, F.M. (eds.), Surface Science of Catalysis, American Chemical Society, Washington, D. C., vol. 482, p. 24..
36. Weinberg, W. H. (1991), in Rettner, C.T. and Ashold, M.N.R. (eds.), Dynamics of Gas-Surface Interactions, The Royal Society of Chemistry, Cambridge, p. 171.
37. Beebe, d, T. P. , Goodman, W., Kay, B. D. and Yates Jr., J. J. (1987), J. Phys. Chem., **87**, 2305.
38. Schoofs, G. R., Arumaninayagam, C. R., McMaster, M. C. and Madix, R. J. (1989), Surf. Sci. **215**, 1.
39. Beckerle, J. D., Yang, Q. Y., Johnson, A. D. and Ceyer, S. T. (1987), J. Chem. Phys., **86**, 7236.
40. Hamza, A. V., Steinruck, H.-P. and Madis, R. J. (1987), J. Chem. Phys., **86**, 6506.
41. Rettner, C. T., Pfnur, H. E. and Auerbach, D. J. (1985), Phys. Rev. Letts., **54**, 2716.

42. C. T. Rettner, H. Stein and Schweizer, E. K. (1988), J. Chem. Phys. **89**, 3337-3341.
43. Alnot, P. and King, D. A. (1983), Surf. Sci. **126**, 359-367.
44. Clavenna, L. R. and Schmidt, L. D. (1970), Surf. Sci., **22**, 365.
45. Singh-Boparai, S. P., Bowker, M. and King, D. A. (1975), Surf. Sci. **53**, 55.
46. Rettner, C. T. and Stein, H. (1987), Phys. Rev. Lett., **59**, 2768.
47. Ertl, G., Lee, S.B., and Weiss, M. (1982), Surf. Sci., **114**, 515.
48. Rettner, C. T. and Mullins, C. B. (1991), J. Chem. Phys. **94**, 1626.
49. Bond, G. C. (1962), Catalysis by Metals, Academic Press, New York.
50. Mullins, C. B. and Weinberg, W. H. (1990), J. Chem. Phys., **92**, 4508.
51. Mullins, C. B. and Weinberg, W. H. (1990), J. Chem. Phys., **92**, 3986..
52. Hamza, A. V. and Madix, R. J. (1987), Surf. Sci., **179**, 25.
53. Sun, Y.-K. and Weinberg, W. H. (1990), J. Vac. Sci. Technol. **A8**, 2445 (1990).
54. Bernard, J. R. (1980), in Rees, L.W. (ed.)Proc. 5th Intern. Conf. Zeolites, Heyden, London, p. 686.
55. Mielczarski, E., Hong, S. B., Davis, R. J. and Davis, M. E. (1992), J. Catal., **134**, 349.
56. Gault, F. G. (1981), Adv. Catal., **30**, 1.
57. Derouane, E. G. and Vanderveken, D. J. (1988), Appl. Catal., **45**, 215.
58. Tauster, S. J. and Steger, J. J. (1990), J. Catal., **125**, 387.
59. DeMallman, A. and Barthomeuf, D. (1986), Stud. Surf. Sci. Catal., **28**, 609.
60. Kustov, L. M., Ostgard, D. and Sachtler, W. M. H. (1991), Catal. Lett. **9**, 121.
61. Newsam, J. M. (1989), J. Phys. Chem., **93**, 7689.
62. Newell, P. A. and Rees, L. V. C. (1983), Zeolites, **3**, 22.
63. Newsam, J. M. (1987), J. Chem. Soc. Chem. Commun. , 123.
64. Rabo, J. A. and Gajda, G. J. (1989), Catal. Rev.-Sci. Eng., **31**, 385.
65. van Santen, R. A. (1991), Theoretical Heterogeneous Catalysis, World Scientific, Singapore.
66. Okamoto, Y., Ogawa, M., Maezawa, A. and Imanaka, T. (1988), J. Catal., **112**, 427.
67. Olah, G. (1974), Carbocations and Electrophilic Reactions, Wiley, New York.
68. Haag, W. O., Dessau, R. M. and Lago, R. M. (1991), Proceedings of the International Symposium on Chemistry of Microporous Crystals, Kodansha Ltd., Tokyo, 1991).
69. Yang, R. and Wong, C. (1981), J. Chem. Phys., **75**, 4471.
70. Keiken, L. and Boudart, M. (1992), Proc. 10th Intern. Congr. Catal., Budapest. (to be published).
71. Gates, B. C., Katzer, J. R. and Schuit, G. C. A. (1979), Chemistry of Catalytic Processes, McGraw-Hill Book Co., New York.
72. Satterfield, C. N. (1970), Mass Transfer in Heterogeneous Catalysis, M.I.T. Press, Cambridge.
73. Larsen, G. and Haller, G. L. (1991), in Yoshida, S, Takezawa, N. and Ono, T (eds.), Catalytic Science and Technology, Kodansha, Tokyo, vol. 1, p. 135.
74. Chen, B., Falconer, J. L. (1992), J. Catal., **134**, in press.
75. Robbins, J. L. and Marucchi-Soos, E. (1989), J. Phys. Chem., **93**, 2885.
76. Vannice, M. A. and Twu, C. C. (1983), J. Catal., **82**, 213.
77. Goodman, D. W. (1982), Surf. Sci., **123**, L679.
78. Resasco, D. E. and Haller, G. L. (1983), J. Catal., **82**, 279.
79. Sinfelt, J. H. (1972), J. Catal., **27**, 468.

80. Wu, X., Gerstein, B C. and King, T. S. (1990), J. Catal., **121**, 271.
81. Logan, A. D., Sharoudi, K. and Datye, A. K. (1991), J. Phys. Chem,. **95**, 5568.
82. Foger, K. (1984), in Anderson, J.R. and Boudart, M. (eds.), Catalysis: Science and Technology, Springer-Verlag, Berlin, vol. 6, p. 227.
83. Glugla, P. G., Bailey, K. M. and Falconer, J. L. (1988), J. Phys. Chem., **92**, 4474.
84. Haller, G. L. and Resasco, D. E. (1989), Adv. Catal., **36**, 173.
85. Sadeghi, H. R. and Henrich, V. E. (1984), J. Catal., **87**, 279.
86. Iglesia, E. and Baumgarten, J. E. (1992), in Proc. 10th Intern. Congr. Catal. to be published, Budapest.
87. Chen, N. Y. and Garwood, W. E. (1978), J. Catal., **52**, 453.
88. Siegel, S. (1966), Adv. Catal., **16**, 123.
89. Horiuti, J. and Polanyi, M. (1934), Trans. Faraday. Soc., **30**, 1164.
90. Knowles, W. S. (1983), Acc. Chem. Res. **16**, 106.
91. Collman, J. P., Hegedus, L. S., Norton, J. R. and Finke, R. G. (1987), Principles and Applications of Organotransition Metal Chemistry, University Science Books, Mill Valley, California.
92. Izumi, Y. (1971), Angew. Chem., **83**, 956.
93. Groenewegen, J. A. and Sachtler, W. M. H. (1977), in Bond, G.C., Wells, P.B. and Tompkins, F.C. (eds.), Proc. 6th Intern. Congr. Catal., The Chemical Society, London, vol. 2, p. 1014.

ELECTRONIC STRUCTURE THEORY FOR TRANSITION METAL SYSTEMS: A SURVEY

M. C. ZERNER
Quantum Theory Project
University of Florida
Gainesville, Florida 32611
USA

ABSTRACT. Theoretical models in common use today for studying the electronic structure and spectroscopy of transition metal containing systems are surveyed. Comparisons are made between the methods that demonstrate that transition metal complexes are "different" and results from common molecular orbital-based theories might be less accurate and more difficult to interpret than they are for molecules without transition metals. A simple interpretation of properties based on orbital arguments, for example, are often misleading. An evaluation of the models and their promise is made throughout the text.

1. Introduction

Perhaps the first modern look at systems containing electrons in d and f orbitals was that of Bethe [1] and Van Vleck [2]. This theory, crystal field theory, nicely summarized by Ballhausen [3], accounted for many of the properties exhibited by these systems, and we briefly discuss this below.

The crystal field Hamiltonian for a one metal atom complex is written as

$$H = \{ -h^2/2m \sum_i \nabla_i^2 - \sum_i Ze^2/r_i + 1/2 \sum_{i \neq j} e^2/r_{ij} +$$

$$+ \sum_i \xi_i l_i \cdot s_i \} + V \qquad (1)$$

$$= H_0 + V$$

and further classified into types according to:

$$V \geq \sum_{i<j} e^2 / r_{ij} \qquad \text{Covalent Compounds}$$

$$\sum_i \xi_i l_i \cdot s_i < V < \sum_{i<j} e^2/r_{ij} \qquad \text{First Transition Series}$$

$$V < \sum_i \xi_i l_i \cdot s_i \qquad \text{Lanthanides and Actinides}$$

D. R. Salahub and N. Russo (eds.), Metal-Ligand Interactions: from Atoms, to Clusters, to Surfaces, 101–123.
© *1992 Kluwer Academic Publishers. Printed in the Netherlands.*

The "undemocratic" breakdown of the Hamiltonian of equation 1 into metal ion and all the rest is rather different in philosophy from most work today, and such a breakdown is of limited utility in studying, for example covalent compounds. For systems and problems in which the metal is of particular importance this breakdown, however, is convenient.

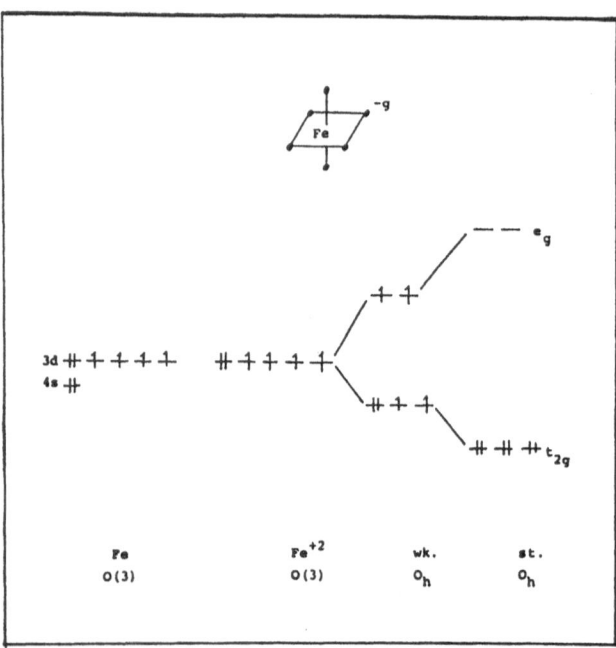

Figure 1. The crystal field view of an octahedral complex of Fe(II).

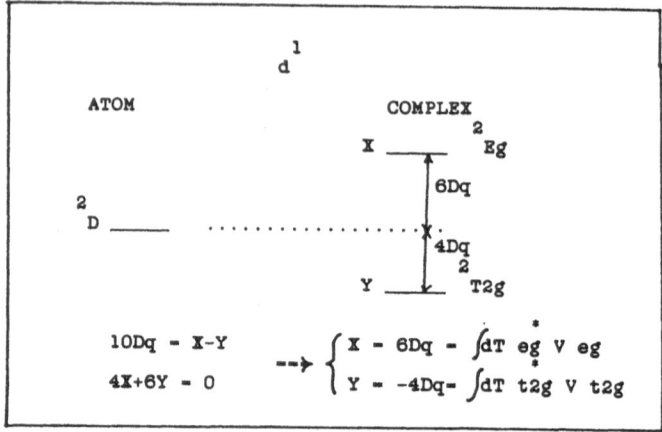

Figure 2. Calculation of the crystal field splitting of a single d electron in an octahedral field.

The perturbation V is then expanded in terms of symmetry components as

$$V = \sum_i \sum_l \sum_m Y_l{}^m (\theta_i \phi_i) R_{n_1} (r_i) = V_O + V_R \qquad (2)$$

in which $Y_l{}^m(\theta\phi)$ are spherical harmonics, V_O the l=0 component, and V_R, the rest. All molecular complexes would have, for example, a spherical component of V, and the largest correction to an octahedral complex would be the l=2 quadrupolar term.

A pictorial way of viewing equation 1 and 2 is given in Figure 1, the well known crystal field splitting for an octahedral complex. On the left side of this figure are the frontier atomic orbitals of the iron atom. In this model one then prepares a common ion of the complex by removing one or more electrons from the atom to a sphere surrounding

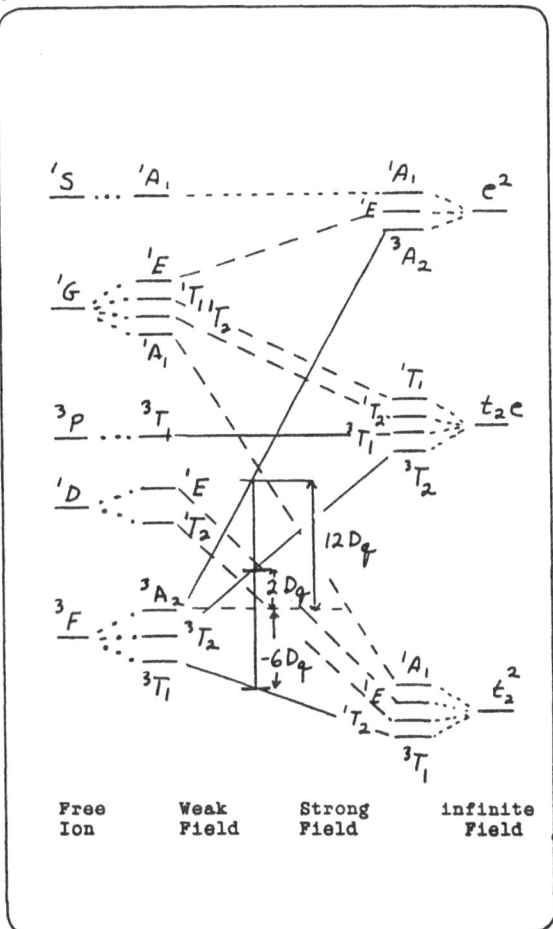

Figure 3. The crystal field splittings of 2 electrons in d orbitals under an octahedral perturbation.

the atom, in this case d^6 Fe(II). Then the charge placed on the sphere is localized at points on the sphere to simulate negative ligands. Depending on the amount of charge and the radius of the sphere (representing metal-to-ligand bond distance) a weak-field high-spin situation results, or a strong-field low-spin situation results. The determination of spin state is a function of the orbital split verses the stability due to Hund's rules: states of highest multiplicity lie lowest in energy. This concept can be made quite quantitative by using Slater-Condon factors [3-6] or Racah coefficients [3] to estimate the energy of the various spin states that result from a given electronic configuration.

In the case of the octahedron, two orbitals go up in energy, for occupying these orbitals with electrons causes repulsion between metal electron and ligand, in this case between $d(x^2-y^2)$ and $d(z^2)$ and ligands lying on the x, y, and z axes. By Gauss' theorem [7] ten electrons in d orbitals, two in each orbital, will lead to no change in the system energy regardless of the interaction. This implies that there is a relationship between the energy in which these two orbitals are raised from that found in the ion, and the amount of energy that the other three orbitals are depressed. This is shown in Figure 2.

The energy splitting of the orbitals defines $=10Dq$. Recall that V_2 is the quadrupolar correction to the Hamiltonian of equation 1. This idea can be, for example, extended for two electrons in d orbitals, as shown in Figure 3.

If one plots the energy of the system relative to the ground state energy one obtains the well known Tanabe-Sugano diagrams, and this is given for the d^7 system in Figure 4.

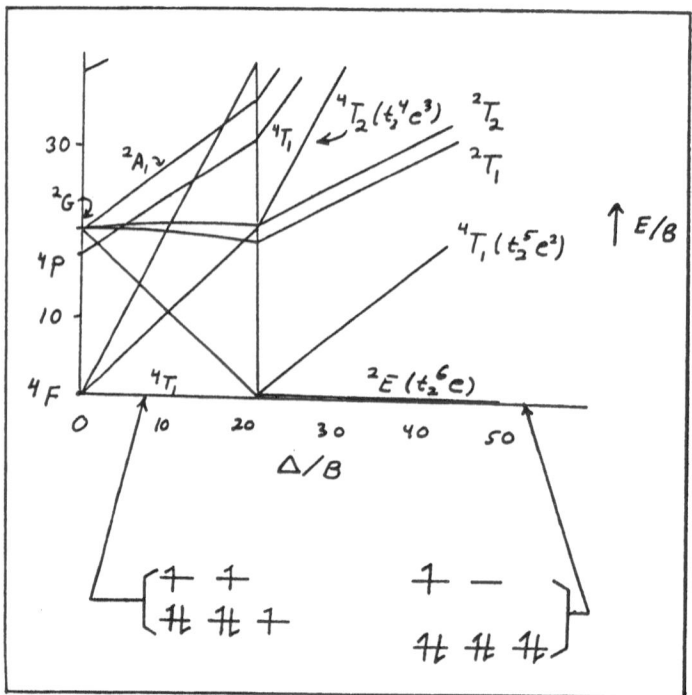

Figure 4. The Tanabe-Sugano Diagram for d^7 in an octahedral environment. The case shown is that for Co(III) with Racah B coefficient $= 970$ cm-1.

These diagrams are of great use, either directly, in explaining observed phenomena, or important in beginning more quantitative calculations as described below, or in the interpretation of more quantitative calculations.

There are several problems with crystal field theory. Foremost among these is that such a theory predicts atomic or ionic d orbitals in the electrostatic field of the ligands. No account is taken of the fact that d orbitals might mix "covalently" with the ligand orbitals. This is at variance with, for example, EPR experiments on paramagnetic systems that show unpaired spin density not only on the central metal ion, but also on the ligands. Such observations strongly point to a somewhat delocalized "d" orbital.

Early attempts to correct for this deficiency quickly led to what is now called "Ligand Field Theory". Although the exact definition of Ligand Field Theory is somewhat obscured, all such models have covalency as a common feature, and might be considered as higher order perturbation theory assuming a Hamiltonian of the form of equations 1 and 2 [3,8].

Figure 5 shows how this might proceed. This again shows the case of an octahedron. In this group the five d orbitals transform according to the e_g and t_{2g} irreducible representation. On the left side of this figure the sigma bonds of the ligands are shown to transform according to a_{1g}, e_g and t_{1u}.

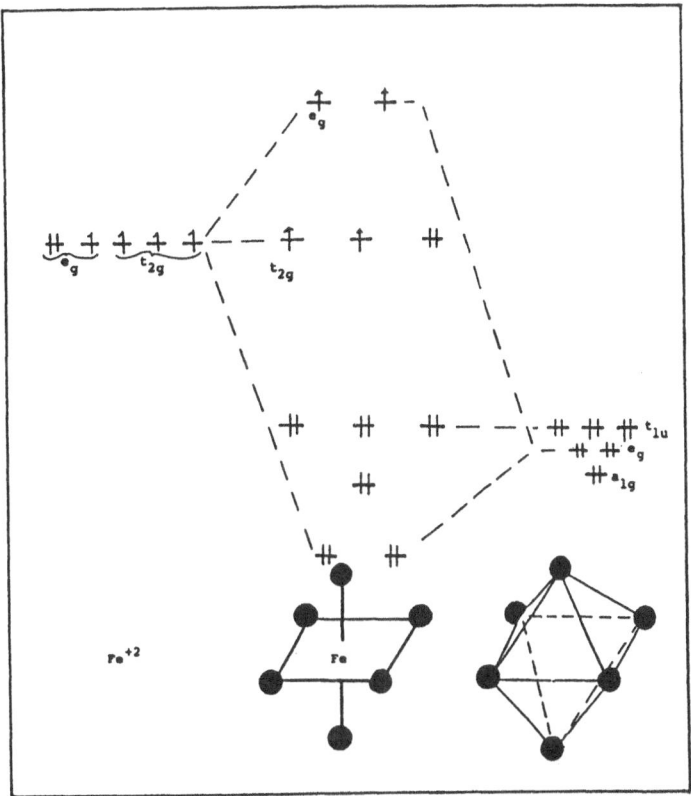

Figure 5. The ligand field picture for an octahedral field.

These ligand-based orbitals are assumed doubly occupied on chemical grounds. In forming the complex, only orbitals of like symmetry can interact. Those orbitals that are mostly d again split into two orbitals of higher energy, e_g, than the remaining three, t_{2g} (compare with Figure 1). But there is a qualitative difference in these two models. The elevation of the e_g "d" orbitals is caused by interaction with the e_g molecular orbitals that are ligand. This results in d orbitals that have ligand character. Electrons assigned to these orbitals are not localized to the metal. In a similar fashion the e_g molecular orbital that originates from the symmetry-adapted combination of ligand sigma bonding orbitals have contributions from the metal e_g atomic orbitals. Occupying these orbitals contributes to the ligand-to-metal charge transfer.

Crystal field and ligand field concepts such as those I have briefly reviewed here are very important in transition-metal chemistry, and probably account for at least 95% of all thinking on this subject. They are used to explain spectroscopy as well as chemistry using frontier molecular orbital concepts. But these models are again limited. They are not generally quantitative. In general, to extract quantitative information from these models requires experimental parameters from the complex of interest. For examples, given the position of one spectroscopic line in a complex, predictions on the location of other transitions can be made. Given the electronic spectrum of a complex, and experimental EPR g values, covalency can be estimated using perturbation theory.

2. Extended Huckel Theories

In order to free these theories from experimental parameters taken from the complex of interest, Wolfsberg and Helmholtz in 1953 [9] borrowed ideas from Mulliken [10] and introduced a Hamiltonian of the ligand-field type in which the interaction between d and ligand orbitals was set proportional to the orbital overlap. This idea also became central to the Extended Huckel models of Libscomb and Lohr [11] where it was applied to boron hydrides, and Hoffmann [12] who coined the name "Extended Huckel". The extended Huckel model was generalized to transition-metal complexes by Ballhausen and Grey [13] and Zerner and Gouterman [14], and is now heavily used by many groups [15,16].

These models proceed to solve a matrix secular equation [17]

$$H\,C = S\,C\,E \tag{3}$$

where the molecular orbitals, ϕ_i are given by

$$\phi_i = \sum_u C(i,u)\,X_u \tag{4}$$

and X_u are the atomic orbitals.

In the simplest forms of this theory,

$$H(uu) = \text{Valence State Ionization Potential, from experiment} \tag{5a}$$

and

$$H(uv) = K(uv)(H(uu)+H(vv))\,S(uv)/2 \tag{5b}$$

and $S(uv)$ is the overlap between atomic orbitals X_u and X_v.

$$S(uv) = (X_u \mid X_v) \qquad (6)$$

K(uv) is a parameter, generally set to 1.75 to 2.0 regardless of u and v, although other choices have been examined [14,18].

This simple model is one of the most useful in quantum chemistry. Parameters are chosen from atoms and not from complexes directly. Many refinements have been made. Perhaps the most important concerns charge iteration [13,14]. In the scheme above too much charge might localize on the metal or ligand atoms, and it is unrealistic to assume a potential for a neutral atom. The iterative or self-consistent charge schemes generally extrapolate parameters from neutral atoms and appropriate ions. The result is far less charge separation in the calculated results than found in the simple non-iterative procedure with parameters obtained only from neutral atoms.

The results of Extended Huckel calculations on transition-metal systems look very much like those obtained from ligand field theory: weakly perturbed metal orbitals in a molecular environment. The orbitals obtained are "physical" orbitals, see Figure 6, in the sense that they are easy to interpret.

The metal d based orbitals are generally the frontier orbitals. The energies of occupied orbitals refer to the negative of ionization potentials, and orbital energy gaps refer to average spectroscopic transitions [14,19]. In this scheme, the energy of the unoccupied orbitals do not refer to electron affinities as this would be inconsistent with their role in spectroscopy.

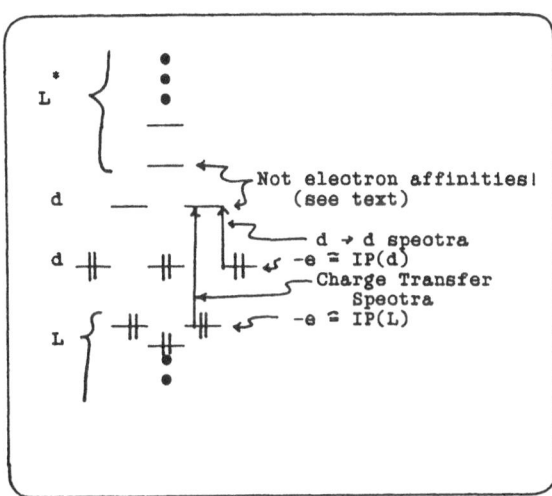

Figure 6. "Physical Orbitals", or those that are easiest to correlate directly to physical observables. Note that either the virtual, or empty, orbitals correspond to electron affinities, or can be related directly to spectroscopy, but not both. This is a consequence of the fact that orbitals cannot be related directly in an exact fashion to any physical observable, see text.

As useful as the Extended Huckel methods have proven, they are at best semi-quantitative. They cannot, for example, yield detailed spectroscopy nor can one obtain structure from first principles. For example, is $CuCl_4^{-2}$ tetrahedral, distorted tetrahedral or square planar? What is the bond length? What are the vibrational frequencies?

3. Molecular Orbital Theory

One might view the progression of theories described above, crystal field theory, ligand field theory, extended Huckel theory, as a progression toward molecular orbital theory, and one might ask "Why not do molecular orbital calculations on transition-metal complexes?" After all, this is the 1990's.

Such calculation started to appear regularly in the 1970's. At first I think it is fair to say that these calculations were disappointing both quantitatively and qualitatively. Computer technology and our knowledge of the problems that arise in such studies just weren't up to quantitative accuracy, and the results obtained often did not coincide with the physical pictures we had developed using crystal and ligand field concepts. As time progressed, however, such calculations became quite capable of making quantitative predictions as exemplified in the subsequent chapters of this book, but often the qualitative pictures that we had developed using simpler models were not obtained. Metal based d orbitals are seldom the highest occupied, transitions between orbitals cannot be based upon orbital energy differences, and often the aufbau principle itself cannot be used to assign electrons to molecular orbitals. Crystal field, ligand field and Extended Huckel Molecular Orbitals are not self-consistent field molecular orbitals as shall be shortly demonstrated, and these orbitals can all, in turn, be quite different from the Kohn-Sham orbitals we discuss later in this chapter.

We briefly review the essential ingredients of molecular orbital theory, and stress especially those aspects of this theory that prove problematical. Table 1 summarizes the model.

In the first step we assume the time independent fixed nuclei, non-relativistic "electronic" Hamiltonian [18]. In general these assumptions are well justified, and can be corrected by perturbation theory. The electronic Hamiltonian is then written in full form

$$H = -\hbar^2/2m \sum_i \nabla_i^2 - \sum_i \sum_A \frac{z_A e^2}{R_{iA}} + \sum_{i<j} e^2/r_{ij} + \sum_{A<B} \frac{z_A z_B}{R_{AB}} \qquad (7)$$

(compare equation (7) with equation (1)).

The second step is the orbital approximation itself. We assume that a good starting point for the calculation is a product of one-electron functions - orbitals. This product must be antisymmetric with respect to the exchange of any two electrons, and is most often further created to be an eigenfunction of spin [17,20], although Unrestricted Hartree Fock theory (UHF) [21, 22] is also often used.

In the third step the molecular orbitals are expanded as a linear combination of atomic-like orbitals. Often these "basis" sets of orbitals are taken as the solution of the same problem as we outline here, but for atoms. This procedure [17,23,24] recommends

TABLE 1.   The steps in the LCAO-MO-SCF theory as usually
implemented. This procedure is also often referred to as the Hartree Fock
(HF) procedure, although, strictly speaking, HF need not employ a basis
of analytical functions, step 3.

1. The time-independent, fixed nuclei, non-relativistic, problem
$$H\psi_a = E_a \psi_a \qquad (A1)$$

2. Molecular Orbital Approximation (MO)

$$\psi_o = \theta(s) \ A \ \{\phi_1(1)\phi_2(2)\phi_3(3)\phi_4(4) \ldots \phi_N(N)\} \qquad (A2)$$

where A is the antisymmetrizer, insuring the Pauli Exclusion Principle
$\theta(s)$ is a spin projection operator insuring that
ψ_o is an eigenfunction of spin (=1 for UHF)
$\phi_j(i)$ is the j'th orbital assigned to electron i

3. Linear Combination of Atomic Like Orbitals (LCAO)

$$\phi_i = \sum_u X_u \ C(u,i) \qquad (A3)$$

4. Variational Principle

$$W(\psi) = (\psi|H|\psi)/(\psi|\psi) \geq Ex$$

$$\delta W/\delta\psi = 0 \ ---> \ FC=SCE \qquad (A4)$$

where W is the expectation value of the energy represented by H for the
trial wavefunction ψ
Ex is the exact energy for this H
F is the Fock or energy matrix
S is the overlap matrix
C are the molecular orbital coefficients of equation A3.
The i'th column of this square matrix corresponds to ϕ_i.
E are the molecular orbital energies as a diagonal matrix

5. Self Consistent Field (SCF)

$$F(C^0) --> C^1 --> \bar{C}^1 --> F(\bar{C}^1) --> C^2 --> \bar{C}^2 \ F(\bar{C}^2) \ etc.$$

where C^0 is the initial case, and $\bar{C}^n$ are extrapolated mo coefficients

---

itself for several reasons. Foremost among these is the fact that an atom in a molecule is
principally an atom. An x-ray picture of a molecule shows weakly perturbed atoms, and
the binding energy is most often a very small fraction of the total energy of the system
(for example, in the triple bonded N_2 molecule, the binding energy is about 0.3% of the
total energy, and this is a strong bond). To do a good job in molecular calculations, atoms
must be well represented.

The fourth step is the assumption of the variational principle that states that the expectation value of the energy, W

$$W = (\Psi|H|\Psi)/(\Psi|\Psi) \geq Ex \tag{8}$$

must be greater than the experimentally observed energy of the Hamiltonian assumed. Varying the wave function with respect to its parameters in equation 8 can then only lead to better energies. If we vary W with respect to the molecular orbital coefficients C, we derive the matrix secular equation shown in Table I, and assumed, for example, in equation 3 for the Extended Huckel Model (H = F).

The F matrix, or Fock matrix, of this theory depends on the kinetic energy of the electrons, the nuclear-electronic attraction and the electron-electron interactions. It is this latter term which makes the solution of equation A4 of Table I non-linear and iterative in nature. We must estimate where the electrons are, calculate the electron-electron term, reform F, solve equation A4, Table I, reevaluate the electron-electron term, reform F, and continue this procedure until the electron field is "self-consistent" with that which was used to obtain it to within a given tolerance.

This is certainly a short description for a rather involved procedure [17], but the steps are correct, if briefly described. Two aspects of this theory are important to stress, especially in the context of its application to problems in transition-metal chemistry.

The first of these is associated with the LCAO approximation, step 3 of Table 1. Although such a procedure admits to the observation that an atom in a molecule is still an atom, and it is important to describe atoms correctly, certain aspects of atomic characterization are lost by this construction, namely the rather unique spin and spacial orbital angular momentum that characterize an atom or an ion. Now this would seem to be of little importance in many cases: for example, we do not really care whether the carbon atom in benzene was a ^3P or a ^1S or a ^1D state when we formed the molecule. The bonding is sufficiently strong so as to uncouple all the angular momenta that the atoms had when isolated. But this is not the case with a transition element. We do care whether the Co(III) ion of Figure 4 was ^4F or ^4P. The very successes of ligand field theory and the utility of the Tanabe-Sugano diagrams depend on the fact that the transition atom is weakly perturbed. In the molecular orbital theory this information is discarded, molecular orbitals created, and electrons assigned to orbitals, generally two at a time, usually in order of increasing orbital energy.

Secondly, it is interesting to note that we, as chemists, often place the highest physical significance on the molecular orbitals and their orbital energies. Yet they have no real meaning other than that ascribed to them through Koopmans' approximation - that in the absence of relaxation (frozen orbital approximation) the molecular orbital energy approximates the negative of ionization energies. For localized orbitals, such as d and f orbitals, this is a poor approximation with relaxation energies in the tens of electron volts. The orbital approximation is a mathematical convenience in solving for the total energy E, and the state wave function, ψ [18, 25]. These are the only quantities with well defined physical significance.

Shortcomings in the LCAO-MO-SCF procedure, (or Hartree-Fock, HF, procedure) are corrected by introducing electron correlation into the calculation. Such procedures are complex and the subject of much intense study today [17, 25, 26]. But there is one aspect of this problem that should be demonstrated here. How does one restore atomic characteristics to an atom in a molecule?

For this we examine the simplest of molecular problems, that of H_2. The usual description of the electronic structure of this molecule is the configuration that assigns two electrons to the bonding orbital as

$$\psi(I) = |\phi\bar{\phi}| \tag{9a}$$

with

$$\phi = (X_a + X_b) / \{2(1 + S)\}^{1/2} \tag{9b}$$

where S is the overlap between X_a and X_b, the (linear combination of) atomic orbitals located on atom "a" and "b". Expanding the determinant of equation 9a yields

$$\psi(I) = \{X_a(1)X_a(2) + X_b(1)X_b(2) + X_a(1)X_b(2) + \\ + X_b(1)X_a(2)\}O(spin) \tag{9c}$$

where the first two terms represent ionic terms, two electrons on atom a, none on b, and two on b, none on a, and the last two terms are covalent terms.

This, by symmetry, is the description of H_2 at all internuclear separations. Although a reasonably good description at equilibrium, the appearance of the ionic terms at large separation causes problems, as H_2 does not dissociate into a 50%-50% mixture of two H atoms and a proton and H anion, as this wave function would suggest. At large internuclear separations we must restore the atomic nature to the H atoms.

This is done through configuration interaction, a well established procedure. In its simplest form we mix with the $|\phi\phi|$ determinant an electronic configuration with two electrons assigned to the antibonding partner of equation 9b

$$\psi(II) = |\phi^*\bar{\phi}^*| \tag{9d}$$

$$\phi^* = (X_a - X_b) / \{2(1 - S)\}^{1/2} \tag{9e}$$

Expanding yields

$$\psi(II) = \{X_a(1)X_a(2) + X_b(1)X_b(2) - X_a(1)X_b(2) - \\ - X_b(1)X_a(2)\}\,O(spin) \tag{9f}$$

Using the variational principle on a wave function of the form

$$\psi = \cos(t)\,|\phi\bar{\phi}| - \sin(t)\,|\phi^*\bar{\phi}^*| \tag{9g}$$

yields the result that t approaches 45 degrees as the internuclear separation increases. Although this observation may at first seem unreasonable, recall that the node between the two H atoms in ϕ^* has decreasing influence as R increases, and the energy of the bonding and antibonding orbital approach one another. At $t = 45^\circ$ the ionic terms of equation 9c and 9f cancel, leading to a proper description of the dissociation process. In other words, we have restored the atomic nature to the H atoms.

Now what relevence has this to the problems associated with solving the Schrödinger equation for transition-metal complexes?

The role of the second configuration in the H_2 problem is somewhat important already at equilibrium, and becomes absolutely essential when the orbital overlap $S=(X_a|X_b)$ becomes less than about 0.2. The typical overlap between a transition-metal d orbital and the orbitals of chelating atoms are generally much smaller than this. We might anticipate that this fact may render many simple Hartree Fock calculations of limited value.

Since we suspect that the atomic nature of the central metal atom is important, which is, after all, the reason that such models as crystal field theory are useful, we will need to insure that the level of theory we employ is sufficient to guarantee a proper representation of the metal atom. Even so, we cannot guarantee that the simple orbital pictures that result from crystal field theory will result from molecular orbital theories, even corrected by configuration interaction.

Consider the simple case of $CuCl_2$ given in Figure 7. Crystal-field considerations yield an orbital level diagram with the $d(z^2)$ "sigma" orbital highest in energy, than the $d(xz)$ and $d(yz)$ "pi" type slightly above the $d(x^2 - y^2)$ and $d(xy)$ "delta" orbitals. Of interest then are the excitations among the d orbitals, and the charge transfer excitations.

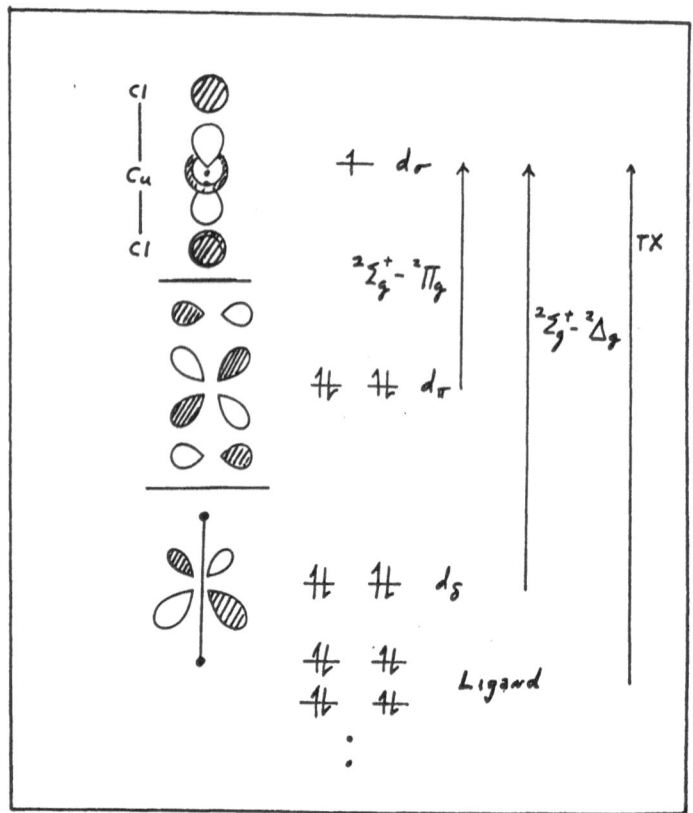

Figure 7. The ligand field construction of orbitals for $CuCl_2$.

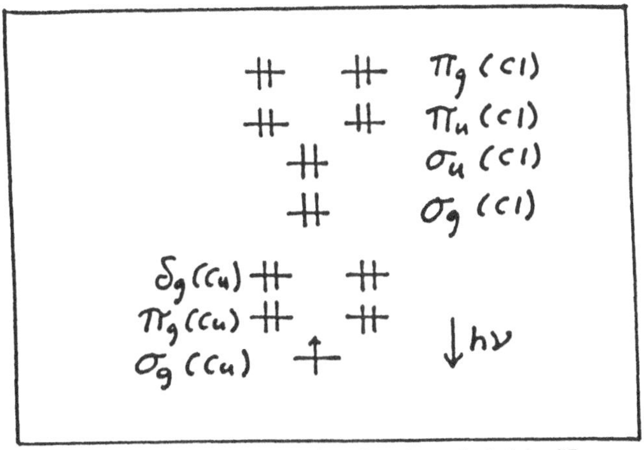

Figure 8. The order of orbitals as calculated using ab-initio Hartree Fock theory. Contrast this with the order given in Figure 7. Note that there is a singly occupied orbital below many doubly occupied ones.

The results of SCF calculations either of the ab-initio or INDO type [27] are given in Figure 8. The highest occupied orbitals are ligand type, not d. The "d" orbital energies are exactly the opposite of that predicted by ligand-field theory, and the singly occupied orbital is the lowest lying d orbital, suggesting at first glance that even the aufbau principle does not apply.

But how do we know such results are meaningful? Removal of an electron from the d(pi) orbitals or d(delta) orbitals and placing this electron into the d(sigma) orbital leads to excited states with excitation energies about 9,000 cm^{-1} higher in energy than the state described in Figure 8. In addition, removal of an electron from the HOMO, a Cl based pi orbital, and placing this in the d(sigma) orbital leads to a charge transfer excitation some 20,000 cm^{-1} higher in energy than the state represented in Figure 8. These observations are consistent with experiment [27]. Excitations down in orbital energy, lead to states of higher energy. Note that these quantitative predictions are consistent with the qualitative crystal-field picture of Figure 7, and, unfortunately, the theory which gives quantitative results (after CI) does not yield an orbital picture of great practical utility. Fortunately, as demonstated in subsequent chapters in this book, this is not always the case. But one must be careful in giving too much significance to orbitals obtained in ab-initio SCF theory, as this extreme case demonstrates.

There is another important point to make here, and one that was alluded to earlier. The crystal field theory was used to assign electrons to orbitals to guide the SCF procedure to convergence. That is, the electron hole was assigned to the d(sigma) orbital to aid SCF convergence. This is often the case: the quantitative calculation is guided by intuition gained through crystal- or ligand-field theory. The SCF procedure would not have converged in this case if the electrons were simply assigned on the basis of the aufbau principle.

As a second example, we consider the case of ferrocene. The ligand-field description of this is summarized in Figure 9. Here we will examine only the ionization spectrum, although we have also examined in some detail the uv-visible spectrum of the neutral [28]. Ab-initio calculations [28, 29] yield the occupied metal-based orbitals some 10eV below the HOMO, a ligand based mo of e$_1$' type. Intermediate Neglect of Differential

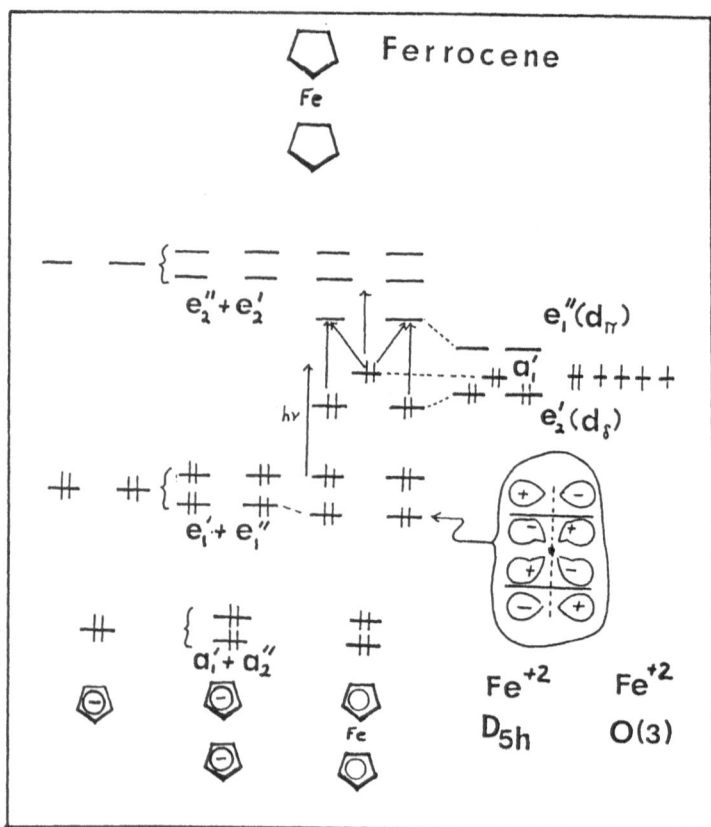

Figure 9. The ligand-field construction of orbitals for ferrocene. Note that both ab-initio and INDO calculations yield the HOMO to be a ligand-based e_1''.

Overlap calculations (INDO/S) [22,30] yield the metal-based orbitals as highest lying, as this method relies heavily on atomic spectroscopy for its parameters and one-center (atomic) relaxation is automatically built-in.

Regardless of this order of orbital energies, all methods predict that the first ionization is best described as loss of an electron from the metal-based $e_2'(d(delta))$ the second from an $a_1'(d(sigma))$, and subsequent ionizations from ligand-based molecular orbitals, as indicated in Table 2.

These results are obtained from delta-SCF calculations, separate calculations on the neutral and positive ions. The orbital energies are not useful for this estimate.

Table 3 presents another interesting aspect of this problem. Here the electron populations are given for each state calculated in this delta SCF procedure. It is most easy to see from the spin density, the numbers in parentheses, where the unpaired electron is. But note that the net population (from a Mulliken population analysis) on the iron atom remains about +2, for the neutral, and for any of the ions, even those that have iron atoms that are formally "ferric".

TABLE 2. The ionization spectrum of ferrocene, in eV. a is reference 37, b reference 29, c reference 28, see text and Figure 9.

Method/state	$^2E_2'$	$^2A_1'$	$^2E_1''$	$^2E_1'$	$^2A_2''$	$^2E_1'$
	|--------metal--------|--------------ligand--------------|					
Experiment(a)	6.9	7.2	8.7	9.4	12.2	13.6
Extended Basis(b) delta-SCF	5.7	7.5	8.8	8.9	13.0	13.6
INDO/S(c) delta-SCF	6.1	5.8	9.6	9.7	13.1	13.9
corrected Koopmans	7.8	8.3	9.6	10.0	13.5	13.9

Considering for example the $^2E_2'$ state, there is 0.92 unpaired spin density on the $d_{x^2-y^2}$ mo, but only 0.72 electron has been lost from this orbital. The loss of electron density from this orbital has been compensated by donation from the ligands into the other d based mo's, with the result that only a net of 0.13 electron is lost from iron with the loss of this electron formally associated with the iron atom. This large donation from ligand to metal upon the loss of an electron in a metal-based orbital is what renders the frozen-orbital approximations (Koopmans' approximation) particularly bad in estimating ionization processes.

3. Density Functional Methods

Models using density functional theories have been useful in the study of transition metal systems for some time. Originally based on spherically averaged "muffin tin" potentials, these models proved effective in studying large systems, yielding "physical" orbitals in the sense previously discussed. Estimates of ionization potentials and electron spectroscopy were obtained from Slater's transition state theory [31], generally involving the loss of half an electron or the promotion of half an electron from one orbital to another [31], but the method, like extended Huckel theory, had only limited quantitative capabilities. Recently this has changed dramatically with the introduction of the local density approximation with realistic potentials, and especially with the introduction of non-local corrections.

Modern local density functional theory (LDF) begins with the Hohenberg-Kohn theorem (32) that proves that the energy can be written as a functional of the energy (33)

$$E(total) = E(p) = To + \int \rho(1)Vnuc \, d\tau(1) +$$
$$+ 1/2 \iint \rho(1)\rho(2)/r(1,2) \, d\tau(1)d\tau(2) +$$
$$+ Exc \tag{10}$$

TABLE 3. Calculated electronic and spin populations (from INDO/S) for ferrocene and the first four states of the ferrocenium ion (30).

	$^1A_1{}'$	$^2E_2{}'(d_{x^2-y^2})$	$^2A_1{}'(d_{z^2})$	$^2E_1{}''$	$^2A_2{}''$
Net Charge	0	+1	+1	+1	+1
Fe					
4s	0.05	0.03 (0.00)[a]	-0.06 (0.04)	0.01 (0.00)	0.04 (0.00)
4p (total)	-0.54	-0.43 (0.00)	-0.42 (0.00)	-0.53 (-0.01)	-0.73 (0.07)
$3d_{z^2}$	1.81	1.86 (0.00)	0.99 (0.92)	1.85 (0.00)	1.83 (0.00)
$3d_{xy}$	1.69	1.83 (0.00)	1.78 (0.01)	1.74 (0.00)	1.69 (-0.01)
$1d_{x^2-y^2}$	1.69	0.94 (0.92)	1.78(0.01)	1.74 (0.00)	1.69 (-0.01)
$3d_{xz}$	0.72	0.89 (0.02)	0.95 (0.01)	0.78 (0.00)	0.75 (-0.02)
$3d_{yz}$	0.72	0.89(0.02)	0.95(0.01)	0.49 (0.28)	0.75 (-0.02)
Net	1.87	2.00 (0.95)	2.03 (1.00)	1.93 (0.28)	1.99 (0.00)
C^b					
$\bar{2}p(\pi)$	1.03	0.95 (0.00)	0.96 (0.00)	0.94 (0.06)	0.94 (0.10)
Net	-0.24	-0.20(0.00)	-0.21 (0.00)	-0.20 (0.07)	-0.20 (0.10)
H^b					
Net	0.06	0.10 (0.00)	0.11 (0.00)	0.10 (0.00)	0.10 (0.00)

[a]The numbers in parentheses are spin densities.
[b]Average values for each carbon and each hydrogen atom.

In this, To represents the kinetic energy of non-interacting electrons, and Exc contains all exchange and correlation, as well as corrections to To. Variation of this expression can be made to yield the one electron Kohn-Sham equation [34],

$$\{-h^2/2m\nabla^2 + v^\sigma(ks)\}\phi_i = \varepsilon_i\phi_i \tag{11}$$

in which $v^\sigma(ks)$ can be decomposed into

$$v^\sigma(ks) = v(nuc) + v(el) + v^\sigma(xc) \tag{12}$$

TABLE 4. Some density functionals [46].

Kohn-Sham exchange only (Xα with α=2/3) [34]	$\rho^{1/3}$	exchange only for electron gas
Xα [38]	$\alpha\,\rho^{1/3}$	KS exchange weighted by 3 α/2
VWN (Vosko, Wilk, Nusair)[39] PZ (Perdew, Zunger) [40]	$\rho^{1/3}$+correl.	electron gas exchange and correlation from accurate Quantum Monte Carlo calcs of Ceperley and Alder [45]
LM (Langreth-Mehl) [41]	$\rho^{1/3}$+correl. +F($\nabla\rho$)	the first "modern" gradient corrected funct.
Becke [42] De Pristo-Kress [43] Perdew-Yue [44]	"	"semi-empirical" gradient functionals

a nuclear part, a classical electrostatic part, and all else.

$$v^\sigma(xc) = \delta Exc/\delta p = v^\sigma(x) + v(cor) + v^\sigma(xc)' \qquad (13)$$

v(x), the exchange part, is most often taken from free electron theory [35,36,33]. There are several formulations of these "density functionals" and some of them are summarized in Table 4.

In the local density functional methods bond distances and force constants seem to be well reproduced, although there is a strong tendency to overbind atoms in a complex. The non-local corrections, see Table 4, tend to correct this and yield a method very capable of yielding good bond lengths as well as good thermochemistry. Advantages of these methods are that they scale as N^3, with N related to the number of atoms. This should be contrasted with conventional ab-initio Hartree Fock theory, that scales with N^4, and correlated methods that scale as higher powers of N. Unsolved problems in density funtional theory include the lack of a theory for excited states, and a lack of a theory for multiplet structure.

Subsequent chapters in this book present many examples of calculations based on density functional methods.

4. A Final Survey

In Table 5 is presented a list of levels of theory. The correlated methods in this Table, if taken to high enough order, are capable of solving the time-independent non- relativistic Schrödinger Equation 7.

TABLE 5. Levels of theory [46].

1.	Hartree Fock HF or SCF [17,23,24,25]	one spin adapted detor build from mo's	Work N**4, Mo's adjusted iteratively to obtain minimum energy, E is variational and size extensive
2.	Configuration Interaction CI [17,47,48]	many detors $\Psi = \sum_I \Psi(I)C(I)$	Work N**5 or greater, variational, slow to converge not size extive.
3.	Multiconfiguration SCF MCSCF [48]	many detors same as CI	CI coefficients and orbital optimized simultaneously, better with shorter CI list variational, not size extensive, often difficult to converge.
4.	Complete (Full) Active Space SCF CAS SCF [48]	special case of MCSCF	All possible detors from set of "active" orbitals are in the the MCSCF, variational, size extensive, CI expansion list rapidly gets large.
5.	Generalized Valence Bond GVB [49]	special case of MCSCF	Combination of detors representing VB functions are used to fix CI coefficients, simplest is perfect pairing.
6.	Many Body (Moller-Plesset) Perturbation Theory MBPT, MP2, MP3.. [17,50]	many detors same as CI	CI coef. chosen by perturbation theory, non-variational, size extensive, second order quite tractable.
7.	Coupled Cluster CPMET, CCD, CCSD, CCSDT [1] [51,52]	many detors same as CI	CI coef. chosen by exponential cluster ansatz, non-variational, size extensive.
8.	Valence Bond [53]	similar to GVB	Non-orthogonal VB detors, variational, small systems, easy "chemical" interpretation.
9.	Quantum Monte Carlo	quantum simulation HF or CI "guiding" funct.	Small systems only. difficult for excited states.

TABLE 6. Some semi-empirical methods used for d- and f-electron systems [46].

1.	Extended Huckel EHT, REX (Relativistic EHT) [12,14,55]	h (ii) related to ionization potentials, h(ij) set proportional to overlap S, S is included in secular equation. (H-E(i) S) C (i)=0	Pi and sigma elect, large systems easy, not self consistent, relativistic version
2.	Iterative EHT IEH, SCCEH ITEREX [13,14,56,57]	As above, iterated to charge consistency, usually based on extrap. between I.P.'s of atomic ions.	as above, relativistic version
3.	Fenske-Hall [54]	As above: h(ij) with terms proportional to S as well as kinetic energy terms, most electrostatics included.	relatively easy, iterative, SCF level
4.	Zero Differential Overlap, ZDO CNDO, INDO, NDDO (AM1) INDO/S, CNDO/S SINDO [57-60,22,30,61]	One center part of h(ii) from IP's. nuclear attraction included. CNDO contains all <i,j|i,j>: INDO as well contains all one-center <i,j|k,l>. NDDO in addition includes all remaining two-center Coulomb type. h(i,j) usually proportional to S. Two-electron integrals often parameterized.	Large complexes, iterative SCF, CI, MBBT, MCSCF versions. Useful for geometry estimate and for electronic spectroscopy.
5.	Partial Retention of Differential Diatomic Overlap PRDDO [57,62]	Attempt is made to fit the one-electron and two-electron pieces of the Fock matrix separately.	An SCF method that simulates MBS ab-initio theory, Good when MBS is good, useful for estimating geom., but no gradient.

Problems that exist are the requirement for very good basis sets [15] and the exponential growth of the CI expansion needed for really high accuracy. Although good size systems are being attacked using such methods, their scaling with the size of the system will limit the utility of these methods to smaller systems.

Some semi-empirical methods are listed in Table 6. These are generally created to mimic ab-initio molecular orbital plus correlation methods. At the SCF level they scale as N^3, and include much of the dynamic correlation through their reliance on empirical information.

Essential correlation is included in many of these methods through small CI's, or limited perturbation theory. These methods have been applied with considerable success to very large systems [63].

120

TABLE 7. Some techniques for solving the Kohn-Sham equations.

1. Scattered-wave (SW) or Multiple Scattering (MS)	Muffin-tin potential partial-wave expansion	Rapid, good one-electron props.,total energies not reliable.
2. Linear Muffin Tin Orbitals (LMTO)	Basis of muffin tin eigenfunctions, linear expansion	Quite rapid,total energy reasonable, basis set choice?
3. Discrete Variational Method (DVM)	Numerical sampling of Slater or numerical basis	Good energy requires fine grid of points.
4. Linear Combination of Gaussian Type Orbitals (LCGTO-LSD)	Gaussian basis fits for the density and exchange-correlation.	Analytical integrals accurate good relative energies. derivative now available, accurate geometries, force constants for good v(xc).

TABLE 8. Critique of modern electronic structure theories for transition-metal complexes.

Crystal Field/Ligand Field

Always useful, often needed to start a calculation, and to interpret a calculation, very good insight, but...

not quantitative.

"Ab-initio" Theory

Correlated theory can get exact answers, but...

expensive, often difficult to interpret, requires real care.

Semi-Empirical and Approximate Methods (LCAO-MO Based)

Inexpensive, truly large systems can be examined, good insight, good starting geometries for other methods, but...

not always reliable, needs some chemical and physical insight not to go astray.

Local Spin Density Methods

Reasonably large systems, good accuracy for geometries and force constants, good insights, easily handles unpaired spins, but...

no theory of multiplet structure (why it handles unpaired spins easily!), no theory for excited states, often overestimates binding.

Some methods of solving the Kohn-Sham equation are presented in Table 7. All have proven quite successful, although the scattered wave method is beginning to disappear in favor of methods utilizing basis sets, either DVM or LCGTO-LSD. These latter two methods with accurate non-local corrections appear to be competitive with good basis set conventional ab-initio calculations at the second-order Moller-Plesset perturbation theory level [17], but scale as N^3, rather than N^4.

Progress in understanding the electronic structure of transition metal systems has progressed enormously in the last decade. Methods used today quantitatively predict structure, reactivity, and spectroscopy, and offer suggestions for experiments not yet done. Our ideas on these systems are evolving on a nearly daily basis as are our ideas on the methodologies needed for these studies. Knowledge and availability of methods to treat transition-metal systems lags considerably behind those routinely available to treat molecules consisting of main group elements, but that is certainly part of the appeal of this area. That, and the fact that transition-metal chemistry makes up such a rich and relatively poorly understood part of our science.

In Table 8 are summarized some of the advantages and disadvantages associated with the models that have been discussed in this survey. This table is for "today" and might well be out of date in as little as a year. We are continuously learning about these models and how they apply to transition-metal systems. Our ideas are changing almost daily.

REFERENCES

1. Bethe, H. (1929), Ann. Physik, 3, 135. See, also, Becquerel, Z. (1929) Physik, **58**, 205.
2. Van Vleck, J. H. (1932), Theory of Magnetic and Electric Susceptibilities, Oxford Univ. Press, Oxford; (1932), Phys. Rev., **41**, 208.
3. Ballhausen, C. J. (1962), Ligand Field Theory, McGraw Hill, New York.
4. Slater, J. C. (1960), Quantum Theory of Atomic Structure, Vol. I and II, McGraw Hill, New York.
5. Griffith, J. S. (1961), The Theory of Transition Metal Ions, Cambridge University Press, Cambridge.
6. Condon, E. U. and Shortly, G. H. (1959), The Theory of Atomic Spectra, Cambridge Univ. Press, Cambridge.
7. See, for example, Panofsky, W. K. H. and Phillips, M. (1962), Classical Electricity and Magnetism, Addison Wesley, New York.
8. Cotton, F. A. (1990), Chemical Applications of Group Theory, John Wiley and Son, New York.
9. Wolfsberg, M. and Helmholz, L. (1952), J. Chem. Phys., **20**, 837.
10. Mulliken, R. S. (1949), J. Chem. Phys., **46**, 497.
11. Lohr, L. L. and Lipscomb, W. N. (1963), J. Chem. Phys., **38**, 1604.
12. Hoffman, R. (1964), J. Chem. Phys., 39, 1397; ibid. **40**, 2047; 40, 2474; 40, 2480; 40, 2745.
13. Ballhausen, C. J. and Grey, H. B. (1962), Inorg. Chem., **1**, 111; ibid. (1964), Molecular Orbital Theory, Benjamin Press, New York.
14. Zerner, M. C. and Gouterman, M. P. (1969), Theoret. Chim. Acta, **4**, 44.
15. See for example, appropriate chapters in Salahub, D. and Zerner, M. C. (1989), The Challenge of d and f Electrons, ACS Sym. Series 394, Washington, D.C.
16. Hoffman, R. (1988), Rev. Mod. Phys., **60**, 601.

122

17. Szabo, A. and Ostlund, N. S. (1982), Modern Quantum Chemistry, McMillan, New York.
18. Newton, M. D., Boer, F. P. and Lipscomb, W. N. (1966), J. Am. Chem. Soc., **88**, 2353.
19. Zerner, M. C., Gouterman, M. P., Kobayashi, H. (1966), Theoret. Chim. Acta, **6**, 363.
20. Pauncz, R. (1979), Spin Eigenfunctions, Plenum Press.
21. Pople, J. A. and Nesbet, R. (1954), J. Chem. Phys., **22**, 571.
22. Bacon, A. D. and Zerner, M. C. (1979), Theoret. Chim. Acta, **53**, 21.
23. Roothan, C. C. J. (1951), Rev. Mod. Phys., **23**, 69; ibid. (1960), **32**, 179.
24. Hall, G. G. (1951), Proc. Roy. Soc., **A205**, 541.
25. McWeeney, R. (1989), Methods of Molecular Quantum Mechanics, Academic Press.
26. Jorgensen, P. and Simons, J. (1981), Second Quantized Based Methods in Quantum Chemistry, Academic Press, New York.
27. Correa de Mello, P., Hehenberger, M., Laisson, S. and Zerner, M. C. (1980), J. Am. Chem. Soc., **102**, 1278.
28. Coutiere, M., Demuynck, J. and Veillard, A. (1972), Theoret. Chim. Acta, **27**, 281.
29. Bagus, P. S., Wahlgren, U. I. and Almlof, J. (1976), J. Chem. Phys., **64**, 2324.
30. Zerner, M. C., Loew, G. H., Kirchner, R. F., and Mueller-Westerhoff, U. T. (1980), J. Am. Chem. Soc., **102**, 589.
31. Slater, J. C. (1974), The Self Consistent Field for Molecules and Solids, Vol. 4, McGraw Hill, New York.
32. Hohenberg, P. and Kohn, W. (1964), Phys. Rev., **136**, B864.
33. Yang, W. and Parr, R. G. (1989), Density Functional Theory of Atoms and Molecules, Oxford Univ. Press, Oxford.
34. Kohn, W., Sham, L. J. (1965), Phys. Rev., **140**, A1133.
35. Lundquist, S. and March, W. H., (eds.). (1983), Theory of the Inhomogeneous Electron Gas, Plenum, New York.
36. Raimes, S. (1972), Many Electron Theory, North Holland, Amsterdam.
37. Rabalais, J. W., Werme, L. O., Berkmark, T., Karlsson, L., Hussain, M. and Siegbahn, K. (1972), J. Chem. Phys., **57**, 1185.
38. Slater, J. C. (1972), Adv. Quantum Chem., **6**, 1; (1974), The Self-Consistent Field Method for Molecules and Solids, Vol. 4, McGraw Hill, New York.
39. Vosko, S. H., Wilk, L.; Nusair, M. (1980), Can. J. Phys., **58**, 1200.
40. Perdew, J. P., Zunger, A. (1981), Phys. Rev. B., **23**, 5048.
41. Langreth, D. C., Mehl, M. J., (1983), Phys. Rev. B., **28**, 1809; erratum (1984), **29**, 2310.
42. Becke, A. D. (1986), J. Chem. Phys., **84**, 4524.
43. De Pristo, A. E., Kress, J. D. (1987), J. Chem. Phys., **86**, 1425.
44. Perdew, J. P., Yue, W. (1986), Phys. Rev. B., **33**, 8800; Perdew, J. P. (1986), Phys. Rev. B., **33**, 8822.
45. Ceperley, D. M., Alder, B. J. (1980), Phys. Rev. Lett., **45**, 566.
46. Adapted from Salahub, D. R. and Zerner, M. C., Chapter 1 in reference 15.
47. Shavitt, I. (1977), Methods of Electronic Structure Theory, Vol. 3, H. F. Shaeffer III, (ed.)., Plenum Press, New York, p. 189.

48. (a), Roos, B. (1975), Computational Techniques in Quantum Chemistry and Molecular Physics, G. H. F. Diercksen, B. T. Sutcliff and A. Veillard, (eds.)., Reidel, Dordrecht, p. 151; (b), Werner, H.-J. (1987), Ab Initio Methods in Quantum Chemistry II, K. P. Lawley, (ed.)., J. Wiley and Sons, New York, p. 1.

49. Bobrowicz, F. W. and Goddard, W. A. (1977), Methods of Electronic Structure Theory, H. F. Schaeffer III, (ed.)., Plenum Press, New York.

50. Bartlett, R. J. (1981), Ann. Rev. Phys. Chem., 32, 359.

51. Bartlett, R. J. (1989), J. Phys. Chem., 93, 1697.

52. Bartlett, R. J., Dykstra, C. E. and Paldus, J. (1984), Adv. Theories and Computational Approaches to the Electronic Structure of Molecules, C. E. Dykstra, (ed.)., Reidel.

53. McWeeney, R. (1990), Intern. J. Quantum Chem., S24, 733.

54. Hall, M. B. and Fenske, R. F. (1972), Inorg. Chem., 11, 768.

55. Lohr, L. L. and Pyykko, P. (1979), Chem. Phys. Let., 62, 333.

56. Pyykko, P. (1983), Chem. Phys., 74, 1.

57. Zerner, M. C. Reviews of Computational Chemistry, K. Lipkowitz and D. Boyd, (eds.)., VCH Publishers, New York, in press.

58. Sadley, J. (1985), Semi-Empirical Methods in Quantum Chemistry, Wiley, New York.

59. Pople, J. A. and Beveridge, D. L. (1970), Approximate Molecular Orbital Theory, McGraw-Hill, New York.

60. Del Bene, J. and Jaffee, H. H. (1968), J. Chem. Phys., 48, 107.

61. Jug, K. and Iffert, R. (1988), J. Comput. Chem., 9, 51; Jug, K. and Schulz, J. (1988), J. Comput Chem., 9, 40.

62. a) Halgren, T. A., Kleier, -., Hall, -., Brown, -. and Lipscomb, W. N. (1978), J. Amer. Chem. Soc., 100, 6595;

    b) Marynick, D. S. and Lipscomb, W. N. (1982), Proc. Natl. Acad. Sci. USA, 79, 1341.

63. See, for example, Thompson, M. and Zerner, M. C. (in press), J. Am. Chem. Soc., who treat over 500 atoms and 1500 electrons at the SCF-CI level of theory.

SELECTIVITY IN CATALYSIS BY METALS AND ALLOYS

A.J. RENOUPREZ
Institut de Recherches sur la Catalyse
2 avenue Albert Einstein
69626 Villeurbanne Cedex
France

ABSTRACT. After a brief review of the mechanism of heterogeneous catalysis on metals, reactions where selectivity is involved are considered. The case of the selective hydrogenation of 1,3 butadiene to butenes is discussed first. From kinetic and spectroscopic evidence, it is shown that selectivity is governed by the competitive adsorption of butadiene and butene. In the last section, hydrogenation reactions of polyfunctional molecules which can follow different pathways are described.

1. Introduction

In the early definition proposed by Berzeluis, 150 years ago, a catalyst is a material which accelerates the rate of a reaction and is recovered unchanged at the end of it. A simple bimolecular reaction like the combustion of hydrogen is feasible at 300 K, since $\Delta G_{300} = -455 \text{ KJmol}^{-1}$.

$$2H_2 + O_2 \longrightarrow 2H_2O$$

But high potential barriers associated with the dissociation of the hydrogen and oxygen molecules have to be overcome. In the gas phase this dissociation would not occur below 3000 K. But in the presence of a transition metal, the dissociative chemisorption of both molecules is exothermic with negative free energy at 300 K. Actually it has been shown recently [1] that these chemisorptions occur already at 130 K and that water formation is possible on platinum at such a low temperature.

In the middle of the 19th Century several catalytic reactions of industrial interest involving the full conversion of the reagents such as the synthesis of ammonia or the oxidation of sulfur dioxide were already in operation. The only requirement for this type of reaction is the obtention of the maximum rate of conversion by choosing the most efficient catalyst, the proper temperature and pressure of reagents.

Now, in 1870 van Hoffmann had already found that silver (and silver oxide) is able to induce the partial oxidation of methanol to formaldehyde and not to carbon dioxide and water. This was the first step toward one of the major achievement of modern catalysis i.e. selectivity. In the simple scheme:

$$A + B \longrightarrow C \longrightarrow D$$

the selectivity for the formation of C is defined as

125

D. R. Salahub and N. Russo (eds.), Metal-Ligand Interactions: from Atoms, to Clusters, to Surfaces, 125–136.
© 1992 *Kluwer Academic Publishers. Printed in the Netherlands.*

$$S_c = \frac{\text{number of moles of C}}{\text{number of moles of C} + \text{D}}$$

Obviously the elucidation of the reason for the specificity of a given catalyst is of primary interest from both practical and scientific view points. For this purpose, a study at the microscopic scale of the various steps of a catalytic reaction is the necessary pathway to the prediction of selectivity.

2. Mechanism of Catalytic Reactions

Before approaching the problem of selectivity in reactions where successive products can be formed, it is necessary to consider first the mechanism of bimolecular reactions leading to a unique product.

A classical example is the oxidation of carbon monoxide on Pd(111) [2]. In the domain of temperature and pressure under study ($400 < T < 700$ K, $P < 10^{-6}$ torr), the adsorption of CO is equilibrated:

$$[S] + CO \underset{k_{-1}}{\overset{k_1}{\rightleftharpoons}} CO_{(a)} \tag{1}$$

where [S] is a surface site and (a) stands for an adsorbed molecule. Conversely, the adsorption of oxygen is an irreversible process:

$$[S] + O_2 \xrightarrow{k_2} O\text{-}O_{(a)} \tag{2}$$

$$[S] + O\text{-}O_{(a)} \xrightarrow{k_3} 2\,O_{(a)} \tag{3}$$

Using pulsed molecular beams experiments, the direct reaction of CO_{gas} with adsorbed oxygen was discarded (Ealy - Rideal mechanism) and the final step would be:

$$CO_{(a)} + O_{(a)} \xrightarrow{k_4} CO_2 + 2[S] \tag{4}$$

This irreversible reaction between adsorbed species is called a Langmuir - Hinshelwood process. The global rate of reaction then reads:

$$r = k_4[O_{(a)}][CO_{(a)}]$$

Actually the experiments have shown that oxygen and carbon monoxide surface coverage are not related in a simple way to the partial pressures and as a function of temperature, the rate exhibits a maximum at 550 K, as shown in Fig. 1.

Below 550K, the adsorption of oxygen is inhibited by adsorbed CO. As the temperature is raised, reaction [1] is displaced to the left and the surface coverage by CO decreases; subsequently the oxygen coverage increases. The rate of CO_2 formation is then:

$$r = k_2 \frac{P_{O_2}}{K_1 P_{CO}} \quad \text{with} \quad K_1 = \frac{k_1}{k_{-1}}$$

Above 550 K, oxygen was shown by LEED to form islands on the metal and the formation of CO_2 takes place at the frontier of these islands. Since the Pd-O bond is stronger than the Pd-CO, the CO molecules have to diffuse toward the islands' edges to react. In this temperature range the reaction rate does not depend upon oxygen coverage and $[O_{(a)}]$ is considered as a constant. The expression of the rate is:

$$r \approx k_4 K_1 [P_{CO}]$$

The rate constant k_4 was determined experimentally and the activation energy E_4 of step [4] was found to be 102 KJmole^{-1}.

For the whole reaction, at high temperature, an apparent activation energy $\Delta H_{adsCO} + E_4$ of -40 KJmole^{-1} was found which accounts for the decrease of the reaction rate upon increasing temperature above 550 K.

3. Selectivity in Reactions Following a Unique Pathway

A classical example of this type of reaction is the hydrogenation of 1,3 butadiene to butenes and butane,which takes place on transition metals.

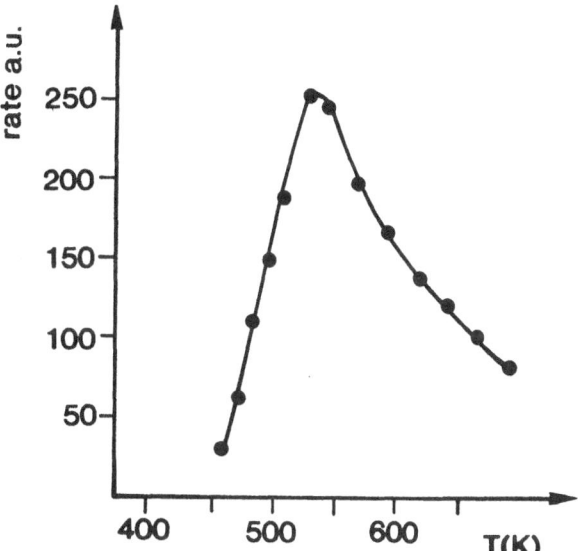

Figure 1. Variation of the rate of oxidation of CO with the temperature (from ref. 2)

$$\begin{array}{cc}
C_4H_6\,{}_{(g)} & C_4H_8\,{}_{(g)} \\
\updownarrow & \updownarrow \\
C_4H_6\,{}_{(a)} + H_2 & C_4H_8\,{}_{(a)} + H_2 \;\rightarrow\; C_4H_{10}
\end{array} \qquad (5)$$

Both the rate and the selectivity for the formation of butenes vary with the nature of the metal. The reactivity of platinum, palladium and nickel have been recently compared by J. Massardier et al. [3].

3.1. REACTION RATES

As shown in Table 1, Pd is one to two order of magnitudes more active than Pt or Ni and the rate of hydrogenation is comparable on Pd supported catalysts and on Pd(111). However on the catalysts, an important decrease of rate (deactivation) with the reaction time is observed whereas it remains constant on single crystal faces. An explanation was found by performing Auger and XPS experiments.

TABLE 1. Reactivity of platinum, palladium and nickel for the hydrogenation of 1,3 butadiene

Sample	rate s^{-1}	selectivity for butenes	order $/H_2$	/H.C.
Pd(111)	0.15	1	1	0
Pd(110)	1.2	1	1	0
Pd/SiO2	0.6 —> 0.2*	0.4 —> 0.9	1	0-0.5
Pt(111)	0.01	0.6	1	0
Ni(111)	0.01	1	1	0

* after deactivation

On Pd(111) and (110), only half a monolayer of carbon is found after the reaction, but on the catalyst composed of 15 Å particles, a thick layer is progressively built up. On the clean surface, the rate was found to follow the relation:

$$r = K(P_{C_4H_6})^0 * (P_{H_2})^1 \exp(-50/RT) \qquad (6)$$

The zero order with respect to butadiene means that the competition of adsorption is in favor of the hydrocarbon. When the surface is deactivated, this order with respect to butadiene is no longer 0 but becomes slightly positive, confirming that the access of butadiene to the adsorption sites is limited. Table 1 also shows that the crystallography of the surface is an important parameter, since (110) faces are one order of magnitude more active than (111). In this case the rate is probably controlled by the adsorption of hydrogen since the sticking coefficient is 0.4 on (110) faces and only 0.1 on (111) faces.

3.2. THE SELECTIVITY

The selectivity for the formation of butenes is 1 for Pd(111) and only 0.6 for Pt (111). A possible explanation could be found in the rake like scheme [5]: butenes would be evolved in the gas phase as soon as formed, on Pd (111) whereas they remain chemisorbed on Pt(111), allowing the reaction to proceed to butane. To corroborate this assumption, a kinetic study of the competitive hydrogenation of butadiene and butene on both metals [4] was performed. However, to distinguish between the ethylenic hydrocarbon formed during the reaction and the molecules present as reactants, the following sequence was followed:

(i) The hydrogenation of a butadiene-propene mixture, then (ii) the hydrogenation of a propene-butene mixture. Actually the ratio of the rates reads:

$$R_{buta}/R_{bute} = (k_{buta}A_{buta}/k_{bute}A_{bute}) (P_{buta}/P_{bute})$$

where k, A and P are the rate constants, the adsorption coefficients and the partial pressures. A set of experiments in which the partial pressure of one of the reactants is kept constant while the other is varied has led to the figures reported in Table 2.

TABLE 2. Butadiene-butene competitive hydrogenation

Sample	A_{buta}/A_{bute}	k_{buta}/k_{bute}	R_{buta}/R_{bute}
Pd/SiO$_2$	12	0.33	4
Pt/SiO$_2$	0.95	0.8	0.75

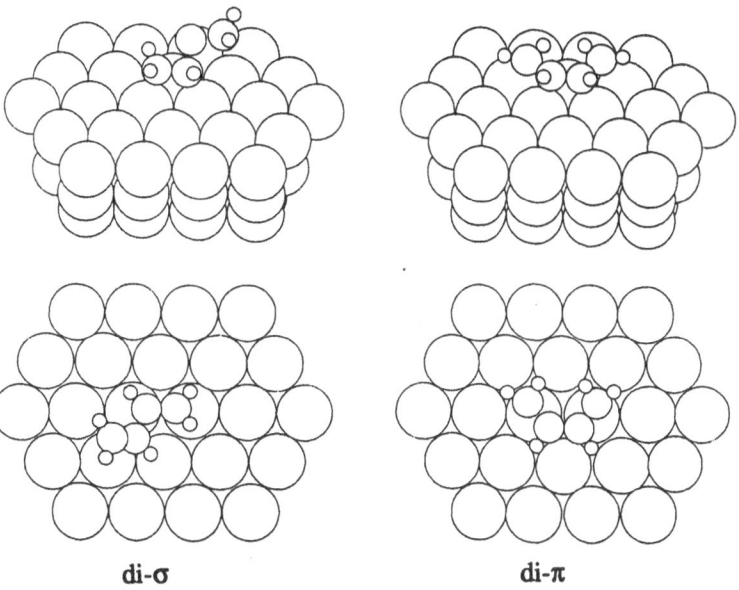

di-σ   di-π

Figure 2. Side and top view of di-σ and di-π adsorption of butadiene on a 49 atom cluster modelling a (111) surface (from ref. 5).

130

TABLE 3. Adsorption of ethylene

Conformation	Pt(111) E Kcalmole⁻¹	Pd(111) E Kcalmole⁻¹
di-σ	15	19
π	8	17

TABLE 4. Adsorption of butadiene

Conformation	Pt(111) E Kcalmole⁻¹	Pd(111) E Kcalmole⁻¹
di-σ	15	20
di-π	18	36

Obviously, the better selectivity of palladium which derives from the ratio of the rate constants, is governed by the adsorption constants rather than by the kinetic constants. A proof of this stronger adsorption of butadiene than of butene was also provided by spectroscopy. The infrared spectra show that butadiene and butene are both diσ-bonded on platinum, probably by the opening of a single double bond. On the contrary, on palladium, butadiene is π bonded and butene is weakly adsorbed.

A theoretical study of the comparative adsorption of butadiene and ethene on Pt and Pd was also performed by Sautet et al. [5] by extended Hückel calculations. The (111) surface was modelled by a 49 atom cluster represented in Fig. 2.

The adsorption of ethylene is controlled by a balance between attractive 2-electron interactions and 4-electron repulsions coming from occupied orbitals of the molecule and the occupied part of the d band as shown in Fig. 3

The resulting adsorption energies for various configurations can be compared in Table 3 and 4.

On platinum the energies of the di-σ and di-π modes are comparable for both molecules. On the contrary, on palladium butadiene adsorption in the di-π configuration is largely favored and has an adsorption energy nearly twice that of ethylene. This confirms the experimental observations of a stronger adsorption of dienes than ethylenic hydrocarbons on palladium and of similar bonding energies on platinum.

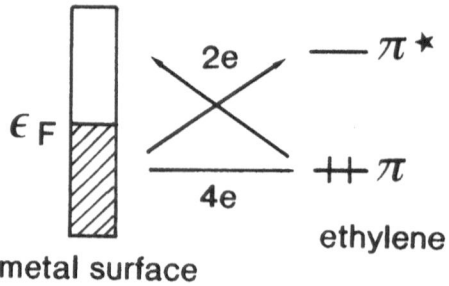

Figure 3. 2-electron and 4-electron interaction between the metal and the ethylene molecule (from ref. 5).

4. Selectivity in Reactions with Several Pathways

4.1. CONVERSION OF HYDROCARBONS ON NICKEL AND RUTHENIUM ALLOYS

It is known that metals such as nickel or ruthenium have a high reactivity for reactions involving a cracking of C-C bonds. Above 600 K, in the presence of hydrogen, hydrogenolysis reactions like [6] generally take place:

$$C_2H_6 + H_2 \longrightarrow 2\ CH_4 \tag{6}$$

From kinetic experiments and magnetic measurements, Martin et al. [6] have shown that this reaction needs large ensembles of 12 contiguous nickel atoms to take place. They derived a simple picture for the chemisorbed ethane molecule stripped of its hydrogen with each carbon atom bonded to three nickel atoms. This was supported by comparing the reactivity of Ni/SiO_2 catalysts with that of nickel-copper alloys. Actually copper is inactive in the chemisorption of hydrogen and hydrocarbons and only plays the role of a diluant of the active metal on the surface. Also only a limited segregation of copper at the surface is observed. Fig.4 shows that the rate of ethane hydrogenolysis is decreased by three orders of magnitude when the copper concentration is 20%. This is consistent with the assumption of an adsorption site for the hydrocarbon composed of a large number of nickel atoms. On the contrary for benzene, the adsorption of the reactants only needs 3 metal atoms. As a consequence the site dilution is less effective and the rate of hydrogenation of this molecule is only decreased by a factor of 3 for the same Cu concentration.

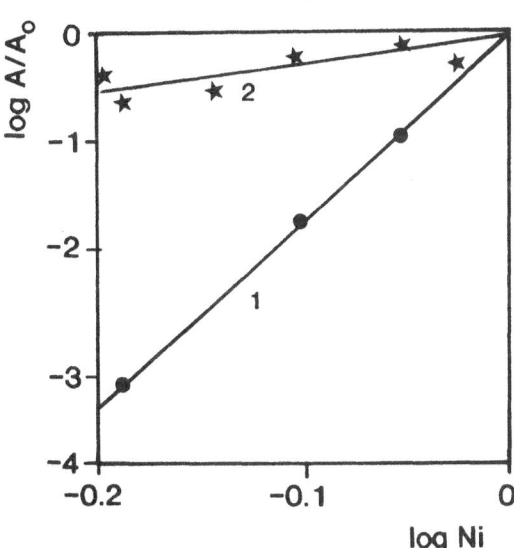

Figure 4. Compared variation with the Cu concentration of the rates of hydrogenolysis of butane (1) and hydrogenation of benzene (2) on Ni-Cu alloys (from ref. 6)

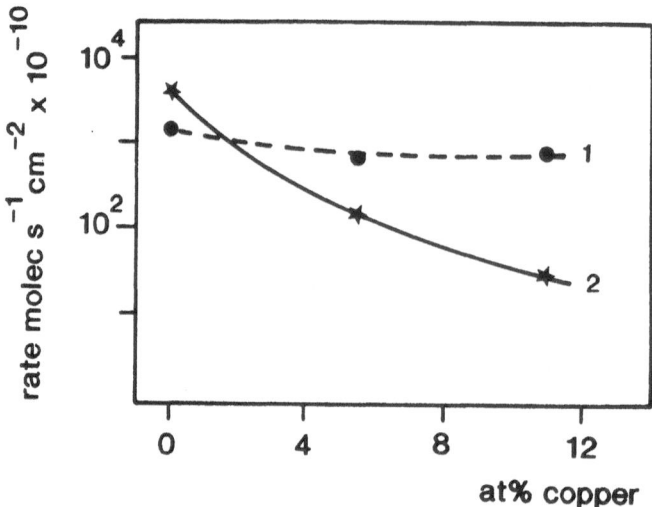

Figure 5. Compared effect of Cu concentration in Ru-Cu catalysts on the rates of dehydrogenation and hydrogenolysis of cyclohexane.

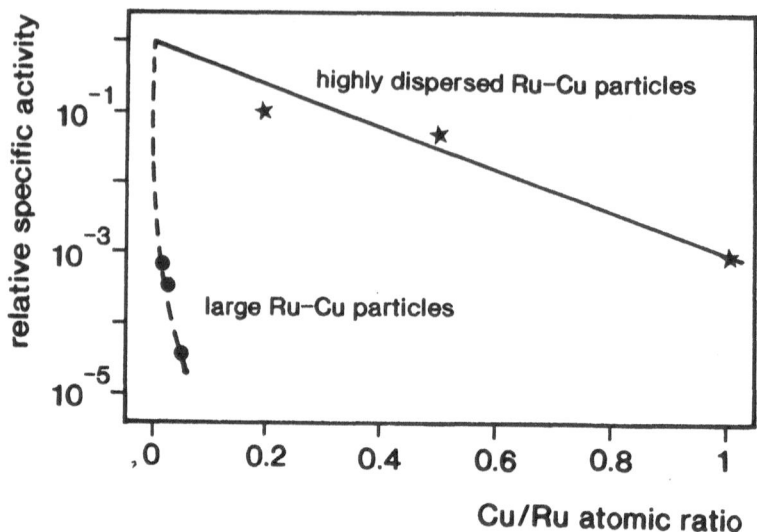

Figure 6. The effect of particle size i.e. surface copper concentration, on the rate of hydrogenolysis of ethane.

Sinfeld et al. [7] have studied the transformation of cyclohexane on a series of supported Ru-Cu supported catalysts. The reaction scheme is the following:

$$C_6H_{12} \longrightarrow C_6H_6 + H_2 \tag{7}$$

$$\longrightarrow C_1 + C_5 \longrightarrow C_1 + C_4 \longrightarrow \tag{8}$$

It is known that in this system, unlike in Ni-Cu, copper has a marked tendency to segregate to the surface of the particles. Hence, even for low I B metal concentration in the bulk, a quasimonolayer of this metal can be present on the surface and as expected, Fig. 5 shows that the rate of reaction [8] is more strongly affected by the presence of Cu atoms on the surface than that of reaction [7].

Now if the particle diameter can be increased from 15 Å to 400 Å, the ratio of surface to bulk atoms decreases from 50 % to 1% and the site dilution effect by unreactive metal should increase strongly. Fig. 6 shows this effect of particle size for the case of the hydrogenolysis of ethane.

4.2. HYDROGENATION OF POLYFUNCTIONAL MOLECULES ON PLATINUM ALLOYS

No electronic structure modification of the active metal is put forward for the interpretation of the reactivity of the catalytic systems mentioned in section 4.1. A different behaviour can be expected if two active metals like platinum and nickel or iron are alloyed. Also, it is more fruitfull to investigate the reactivity of this type of bimetallic association in the catalytic transformation of polyfunctional molecules. Among the reactions which can follow parallel pathways, one of the most interesting is the selective hydrogenation of α,β ethylenic aldehydes:

$$R\text{-}CH=CH\text{-}CHO + H_2 \overset{\displaystyle \overset{A}{RCH=CH\text{-}CH_2OH}}{\underset{\displaystyle \underset{B}{RCH_2CH_2CHO}}{\Large\searrow}} \longrightarrow RCH_2CH_2CH_2OH \tag{9}$$

Indeed ethylenic alcohols are important intermediates in the synthesis of vitamin A or in the perfume industry, but the C-C double bond is more reactive than the carbonyl function and on most catalysts such as palladium or nickel the reaction would follow pathway B or proceed to complete hydrogenation of the molecule. Recently Goupil et al. [8] have studied the hydrogenation of cinnamal on carbon supported bimetallic platinum-iron catalysts. As shown on Fig. 7, the selectivity for the production of cinnamol increases from 70 to 90% when the iron concentration is increased from 0 to 50 %. Also, the reaction rate is 50 times larger on the $Pt_{80}Fe_{20}$ catalyst than on platinum.

This reaction was performed in the liquid phase, under high hydrogen pressure. The influence of the alloy composition is thus difficult to draw out. A similar reaction was thus studied in the gas phase on well defined Pt and Pt-Fe single crystal faces exhibiting a (111) orientation [9]. The $Pt_{80}Fe_{20}$ alloy was shown by X-Ray diffraction in grazing incidence to have an ordered Pt_3Fe structure. The analysis of the LEED diagram also indicates that the first layer is composed of pure platinum whereas the second layer

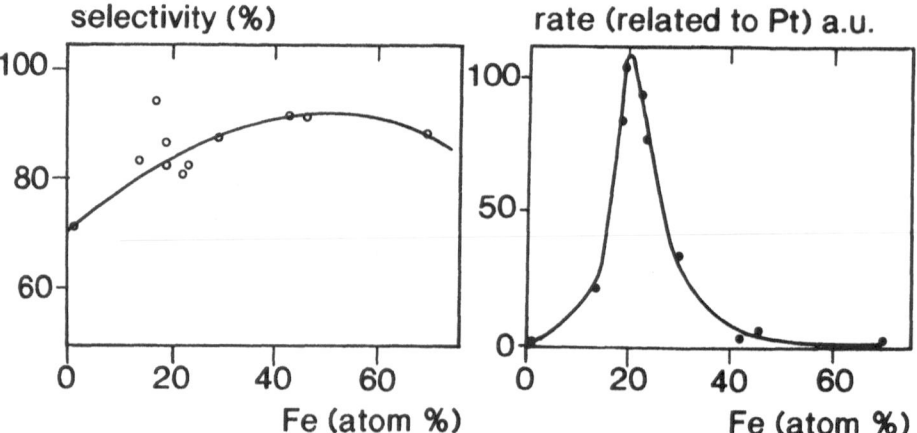

Figure 7. Activity and selectivity of Pt-Fe/carbon bimetallic catalysts in the hydrogenation of cinnamaldehyde (from ref. 8).

contains 12 at % Fe. Consequently, as shown on Fig. 8, two types of sites are present: C_{3v} sites composed of 3 Pt atoms in the first layer and either a Pt or an Fe underneath.

The electronic properties of this surface were studied by UPS using synchrotron radiation [10]. As shown in table 5, on pure platinum, one can distinguish between surface and bulk levels. Moreover the Pt $4f_{7/2}$ surface levels of the alloy are split into two lines separated by 0.4 eV. The level located at 70.5 eV corresponds probably to the sites of type 1 composed only of platinum atoms. The presence of iron in the sites of type 2 can explain the increase of binding energy of the second line.

The intrinsic effect of alloying was then studied in the hydrogenation of two simple molecules: butenal and methylbutenal. As can be seen in table 6, the alloy is 3 to 6 times more active than platinum and slightly more selective for the formation of ethylenic alcohol.

Obviously the presence of the methyl group has a large influence on the selectivity. Its steric hindrance probably prevents the molecule from being adsorbed by the double bond which allows a preferential activation of the carboxylic function.

The composition of the catalyst also has an undeniable effect on the reactivity but the rate increase observed in liquid phase for the alloy is one order of magnitude larger than the increase measured on single crystals, for similar molecules.

TABLE 5. Pt $4f_{7/2}$ core levels in Pt(111) and $Pt_{80}Fe_{20}$ (111)

	bulk level eV	surface levels eV
Pt (111)	70.95	70.6
PtFe (111)	71.25	70.9 70.5

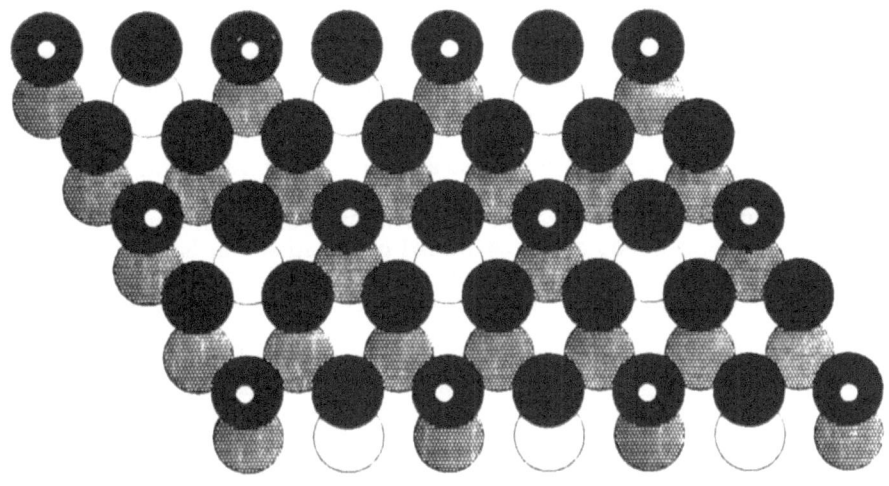

Figure 8.   The two types of sites of the $Pt_{80}Fe_{20}$ (111) surface. Dark circles are surface Pt atoms; white and grey circles are respectively Fe and Pt atoms of the second layer.

TABLE 6. Hydrogenation of butenal and methylbutenal

| | | rates s^{-1} x 10^2 | |
		Pt(111)	PtFe(111)
butenal	butanal	6	20
	butenol	6	29
selectivity % alcohol	10	16	
CH_3butenal	CH_3butanal	2	5.5
	CH_3butenol	5.5	8
selectivity % alcohol		57	70

5. Conclusions

All the recent studies show that selectivity is strongly influenced by the nature of the catalyst. When pure metals are compared to alloys, the modifications of selectivity originate from a change in the adsorption competition between the reagents and the various reaction products. In most cases this is associated with modifications of the electronic properties of the surface of these alloys. A slight shift of the core levels and small changes in the shape of the valence band of the noble metal is generally observed. More intriguing is the strong effect on the reactivity observed in some cases for well defined compositions of the alloy. A partial explanation can be looked for in the selectivity: one of the reaction products can be a strong poison of the surface sites and if its formation is inhibited on the alloy, the reactivity can increase by several orders of magnitude.

During a long period of time, the concept of site dilution by an inactive metal was considered as the only explanation of reaction rate modifications upon alloying. Although this is still a reasonable assumption in reactions involving the rupture of a molecule in several fragments which requires a large number of contiguous adsorption sites, this cannot explain an important increase of the reaction rate observed in other types of reactions. Two assumptions can thus be put forward: first when a Langmuir-Hinshelwood mechanism is postulated, the rate would increase if the surface diffusion of one of the reagent increases. Secondly, as was shown in the oxidation of CO, a maximum of rate is observed if the surface coverage of both reagents has the same magnitude. If the sticking coefficient of a weakly adsorbed reagent is increased, the rate would also be augmented.

REFERENCES

1. Gland, J.L., Fisher, G.B. and Kollin, E.B. (1982), J. Catal., **77**, 263.
2. Engel, T. and Ertl, G. (1979), Adv. Catal., **28**, 1.
3. Massardier, J., Bertolini, J.C. and Renouprez, A. (1988), in Phillips, M.J. and Terman (eds), Proceed. 9th intern. congr. catal., Phillips, Ottawa, **3**, 1222.
4. Ouchaib, T., Massardier, J. and Renouprez, A. (1989), J. Catal., **119**, 517.
5. Sautet, P. and Paul, J.F. (1991), Catalysis Letters, **9**, 245.
6. Martin, G.A. (1988), Catal. Rev., **30**, 519.
7. Sinfeld, J.H. (1991), Catalysis Letters, **9**, 159.
8. Goupil, D., Fouilloux, P. and Maurel, R. (1987), React. Kinet. Catal. Lett., **35**, 185.
9. Beccat, P., Bertolini, J.C., Gauthier, Y., Massardier, J. and Ruiz, P. (1990), J. Catal., **126**, 451.
10. Barret, N., Guillot, C., Bertolini, J.C., Massardier, J. and Khanra, B.C. (1991), Surf. Sci. Letters, in press.

METAL - CO INTERACTIONS : WELL-DEFINED SURFACES AND SUPPORTED PARTICLES

A. GOURSOT
Laboratoire de Chimie Organique Physique et Cinétique
Chimique Appliquées, UA. 418 CNRS
Ecole de Chimie, 8 rue de l'Ecole Normale
34053 Montpellier Cédex 1
France

ABSTRACT. Catalysis on supported noble metals such as Rh, Pd, Pt occupies the major part of automobile exhaust gas conversion and plays a great role in many other industrial processes of petroleum and fine chemistries. CO adsorption is the initial step of hydrogenation reactions, yielding alcohols and alkanes, but is also of major importance for the basic understanding of the mechanisms of catalysis. In that context, the determination of metal surface properties is essential. The accelerated development of surface science techniques allows the detailed investigation of the surface structure of the adsorbed overlayer on single-crystal clean surfaces, yielding their ordering, the geometry of adsorbed molecules, their vibrational and bonding properties at different metallic sites and various coverages. These informations, together with those from quantum chemical calculations are of fundamental importance for the comprehension of the adsorption properties of CO molecules on the more complex real catalysts, namely supported metal particles. The nature of the support and the size of the particles are two new factors which have to be studied. Electronic and geometric factors have to be taken into account to explain the changes in binding energy and adsorptive capacity of CO on supported particles with varying metal and size, and also when additives have been incorporated in the surface or when other molecules are coadsorbed.

1. Introduction

Catalysis on metals plays an important role in many industrial processes, particularly in petroleum chemistry, oil refining and automobile exhaust gas conversion. Noble metals like Rh, Pd, Pt are of major interest. Unsupported metals were the first catalysts to be prepared and they are still used as gauze (PtRh), powder (Ni), or films. However, in order to increase the ratio of active surface atoms with respect to their total number, supported metal particles are now generally used. The support is a porous oxide of Al, Si, Mg, Ti, etc, exhibiting a network of pores. Small metallic particles, ranging in size from 10 to 500 Å, are fixed to the walls of these pores by oxygen-metal bonds.

It is not always easy to delineate the influence of various factors, such as the size and shape of the metal particles or the nature of the support on the activity and selectivity of catalytic reactions. Single-crystal metal surfaces can then be regarded as model systems for chemisorption reactions. Moreover, the recent development of surface science

D. R. Salahub and N. Russo (eds.), Metal-Ligand Interactions: from Atoms, to Clusters, to Surfaces, 137–153.

techniques provide much valuable information about the bonding modes of adsorbates with respect to the surface.

The adsorption of carbon monoxide on metal surfaces and supported particles is one of the most extensively studied for many years.

* CO hydrogenation (Fischer-Tropsch catalysis) has been known for decades to produce organic compounds according to two different pathways (i) dissociation of CO leads to adsorbed C and O species and gives rise to hydrocarbons; (ii) a non-dissociative reaction yields first methanol and then higher alcohols. The nature of the metal, of the active site, of the support are, among others, possible variables which are able to orientate the reaction pathway. For example, the first reaction occurs on Ni, whereas the second one prevails on Pd or Pt particles [1,2]. Both reactions are observed in the case of Rh, with a preference for the formation of hydrocarbons in the case of small Rh clusters on acidic support, while the selectivity in alcohols is enhanced on basic supports [3]. This shows that Rh is a very versatile and thus very interesting catalyst.

* a major use of Pd, Rh and Pt supported catalysts is the conversion of automobile exhaust gases [4]. Pd and Pt are the main catalytic components used for the oxidation of CO and hydrocarbons. Rh is principally used to reduce NO to N_2 but is also effective for catalyzing CO oxidation [5]. In catalytic converters, metals are present on alumina in the form of ultra-dispersed particles, with few metal atoms, all of which are probably in close contact with the support. It is thus important to identify the active forms of the metal supported particles and to analyze their state after adsorption of CO.

* finally, CO is a very convenient small adsorbate, which is relatively easy to study by various methods, including quantum chemical calculations. For a long time, CO has been used as a probe in order to characterize the metal "surface". This is particularly the case for infrared spectroscopy investigations, since the C-O stretching vibrational mode has a sufficiently high extinction coefficient when it is adsorbed on metal surfaces. Moreover, its adsorption strength at the various possible sites is influenced by the direct environment and is related to the C-O vibrations [6].

In an attempt to understand the behavior of supported metal catalysts, ie their activity, selectivity, poisoning, a large number of investigations have been performed.

Single-crystal surfaces have received great attention because their structures are well-defined and because of the arsenal of analytical techniques which can be applied. However, their properties are not easy to correlate with those of real catalysts, especially when reactions are sensitive to the size of the metal particles. For this reason, model supported particles have been studied under UHV conditions, combining surface science and chemical methods. Concurrently, the thermal, vibrational and chemical behavior of carbon monoxide adsorbed on commercial catalysts has also been extensively investigated.

In parallel, the bonding interactions of CO with model clusters or surfaces have been studied by a great number of theoretical investigations. The first description was given by Blyholder in a model involving a 5σ CO donation to the metal and a backdonation to the empty $2\pi^*$ CO orbital [7]. Since then, the adsorption of CO on clusters modeling different sites has been studied with various quantum chemical methods. The most sophisticated of these methods can account for the structural properties of the adsorbed species, including the vibrations [8-22].

The character of the chemisorption bond, which emerges from most of the recent calculations is consistent with the simple Blyholder model, although some adjustments have to be made. CO chemisorption is accompanied by an initial σ repulsion which is followed by a rehybridization of the metal s, pσ and dσ orbitals and a polarization of the 5σ CO orbital (mainly C lone pair), which mixes with 4σ (mainly O lone pair). The populated bonding and antibonding σ– and π– type levels, describing the substrate-CO interactions, differ with the metal and site of adsorption.

Multibonded sites correspond to enhanced metal to $2\pi^*$ CO backdonation, because of the participation of this antibonding orbital to more molecular orbitals of the cluster. This increased population of an antibonding CO orbital induces the lengthening of the C-O bond and thus a decrease of the corresponding stretching frequency, from top to bridge to 3-fold sites. This effect can be quantitatively compared with experiment for Rh and Pd models [8,11,22]. In the same way, the observed higher CO stretching frequencies on Rh substrates with respect to Pd ones can be related to a stronger chemisorption bond with Rh, mainly due to a favored rehybridization of the σ type orbitals for Rh atoms [22]. Although many aspects of the CO chemisorption on noble metals have been explored on the basis of quantum chemical calculations, much effort has still to be made in order to obtain quantitative predictions of the adsorption energies. High accuracy calculations are now needed to account for coverage and size effects.

The main characteristics of the chemisorptive CO bond will be analyzed in more detail in another paper [22]. The different descriptions obtained from calculations based on various methods will then be compared. As a first step, the present article is devoted to the very rich experimental information which is at the basis of all studies on models of chemisorption. The main properties of CO chemisorption on some well-defined surfaces and on supported particles of Rh, Pd and Pt will be reviewed. Finally,the influence of additives and coadsorbed species will be briefly analyzed.

2. Well-Defined Surfaces

During the last ten years, there has been a rapid improvement of surface science techniques.The structure and bonding of single-crystal surfaces and monolayers of adsorbed molecules have been extensively investigated, as a function of temperature, adsorbate coverage and surface composition. The structure of clean metal surfaces and CO overlayers have been studied by a large variety of techniques including low energy electron diffraction (LEED), electron energy loss spectroscopy (EELS), without or with high resolution (HREELS), extended X-ray absorption fine structure (EXAFS), infrared spectroscopy (IR), infrared reflection absorption spectroscopy (IRAS), ultra-violet photoelectron spectroscopy (UPS), Auger electron spectroscopy (AES), thermal desorption spectroscopy (TDS) or temperature programmed desorption (TPD), workfunction measurements. An exhaustive review of surface science techniques is given in ref.[23], together with applications.

These investigations have allowed assignments of the CO adsorption sites as a function of coverage (θ). The geometries of the adsorbed molecules are often provided for given values of θ, related to specific LEED patterns. The measured frequencies of surface vibrational spectra (IR, HREELS, IRAS) show the dependence of the C-O stretching modes on the adsorption site and coverage values.

Some insight into the differences in binding energies between various sites are given by IRAS measurements with varying temperatures and by TPD experiments. From the IR frequencies of the C-O stretching bands and by comparison with organometallics, CO molecules have been found adsorbed at top, bridge or 3-fold sites, depending on the surface and θ values. When the coverage increases, the C-O stretching band shifts to higher frequencies. This shift can be related to dipole-dipole coupling and indirect (through metal) interaction (reduced metal backdonation to the CO $2\pi^*$ orbitals). For high coverages, direct intermolecular repulsions between adjacent molecules raise the energy of the $2\pi^*$ CO orbitals and thus decrease the amount of backdonation.

When coverage increases, the TPD peaks are shifted to lower temperatures, in relation with a decrease of the metal-CO bonding energies, due to the larger intermolecular repulsions. As expected, the strongest metal-CO species correspond to the first site to be occupied upon adsorption, or the last one upon desorption.

2.1. RH SURFACES

Most data concern CO chemisorption on Rh(111) and Rh(100) surfaces, including LEED [1,24-30,36-38], TDS [1,25,33,35,37], EELS [27,31-34,36], AES[1,24], XPS[30], IRAS [38] results.

2.1.1. *Rh(111)*.
The adsorption of CO at 300K produces a series of well-ordered LEED patterns :

-a ($\sqrt{3}$x$\sqrt{3}$)R30° pattern is observed up to a coverage of one-third (θ=1/3). HREELS results in combination with LEED ones show that CO adsorbs at top sites with the C end down [27,28,31].

-at higher coverages (1/3≤θ≤3/4), a split (2x2) LEED pattern is observed, which is explained by a double diffraction from an hexagonal CO overlayer. This corresponds to the population of top and bridge sites [25,27,29,32].

-finally, at θ = 3/4, a (2x2) LEED pattern is visible, where twice as many top as bridge sites are occupied [29].

The order-order transition of the ($\sqrt{3}$x$\sqrt{3}$)R30 to the split (2x2) pattern was shown to be reversible [25]. This transition is reported to occur at a higher θ value (0.5) by De Louise et al. [30] and is correlated with the appearance of a new XPS O1s band, characteristic of a bridge-bonded CO, in addition to the on-top one.

At low coverage, the on-top CO is found to be perpendicular to the surface, with Rh-C = 1.95±0.1 Å and C-O = 1.07±0.1 Å [28]. At the saturation coverage (θ = 3/4), the on-top CO molecules are tilted (16°) from the ideal top site, due to very close bridged-bonded CO molecules, located at a distance of 2.85 Å. The Rh-C bond lengths are then estimated at 1.93±0.1 Å and 2.03±0.07 Å for the near-top and bridge sites respectively, and the C-O bond is 1.15±0.1 Å [29].

The C-O stretching frequency of the on-top molecules increases with coverage, from 1990 cm^{-1} to 2070 cm^{-1} (θ = 3/4), whereas the Rh-C stretching frequency decreases from 460 to 420 cm^{-1} [27,34]. At intermediate coverage (θ = 1/3), the on-top C-O stretching shifts to 2000 cm^{-1}, while the bridged C-O band appears at 1870 cm^{-1} [27,34].

Finally, TDS measurements [27] show that the adsorbed CO at bridge sites have an approximately 4 kcal mole^{-1} lower binding energy to the Rh surface than those at top sites.

2.1.2. *Rh(100)*.
-The interaction of CO with a Rh(100) surface at 300K reveals no ordering in the CO overlayer for θ lower than 0.2 [25,36-38]. At this coverage, both top and bridge sites are occupied. The first one is related to a C-O stretching frequency at 2005 cm^{-1}, while the second one corresponds to a 1880 cm^{-1} value [38].

-A c(2x2) LEED pattern becomes apparent from θ = 0.2 up to θ≈0.6 [25,36-38]. Both bridge and top sites remain occupied and the related C-O stretching frequencies shift towards higher values. Above θ≈0.5, the rate of frequency shift decreases for both bands [38].

-At θ greater than 0.6, the LEED pattern becomes more complicated and a p(4√2x4√2)R45° pattern is found at saturation coverage [25,36-38]. At $\theta = 0.6$, the C-O stretching bands are located at 2060 and 1940 cm^{-1} for the top and bridge sites [38].
From a series of temperature dependent IRAS measurements [38], it has been shown that the bridge-bonded CO molecules are slightly more stable than the on-top ones, with an energy difference ranging from ca. 0.1 to 0.4 kcal mole^{-1} for $\theta = 0.2$-0.5.

2.2. PD SURFACES

The chemisorption of CO on Pd(111) and, to a lesser extent, on Pd(100), has been widely studied by various techniques. Among them are found LEED [39-46,51], EELS [39,46,51], IRAS [40,50], TDS [41,47,49] workfunction measurements [39,41], UPS [42].

2.2.1. Pd(111).
–Under one-third monolayer coverage ($\theta = 1/3$), the surface structure of Pd(111) with the CO overlayer has a (√3x√3)R30° periodicity. LEED results, in conjunction with IRAS measurements, show that CO molecules are adsorbed at 3-fold sites, through the C atom [39-46].The C-O stretching frequency is 1820 cm^{-1} at very low coverage, with a gradual shift to 1840 cm^{-1} at $\theta = 1/3$.
-For intermediate coverages ($1/3 \leq \theta \leq 1/2$), the LEED pattern remains unchanged, but 3-fold and bridge sites are now occupied. Indeed, an additional band, related to the bridged species, appears at 1880 cm^{-1} ($\theta = 1/3$) in the surface IR spectrum [40] and a second peak also appears at lower temperature in the thermal desorption spectrum [42,45,47,48]. But the broadness of this new band, located at 1850 cm^{-1} for $\theta \approx 0.32$-0.40, suggests that, due to the high mobility of CO on the [111] surfaces [49], some CO molecules move from 3-fold to bridge sites, whilst remaining in the same LEED pattern. A further increase in coverage changes the LEED pattern which corresponds to a c(4x2)R45° structure at $\theta = 0.50$ [40]. Only bridge sites are now occupied, related to a unique C-O stretching band at 1940 cm^{-1}.
-In the compression stage above $\theta = 0.50$, a second peak is observed above 2000 cm^{-1}, assigned to CO at top sites [40].
The geometry of the CO/Pd(111) system has only been investigated in the (√3x√3)R30° structure [45]. The Pd-C and C-O bond lengths are estimated as 2.05±0.04 Å and 1.15±0.05 Å, respectively. It is worth noting that the CO molecules are then favored to adsorb at hollow sites of fcc type (with no atom in the second layer), preferentially to hcp ones.

2.2.2. Pd(100).
-The LEED patterns, for every θ value, correspond to a c(4x2)R45° structure [39,51]. Since the IR spectra at various coverages show only one band, the observation of the LEED results means that only bridge sites are occupied. The C-O stretching frequencies vary from 1890 cm^{-1} for $\theta = 0.03$ to 1930 cm^{-1} at $\theta = 0.50$, where the Pd-C stretching frequency amounts to 330 cm^{-1} [39,50,51].
-At $\theta = 0.50$, the geometry of the adsorbate complex is evaluated as follows [39] : Pd-C = 1.93±0.07 Å and C-O = 1.15±0.1 Å. The CO adsorption energy on this surface is close to 40 kcal mole^{-1} [52]. A very similar value is expected for Pd(111) [53].

2.3. PT SURFACES

Carbon monoxide on Pt seems to be one of the most intensively studied chemisorption systems. A large variety of surfaces have been investigated, with a particular attention to Pt(111). Numerous techniques have been applied to these systems, including LEED [54-64], TDS [54,56,63,73-78], NEXAFS [79], IRAS [65,73,74], EELS [56,60,69,72] and IR [63,66-68,70,72].

Among these studies, a large number is devoted to the structural orientations of the CO molecules, revealing their energetically favorable tilting on stepped surfaces or on the atomic ridges of fcc Pt(110) substrates [59-63,78,79].

2.3.1. *Pt(111)*.

-Up to a coverage value of one-third ($\theta = 1/3$), a $(\sqrt{3}x\sqrt{3})R30°$ LEED pattern is obtained [54-58]. The observed C-O stretching vibration, 2050 cm^{-1} at low coverage, is characteristic of CO at the top site [76].

-As θ increases to 0.50, a c(4x2) overlayer is formed, which comprises equal numbers of top and bridge occupied sites [57,58,65-69,76]. The C-O stretching band at 1850 cm^{-1} is assigned to bridge-bonded CO. There has been a controversy about a band at 1810 cm^{-1}, attributed to a 3-fold adsorbed CO [76]. However, it appears that the adsorption of CO is strongly dependent on surface conditions [66]. In a more recent work about CO adsorption on Pt electrode surfaces, the observed C-O stretching band at ≈ 1780 cm^{-1} has been assigned to a 3-fold species [72].

In the c(4x2) overlayer, the Pt-C bond lengths have been evaluated as 1.85 and 2.05 Å for the top and bridge sites respectively [58].

-At a coverage higher than 0.6, the top CO molecules are reported to be tilted 6° away from the surface normal [55].

The CO binding energy is estimated at about 33 kcal mole^{-1} for the top site, this value being essentially coverage independent [54,77]. For coverage values close to one-third, the bridge site is less stable than the top one, with a very small difference of 1.5 [57] or 7 kcal mole^{-1} [77].

2.3.2. *Pt(110)*.

This surface has received considerable attention, since chemisorbed CO molecules on its atomic ridges are tilted, due to an excessive close-packing.

-Vibrational studies have shown that, for a coverage range from 0 to 0.50, the observed (1x2) LEED pattern suggests the formation of islands of tilted CO molecules [59,60]. A single thermal desorption peak at coverages below 0.50 is related to the occupation of a single adsorption site, namely a top site (C-O stretching vibration ranging from 2080 to 2130 cm^{-1}) [60,63].

-At $\theta = 0.50$, an additional desorption peak appears at a lower temperature [59,60,63]. However, only a single linearly bonded CO is observed in vibrational spectroscopy throughout the coverage range. This second desorption peak can be interpreted in terms of a phase transition in the CO overlayer, involving the tilting of the CO molecules [59-63]. Values of 20° [62] and 26° [59] have been proposed for the angle of tilt away from the surface normal.

3. Supported Particles

The commercial supported metal catalysts differ widely from clean well-defined surfaces. They are more generally formed by impregnation of metal salts on the support, i.e. alumina, silica, titania, magnesia, etc, followed by evaporation and reduction [80,81]. A slow evaporation and a further reduction at high temperature yield particles larger than ca. 100 Å. A quick evaporation associated with lower reduction temperatures provides particles ranging from 20 to 50 Å [82]. The very small particles ($\leq$15 Å) are advantageously prepared by ligand exchange [83].

The size of the particles is evaluated by transmission electron microscopy (TEM). Their structure before and after CO chemisorption has been the most extensively studied by IR [3,84-94,125-127] and by TPD [95-99,102,127]. Other techniques, including EXAFS [3,100-102], electron spin resonance (ESR) [102,103], XPS [102,104,105] and nuclear magnetic resonance (NMR) [103,106] have also been employed.

Model supported particles are also very useful, because they closely resemble supported catalysts, but a complete morphological, physical and chemical characterization may be obtained. These systems consist of metal particles deposited onto a flat support, most commonly by vacuum evaporation. This method is used to produce epitaxial thin films [107]. Different types of metallic particles can be obtained, depending on experimental conditions [108,109]. These model samples are characterized by TEM. Among other studies, the chemical bonding and vibrational analysis of chemisorbed CO have been investigated by inelastic electron tunneling spectroscopy (IETS) [110,111], EELS [110,112] and UPS [105], while the adsorption energetics have been obtained by TPD measurements [97,99,112].

Comparisons between the vibrational, electronic and thermal spectra of CO chemisorbed on supported metal particles and well-defined surfaces show that the particles display a distribution of several crystal planes : their number may vary with size and be responsible for size dependence in structure sensitive reactions [113-115].

Moreover, it has been shown that thermal treatments with oxygen and cycles of CO exposure modify the morphology of the particles [108,116]. As a consequence of the multiplicity of orientations, IR and TPD spectra of supported metal particles show that, generally, CO adsorbs at all possible sites and that the relative stabilities of the adsorbed species can then be different from their clean surface counterparts. This is particularly true for metals on which CO adsorption exhibits a strong dependence on crystallographic orientations. In that case, the vibrational and thermal features in IR and TPD spectra will differ from those obtained on crystal surfaces. For example, on-top CO is observed on Pd(111) only at a coverage higher than 0.5 (40). For Pd/SiO$_2$, this species, related to a C-O stretching frequency at 2060-2080 cm^{-1}, appears already at low coverages [92,117,118,126]. In the same way, IR spectra of CO on Pt/Al$_2$O$_3$ during desorption show that CO is bonded to both 3-fold and bridge sites, in addition to top sites [89,127], whereas coordination at 3-fold sites has not been proven on Pt(111).

As expected, coverage effects are as important as on single crystal surfaces, although they appear through different experimental results, because of the different distribution of sites. However, the higher energy shift of CO stretching vibrations with respect to coverage remains a common behavior to surfaces and supported particles [84,85,88-90,99,118].

It should be noted that comparison between the CO vibrational frequencies on monocrystals and metal particles must be singleton values, i.e. frequencies at zero coverage. In that case, the adsorbed CO molecules do not interact with their neighbors. Singletons are obtained by performing adsorption of very small amounts of CO or desorption at increasing temperatures, in order to achieve a uniform CO distribution [119,120]. The singleton values are important characteristics, since they are determined only by the binding of CO with metal atoms. In fact, true singletons are very difficult to obtain. This may explain the discrepencies between values given by different authors, especially for supported Pd particles [91,94]. In most cases, the reported singleton frequencies are close to those for single-crystal surfaces, showing only a small influence of the support [93,94,127].

However, this conclusion should not hold for very small particles where most metal atoms are in close contact with the support. Indeed, it has been shown that the nature of the support is the most effective factor in the hydrogenation reaction of CO on very dispersed particles : TiO_2 was found to be far more active than SiO_2 [3,121-123] and the selectivity was found to depend on the acidic properties of the support [3].

The size effect for Rh, Pd, Pt particles may originate from several reasons : larger interactions with the support for small particles, different crystallographic orientations with different sites, changes in the morphology of the particles after CO adsorption.

The latter possibility seems to be the real one for very dispersed Rh particles, which have a specific behavior upon CO adsorption, since the CO to Rh ratio reaches 2.0, yielding dicarbonyl Rh species.

For medium or large Rh particles (≥ 30 Å), the TPD curves resemble those obtained for single crystals, with a CO desorption peak at 500 K and a shoulder at 425 K, related to top and bridged CO. Since these sites are populated on various Rh surfaces, the TPD results are independent of crystal plane [25]. Thus, there is no observable size effect for these particles [96].

This result implies that there should be a size effect for Pd and Pt particles, for which the nature of the adsorption sites is strongly dependent on crystallographic orientation. Indeed, a study of the IR spectra of Pd/SiO_2 particles as a function of their size shows that the ratio of multibonded CO to on-top CO is decreasing when dispersion increases, favoring top sites on small particles [99,117,124].

In the same way, the TPD curves for CO from Pt particles on Al_2O_3 show a very clear size dependence [96,127]. The interpretation of these curves in terms of changes in the distribution of sites is not obvious. Indeed, Pt(110) exhibits two desorption peaks, whereas only on-top tilted CO are observed in vibrational spectroscopy at saturation coverage [63]. Moreover, recent studies on supported Pt particles with varying sizes have shown that a new band (or shoulder) appears in the on-top region, only in the case of very small particles (≤ 10 Å) [128,129]. This band, which is very sensitive to the basicity of the support, has been attributed to CO molecules adsorbed on Pt sites in close contact with the support [129].

The C-O stretching frequencies measured for Rh/Al_2O_3, Pd/SiO_2 and Pt/Al_2O_3 are reported in table 1. They correspond to large particles. Somewhat different frequencies can be found in the literature, due to different reduction conditions, sizes or supports. However, the discrepencies are less than 15 cm^{-1} for values at full coverage ($\theta=1$). Below full coverage, variations can be larger, especially for Pd particles, reaching their maximum for singleton values.

Complete descriptions of the IR spectra are given in refs.[84-87,125] for Rh, in refs.[91,92,94,116,126] for Pd and in refs.[89,93,127,130] for Pt particles.

Highly dispersed Rh/Al_2O_3 particles, which have been characterized and studied for twenty years because of their industrial applications, behave differently from Pd and Pt supported particles. Indeed, the latter ones always display single bonded CO molecules, whilst IR spectra of small Rh particles comprise C-O stretching bands characteristic of $Rh(CO)_2$ species. These bands, absent from single-crystal spectra, appear at 2030 and 2100 cm^{-1} and are coverage independent.

Extensive experimental studies, including all techniques have been performed on this subject [85-88,90,100-103]. The more recent and justified explanation is that CO adsorption disrupts the Rh-Rh bonds of the small supported crystallites, increasing the distances between the $Rh(CO)_2$ sites, which can be viewed as $Al-O-Rh(CO)_2$.

TABLE 1. CO stretching frequencies (cm^{-1}) for Rh/Al$_2$O$_3$, Pd/SiO$_2$ and Pt/Al$_2$O$_3$ particles

	C-O stretching frequency		site assignment
	$\theta = 1$	singleton	
Rh/Al$_2$O$_3$	2070	1975 (94) { 2015 (125)	top[a]
	1900[b]	≈1800	bridge
Pd/SiO$_2$	2095	2060 1960 (91)	top
	1965	{ 1895 (94)	bridge[a]
	1910	1920 (91) { 1805 (94)	3-fold[a]
Pt/Al$_2$O$_3$	2090	2030 (93) {2040 (127) 2070 (128)	top[a]
	2040	2040 (128)	top[c]
	1850 to 1750	≈1860	bridge
		≈1760	3-fold

(a) more stable site(s) upon desorption
(b) this band is shifted to 1870 cm^{-1} for small particles
(c) assigned to steps or support effects

Surface OH groups, present on the Al$_2$O$_3$ substrate play a determinant role in the formation of these dicarbonyl Rh species [133]. Their role has been demonstrated by addition of (CH$_3$)$_3$SiCl, which interacts with the surface OH groups on Al$_2$O$_3$, leading to the almost complete suppression of the geminal dicarbonyl species [136].

It should be mentioned that modifications in CO chemisorption due to support effects also occur for supported metal thin films. Much available information concerns thin films of Pd on Nb(110) and Ta(110) [131-134]. The Pd monolayer undergoes strong alteration of its electronic structure, which involves a substantial decrease of the Pd-CO bond strength. Indeed, the heat of adsorption of CO on Pd/Ta(110) is only 14 kcal mole^{-1} [134], instead of 38-40 kcal mole^{-1} for CO on Pd(110) and Pd(111) [52,53]. Very close values were found for Pd/SiO$_2$, with activation energies for CO adsorption estimated at 35 kcal mole^{-1} for large particles ($\geq$50 Å) and 38 kcal mole^{-1} for small particles [99].

4. Additives and Coadsorbants

It is possible to modify the properties of metal catalysts by adding a second transition metal to get bimetallic catalysts or alloys, or by adding on the metal surface promoters or poisons, such as alkalis, chalcogens, metallic ions.

The addition of promoters to metallic catalysts is of general use for various reactions, including the Fischer-Tropsch synthesis [137]. Moreover, the development of the use of bimetallic catalysts in the industrial hydrogenolysis reactions has intensified the studies of the catalytic behavior of alloys [138,139].

Additives (or modifiers) are generally called promoters when they increase the selectivity in desired products or poisons if they reduce the activity of a given catalytic reaction. The borderline between alloying and promoter effects is not very easy to delineate in the case of metallic additives, since their chemical state on the surface is more or less well-defined.

Two different concepts have been used to explain the role of these additives (or coadsorbants). The first one assumes modifications of the electronic properties of the metal atoms, due to the so-called "ligand effect". The second interpretation is based on the theory of ensembles [140-143]. In this representation, a reaction requires an ensemble of neighboring atoms, then the dilution of the surface by the additive should be taken into account.

In terms of the modifications in the local surface electron density induced by additives, we can say that those which are electron donors to the metal promote backdonation to the empty CO $2\pi^*$ orbital. This increase in backdonation weakens the C-O bond, lowers the C-O stretching frequency and strengthens the metal-CO bond, which can increase the propensity for CO dissociation. The inverse effect is produced by electron acceptor species.

These electronic modifications have been observed in both cases by UPS. Alloying Pd with a more electropositive element such as Zr (containing d electrons) shifts the Pd 4d orbital to higher energy with respect to pure Pd and lowers the activation energy for CO dissociation [144]. On the contrary, the presence of electron acceptor atoms, coadsorbed with CO has been shown to decrease the CO binding energy on Ni surfaces [145].

However, many experiments have demonstrated that geometric factors, such as dilution and size effects [125-127,146] or blocking of sites [147,148] are also of major importance. Direct electrostatic interactions between CO and the additives also have to be taken into account, in addition to the indirect ones through the metal substrate [118,126,147]. Finally, other cooperative effects, such as the stabilization of intermediate products on the alloys, have been suggested for PtZn samples [149].

The balance of these factors depends on the reaction which is considered and on the nature of the slow step of the reaction mechanism [150].

In any case, the role of the additives in the chemisorption varies with the nature of the metal-additive couple on the surface. For instance, addition of Cu to Pd induces no red shift of the C-O stretching frequency, whereas addition of Ag, Sn and especially Pb shows a decrease of this frequency [118,151]. The same larger effect for Pb is also reported in the case of PtPb alloys, with respect to PtCu or PtAg [146,152,153].

The UPS study of ZrPd, SiPd and AlPd alloys showed that the Pd-CO bond is weakened for alloys with elements containing no d electrons (Al, Si) and that CO dissociation is promoted only in the case of Zr addition [144].

It is well-known that Rh catalysts are able, depending the nature of the support, to catalyze the conversion of CO + H_2 to aldehydes and alcohols, and also to dissociate CO, leading to the formation of alkanes [3,154]. Addition of Mn, Ti, Zr to Rh/SiO$_2$ lowers the temperature of alkane formation [147]. This indicates that CO dissociation is enhanced and that the rate of CO conversion on Rh is increased. Moreover, further evidence of the C-O bond weakening is obtained from the IR spectra. Indeed, the bridged C-O stretching frequency, which occurs at ca. 1870-1900 cm^{-1} on Rh/Al$_2$O$_3$ is decreased to 1760 cm^{-1} on Rh+Ti, 1670 cm^{-1} on Rh+Zr and 1520 cm^{-1} on Rh+Mn. These results are interpreted in terms of a coordination of both C and O [155].

Addition of Fe and Zn to SiO$_2$ increases the temperature of alkane formation, which is interpreted as a sign of decreased CO dissociation. Additionally, the amount of bridged chemisorbed CO decreases with increasing Fe content of the RhFe samples [155]. In the case of Zn addition, the IR spectra are strongly altered, suggesting that the bridge sites are blocked with Zn [153,155].

Alkalis have been used as promoters to increase the selectivity towards heavy alkanes in CO + H$_2$ reactions. Their effect can be correlated with their electron donor character, making a dissociative adsorption more probable for CO [156]. Among other techniques, IR spectroscopy has been used to study the influence of alkali promotion on the adsorption of CO, and particularly that of potassium. Deposition of K induces the presence of a new absorption band in the region of the lowest C-O stretching frequencies [125,126,157-159]. This band certainly reflects local interactions of CO with the promoter and can also indicate that CO changes its site from top to multibonded positions.

Adsorption of Carbon ("coking"), sulfur, halogens on metal catalysts has been the subject of extensive studies since it is the most serious cause of catalyst poisoning. The presence of these additives blocks the reaction sites, but also induces changes in the surface morphology and modifies metal-support and metal-adsorbate interactions [160].

Most experiments have shown that their presence on metal surfaces reduces the adsorption rate of CO, its binding energy to the metal, its adsorptive capacity and its dissociation ability.

At low S (Se) coverage, CO adsorption induces a rearrangement of the chalcogen overlayer on Pt(111) and Pt(110), while no effect has been observed on Pt(100) [160,161]. Moreover, recent results underline that the chalcogen modifiers stiffen the CO vibrational modes, which can be attributed to direct repulsive interactions [161].

Coadsorption of CO with molecules, such as O$_2$, H$_2$, NO, H$_2$O, C$_6$H$_6$, has also been intensively investigated. The observed IR frequency shifts have been analyzed in terms of "ligand" and geometric effects. Generally, coadsorption is accompanied by significant changes in the adsorption sites, but, once again, the results depend on the metal. For example, on Rh(111), NO migrates from bridge to top site in the presence of CO [162], whereas, CO is probably displaced from top to 3-fold sites if water is coadsorbed [163]. When CO and benzene are coadsorbed, CO migrates from top to bridge sites and benzene prefers 3-fold rather than bridged coordination [164]. On the contrary, on Pt(111), there is no clear migration of CO or benzene when they are coadsorbed, and the presence of water does not induce changes in CO adsorption sites, different from coverage effects.

The structure of the overlayer of coadsorbed species can be different according to the sequence of adsorption. For example, CO and NO form separate islands if CO is adsorbed first, but CO adsorbs dispersively throughout the NO layer [162].

Due to direct and indirect interactions between coadsorbed species, one can expect that their binding energies to the metal surface may be modified. In fact, this depends on the system studied.

The Rh-CO bond is weakened in well-mixed CO + NO adlayers and this has been attributed to repulsive NO-CO interactions through the Rh substrate. However, the NO chemisorption is not measurably affected by coadsorbed CO [162]. An opposite behavior on Rh(111) has been reported for CO-water interactions, which are attractive for adjacent sites since the temperature of water desorption increases in the presence of coadsorbed CO [163]. The behavior of the coadsorbates CO and H_2O is different on Pt(111), where the temperature of water desorption decreases for all CO coverages. In that case, repulsive CO-H_2O interactions provide migration of the two species to separate islands [163].

5. Conclusion

The adsorption of CO on single-crystal transition metal surfaces and supported particles has received considerable attention during the last decade. Not only is chemisorbed CO the much used as a probe of the metal "surface", but its industrial importance is great in catalysis of CO hydrogenation to produce hydrocarbons and of CO oxidation to convert automobile exhaust gases.

The development of surface science techniques has favored extensive studies of ordered layers of adsorbed molecules on clean metal surfaces. The structure and bonding at the surface and their changes with coverage and temperature are essential informations in order to understand the more complex behavior of CO on the real metal catalysts. The geometry of the adsorbed species at different sites and their vibrational modes determined on several metal surfaces are helpful data for the study of the adsorption properties on metal supported particles. Surface studies have also shown how changes can occur in the ordered overlayer with increasing coverage or with coadsorbed species, including possible reconstruction of the metal surface itself.

Chemisorption on supported metal particles is less easy to define, because of the great variety of sizes (from a few to thousands of atoms), because of the various interactions with the supports and, also, the three dimensional shape of the particles, which implies the presence of several crystal planes. Comparison of chemisorption on single-crystal planes and on metal particles in terms of the vibrational properties of the adsorbates is hindered by the technical difficulties in obtaining values at zero-order coverage. It would be very interesting, for comparisons with calculations on model clusters, to have reliable singleton values, measured for particles of varying sizes and with different supports.

The influence of additives or coadsorbed species on CO chemisorption is a balance of geometric and electronic effects, as shown by a large number of experimental results. Many interesting and sometimes puzzling results are available in this field of catalysis, where the activity and selectivity of a metal catalyst are modified by addition of other species in the metal surface or at the surface. Theoretical methods would be of great help to analyze the electronic factors for different additives (coadsorbants) and to evaluate their incidence on the adsorption properties (geometries, vibrations, energies).

The richness of the experimental results on Rh, Pd, Pt surfaces and supported particles shows how these metals can behave differently. It is thus a real challenge for theoretical approaches to understand the adsorbate-metal chemical bond and how it is influenced by the supports, the geometry of the catalyst and the modifiers.

REFERENCES

1. Sault, A.G. and Goodman, D.W. (1989), in K.P. Lawley (ed.), Model Studies of Surface Catalyzed Reactions, Advances in chemical Physics, John Wiley and Sons, New York, p. 153.
2. Poustna, M.L., Elek, L.F. , Ibarbia, P.A., Risch, A.P. and Rabo, J.A. (1978), J. Catal. **52**, 157.

3. Katzer, J.R., Sleight, A.W., Gajardo, P. ,Michel, J.B., Gleason, E.F. and S. Mc Millan, S. (1981), Disc. Faraday Soc. **72**, 121.
4. Taylor, K.C. (1984), Automobile Catalytic Converters, Springer, New York.
5. Oh, S.H., Fischer, G.B., Carpenter, J.E. and Goodman, D.W. (1986), J. Catal. **100**, 360.
6. Ryberg, R. (1989), in K.P. Lawley (ed.), Infra-Red Spectroscopy of Molecules Adsorbed on Metal Surfaces, Advances in Chemical Physics, John Wiley and Sons, New York, p.1.
7. Blyholder, G. (1964), J. Phys. Chem. **68**, 2772.
8. Ray, N.K. and Anderson, A.B. (1982), Surface Sci. **119**, 33.
9. Anderson, A.B. and Awad, M.K. (1985), J. Am. Chem. Soc. **107**, 7854.
10 Andzelm, J., Salahub, D.R. (1986), Int. J. Quantum Chem. **29**, 1091.
11. Andzelm, J., Salahub, D.R. (1987), in Jena, P., Rao, B.K. and Khanna, S.N. (eds.), Physics and Chemistry of Small Clusters, Nato ASI Series B158, Plenum New York, p.867.
12. Pacchioni, G. , Koutecky, J. (1987), J. Phys. Chem. **91**, 2658.
13. Blomberg, M.R.A., Lebrilla, C.B., Siegbahn, P.E.M. (1988), Chem. Phys. Lett. **150**, 522.
14. Caffarel, M., Claverie, P., Mijoule, C., Andzelm, J. and Salahub, D.R. (1989), J. Chem. Phys. **90**, 990.
15. Pacchioni, G., Bagus, P., S. (1990), J. Chem. Phys. **93**, 1209.
16. Bagus, P.S., Pacchioni, G. (1990), Surface Sci. **236**, 233.
17. de Koster, A., Van Santen, R.A. (1990), Surface Sci. **233**, 366.
18. Wong, Y.T., Hoffmann, R. (1991), J. Phys. Chem. **95**, 859.
19. Mains, G.J., White, J.M. (1991), J. Phys. Chem. **95**, 112.
20. Smith, G.W., Carter, E.A. (1991), J. Phys. Chem. **95**, 2327.
21. Papai, I., Goursot, A., St-Amant, A. and Salahub, D.R. (1991), submitted.
22. Goursot, A., Papai I. and Salahub, D.R. (1991), in press.
23. Van Hove, M.A., Wang, S.W., Ogletree, D.F., Somorjai, G.A., (1989), in P.O. Löwdin (ed.), The State of Surface Structural Chemistry : Theory, Experiment, Results, Advances in Quantum Chemistry, Academic Press Inc., New York, p. 1.
24. Grant, J.T., Haas, T.W., Somorjai, G.A. (1970), Surface Sci. **21**, 76.
25. Castner, D.G., Sexton, B.A., Somorjai, G.A. (1978), Surface Sci. **71**, 519.
26. Thiel, P.A., Williams, E.D., Yates, J.T. and Weinberg, W.H. (1979), Surface Sci. **84**, 59.
27. Dubois, L.H., Somorjai, G.A. (1980), Surface Sci. **91**, 514.
28. Koestner, R.J., van Hove, M.A., Somorjai, G.A. (1981), Surface Sci. **107**, 439.
29. van Hove, M.A., Koestner, R.J., Frost, J.C., Somorjai, G.A. (1983), Surface Sci. **129**, 482.
30. de Loiuse, L.A., White, E.J., Winograd, N. (1984), Surface Sci. **147**, 252.
31. Ibach, H., Mills, D.L. (1982), Electron Energy Loss Spectroscopy and Surface Vibrations, Academic Press, New York.
32. Crowell, J.E., Somorjai, G.A. (1979), Appl. Surf. Sci. **19**, 73.
33. Dubois, L.H., Hansma, P.K., Somorjai, G.A. (1980), J. Catal. **65**, 318.
34. Root, T.W., Fischer, G.B., Schmidt, L.D. (1986), J.Chem. Phys. **85**,4687.
35. Richter, L.J., Gurney, B.A., Ho, W. (1987), J. Chem. Phys. **86**, 477.
36. Gurney, B.A., Richter, L.J., Villarubia, J.S. and Ho, W. (1987), J. Chem. Phys. **87**, 6710.
37. Kim, Y., Peebles, H.C., White, J.M. (1982), Surface Sci. **114**, 363.

38. Leung, L.W.H., He, J.W. and Goodman, D.W. (1990), J. Chem. Phys. **93**, 8328.
39. Behm, R.J., Christmann, K., Ertl, G. (1980), J. Chem. Phys. **73**, 2984.
40. Bradshaw, A.M., Hoffmann, F.M. (1978), Surface Sci. **72**, 513.
41. Ertl, G., Koch, J. (1972), in F. Ricca (ed.), Adsorption-desorption phenomena, Academic Press, New York.
42. Conrad, H., Ertl, G., Kuppers, J. (1978), Surface Sci. **76**, 323.
43. Biberian, J.P., van Hove, M.A. (1984), Surface Sci. **138**, 361.
44. Mirana, R., Wandel, K., Rieger, D. and Schnell, R.D. (1984), Surface Sci. **139**, 430.
45. Ohtani, H., van Hove, M.A. and Somorjai, G.A. (1987), Surface Sci. **187**, 372.
46. Netzer, F.P. and El Gomati, M.M. (1984), Surface Sci. **124**, 26.
47. Noordermeer, A., Kok, G.A. and Nieuwenhuys, B.E. (1986), Surface Sci. **172**, 349.
48. Brown, A. and Vickerman, J.C. (1983), Surface Sci., **124**, 267.
49. Doyen, G and Ertl, G. (1974) Surface Sci. **43**, 197.
50. Yoshioka, K., Kitamura, F., Takeda, M., Takahashi, M and Ito, M. (1990), Surface Sci. **227**, 90.
51. Behm, R.J., Christmann, K., Ertl, G, van Hove, M.A., Thiel, P.A. and Weinberg, W.H. (1979), Surface Sci. **88**,L59.
52. Tracy, J.C., Palmberg, P.W. (1969), J. Chem. Phys. **51**, 4852.
53. Ladda, S., Poppa, H. and Boudart, M. (1981), Surface Sci. **102**, 151.
54. Ertl, G., Neumann, M and Streit, K.M. (1977), Surface Sci., **64**, 393.
55. Kiskinova, M., Szabo, A. and Yates, J.T. (1988), Surface Sci. **205**, 815.
56. Steiniger, H., Lehwald, S., Ibach, H. (1982), Surface Sci. **123**, 264.
57. Schweizer, E., Persson, B.N., Tushaus, M., Hoge, D., Bradshaw, A.M. (1989), Surface Sci. **213**, 49.
58. Ogletree, D.F., van Hove, M.A. and Somorjai, G.A. (1986), Surface Sci. **173**, 351.
59. Hoffmann, P., Bare, S.R., Richardson, N.V., King, D.A. (1982), Solid State Commun. **42**, 645.
60. Bare, S.R., Hoffmann P. and King, D.A. (1984), Surface Sci., **144**, 347.
61. Biberian, J.P. and van Hove, M.A. (1984), Surface Sci. **138**, 361.
62. Rieger, D., Schnell, R.D. and Steinmann, W. (1984), Surface Sci. **143**, 157.
63. Hayden, B.E., Robinson, A.W. and Tucker, P.M. (1987), Surface Sci. **192**, 163.
64. Baetzold, R.C. (1985), J. Chem. Phys. **82**, 5724.
65. Malik, I.J., Trenary, M. (1989), Surface Sci. **214**, L240.
66. Tobin, R.G., Phelps, R.B. and Richards, P.L. (1987), Surface Sci. **183**, 427.
67. Tobin, R.G. and Richards, P.L. (1987), Surface Sci. **179**, 387.
68. Krebs, H.J. and Luth, H. (1977), Appl. Phys. **14**, 337.
69. Frointzheim, H., Hopster, H., Ibach, H. and Lehwald, S. (1977), Appl. Phys. **13**, 147.
70. Crossley, A. and King, D.A. (1980), Surface Sci. **95**,131.
71. Avery, N.R. (1981), J. Chem. Phys. **74**, 202.
72. Kitamura, K., Takahashi, M. and Ito, M. (1989), Surface Sci. **223**, 493(1989)
73. Greenler, R.G., Burch, K.D., Kretschnar, K., Klauser, R. and Hayden, B.E. (1985), Surface Sci. **152/153**, 338.
74. Hayden, B.E., Kretschner, K., Bradshaw, A.M., Klauser, R. and Greenler, R.G. (1985), Surface Sci. **149**, 394.

75. Collins, D.M. and Spicer, W.E. (1977), Surface Sci. **69**, 85.
76. Hayden, B.E. and Bradshaw, A.M. (1983), Surface Sci. **125**, 787.
77. Frointzheim, H. and Schulze, M. (1989), Surface Sci. **213**, 837.
78. Siddiqui, H.R., Guo, X., Chordenkoff, I. and Yates, G.T. (1987), Surface Sci. **191**, L813.
79. Somers, J.S., Lindner, T., Surman, M. and Bradshaw, A.M. (1987), Surface Sci. **183**, 576.
80. Prestridge, E.B., Via, G.H. and Sinfelt, H. (1977), J. Catal. **50**, 115.
81. Paul, D.K., Ballinger, T.H., Yates, J.T. (1990), J. Phys. Chem. **94**, 4617.
82. Marcilly, C. (1984), in B. Imelik, G.A. Martin, A. Renouprez (eds.), Catalyse Par les Métaux, Editions du CNRS, p 121.
83. Boitiaux, J.P., Cosyns, J. and Vasudevan, S. (1983), in G. Poncelet, P. Grange, P.A. Jacobs (eds.), Preparation of Catalysts III, Elsevier, Amsterdam, p. 123.
84. Yang, A.C. and Garland, C.W. (1957), J. Phys. Chem. **61**, 1504.
85. Yates, J.T., Duncan, T.M., Worley, S.D. and Vaughan, R.W. (1979), J. Chem. Phys. **70**, 1219.
86. Cavanagh, R.R. and Yates, J.T. (1981), J. Chem. Phys. **74**, 4150.
87. Rice, C.A., Worley, S.D., Curtis, C.W., Guin, J.A. and Tarrer, A.R. (1981), J. Chem. Phys. **74**, 6487.
88. Solymosi, F., Pasztor, M. (1985), Phys. Chem. **89**, 4789.
89. Haaland, D.M. (1987), Surface Sci. **185**, 1.
90. Dictor, R. and Roberts, S. (1989), J. Phys. Chem. **93**, 2526.
91. Hendrickx, H.A.C.M., Des Bouvrie, C. and Ponec, V. (1988), J. Catal. **109**, 120.
92. Palazov, A., Chang, C.C., Kokes, R.J. (1975), J. Catal. **36**, 338.
93. Primet, M. (1984), J. Catal. **88**, 273.
94. Bredikhin, M.N. and Lokhov, Yu A. (1989), J. Catal. **115**, 601.
95. Altman, E.I. and Gorte, R.J.(1986), Surface Sci. **172**, 71.
96. Altman, E.I. and Gorte, R.J. (1988), Surface Sci. **195**, 392.
97. Ladda, S., Poppa, H. and Boudart, M. (1981), Surface Sci. **102**, 151.
98. Doering, D.L., Dickinson, J.T. and Poppa, H. (1982), J. Catal. **73**, 104.
99. Gillet, E., Channakhone, S. and Matolin, V. (1986), J. Catal. **97**, 437.
100. van't Blik, H.F.J., van Zon, J.B.A.D., Huizinga, T., Koningsberger, D. and Prins, R. (1983), J. Phys. Chem. **87**, 2264.
101. van't Blik, H.F.J., van Zon, J.B.A.D., Huizinga,T., Koningsberger, D. and Prins, R. (1984), J. Mol. Catal. **25**, 379.
102. van't Blik, H.F.J., van Zon, J.B.A.D., Huizinga,T., Koningsberger, D. and Prins, R. (1985), J. Am. Chem. Soc. **107**, 3139.
103. Robbins, J.L. (1986), J. Phys. Chem. **90**, 3381.
104. Buchanan, D.A., Hernandez, M.E., Solymosi, F., White, J.M. (1990), J. Catal. **125**, 456
105. Takasu, Y., Unwin, R., Tesche, B., Bradshaw, A.M. and Grunze, M. (1978), Surface Sci. **77**, 219.
106. Duncan, T.M., Zilm, K.W., Hamilton, D.M. and Root, T.W. (1989), J. Phys. Chem. **93**, 2583.
107. Thomas, M., Poppa, H. and Pound, G.M. (1979), Thin Solid Films **58**, 273.
108. Gillet, M.F. and Channakhone, S. (1986), J. Catal. **97**, 427.
109. Robinson, F., Gillet, M.F. (1982), Thin Solid films **98**, 179.
110. Kroeker, R.M., Kaska, W.C., Hansma, P.K. (1979), J. Catal. **57**, 72.

152

111. Dubois, L.H., Hansma, P.K. and Somorjai, G.A. (1980), Appl. Surface Sci. **6**, 173.
112. Chen, J.G., Colaianni, M.L., Chen, P.J. and Yates, J.T. (1990), J. Phys. Chem. **94**, 5059.
113. Goodman, D.W., Kelley, R.D., Madey, T.E. and Yates, J.T. (1980), J. Catal. **63**,226.
114. Kelley, R.D. and Goodman, D.W. (1982), Surface Sci. **123**, 743.
115. Goodman, D.W. (1982), Surface Sci. **123**, 679.
116. Hicks, R.F., Qi, H., Kooh, A.B. and Fischel, L.B. (1990), J. Catal. **124**, 488.
117. Sheu, L.L., Zharpinski, Z. and Sachtler, W. (1989), J. Phys. Chem. **93**, 4890.
118. Hendrickx, H.A.C.M. and Ponec, V. (1987), Surface Sci. **192**, 234.
119. Primet, M., De Menorval, L.C, Ito, T. and Fraissard, J. (1985), J. Chem. Soc. Far. Trans.1 **81**, 2866.
120. Eischens, R.P., Francis, S.A. and Pliskin, W.A. (1956), J. Phys. Chem., **60**, 194.
121. Solymosi, F., Erdohelyi, A. and Bansagi, T. (1981), J. Catal. **68**, 371.
122. Solymosi, F., Tombacz, I. and Kocsis, M. (1982), Catal. **75**, 78.
123. Solymosi, F., Tombacz, I. and Koszta, J. (1985), J. Catal. **95**, 578.
124. van Hardeveld and R., Hartog, F. (1972), Adv. Catal. **22**, 75.
125. Angewaare, P.A.J.M., Hendrickx, H.A.C.M. and Ponec, V. (1988) J. Catal. **110**, 18.
126. Angewaare, P.A.J.M., Hendrickx, H.A.C.M. and Ponec, V. (1988), J. Catal. **110**, 11.
127. Barth, R., Pitchai, R., Anderson, R.L. and Verykios, X.E. (1989), J. Catal. **116**, 61.
128. De Menorval, L.C. (1991), in press.
129. de Mallmann, A. and Barthomeuf, D. (1989), in J. Weitkamp (eds), Zeolites as Catalysts, Sorbents and Detergent Builders, H.G. Karge, p.429.
130. Toop, F.S., Toolenaar, J.C.M and Ponec, V. (1982), J. Catal. **73**,50
131. Neiman, D.L. and Koel, B.E. (1988), in D.M. Zehner, D.W. Goodman (eds), Physical and Chemical Properties of Thin Metal Overlayer and Alloy Surfaces, Materials Research Society, Pennsylvania.
132. Ruckman, M.W. and Strongin, M. (1984), Phys. Rev. **B29**, 7105.
133. Ruckman, M.W. and Strongin, M. (1987), Phys. Rev. **B35**, 487.
134. Koel, B.E., Smith, R.J. and Berlowitz, J.P. (1990), Surface Sci. **231**, 325.
135. Basu, P., Panayotov, D. and Yates J. T. (1987), J. Phys. Chem. **91**, 3133
136. Paul, D.K., Ballinger, T.H. and Yates, J.T. (1990), J. Phys. Chem. **94**, 4617.
137. Storch, H.H., Golumbig, N.G. and Anderson, R.B. (1951), The Fischer-Tropsch and Related Syntheses, J. Wiley, New York.
138. Sinfelt, H. (1977), in J.F. Bunnet (ed.), Catalysis by Alloys and Bimetallic Clusters, American Chemical Society, p. 15.
139. Sinflet, H. (1983), Bimetallic Catalysts : Discoveries, Concepts and Applications, J. Wiley and Sons, New York.
140. Soma-Noto Y. and Sachtler, (1974), J. Catal. **34**, 162.
141. Sinfelt, J.H. (1977), Acc. Chem. Res. **10**, 15.
142. Ponec, V. (1983), Advances in Catalysis **32**, 149.
143. Martin, G.A. (1988), Catal. Rev. Sci. Eng. **30**, 519.
144. Hauert, R., Oelhafen, P. and Guntherodt, H.J. (1989), Surface Sci. **220**, 341.
145. Ertl, G. (1977), J. Vac. Sci. Technol. **14**, 435.
146. Toolenaar, F.J.C.M., Ponec, V. (1983), J. Catal. **83**, 251.

147. Sachtler, W.M.H. and Ichikawa, M. (1984), J. Phys. Chem. **89**, 1564.
148. Jen, H.W., Zheng, Y., Shriver, D.F., Sachtler, W.M.H. (1989), J. Catal. **116**, 361.
149. Boccuzzi, F., Chiorino, A., Ghiotti, G., Pinna, F., Strukul, G. and Tessari, R. (1990), J. Catal. **126**, 381.
150. Figueras, F. and Coq, B. , Catalysis To-day, (1991), in press.
151. Kharson, M.S., Kadinov, G.B. and Palazov, A.N. (1979), React. Kinet. Catal. Lett. **10**, 267.
152. Palazov, A., Vonev, C., Kadinov, G., Shopov, D., Lietz, G. and Volter J. (1981), J. Catal. **71**, 1.
153. Bastein, A.G.T.M. and Toolenaar, F.J.C.M. (1982), J. Chem. Soc. Chem. Commun. **627**.
154. Ichikawa, M. (1982), Chem. Tech. **674**.
155. Ichikawa, M. and Fukushima, T. (1984), J. Phys. Chem. **89**, 1564.
156. Martin, G.A. (1982), in B. Imelik et al. (eds), Metal-Support and Metal-Additive Effects in Catalysis, Elsevier, Amsterdam, p. 315.
157. Garfunkel, E.L., Crowell, J.E. and Somorjai, G.A. (1982), J. Phys. Chem. **86**, 310.
158. Crowell, J.E., Garfunkel, E.L. and Somorjai, G.A. (1982), Surface Sci. **121**, 303.
159. Crowell, J.E. and Somorjai G.A. (1984), Appl. Surf. Sci. **19**, 73.
160. Oudar, J. (1985), in J. Oudar, H. Wise (eds), Deactivation and Poisoning of Catalysts , Marcel Dekher Inc., New York, p. 51.
161. Kiskionva, M.P., Szabo, A. and Yates, J.T. (1990), Surface Sci. **226**, 237.
162. Root, T.W., Fisher, G.B. and Schmidt, L.D. (1986), J. Chem. Phys. **85**, 4687.
163. Wagner, F.T., Moylan, T.E. and Schmieg, S.J. (1988), Surface Sci. **195**, 403.
164. Ogletree, D.F., van Hove, M.A. and Somorjai, G.A. (1987), Surface Sci. **183**, 1.

LINEAR SEMIBRIDGING CARBONYLS 5. THE STRUCTURE AND BONDING OF THE CHROMIUM CYCLOPENTADIENYL DICARBONYL DIMER.

R.L. WILLIAMSON, A. A. LOW, M. B. HALL
Department of Chemistry, Texas A&M University
College Station, Texas 77843-3255, USA

M. F. GUEST
SERC, Daresbury Laboratory
Warrington, WA4 4AD, U.K.

ABSTRACT. The structure and bonding of $[Cr(Cp)(CO)_2]_2$ was examined by optimizing the geometry of the dimer, fragment moiety, and analogous stable monomers. The optimized structure of $Cr(Cp)(CO)_2N$ has a "piano stool" geometry with a fragment moiety geometry nearly identical to the geometry of the fragment moiety in the dimer. Examination of x-ray and theoretical geometries of monomers with the fragment moiety bonded to a single ligand shows that the geometry of the fragment moiety is insensitive to the ligands bonded to it. Van der Waals calculations on $[Cr(Cp)(CO)_2]_2$ predict a geometry with the fragment moiety tilted away from a semibridging orientation to an "ethane-like" geometry. Perfect pairing GVB and CASSCF calculations predict a metal-metal triple bond made up of a σ and two π bonds. Because correlation was added to the primarily metal-metal bonding orbitals the fragment moieties are still significantly tilted away from the experimental geometry and toward the "ethane-like" geometry. Additional correlation of the metal-carbonyl bonding orbitals favors the fragment moiety tilting down toward the experimental geometry. Analysis of the transformed SCF wavefunction shows that the semibridging carbonyl ligands are acting as π acceptors from filled distal-metal d-orbitals.

1. Introduction

Recent X-ray crystal structures [1-4] of four metal dimers with "linear" semibridging carbonyls, $[(\eta^5-C_5R_5)M(CO)_2]_2$ (M=Cr,Mo; R=H,Me), have given rise to questions about the bonding of the carbonyls and investigations as to why the Cp-M-M-Cp (Cp=$\eta^5-C_5H_5$) moiety is linear in only the molybdenum dimer (**1**). As defined by Crabtree and Lavin [5] a linear semibridging carbonyl has a M-C-O angle characteristic of a terminal carbonyl (160-180°) and a M-M-C angle characteristic of a semibridging one (40-75°). The four complexes above have M-C-O angles (167-173°) and M-M-C angles (66-73°) well within the limits of linear semibridging carbonyls.

The controversy concerning the nature of the bonding in linear semibridging carbonyl ligands has spawned three hypotheses. Klingler, Butler, and Curtis [1] have suggested that the carbonyls are acting as 4-electron donors into empty metal d-orbitals. Two electrons are donated from the carbonyl 5σ orbital to one metal atom and two electrons are

D. R. Salahub and N. Russo (eds.), Metal-Ligand Interactions: from Atoms, to Clusters, to Surfaces, 155–173.

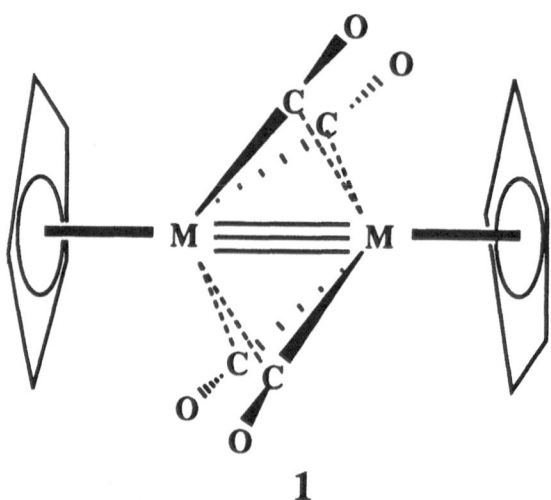

1

donated from the carbonyl 1π orbital to the other metal atom. They further explain that this σ and π donation from the carbonyl ligand would tend to lengthen the metal-metal bond. The metal-metal bond distances for these molybdenum dimers are longer than other known molybdenum-molybdenum triple bond distances.

A second model, first proposed by Cotton [6] to explain normal (bent) semibridging carbonyls, describes the carbonyls as acceptor ligands rather than donor ligands. The results of extended Hückel calculations on these systems by Jemmis, Pinhas, and Hoffmann [7] led them to suggest that the carbonyls are accepting electron density from metal d-orbitals into the empty CO 2π orbital, however, they did not explain why the carbonyl ligands are linear. They further stressed that the metal-metal bond order is diminished by semibridging carbonyls accepting electron density from metal orbitals of both metals. Morris-Sherwood, Powell and Hall [8], using Fenske-Hall molecular orbital calculations, also conclude that the carbonyls are accepting density from the metals. They explain the linearity of the carbonyl ligands by showing that the carbonyl ligand must remain linear in order to achieve maximum overlap of the empty CO 2π orbital with the filled metal-metal π-orbital. This results in the formation of a 3-center, 2-electron M-C(O)-M bond.

The third hypothesis was proposed by Colton and McCormick [9] and is based on the geometric parameters and IR carbonyl stretching frequencies. They suggested that the carbonyls are not truly semibridging but rather almost terminal carbonyls, and classified them as "borderline" semibridging. Although these models might appear to be mutually exclusive they might all be correct to one degree or another. The true picture of the bonding in linear semibridging carbonyls could include electron donation from the distal carbonyl 1π orbital to the metal (first model) as well as electron donation from the metal to the distal carbonyl 2π orbital (second model). If these donor-acceptor interactions are relatively weak, the carbonyls would appear more like terminal ligands and less like bridging ligands (third model).

The factors affecting the geometry of the Cp-M-M-Cp moiety are even less well understood than the factors affecting the geometry of the linear semibridging carbonyls. Curtis et al. [3] suggested that the linear geometry of the Cp-M-M-Cp moiety arises because electron density is donated from the Cp-ring orbitals into empty metal-metal antibonding d-orbitals. When the moiety is linear, the metal-metal π^*-orbital can accept density from the Cp ring, but when the Cp rings are bent off of the metal-metal bond axis such that the M-M-Cp angle is 135°, the metal-metal π^* orbital can no longer accept density from the ring. However, in this bent geometry the metal-metal σ^* orbital can accept density from the ring. The M-M-Cp angle which minimizes the energy for a particular complex would depend upon the relative energies of the metal orbitals and their interaction with the Cp ring orbitals. Although this model offers a reasonable rationalization of the electronic structure, it does not explain the geometric structure.

Although much study has been done to understand the bonding in the "linear semibridging" carbonyl chromium and molybdenum dimers, the semi-empirical and approximate molecular orbital (MO) calculations have left some questions unanswered. Here, we use *ab-initio* MO calculations to investigate the chromium cyclopentadienyl dimer to determine the nature of the linear semibridging carbonyls bonding, the nature of the metal-metal bond, and the geometry of the Cp-Cr-Cr-Cp moiety.

2. Computational Details

Using analytical gradient techniques, we did complete geometry optimizations with SCF wavefunctions and partial geometry optimizations with higher level wavefunctions. The higher level calculations include perfect-pairing generalized valence bond (GVB) and complete-active-space self-consistent-field (CASSCF). In the perfect-pairing scheme for the metal-metal bond each GVB pair involves one occupied metal-metal bonding orbital with a corresponding unoccupied metal-metal antibonding orbital. Two CASSCF calculations were done with active spaces of six electrons in six orbitals (6/6) and six electrons in ten orbitals (6/10). The van der Waals repulsion energies were calculated with the graphics program CHEMX [10] in which the repulsions between atoms that are bonded to each other are not included in the van der Waals energy. The carbonyl ligands were treated both as terminal carbonyls (with the carbon bonded to only one chromium atom) and as bridging carbonyls (with the carbon atoms bonded to both chromium atoms). Each of the carbon atoms in the Cp ring was bonded to a chromium atom. Sulfur was substituted for chromium because CHEMX does not have atomic parameters for transition metals. All calculations of single metal and dimer complexes were done in C_S and C_{2h} symmetry, respectively.

The basis set used for all calculations is a 3-21G basis set developed in a previous study [11]. The wavefunctions of several complexes were transformed from a basis set of atomic functions to a basis set of molecular functions in which the molecular functions are the molecular orbitals for the individual ligands or fragments. This allows the metal-ligand or fragment-fragment interactions to be separated out from the intra-ligand interactions.

Electron deformation densities were calculated by subtracting the electron density of the promolecule from the total electron density of the molecule. The electron density of the promolecule is the sum of the electron densities of each part, with the parts arranged in the geometry of the molecule. For all experimental and most theoretical deformation density plots, each part is a single atom whose density is the averaged spherical electron density; however, it is also informative to construct the promolecule with calculated densities of isolated ligands for monomers or metal-ligand fragments for cluster compounds [12].

Steigerwald and Goddard [13] have suggested that a good model for cyclopentadienyl is chloride. They base their suggestion on empirically determined ionization enthalpy (IE)

158

and metal-Cp bond distances. Because the number of basis functions for a Cl⁻ ligand is much smaller than the number of basis functions for a Cp⁻ ligand, this is very appealing to computational chemists who wish to calculate large complexes with Cp⁻ ligands. Since no work has been done to find a good substitute for Cp⁻ other than the study above, we looked at several model ligands including Cl⁻ to either confirm Cl⁻ as a good model for Cp⁻, or find a ligand that models Cp⁻ better than Cl⁻.

TABLE 1. Eigenvalues of σ and π Molecular Orbitals and Ionization Enthalpy (IE) of Cp⁻, OH⁻, SH⁻, and Cl⁻

Ligand	Energy (au)			IE
	σ orbital	π orbitals	σ-π	(kJmol⁻¹)
Cp⁻	-0.2602	-0.0465	0.2137	172[a]
Cl⁻	-0.1287	-0.1287	0.0	349[b]
OH⁻	-0.1217	0.0072	0.1289	177[a]
SH⁻	-0.2251	-0.0761	0.1490	220[a]

[a]Reference 22
[b]Reference 23

In addition to Cl⁻, we examined OH⁻ and SH⁻. The M-O-H and M-S-H angles were set at 180° in order to orient the ligand π orbitals to model the orientation of the π orbitals in Cp⁻. We considered OH⁻ because its IE is close to the IE of Cp⁻ and because the σ and π molecular orbitals for OH⁻, as in Cp⁻, are not degenerate. We also considered SH⁻ because the energy difference between the σ and π orbitals for SH- is larger than the energy difference for OH⁻, however, the IE of SH⁻ is higher than the IE of Cp⁻. By comparison, the Cl⁻ ligand has a much higher IE than SH⁻ and it has σ and π orbitals which are degenerate. The calculated energy differences between the σ and π orbitals and IE values for Cp⁻, OH,⁻ SH⁻, and Cl⁻ are shown in Table 1. The Cp⁻ ligand has the largest splitting of 0.213 au. As expected, the orbital splitting for OH⁻ is slightly smaller than the splitting of SH⁻.

To further test the model Cp⁻ ligands, we optimized the geometries of complexes with Cp⁻ ligands and compared those geometries with the optimized geometries of complexes for which the Cp⁻ ligand was replaced with Cl⁻ ,OH⁻, and SH⁻. The results of the optimized geometries of $Cr(L)(CO)_2N$, $Mn(L)(CO)_3$, and $Ti(L)_2H_2$ (L = Cp, Cl, OH, SH) are listed in Table 2 and show that the optimized geometries of the OH chromium and manganese complexes are most like the optimized geometries of the analogous Cp complexes; however, the optimized geometry of the SH titanium complex is closest to the analogous Cp complex.

TABLE 2. Selected Optimized Bond Lengths and Bond Angles for $Cr(L)(CO)_2N$, $Mn(L)(CO)_3$, and $Ti(L)_2H_2$ (L = Cp, OH, SH, Cl).

	L			
	Cp	Cl	OH	SH
$Cr(L)(CO)_2N$				
Cr-N	1.54	1.48	1.51	1.48
Cr-CO	1.91	2.06	2.01	2.06
OC-Cr-CO	93.3	95.5	97.0	95.6
L-Cr-$(CO)_2$	133.7	119.0	127.0	119.2
N-Cr-$(CO)_2$	95.6	106.9	101.1	104.0
$Mn(L)(CO)_3$				
Mn-CO	1.88	1.95	1.89	1.94
OC-Cr-CO	96.8	96.0	95.8	97.0
Mn-C-O	174.3	176.1	174.8	175.7
L-Mn-CO	120.3	120.9	121.0	120.1
$Ti(L)_2H_2$				
Ti-H	1.67	1.65	1.69	1.65
L-Ti-L	141.3	130.0	125.0	134.8
H-Ti-H	78.0	84.7	101.0	83.8

Because we have optimized the geometries of only three model complexes and OH is not the best model for all three complexes it is premature to suggest that OH is a good model for Cp in all cases. However, the optimized geometry of the OH chromium complex is closest to the optimized geometry of the Cp chromium complex, the IE of OH is closest to Cp, and the σ and π orbitals of OH are not degenerate. Thus, for this study, we decided to use OH as a model ligand for the Cp ligand in $[Cr(Cp)(CO)_2]_2$. Although the σ and π orbital splitting in SH⁻ is closest to the splitting in Cp⁻, the difference between the splitting in the OH⁻ and SH⁻ ligands is only one fourth of the difference in splitting between the OH⁻ and Cp⁻ ligands. Examination of the deformation density also suggested that OH⁻ was the best model in these systems [14].

All calculations were done with GAMESS (Generalized Atomic and Molecular Electronic Structure System, SERC Daresbury Laboratory) on an IBM 3090/600 and an

FPS-264 at the Cornell National Supercomputer Facility, an IBM 3090/200 at Texas A&M University, and on an FPS-264 at the Department of Chemistry of Texas A&M University. The contour maps were drawn with Precision Visuals DI3000 and the density maps were calculated with MOPLOT on a VAX 11/780 and MicroVAXII at the Department of Chemistry of Texas A&M University.

3. Results and Discussion

3.1. VAN DER WAALS ENERGIES OF Cp AND OH DIMERS

Point by point van der Waals energies were calculated as the angles were varied in both the Cp and OH dimers. The carbonyls were treated as terminal ligands which allows the metal to distal carbonyl van der Waals interaction to be included in the total energy. The van der Waals energy curves in Figure 1 show the Cp dimer to be sterically crowded and the OH dimer to be relatively uncrowded. When the Cp-Cr-Cr-Cp moiety is fixed at the X-ray diffraction geometry for the Cp dimer and the carbonyls rocked by varying the Cr-Cr-$(CO)_2$ angle, the van der Waals minimum Cr-Cr-$(CO)_2$ angle is 12° larger than experiment. The Cp-Cr-$(CO)_2$ angle is the angle of the Cp centroid, metal atom, and $(CO)_2$ plane. The $(CO)_2$ plane is the plane containing the metal atom and the two carbonyl ligands. Likewise for the OH dimer, the Cr-Cr-$(CO)_2$ angle is 26° larger than experiment. When the Cp ligand is rocked and the $Cr_2(CO)_4$ moiety is fixed at the experimental geometry, the van der Waals minimum Cr-Cr-Cp angle is 6° smaller than experiment. Similarly, the van der Waals minimum Cr-Cr-OH angle for the OH dimer is 22° smaller than experiment. It appears that the OH dimer prefers a geometry that orients the CO and OH ligands similar to the orientation of the hydrogen atoms in staggered ethane. However, the bulky Cp ligands in the Cp dimer prevent the individual ligands from rearranging into an "ethane-like" geometry.

We also calculated point by point van der Waals energies of the Cp and OH dimers as both Cr(L)$(CO)_2$ fragments (L=OH,Cp) were tilted concertedly clockwise and counterclockwise. The Cp-Cr-$(CO)_2$ and HO-Cr-$(CO)_2$ angles were fixed and the energies calculated by two methods. Firstly, the carbonyl ligands were bonded as "terminal" carbonyls such that the metal to distal-carbonyl van der Waals interaction was included, and secondly, the carbonyls were bonded as "semibridging" carbonyls such that metal-carbonyl van der Waals interactions were not included. The tilt was measured by the Cr-Cr-$(CO)_2$ angle. Unlike the van der Waals calculations above for the individual ligands, the van der Waals energy curves shown in Figure 1 are relatively flat near the minimum. This is because as the one fragment tilts, its carbonyls move away from the approaching Cp ligand of the other fragment, thus the inter-ligand distances between fragments remain virtually unchanged.

The van der Waals minimum Cr-Cr-$(CO)_2$ angles for the Cp and OH dimers with "terminal" carbonyls are 31° and 46° larger, respectively, than the experimental structure of the Cp dimer. This suggests that an attractive interaction is tilting the fragments into the sterically unfavorable experimental geometry. If there is bonding between the metal and the distal carbonyl carbon one would expect the van der Waals minimum to move closer to the experimental geometry when the metal-distal carbonyl interaction is dominated by an attractive interaction (bonding) rather than a repulsive interaction (van der Waals). Therefore, as mentioned above for the second method, we removed the metal

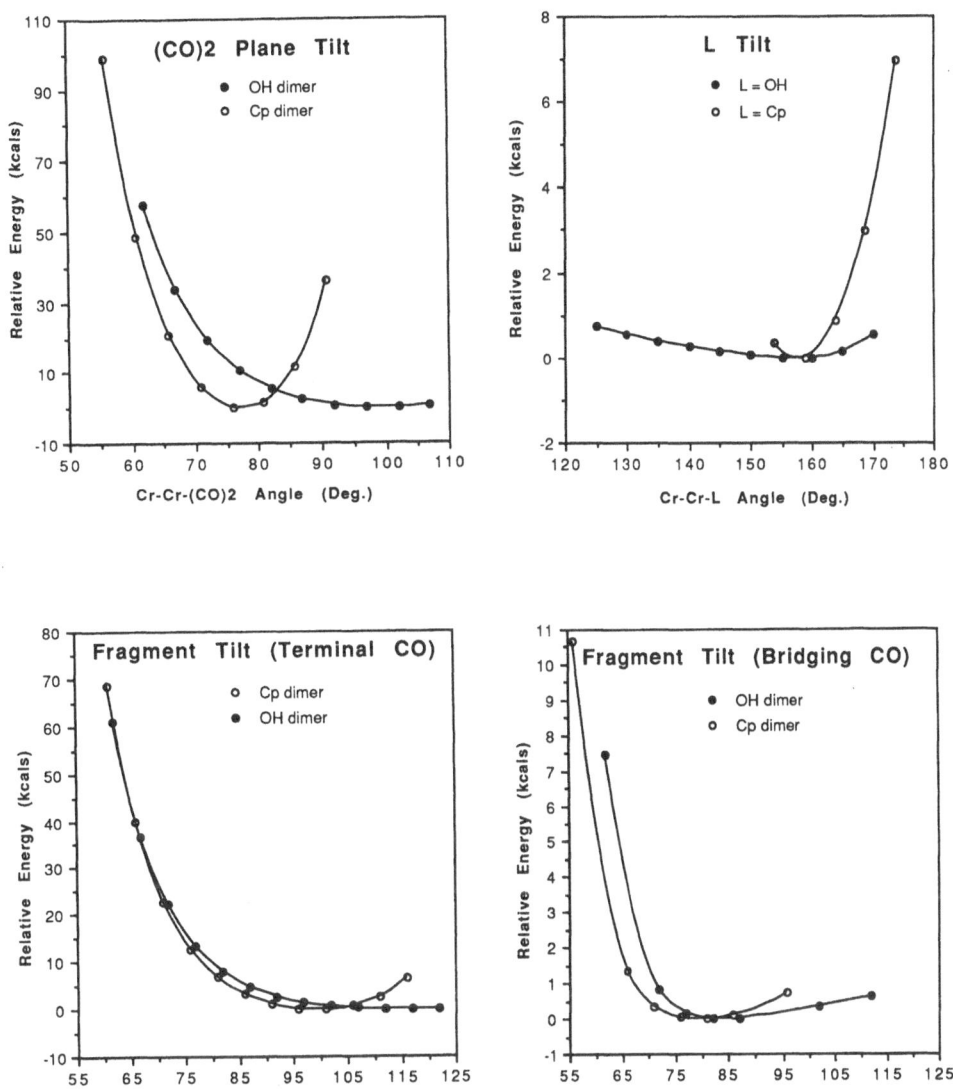

Figure 1. Van der Waals energy curve of ligand bending and fragment tilting in [Cr(Cp)(CO)₂]₂ and [Cr(OH)(CO)₂]₂.

to distal carbonyl van der Waals interaction to simulate a dominant bonding interaction. The resulting van der Waals minimum Cr-Cr-(CO)₂ angles for the Cp and OH dimers decreased to 13° and 18° larger, respectively, than experiment. This large shift of the van der Waals minima toward the experimental Cr-Cr-(CO)₂ angle supports the suggestion of an attractive interaction (or lack of a repulsive one) between the distal carbonyls and the metal.

3.2. STABLE MONOMERS

The chromium dimer can be thought of as two $Cr(Cp)(CO)_2$ fragments held together through metal-metal and metal-semibridging carbonyl bonds. By replacing one of the fragments with ligands which have no propensity to bond with the carbonyls, we can get some idea of the relative importance of the metal-carbonyl and metal-metal bonds on the structure. We examined x-ray crystal structures of eight complexes which contained the fragment moiety bonded to a ligand. Table 3 lists the complexes and their experimental $Cp-Cr-(CO)_2$ angles. The Cp ring in the complex with L=CO is $C_5H_4S(CH_3)_2$ and in the other eight complexes it is C_5H_5. The complexes with the smallest $Cp-Cr-(CO)_2$ angles, 129°(1), are the chromium dimer and the dimer-like complexes which have $Cr-S_2-Cr$ and $Cr-Se_2-Cr$ linkages bonding the fragments together. The complex with the largest $Cp-Cr-(CO)_2$ angle, 143.5°, is the CO complex with the sulfur substituted Cp ring. The $Cp-Cr-(CO)_2$ angles for the five other complexes range from 132 to 138° and the average $Cp-Cr-(CO)_2$ angle for all the complexes is 134.6°.

We then optimized the geometries of the model complexes $CpCr(CO)_2N$ and $CpCr(CO)_2(CH)$ (see Table 2). We picked these complexes because the fragment moiety $(Cr(Cp)(CO)_2)$ is bonded to the added ligand by a triple bond. Thus, these represent possible electronic analogues of the metal-metal dimer system which, according to the $18e^-$ rule, is predicted to have a $Cr≡Cr$ triple bond.

When a nitrogen ligand is bonded to the fragment moiety and the geometry is optimized, the $Cp-Cr-(CO)_2$ angle found to be 133.8°, 4.8° larger than the $Cp-Cr-(CO)_2$ angle of the dimer. The $N-Cr-(CO)_2$ angle of 95.6° is 28.6° larger and the $Cp-Cr-N$ angle of 130.6° is 33.4° smaller than the analogous angles in the dimer. The optimized geometry of the $CpCr(CO)_2N$ complex has a "piano stool" geometry with near equivalent $Cp-Cr-CO$ and $Cp-Cr-N$ bond angles. The $Cp-Cr-(CO)_2$ angle of the optimized

TABLE 3. X-Ray Crystal Structure $Cp-Cr-(CO)_2$ Angles for $Cr(Cp)(CO)_2L$.

	L	$Cp-Cr-(CO)_2$ Angle (Deg.)
1.	CO	143.5[a]
2.	PPh₃	138.0[b]
3.	NO	136.8[c]
4.	NS	136.3[d]
5.	$SCr(Cp)(CO)_2$	135.4[e]
6.	$SeCr(Cp)(CO)_2$	133.6[f]
7.	$S_2Cr(Cp)(CO)_2$	129.8[g]
8.	$Se_2Cr(Cp)(CO)_2$	129.4[f]
9.	$Cr(Cp)(CO)_2$	129.0[h]

[a]Reference 15	[b]Reference 16	[c]Reference 17
[d]Reference 18	[e]Reference 19	[f]Reference 20
[g]Reference 21	[h]Reference 3	

CpCr(CO)$_2$(CH) complex, which has a Cr≡C triple bond, is almost identical to the optimized geometry of the nitrogen complex. In general, the complexes, including those with experimental crystal structures, have Cp-Cr-(CO)$_2$ angles between 129° and 144°. This result would suggest that the fragment geometry itself is insensitive to changes in the added ligand.

3.3. OH DIMER [CR(OH)(CO)$_2$]$_2$

We optimized the geometry of the model OH dimer [Cr(OH)(CO)$_2$]$_2$ at several fixed Cr-Cr bond distances. A potential energy curve of the OH dimer at the various Cr-Cr bond distances (see Figure 2) has a minimum at 2.20 Å and is 119 kcals/mol above the energy of the two optimized fragments. The energy of the optimized bare fragment with OH substituted for Cp was used to determine the energy of the dissociated fragment.

The optimized geometry at 2.20 Å is in good agreement with the experimental geometry of the Cp dimer considering the geometric parameters of the fragment moiety. The OC-Cr-CO and HO-Cr-(CO)$_2$ angles are 6.2° and 1.1° larger, respectively, and the Cr-CO bond distance is 0.01 Å longer than the experimental geometry of the Cp dimer.

OH Dimer - Fragment Dissociation

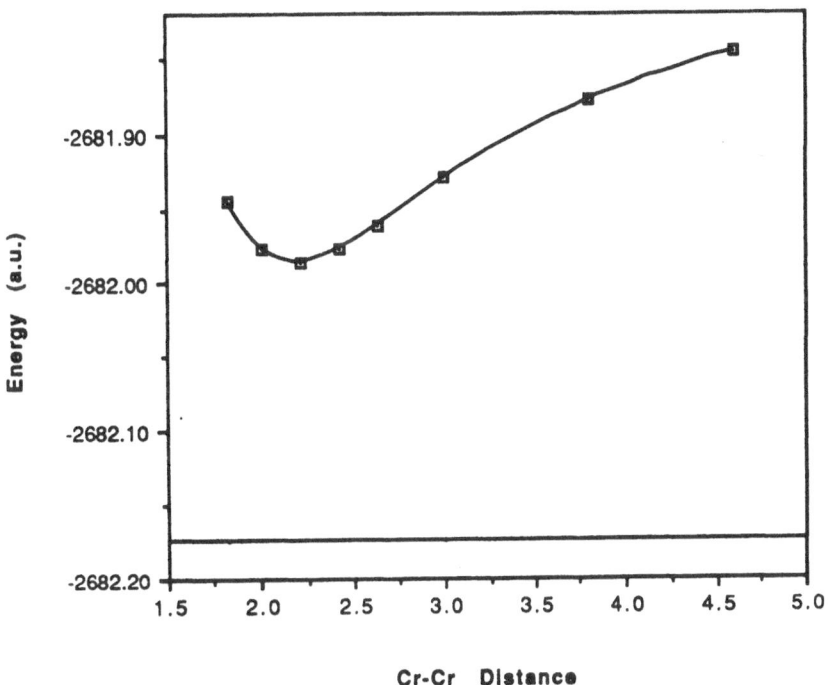

Figure 2. Potential energy curve for the dissociation of [Cr(OH)(CO)$_2$]$_2$ into Cr(OH)(CO)$_2$ fragments. The Cr-Cr distance is in Å.

164

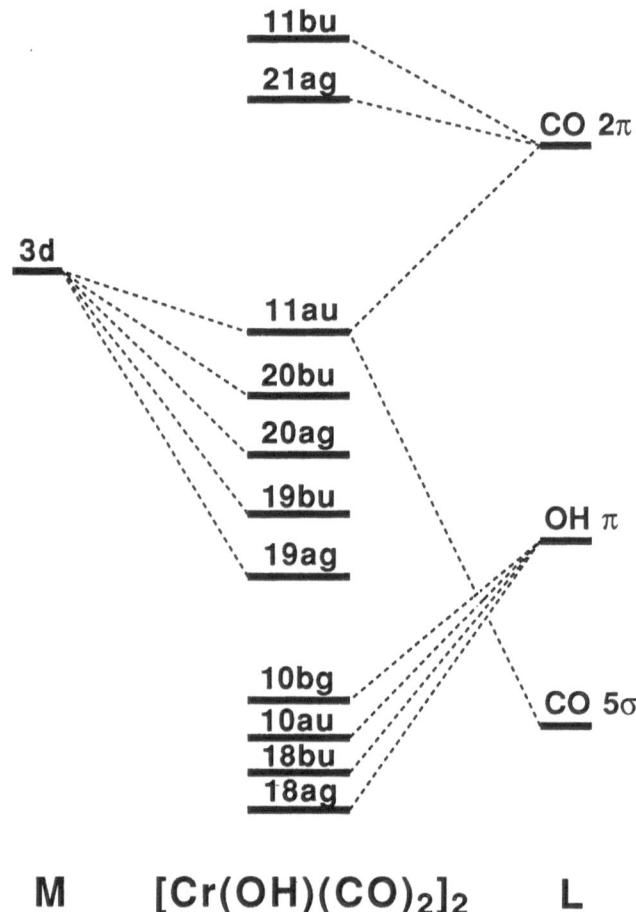

Figure 3. Molecular orbital diagram of [Cr(OH)(CO)₂]₂.

Although the Cr-Cr-(CO)₂ angle, which describes the degree to which the carbonyls of one fragment moiety bend over toward the metal of the other fragment, is 10.6° larger than experiment, it is still much smaller than would be expected for carbonyls which are not linear semibridging. The fragment with a nitrogen ligand, which has no carbonyl-nitrogen interactions has an N-Cr-(CO)₂ angle 28.6° larger than the analogous angle in the dimer.

A molecular orbital diagram for the OH dimer at the optimized geometry is shown in Figure 3. The 11a$_u$ is the highest occupied molecular orbital (HOMO) and has metal-metal π bonding and metal-carbonyl-metal bonding interactions. The 19a$_g$ and 20a$_g$ orbitals are metal-metal δ and σ bonding, respectively, and the 19b$_u$ and 20b$_u$ orbitals are metal-metal δ^* mixed with π character and metal-metal π mixed with σ^* character, respectively. The a$_g$ and b$_u$ orbitals also have metal-carbonyl bonding character between the metal of one fragment and the carbonyl of the other fragment. Because of the large amount of bonding-antibonding mixing in the a_g and b_u orbitals, the net bonding of the metal-metal bond is difficult to determine. Contour plots of the metal-based orbitals are shown in Figure 4.

We considered the effects of tilting the fragments of the dimer by calculating SCF energies of the OH dimer point by point at various fragment tilt angles (similar to van der

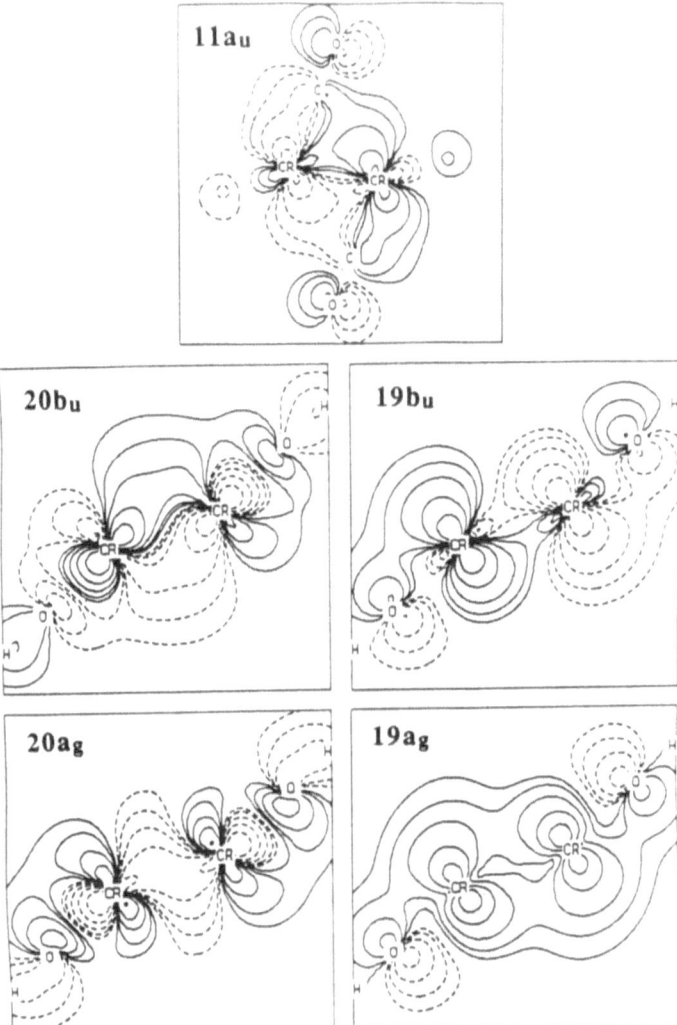

Figure 4. Contour plots of metal-based orbitals of the OH dimer with a Cr-Cr-(CO)$_2$ angle of 71.9°. Plot 11a$_u$ is in the plane of the carbonyls and the remaining four plots are in the symmetry plane. Contours are geometric beginning at 0.02 au and increase in value by doubling the value of the previous contour.

Waals calculations above). The results are shown in Figure 5 (SCF curve). The first and lower energy minimum has a Cr-Cr-(CO)$_2$ angle at the optimized geometry as expected, however, there is a second minimum 4.6 kcals/mol higher in energy with a Cr-Cr-(CO)$_2$ angle 36° larger than the optimized value. The geometry at the second minimum has a ligand orientation analogous to the orientation of the hydrogens in ethane. When the ligands at this "open" geometry were allowed to relax by optimizing the angles and bond lengths of the ligands and keeping the Cr-Cr distance fixed, the dimer did not return to the "semibridging" structure but remained an "open" structure with a total energy now lower than the first minimum.

We then did perfect pairing GVB calculations of the OH dimer at three different geometries: the "semibridging" geometry (Cr-Cr-(CO)$_2$ = 71.9°), the "mid-point" geometry (Cr-Cr-(CO)$_2$ = 91.9°), and the "open" geometry (Cr-Cr-(CO)$_2$ = 111.9°). The three GVB pairs consisted of bonding-antibonding orbital pairs, one of σ symmetry (20a$_g$, 21b$_u$) and two π symmetry (20b$_u$,21a$_g$ and 11a$_u$,11b$_g$). The results in Figure 5 (GVB(3) curve) now show a single minimum at the "mid-point" geometry. Although the energy has decreased significantly the complex is still unbound with respect to the optimized fragments.

Contour plots of the GVB orbitals show the strongly occupied metal-metal π and σ bonding orbitals and the corresponding weakly occupied antibonding orbitals (see Figure 6). The extensive mixing seen in the SCF wavefunction is not evident in the correlated wavefunction. The two metal-based orbitals not included in the GVB space, which were a mixture of δ, π and σ character at the SCF level, developed almost pure $\delta*$ and δ character. Although the GVB wavefunction does well in correlating the metal-metal triple bond, it does not correctly predict the Cr-Cr-(CO)$_2$ angle found in the experimental Cp dimer which is about 30° smaller than our theoretical CP dimer value. However, it is important to remember that we are studying the model OH dimer. The van der Waals calculations show that the steric effect of the Cp ligand would tilt the fragments toward the experimental geometry. In addition, differences in the electronic effects of the Cp and OH ligands, particularly the stronger "π" bonding of Cp, might contribute to a difference in the structures.

We thought the Cr-Cr-(CO)$_2$ angle might decrease if the number of GVB pairs increased. We added the two occupied metal-metal δ and $\delta*$ orbitals and two virtual orbitals which had significant metal distal-carbonyl bonding character to the GVB(PP) space. The results for five GVB pairs (GVB(5)) are shown in Figure 5. The energy decreases, but the complex is still not bound with respect to the optimized fragments. The character of the three GVB pair orbitals and the δ and $\delta*$ orbitals does not change from their analogues in the prior GVB(3) calculation . However, the remaining two weakly occupied counterparts of the δ and $\delta*$ orbitals, which were chosen because they had significant metal-carbonyl bonding character, change significantly to orbitals with no metal-carbonyl bonding character. All the orbitals have been used to provide correlation for the metal pairs, as these are more important for the energy.

Increasing the amount of electron correlation with a CASSCF of six electrons in six orbitals (CAS(6/6)) lowered the energy so that the dimer is now bound with respect to the optimized fragments. However, the fragment tilt angle at the energy minimum is still far from the experimental value for the Cp dimer. The six orbitals in the active space are the same as the GVB pair orbitals in the GVB(3) wavefunction. The orbitals discussed below have been transformed into natural orbitals.

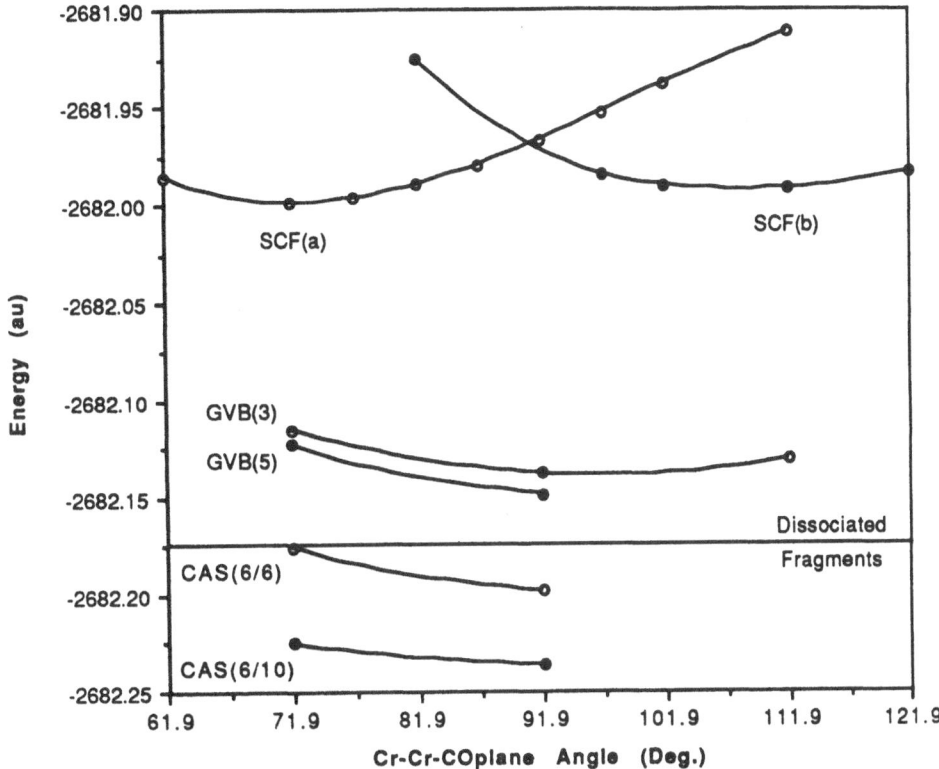

Figure 5. Potential curve of SCF, GVB, and CASSCF wavefunctions as fragments of $[Cr(OH)(CO)_2]_2$ are tilted.

The leading configuration is the reference configuration with a coefficient of 0.84. The configuration with the next highest coefficient, a value of 0.23, is the double excitation of two electrons from the π-bonding orbital (a_u) to the π-antibonding orbital (b_g). Other configurations with coefficients above 0.10 include two configurations that are double excitations from the other two metal-metal bonding orbitals (a_g and b_u) into their antibonding counterparts (b_u and a_g) and two configurations that are each two single excitations from two metal-metal bonding orbitals into metal-metal antibonding orbitals.

At the "semibridging" geometry, the leading coefficient is slightly larger than it is at the "mid-point" geometry. Often, as the approximate wavefunction approaches the true wavefunction the value of the leading coefficient will increase. Thus, the "semibridging" geometry wavefunction may be closer to the true wavefunction than the "mid-point" geometry wavefunction.

The results of the CASSCF calculation with six electrons in ten orbitals (CAS(6/10)) is similar to the CAS(6/6) calculation. The energy decreased substantially and the energy minimum is still at the "mid-point" geometry; however, the decrease in energy at the "semibridging" geometry is greater than the decrease in energy at the "mid-point" geometry. This suggests that the minimum energy geometry is moving from the "mid-point" geometry towards the "semibridging" geometry.

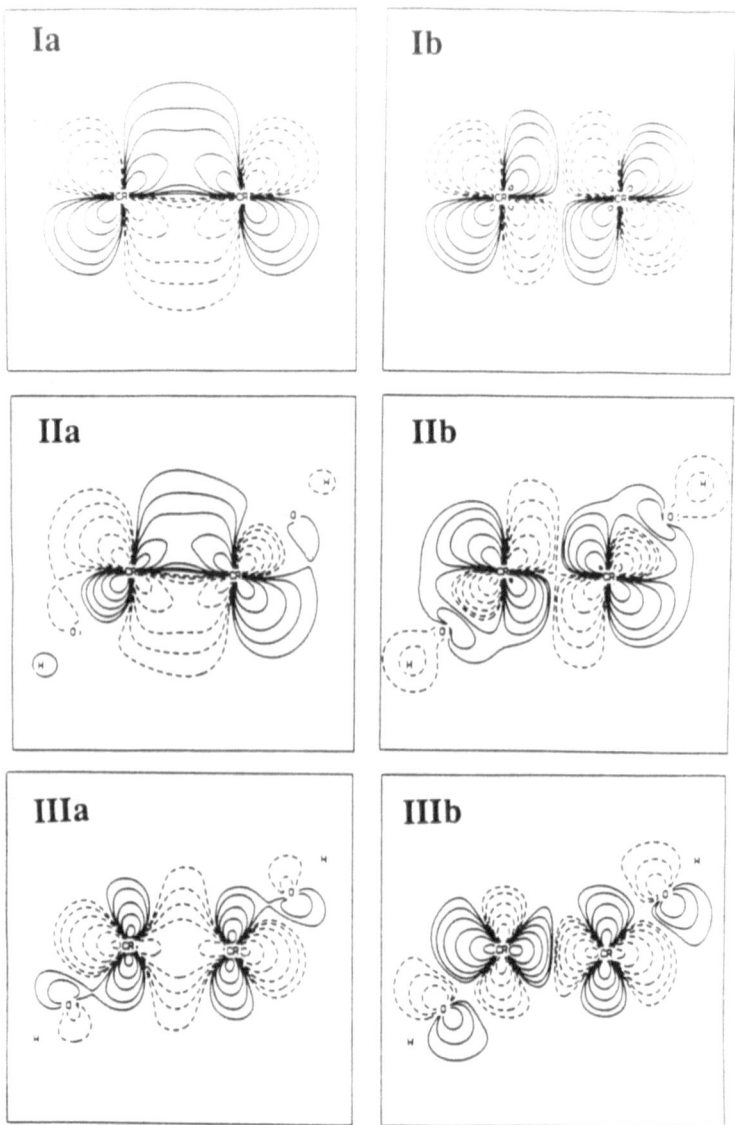

Figure 6. Contour plots of the OH dimer GVB(3) pairs. Pair I is in the XZ plane and pairs II and III are in the XY (symmetry) plane. Contours are geometric beginning at 0.02 au and increase in value by doubling the value of the previous contour.

The coefficient of the reference configuration decreased slightly to 0.83. The other configurations with coefficients above 0.10 showed a similar decrease. Although the active space was increased to include four additional orbitals, none of the larger coefficient configurations included orbitals from the additional space. The only configuration above 0.04 that had orbitals from the additional space had a coefficient of 0.06 and included the two orbitals which are antibonding between the metals and carbonyl ligands. Although no configurations with large coefficients included the metal-carbonyl bonding orbitals, there are several such configurations which are above 0.01 but less than 0.04.

In general, as we have increased the amount of electron correlation from the GVB(3) wavefunction to the CAS(6/10) wavefunction, the energy at the "semibridging" geometry has decreased more than the energy at the "mid-point" geometry. One might hypothesize that an even larger CASSCF or a CI wavefunction might include enough configurations with metal distal-carbonyl bonding to tilt the fragment to the experimental geometry.

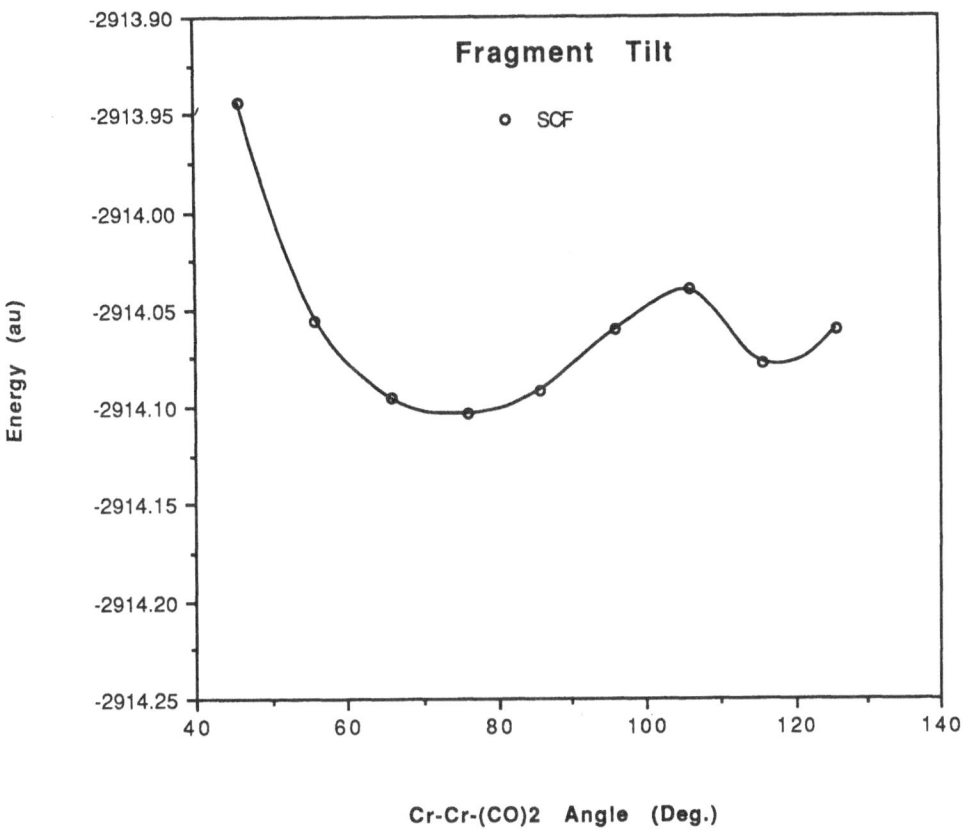

Figure 7. Potential energy curve of $[Cr(Cp)(CO)_2]_2$ as fragments tilt.

3.4. CP DIMER [Cr(C₅H₅)(CO)₂]₂

Because the calculations of the chromium dimer with OH substituted for Cp ligands does not correctly predict the experimental Cp dimer geometry, even with correlation added to the wavefunction, we calculated the potential curve for the fragment tilting of the Cp dimer at the SCF level. The results (SCF curve in Figure 7) show a double minimum similar to the results for the OH dimer with the first and second minima at Cr-Cr-(CO)₂ angles of 73° and 117°, respectively. At the first minimum the Cr-Cr-(CO)₂ angle is only 7° larger than experiment and the dimer is unbound with respect to the optimized fragments by 137 kcals/mol. The second minimum is 16.4 kcals higher in energy than the first minimum which is over three times the difference found in the OH dimer. Preliminary GVB calculations appear to result in a single minimum.

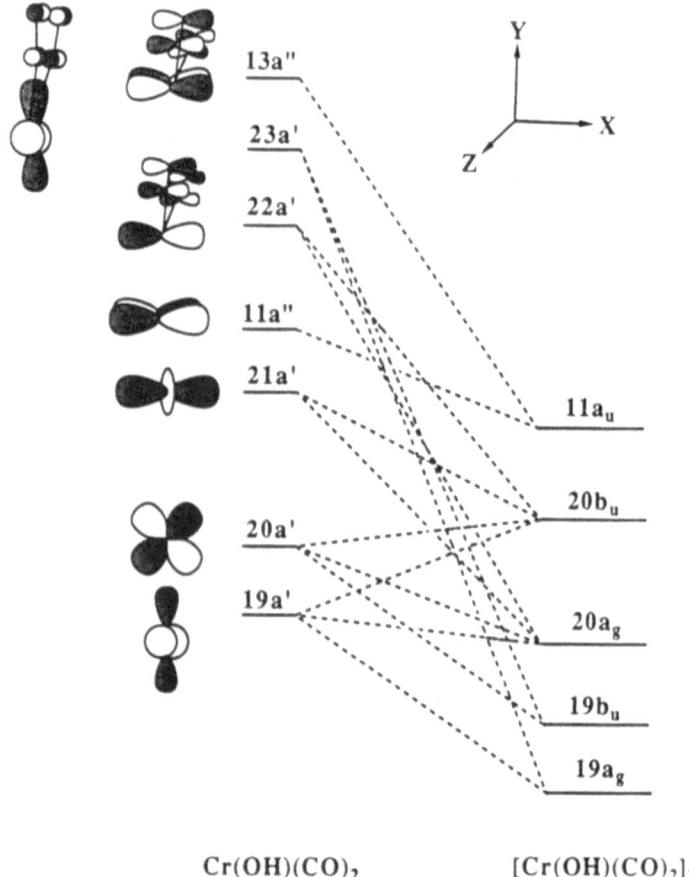

$$Cr(OH)(CO)_2 \qquad\qquad [Cr(OH)(CO)_2]_2$$

Figure 8. Molecular orbital diagram of [Cr(OH)(CO)₂]₂ in terms of the fragment orbitals of Cr(OH)(CO)₂.·

3.5. BRIDGING CARBONYL INTERACTIONS

To help understand the interactions important for fragment-fragment bonding in the dimer at the optimized geometry, the basis set was transformed from the set of atomic orbitals to a set of fragment molecular orbitals. The molecular orbitals of the fragment were calculated with a wavefunction of three singly occupied orbitals and at the geometry of the fragment in the optimized dimer. A molecular orbital diagram of the transformed dimer is shown in Figure 8. The three unoccupied fragment orbitals, 13a", 23a' and 22a' are carbonyl 2π with small admixture of metal. The three singly-occupied fragment orbitals, 11a", 21a', and 20a' are primarily metal d_{xz}, d_{x^2}, d_{xy}, respectively, and the doubly-occupied 19a' is metal $d_{y^2-z^2}$ with some carbonyl bonding.

Previous investigations [3,6-8] of the bonding between the metal orbitals of one fragment and the carbonyl p orbitals of the other fragment have agreed that there are interactions between the metal and carbonyls, however, they do not agree about whether the distal-metal orbitals are donating electron density into carbonyl orbitals or the carbonyl orbitals are donating electron density into distal-metal orbitals. We examined Mulliken populations of the fragment orbitals in both the bare fragment and in the bonded dimer to see how the populations changed. Although Mulliken populations are not reliable for absolute population values, they are dependable for comparison between like orbitals of similar systems. The populations listed in Table 4 show that the 11a", 19a', and 21a' orbitals lose electron density and the 13a" and 22a' orbitals gain electron density. This suggests that the carbonyl 2p orbitals of one fragment are accepting electron density from the metal d-orbitals of the other fragment.

4. Conclusions

When the planar open-shell $Cr(Cp)(CO)_2$ fragment is bonded to a ligand such as N or CH, the geometry of this fragment is very similar to the geometry of the same fragment in the dimer. The Cp-Cr-$(CO)_2$ angles for other known complexes, $Cr(Cp)(CO)_2NO$, $Cr(Cp)(CO)_2PPh_3$, and $Cr(Cp)(CO)NS$ are only 7-9° larger than the Cp-Cr-$(CO)_2$ angle in the dimer. Thus, the geometry of the common fragment is relatively insensitive to the ligand bonded to it.

TABLE 4. Mulliken Populations of $Cr(OH)(CO)_2$ Fragment Molecular Orbitals.

Orbital	Isolated Fragment	Dimer	Difference
13a"	0.000	0.170	+0.170
23a'	0.000	0.073	+0.073
22a'	0.000	0.064	+0.064
11a"	1.000	0.832	-0.168
21a'	1.000	0.431	-0.569
20a'	1.000	1.669	+0.669
19a'	2.000	1.775	-0.225

172

Van der Waals calculations on [Cr(Cp)(CO)$_2$]$_2$ without any bridging carbonyl bonding predict a geometry with the fragments tilted back to an "ethane-like" geometry. *Ab initio* SCF calculations also predict an "ethane-like" geometry; however, a second more stable geometry, which is close to the experimental one, is also predicted. The origin of these two stable geometries may be artifacts of the Hartree-Fock approximation. When electron correlation is included in the wavefunction, the two minima disappear and a single minimum appears at a geometry between the experimental and "ethane-like" geometry.

The metal-metal orbitals of the Hartree-Fock wavefunction contain substantial mixing of σ and π* and of π and σ* which make analysis of the bonding difficult; however, the correlated wavefunction predicts a metal-metal triple bond made up of a σ and two π bonds. Analysis of the transformed Hartree-Fock wavefunction shows that the semibridging carbonyl ligands are acting as π acceptors from filled d-orbitals of the distal metal.

In Part 4 of this series we showed that linear semibridging carbonyls are characteristic of systems where the metal-to-metal bonding is strong and steric interference prevents the carbonyls from being terminal [24]. This reasoning can be applied here to explain why Cp-Cr-Cr-Cp is not linear while Cp-Mo-Mo-Cp is linear, and why the Cp-M-(CO)$_2$ angle is smaller when M=Mo. Because Mo is larger than Cr, Mo complexes generally display a structure that is less dominated by steric (repulsive) forces and more dominated by bonding (attractive) forces. When the metal is made smaller and the M-M bond length shrinks, there is an increase in the steric repulsion between the carbonyls on one metal and the Cp on the other metal. Thus, the Cr-Cr-(CO)$_2$ angle increases and the Cp-Cr-Cr angle decreases, while the Cp-Cr-(CO)$_2$ angle remains fairly constant. Therefore, the whole CpM(CO)$_2$ fragment rotates, a result which is consistent with both steric factors playing a major role in the carbonyl to distal -metal distance and strong metal-to-metal bonding.

5. Acknowledgments

The authors thank the National Science Foundation (Grant No. CHE 86-19420 and CHE 91-13634) and the Robert A. Welch Foundation (Grant No. A-648) for financial support. This research was conducted in part with the use of the Cornell National Supercomputer Facility, a resource for the Center for Theory and Simulation in Science and Engineering at Cornell University, which is funded in part by the National Science Foundation, New York State, and the IBM Corp.

REFERENCES

1. Klingler, R. J., Butler, W. M. and Curtis, M. D. (1978), J. Am. Chem. Soc., **100**, 5034.
2. Potenza, J., Giordano, P., Mastropaolo, D., Efraty, A. and King, R. B. (1972), J. Chem. Soc. Chem. Comm., 1333.
3. Curtis, M. D. and Butler, W. M. (1978), J. Organomet. Chem., **155**, 131.
4. Huang, J. and Dahl, L. F. (1983), J. Organomet. Chem., **243**, 57.
5. Crabtree, R. H. and Lavin, M. (1986), Inorg. Chem., **25**, 805.
6. Cotton, F. A. (1974), Prog. Inorg. Chem., **21**, 1.
7. Jemmis, E. D. and Pinhas, A. R.and Hoffmann, R. (1980) J. Am. Chem. Soc., **102**, 2576.

8. Morris-Sherwood, B. J., Powell, C. B. and Hall, M. B. (1984), J. Am. Chem. Soc., **106**, 5079.
9. Colton, R. and McCormick, M. (1980), J. Coord. Chem. Rev., **31**, 1.
10. CHEMX is an interactive molecular-modeling program marketed by Chemical Design Limited.
11. Williamson, R. L. and Hall, M. B. (1988), Intern. J. Quantum Chem. Symp., **31**, 503.
12. Kok, R. A. and Hall, M. B. (1985), J. Am. Chem. Soc., **107**, 2599.
13. Steigerwald, M. L. and Goddard, W. A., III (1985), J. Am. Chem. Soc., **107**, 5027.
14. Williamson, R. L. (1989), Ph.D. dissertation, Texas A&M University.
15. Andrianov, V. G., Struchkov, Y. T., Setkina, V. N., Zhakaeva, A. Z. and Zdanovitch, V. I. (1977), J. Organomet. Chem., **140**, 169.
16. Cooley, N. A., Watson, K. A., Fortier, S. and Baird, M. (1986), C. Organometallics, **5**, 2563.
17. Atwood, J. L., Shakir, R., Malito, J. T., Jerberhold, M., Kremnitz, W., Bernhagen, W. P. E. and Alt, H. B. (1979), J. Organomet. Chem., **165**, 65.
18. Greenhough, T. J., Kolthammer, B. W .S., Legzdins, P. and Trotter, J. (1979), Inorg. Chem., **18**, 3548.
19 Greenhough, T. J., Kolthammer, B. W. S., Legzdins, P. and Trotter, (1979), J. Inorg. Chem., **18**, 3543.
20. Goh, L. Y., Wei, C. and Sinn, E. J. (1985), J. Chem. Soc. Chem. Commun., 462.
21. Goh, L. Y., Hambley, T. W.and Robertson, G. B. (1987), Organometallics, **6**, 1051.
22. Janousek, B. K. and Brauman, J. I. (1979) in Bowers, M.T. (ed.), Gas Phase Ion Chemistry, Academic Press, New York, Vol. 2, Chapter 10.
23. Hotop, H. and Lineberger, W. C. (1975), J. Phys. Chem. Ref. Data, **4**, 539.
24. Simpson II, C. Q. and Hall, M. B. (1992), J. Am. Chem. Soc., **114**, 000.

CARBON DIOXIDE ORGANOMETALLIC CHEMISTRY: THEORETICAL DEVELOPMENTS

A. DEDIEU, C. BO and F. INGOLD
Laboratoire de Chimie Quantique
UPR 139 du CNRS, Université Louis Pasteur
4 Rue Blaise Pascal, F-67000 Strasbourg
France

ABSTRACT. The theoretical aspects of carbon dioxide chemistry are reviewed:It is first shown how theory can rationalize and predict on the basis of molecular orbital arguments the various coordination mode of CO_2 to transition metal complexes. The activation of coordinated CO_2 and its subsequent reactivity is investigated.The insertion of CO2 into metal hydride and metal carbon bonds is then analyzed as an example of carbon dioxide reactivity towards organometallic systems. The analogy with other organometallic or organic reactions is also briefly studied.

1. Introduction

The transformation of carbon dioxide has been attracting the attention of chemists for many years, not only in order to recover it from the atmosphere or to understand its transformation in biochemical or biological processes, but also to use it as a source of carbon in chemical processes leading to useful products [1].The transformation of carbon dioxide into carbon containing chemicals is hampered however by its intrinsic thermodynamic stability. Activation of CO_2 is therefore needed. This may be achieved with the help of transition metal complexes used as catalysts [2]. The influence of the metal can be on CO_2 itself, by making the attack of an external reagent on <u>coordinated</u> CO_2 easier. Alternatively the coupling of CO_2 and of a substrate can require the coordination of the substrate instead of CO_2 and use CO_2 as an external reagent . A third possibility that one should not forget is to have a simultaneous coordination of CO_2 and of the substrate followed by the coupling in the coordination sphere.

There have been of course many experimental and theoretical studies devoted to a better understanding of these various modes of activation. In this paper, we shall attempt to review the major advances that have been made in the theoretical realm and to show how theory and experiment can successfully interplay. A recent review by Sakaki has focused mainly on the stereochemistry and on the nature of the bonding in transition metal carbon dioxide complexes [3]. We shall also look here at the various coordination modes of CO_2 but we will put the emphasis more on the reactivity aspects of carbon dioxide organometallic chemistry.

For a better understanding of the arguments used in the molecular orbital theory of carbon dioxide coordination and reactivity, it may be useful to recall the valence orbitals of

175

D. R. Salahub and N. Russo (eds.), Metal-Ligand Interactions: from Atoms, to Clusters, to Surfaces, 175–197.

CO$_2$. CO$_2$ is a heterocumulene and as such has two sets of π orbitals orthogonal to each other. These are shown on the left hand side of the Figure 1. They comprise, in ascending order of energy, (i) the completely bonding combinations of the p lobes on the carbon and oxygen atoms (to form the $\pi_{//}$ and $\pi_\perp$ orbitals; the index // denotes the π orbital lying in

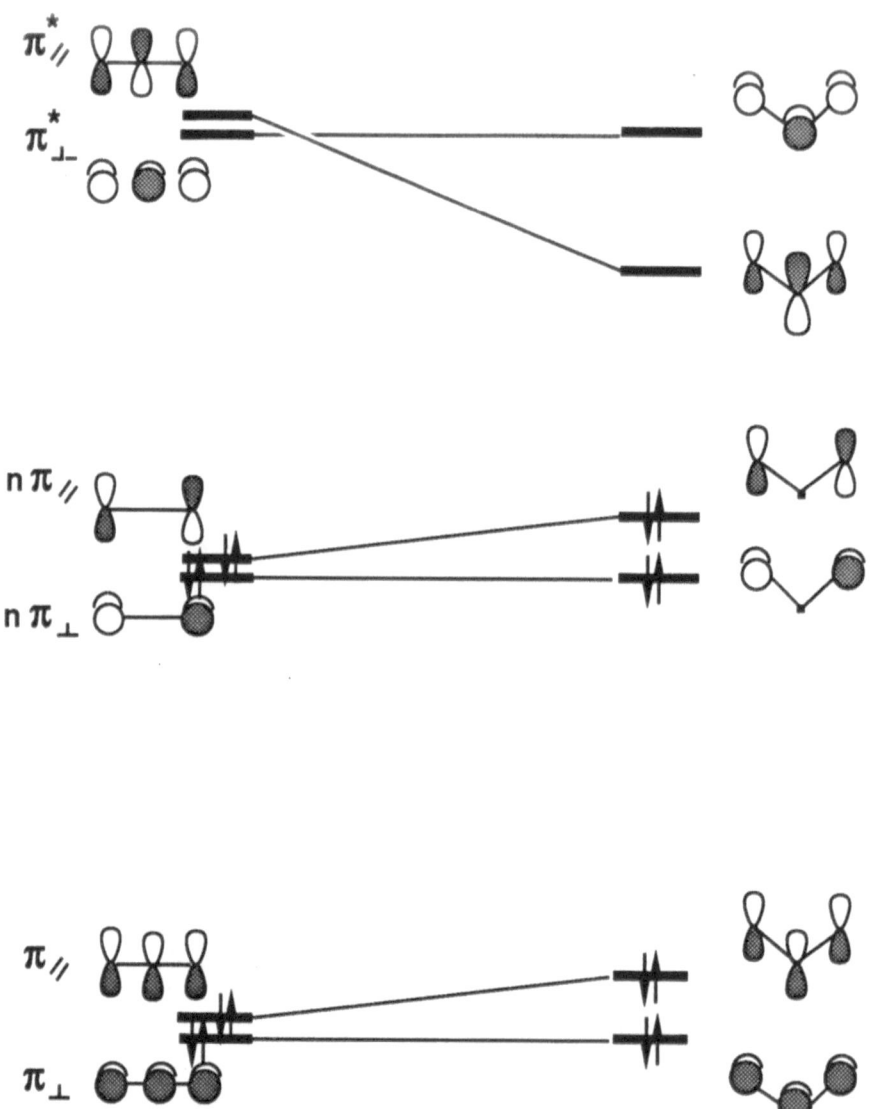

Figure 1: Orbital correlation diagram for the linear to bent deformation of CO$_2$

the CO_2 plane, the index $\perp$ the π orbital perpendicular to this plane); (ii) the non-bonding combination of the p lobes on the oxygen atoms ($n\pi_{//}$ and $n\pi_{\perp}$) and finally (iii) the antibonding combinations ($\pi_{//}^{*}$ and $\pi_{\perp}^{*}$) of the p lobes on oxygen and carbon. Note already the polarization of the π^* orbital which is easily understandable on the basis of the electronegativity difference betweeen carbon and oxygen. The orbital on the left side of the Figure 1 are for a linear ground state geometry. When CO_2 either coordinates or reacts it bends. The degeneracy between the $//$ and $\perp$ orbitals is lifted and one gets the evolution shown schematically on the Figure 1 (for a more quantitative picture we refer the reader to reference [3]. The most prominent changes occur with the $\pi_{//}^{*}$ orbital: it is lowered in energy (due to the release of the repulsive interaction between the carbon and oxygen atoms) and localizes more on the central carbon atom, thus giving to the bent CO_2 a Lewis acid character.

Another feature that is important when dealing with the electronic structure of CO_2 is its charge distribution. The two electronegative oxygen ends make the central carbon atom quite positive . A Mulliken population analysis based on a wave function built from a $(9,5) < 3,2>$ split valence basis set [4], yields atomic charges of $+ 0.75$ and -0.37 e for the carbon and oxygen atoms respectively. One may therefore expect that electrostatic type interactions will play an important role [6] . In that respect it might be worthwhile to note here that a calculation carried out with a polarized basis set [5] increases the charge separation between carbon and oxygen: the respective C and O atomic charges now amount to $+0.91$ and -0.46 respectively Although one should not be too premature when drawing conclusions from Mulliken population analysis results, the role of polarization functions seems far from being negligible and may have been overlooked in all computations that have been carried out so far [8] .

2. The various coordination modes of carbon dioxide to a transition metal complex

The known X-ray crystal structures of carbon dioxide complexes have revealed two preferred coordination modes to a monometallic transition metal complex, either the η^1- C coordination mode **1** [9] or the η^2 side-on coordination mode **2** [10]. But one may think to other possibilities, the most likely ones being η^1-O coordination modes, either **3a** or **3b**. Theoretical studies have been quite successful in assessing the criteria that allow to rationalize or to predict the most stable conformation of the CO_2 ligand for a given transition metal complex. They have also tried to rationalize the current absence -at least up to now- of the η^1 -O coordination mode. The η^1 - C *vs.* η^2 coordination mode dichotomy is under both orbital and electrostatic control [11]. The factors that govern the

1   2

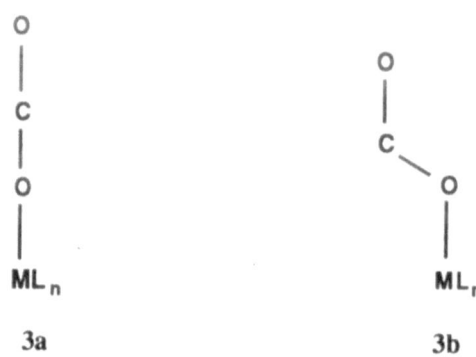

3a

3b

preference for one co_____ally
analyzed [11] and we shall summarize here the most important ones.

In the η^1 - C coordination mode there is a strong two electron stabilizing and charge transfer interaction between a d_{z^2} type orbital (which is doubly occupied) and the π^* orbital of CO_2 (which is empty), see 4. There may be a slight four electron destabilizing interaction between a d_π orbital and $n\pi(CO_2)$ but the overlap is not quite favorable owing to the direction of the lobes on the oxygen atoms, see 5. In the $\eta 2$ side-on coordination mode the orbital interaction pattern is reversed. The d_π orbital interacts in a two electron stabilizing interaction with π^* (CO_2), see 6, but this two electron interaction is somewhat offset by a four electron destabilizing interaction between the d_{z^2} type orbital and $n\pi(CO_2)$, see 7.

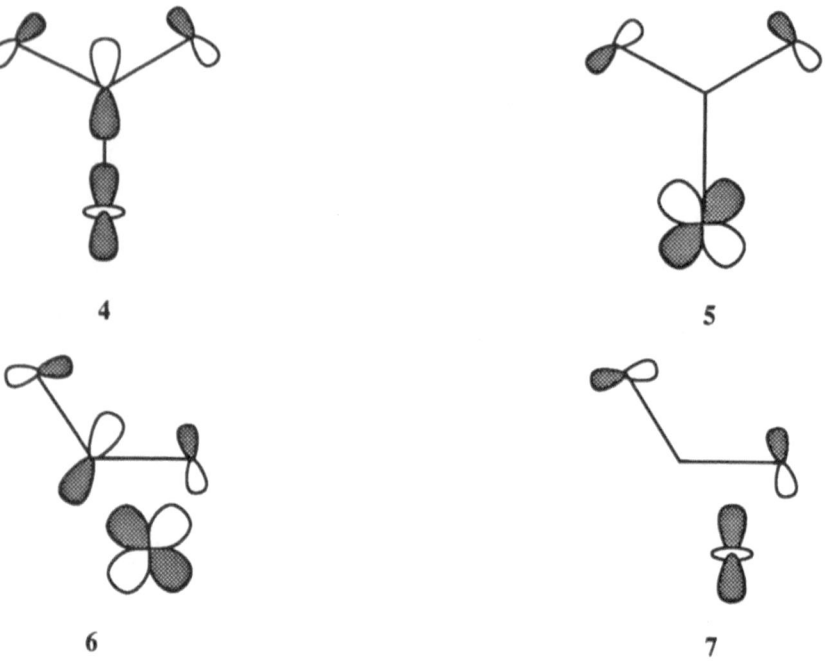

4

5

6

7

In addition to these orbital interactions that have been delineated by other authors and/or other methods [7,12-17] one needs, as said in the introduction, to take into account the electrostatic type interactions. Since the metal atom is usually positively charged, one gets a repulsion in the η^1-coordination mode, see **8**. The η^2 coordination mode is not as unfavorable since it experiences both a repulsive (between C and M) and an attractive (between O and M) electrostatic interaction, see **9**.

Combining therefore both the orbital and electrostatic interactions, one ends up with the following conclusions:

(i) the η^1 - C coordination mode will be most favored when the transition metal ML_n fragment has a doubly occupied d_σ type orbital, relatively high in energy. This high energy will be best achieved if the metal is in a low oxidation state (this also reduces the repulsive electrostatic interaction). Examples following this rule of thumb are Co(salen)(CO_2) - [9a] and RhCl(diars)$_2$$(CO_2)$ [9b]. More generally the η^1 - C coordination mode can be predicted for CO_2 adducts of d^8 square planar or trigonal pyramid transition metal complexes. Adducts of d^8 square pyramid transition metal complexes should also exhibit this coordination mode (note that in this case the 18 electron rule would lead to consider the coordinated CO_2 as a CO_2 $^{2-}$ ligand, therefore emphasizing the charge transfer interaction $d_\sigma \rightarrow \pi^*(CO_2)$).

(ii) the η^2 side on coordination mode is favored by a high lying d_π type orbital. This is indeed the case in bis phosphine nickel complexes [10a]. Fe $(PR_3)_4(CO_2)$ complexes which have a trigonal bipyramid geometry are also expected to have the η^2 side on coordination mode in the equatorial plane [14]. In these systems the hybridization of the d_π orbital, away from the phosphine ligands, increases its overlap (and therefore the stabilizing interaction) with $\pi^*(CO_2)$. Remember also that a stronger stabilization of the η^2 CO_2 complex will be achieved if the d_σ orbital pointing towards the carbon dioxide ligand is empty. Such a situation is encountered in the systems Nb(η-$C_5H_4Me)_2(CH_2SiMe_3)(CO_2)$ [10b], Mo$(PR_3)_4$ $(CO_2)_2$ [10c] , and MoCp$_2(CO_2)$ [10d] .

One may finally wonder about the occurence of the η^1 - O coordination mode (**3a** or **3b**) . The investigations carried out at the SCF level by Sakaki and others [11,12] seem to indicate that in most cases this structure is destabilized quite appreciably with respect to either the η^1 - C or the η^2 side-on structure. This destabilization has been traced to repulsive molecular orbital interactions rather than to electrostatic interactions which, in this case, are favorable. The repulsive interactions occur between the doubly occupied $d\pi$

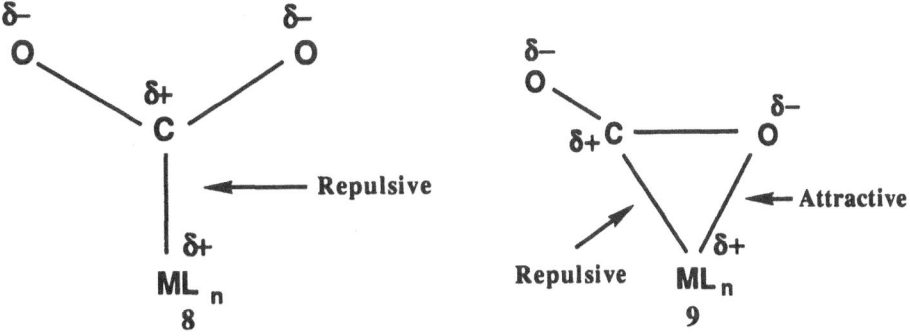

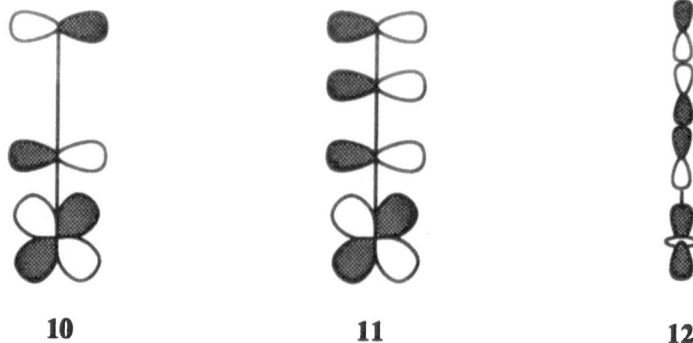

10   **11**   **12**

orbital and the π and $n\pi$ orbitals of CO_2, and/or between a doubly occupied $d\sigma$ orbital and a σ lone pair on the bound oxygen atom of CO_2, see **10-12**.

It should be pointed out, however, that this quite appreciable destabilization has been found with SCF calculations and for a linear geometry of CO_2. The situation might be different if electron correlation is taken into account and/or if CO_2 is allowed to bend. CAS-SCF calculations (which bring non-dynamical correlation effects) have been carried out on the $Ni(NH_3)_2(CO_2)$ complex in the context of a study of the coupling of CO_2 and C_2H_4 [18]. The η^1 structure with a bent CO_2 attached through an oxygen end to the Ni atom was found to be 1.3 kcal/mol only above the η^2 structure. The analysis has been refined by optimizing the O-C-O angle at the CAS-SCF level in both structures and by carrying out a contracted CI calculation to take into account some dynamical correlation effects [19,20]. The corresponding results, shown in Table 1, point clearly to the importance of electron correlation in stabilizing the η^1 structure with respect to the η^2 side on structure. This does not mean that a η^1 - O structure is a stationary point, i.e. a true intermediate on the potential energy hypersurface of $Ni(NH_3)_2(CO_2)$ but that a deformation of the side-on structure towards this structure should not be overlooked in chemical processes . In fact Aresta and coworkers have observed some fluxionality in bis phosphine carbon dioxide complexes in which such structures may play a critical role [20].

TABLE 1.   Relative energies (in kcal/mol) of the η^2 and η^1 O structures of $Ni(NH_3)_2(CO_2)$ [a]

	SCF [b]	CAS-SCF	CI
η^2	0. [b]	0. [d]	0. [f]
η^1	+27.4 [c]	+5.3 [e]	+7. [f]

[a] The geometrical parameters other than the O-C-O angle can be found in reference 19. [b] The optimized SCF O-C-O angle amounts to 133.4°. [c] The optimized SCF O-C-O angle amounts to 130.3°. [d] The optimized CAS-SCF O-C-O angle amounts to 140.0°. [e] The optimized CAS-SCF O-C-O angle amounts to 136.5°; [f] The CAS-SCF optimized O-C-O angle was retained.

TABLE 2. CO_2 binding energies [a] (in kcal /mol) to various transition metal complexes

System	Method			
	SCF	MP2	SD-CI	CAS-SCF+CI
$Ni(NH_3)_2(\eta^2\text{-}CO_2)$[b]	61	-	-	37
$Ni(PH_3)_2(\eta^2\text{-}CO_2)$[c]	17	-30	7	-
$Co(alcn)(\eta^1\text{-}CO_2)$[d]	6	-	-	-
$RhCl(AsH_3)_4(\eta^1\text{-}CO_2)$[e]	-2	27	-	-
$NiF(NH_3)_4(\eta^1\text{-}CO_2)$[f]	22	-	48	-

[a] the binding energies are computed with respect to the fragments in the geometry of the complex. No basis set superposition error correction is applied except for the CAS-SCF+CI calculations of $Ni(NH_3)_2(CO_2)$ where the frozen core approximation was used; [b] For a N-Ni-N angle of 90°, this work; [c] For a P-Ni-P angle of 120°, ref. 16; [d] ref. 11; [e] ref. 24; [f] ref. 17.

The previous considerations are only qualitative in nature. It is interesting to gauge the interaction energy between CO_2 and a transition metal complex. Table 2 summarizes the results which have been obtained so far. The data are too scarce to draw definite general conclusions (moreover most of the calculations have been carried out with different basis sets). Yet from the available SCF data CO_2 seems to be more strongly bound in the η^2 side on coordination mode than in the η^1 - C coordination mode (the Ni fluoride system lies somewhat apart because, at variance with the other systems which are closed-shell species, it involves a Ni(I) d^9 metal atom and is therefore an open shell system). A recent LSD calculation carried out on the bare $PdCO_2$ system [21] is quite interesting in that respect: the atomic ground state of Pd is d^{10}, all d levels are occupied and the system experiences in both coordination modes a 2e stabilizing interaction (either 4 or 6) and a 4e destablizing interaction (either 5 or 7). The energy difference between the two structures (2.5 kcal/mol in favor of the η^1-C coordination mode [21]) therefore corresponds to an *intrinsic* binding energy difference between the two coordination modes.

The data reported for calculations carried out beyond the SCF approximation are even more scarce than the SCF data. They question however about the degree of reliability of the SCF approach for studying this type of problem. It is for instance clear that MP2 type calculations fail to give a correct binding energy in the case of the Ni systems. This is due to the near degeneracy correlation effects which are known to be important in Ni(0) complexes with a low coordination number [22] On the other hand for the d^8 (η^1 - CO_2) Rh and Co complexes, one may expect that near degeneracy correlation effects are less crucial. The increase in the binding energy on going from SCF to MP2 obtained in the case of the Rh complex is probably meaningful and transposable to the Co complex. Nevertheless the scarcity of the results which have been obtained so far calls for a study which should try to assess more comprehensively correlation effects on the binding energy.

Electron correlation has also a quite important effect on the energy difference between the η^1-C and the η^2 side on structures. Sakaki and coworkers have analyzed this problem in detail for the bis phosphine nickel complex $Ni(PH_3)_2(CO_2)$ [16]. Table 3 summarizes some of their findings. The energy difference between the two structures decreases from

TABLE 3. Energy difference (in kcal/mol) between the η^1-C and the η^2 side on structures of $Ni(PH_3)_2(CO_2)$ according to the level of the calculation.(results taken from reference 16

Method	$Ni(PH_3)_2(\eta^2\text{-}CO_2)$	$Ni(PH_3)_2(\eta^1\text{-}CO_2)$
SCF	0.	+24.
est. SD-CI	0.	+10.
est. full CI	0.	+8.4

24 to 8.4 kcal/mol on going from SCF to the estimated full CI limit. We can also link these results to the strong correlation effects that have been also found when comparing the η^1-C and the η^1-O coordination modes in $Ni(NH_3)_2(CO_2)$, see Table 1.

3. The reactivity of the CO_2 ligand

In fact very little is known about the actual activation brought by the coordination of CO_2 to a transition metal complex. The activation towards the O-centered electrophilic addition has been experimentally evidenced in hexacoordinated iridium complexes[9b] and in $MoCp_2(CO_2)$ complexes [23]. Sakaki and coworkers have rationalized and quantified this activation [24]. They have compared the attack of H^+ on CO_2 either coordinated to the $RhCl(AsH_3)_4$ system (taken as a model of the experimental $IrCl(dmpe)_2$ system $(dmpe=(CH_3)_2PCH_2CH_2P(CH_3)_2)$,or coordinated (in the Lewis sense) to NH_3, see below. The stabilization energy is much greater in the Rh case. The attack is found to occur preferentially on the oxygen atom with an optimum angle around 90°. This has been traced, see Figure 2, to a three-orbital-four-electron interaction between d_{z^2}, $\pi_{//}^*$ and $\pi_{//}$ of CO_2, see Figure 2. This interaction polarizes the HOMO on the terminal oxygen atom, and enhances therefore its ability to interact with electrophiles.

(reproduced with permission of ref. 24)

Theoretical studies can also be used to assess in a coupling reaction between CO_2 and a substrate whether or not CO_2 acts as a ligand coordinated to the metal atom. We have looked ourselves to this type of problem in the coupling reaction of CO_2 and C_2H_4 mediated by Ni(0) complexes [18,19]. The experimental procedure makes use of $Ni(cod)_2$ complexes (cod= cyclooctadiene) in the presence of DBU which is an imine type ligand (DBU=1,8-Diazabicyclo[5.4.0]undec-7-ene).[25]. Since the product of the coupling reaction is a bis-DBU metallalactone complex, we have as a first step of our investigation compared two possible reaction channels, either the reaction of CO_2 with a $Ni(NH_3)_2(C_2H_4)$ system or the reaction of C_2H_4 with a $Ni(NH_3)_2(CO_2)$ system.

In order to differentiate between these two channels, we first considered prototypical structures of each of these two pathways, namely $Ni(NH_3)_2(\eta^2\text{-}C_2H_4)$ **13**, $Ni(NH_3)_2(\eta^1\text{-}C_2H_4)$ **14** , $Ni(NH_3)_2(\eta^2\text{-}CO_2)$ **15**, $Ni(NH_3)_2(\eta^1\text{-}CO_2)$ **16** . The dihapto coordination modes are known to be the ground state structures of the ethylene and carbon dioxide NiL_2 complexes. The monohapto structures are neither true intermediates nor true transition states on the potential energy hypersurface of the coupling reaction. But since the geometries of the C_2H_4 and CO_2 ligand in the η^1 complexes were chosen so as to correspond to the structure of the metallalactone product (which is known experimentally [25a]), the energy difference between the monohapto structures and the dihapto ones is the deformation energy of either the $Ni(NH_3)_2(C_2H_4)$ or $Ni(NH_3)_2(CO_2)$ to achieve the metallalactone geometry. In order to get reliable estimates of these energy differences it was found necessary to go beyond the SCF approximation [18,19].

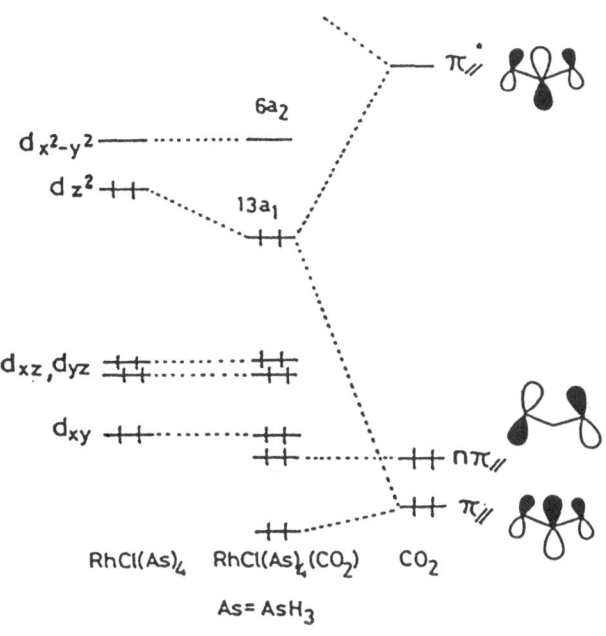

Figure 2. Orbital interaction diagram between the orbitals of $RhCl(AsH_3)_4$ and of CO_2, and the resulting polarization of the HOMO (reproduced with permission from ref. 24)

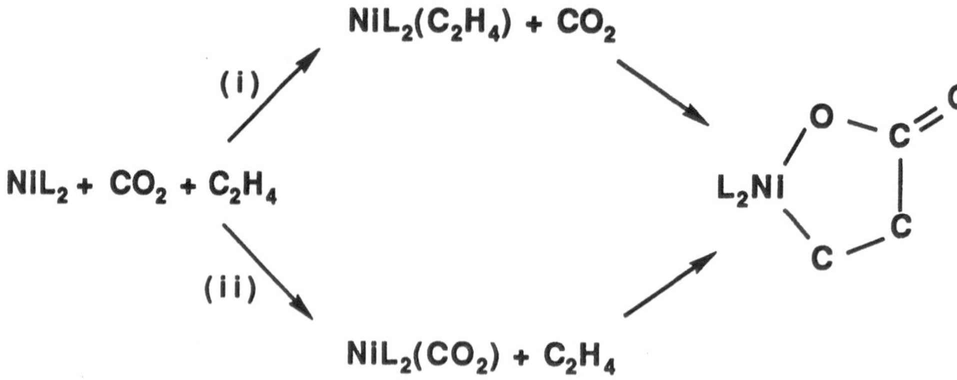

$$NiL_2(C_2H_4) + CO_2$$

(i)

$$NiL_2 + CO_2 + C_2H_4$$

(ii)

$$NiL_2(CO_2) + C_2H_4$$

L=NH3

CASSCF [26] calculations have therefore been carried out on these systems. Table 4 reports the corresponding values. It appears clearly from this Table that the η^2 to η^1

Table 4. CAS SCF relative energies (in kcal/mol) of the Ni(NH3)2(CO2)(C2H4) isomers

System	Relative energies
Ni(NH3)2 +C2H4+CO2 [a]	0.
Ni(NH3)2(η^2-C2H4) + CO2 [b]	-8.5
Ni(NH3)2(η^1-C2H4) + CO2 [b]	+20.5
Ni(NH3)2(η^2-CO2) + C2H4 [b]	-14.8
Ni(NH3)2(η^1-CO2) + C2H4 [b]	-9.5

[a] For the optimum N-Ni-N angle of 180°. [b] For a N-Ni-N angle of 90°

deformation energy is much lower for the CO_2 complex than for the C_2H_4 complex, thus being an indication that the pathway involving the attack of C_2H_4 on coordinated CO_2 is preferred over the attack of CO_2 on coordinated C_2H_4. CAS-SCF calculations have also been carried out for the attack of C_2H_4 on the η^1 CO_2 complex and for the attack of C_2H_4 on the η^1 C_2H_4 complex. As shown on Figure 3 the first pathway is highly preferred over the second one, whatever the level of the calculation (CAS-SCF or CAS-SCF + CI

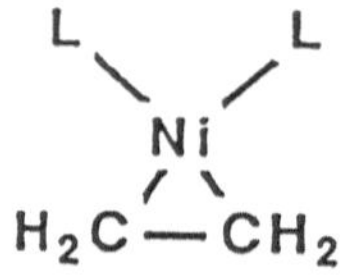

13

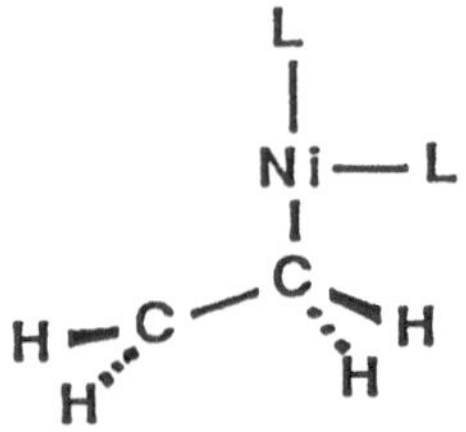

14

15

16

[27]) is. The energy barrier (about 25 kcal/mol at the CAS-SCF + CI level)), although being probably somewhat too high, is reasonable if one considers that the coupling takes place at 40°C and needs 90 hours to be completed.

How can we rationalize the fact that the attack of CO_2 on C_2H_4 is preferred over the attack of C_2H_4 on CO_2? We believe that this is due to the *strong diradical character* that we found for the η^1 CO_2 complex [18,19] stronger at least for η^1 CO_2 than for η^1 C_2H_4. This is best seen from the natural orbital analysis of the CAS-SCF wave functions. The scheme below indicates that these wave functions are essentially characterized by two "frontier"orbitals, made of the bonding and antibonding combination of $d_{x^2-y^2}$ with $\pi^*(CO_2)$ and $\pi^*(C_2H_4)$ respectively.The occupation numbers are 1.297 and 0.704 for the η^1 CO_2 complex, they are much lower for the η^1 C_2H_4 complex, 1.537 and 0.468 [29]. In line with this result, the triplet state $^3A'$ of the η^1 CO_2 complex is computed 4.1 kcal/mol only above the $^1A'$ singlet ground state. Thus this 1,3 diradicaloid centered on Ni and C gives rise to a quite easy [3+2] cycloaddition reaction with ethylene. This feature is in fact reminiscent of an old result of Aresta and coworkers [30], namely the ease of the coupling of O_2 with bis-tricyclohexylphosphine nickel CO_2.

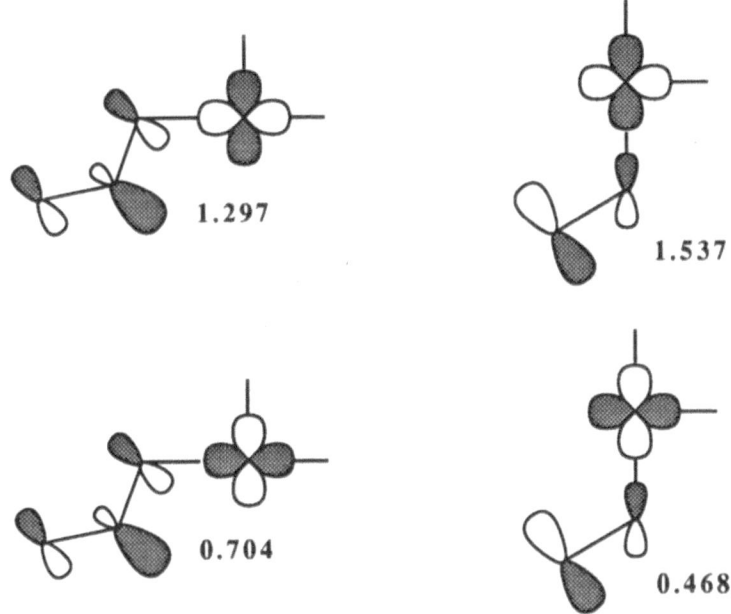

1.297

1.537

0.704

0.468

O_2 is a diradical and as such will easily couple with any system that has a strong diradical character: it reacts for instance easily with strained cycloalkynes. Another interesting resemblance with the [3+2] organic cycloadditions is brought by the isolobal analogy between carbon dioxide and formaldehyde on one hand and by the $Ni(NH_3)_2$ fragment and CH_2 [31] on the other hand. This makes the $Ni(NH_3)_2(\eta^1 CO_2)$ system isolobal to the carbonyl ylid CH_2-O-CH_2. Carbonyl ylids have diradical structures and are known to undergo easily 1,3 dipolar addition with olefins [32,33]. One should be aware however that the most stable structure of CH_2-O-CH_2 is the edge to edge planar one and that the face to face structure, which is the strict analog of the $Ni(NH_3)_2(\eta^1 CO_2)$ complex, is not a true intermediate [33].

4. The reactivity of the CO_2 ligand as an external reagent

CO_2 can of course react with organometallic systems and transition metal complexes [1, 2] and the class of reactions that has been studied from a theoretical point of view is the insertion of CO_2 into the metal hydride bond or the metal methyl bond of a transition metal complex. Studies of this type have been carried out by Sakaki and coworkers [36] and by ourselves [37]. There were several reasons to undertake such studies, the most obvious one being to establish or at least to clarify the reaction mechanism. For instance in the case of the group 6 metal hydride and alkyl anions $RM(CO)_5^-$ (R= H, CH_3; M=Cr, Mo, W) the insertion had been found on the basis of kinetic studies [38] to involve a *dissociative* mechanism in the case of the metal hydride bond and an *associative* mechanism in the case of a metal alkyl bond. The reason for the difference was not known however. Also in the case of $HCr(CO)_5^-$ it was suggested that the dissociation of the ancillary ligand might be triggered by a preassociation with CO_2. It might be interesting to check this hypothesis through some calculations. For the cooper system the reaction mechanism was largely unknown [39].

Table 5 summarizes the energetics of the CO_2 insertion into the metal hydride bond process as obtained at the SCF level from the two sets of studies. In every case the process is thermodynamically favorable, the reaction being always quite exothermic. The $\eta1$-O conformation with the non coordinating oxygen of the carboxylate ligand pointing towards the metal atom 17 is always found, in agreement with the experimental data, more stable than 18 where H or the alkyl ligand points towards the metal. This has been ascribed, for the copper system, to an unfavorable electrostatic interaction in 18 between the metal atom which is positively charged and H (or R) which is also positively charged [36]. The energy difference between 17 and 18 is smaller for the chromium system than for the copper system. The slight negative charge born by the hydrogen atom (according to the SCF calculations [37]) may account for this feature. But one may also think to some

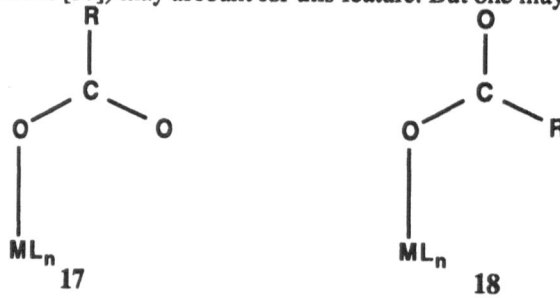

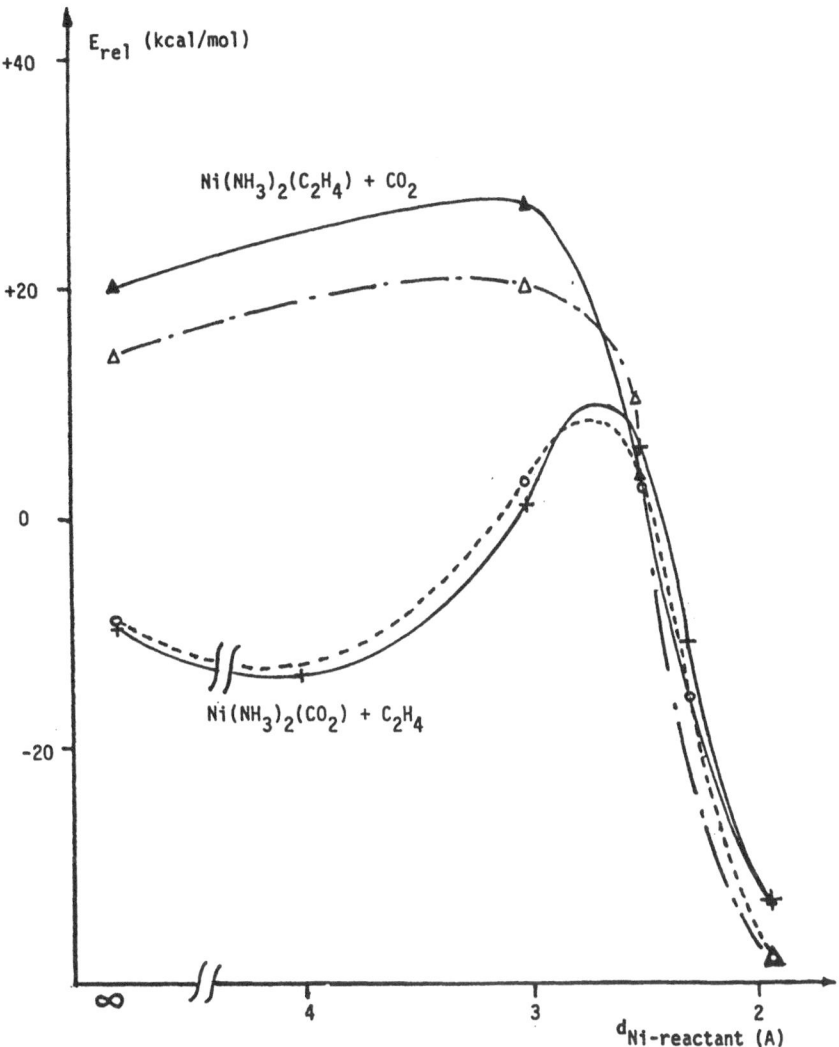

Figure 3. Energy profiles for the attack of C_2H_4 on $Ni(NH_3)_2(\eta^1\text{-}CO_2)$ and of CO_2 on $Ni(NH_3)_2(\eta^1\text{-}C_2H_4)$. The two lowers curves (—— CAS-SCF and - - - o - - - CI) refer to the $Ni(NH_3)_2(\eta^1\text{-}CO_2)$+ C_2H_4 reaction   The two higher curves (—Δ— CAS-SCF and --- Δ --- CI) refer to the $Ni(NH_3)_2(\eta^1\text{-}C_2H_4)$ + CO_2 reaction The zero of energy is for the $Ni(NH_3)_2$ + C_2H_4 + CO_2 separated systems in their equilibrium geometries.

slight stabilizing interaction between one lone pair of the formate oxygen atom and the π^* orbital of the ancillary carbonyl ligand, of the same sort that the one which would occur in a nucleophilic addition reaction. This latter hypothesis has been raised by Darensbourg and coworkers [40]. A Bader type analysis [41] based on the charge density differential ∇

TABLE 5. Thermodynamics (in kcal/mol at the SCF level) of the reaction HML_n + $CO_2 \rightarrow L_nM(O_2CH)$ for various conformations of the $L_nM(O_2CH)$ system

$L_nM(O_2CH)$	HML$_n$		
	CuH(PH$_3$)$_2$ [a]	CuH(PH$_3$)$_3$ [b]	HCr(CO)$_5$$^-$ [c]
H–C=O, M–O	-41.2	-45.5	-41.6
O=C–H, M–O	-55.6	-59.3	-45.6
M, O, C–H, O	-56.8	-57.0	-33.3 [d]

[a] ref [36], not corrected by the basis set superposition error; [b] ref [36], not corrected by the basis set superposition error; [c] ref [37], not corrected by the basis set superposition error; [d] in this case the product is Cr(CO)$_4$(η^2-O$_2$CH) + CO.

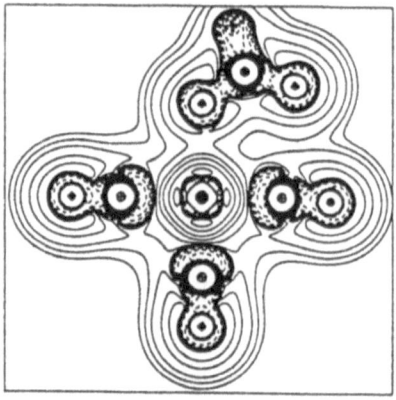

Figure 4. Contour plot of $\nabla^2\rho$ for Cr(CO)5(HCO$_2$). The dashed line denotes regions of charge concentration and the solid line the regions of charge depletion. Starting at zero contour, contour values change in steps of $\pm\,2\times10^n$, $\pm\,4\times10^n$, and $\pm\,8\times10^n$, with n going from -2 to 0 and increasing n in step of unity, up to the ±10 a.u. contours.

ρ and on the Laplacian of the charge density $\nabla^2\rho$ also supports this view: a bond critical point in the charge density of type [3,-1] appears between the distal oxygen atom and the carbon atom of the ancillary carbonyl ligand [42]. The electron density ρ at this point is low, 0.017 a.u.. This value is about twenty times lower than the one found for the C-O carbonyl bond. But in the map of Figure 4 that displays the Laplacian of charge density of the formate system (with the conformation **18**) one maximum of charge concentration of the formate oxygen atom (critical point of [3,-3] type) fits the hole in the valence shell of the carbonyl carbon atom. Although this hole is not a true minima (it is a [3,+1] point in $\nabla^2\rho$) the same type of point was identified as a point of nucleophilic attack in organic systems [43]. In line with this hypothesis, a geometry optimization of the Cr-C-O angle yields a bending of 10° [42], as expected at the beginning of a nucleophilic attack.

A reaction path for the insertion of CO_2 into the copper hydride bond has been determined by Sakaki and coworkers [36], see Figure 5. It is characterized at intermediate distance by the approach of the carbon atom towards the hydride ligand. In the vicinity of the transition state, the Cu-H-C angle begins to decrease as a result of some Cu-O interaction. One then finally yields a monohapto formate structure such as **17**. The energy barrier computed for the insertion process is quite low. It ranges from 8 to 16 kcal/mol for $HCu(PH_3)_2 + CO_2$, depending on the quality of the basis set and on the inclusion of electron correlation effects through MP2 calculations. These last effects are in fact found to be less important than the basis set effects. In particular, the use of a pseudopotential seems to increase the energy barrier [36]. The monohapto structure such as **17** can then rearrange without any noticeable energy barrier, either through a rotation around the O-C bond or through some sliding around the coordinated oxygen atom, to the monohapto structure corresponding to **18**. A final rearrangement to the most stable dihapto structure is easy, the corresponding energy barrier being of the order of a few kcal/mol only [36]. It should be noted however that these two rearrangement processes have beeen artificially separated for the sake of computational simplicity and that they could well proceed simultaneously.

The driving force for the easy insertion step is, according to Sakaki et al. [36], dual: One finds first a quite strong charge transfer interaction between $Cu(H)(PH_3)_n$ and CO_2, essentially between the Cu-H σ orbital and the π^* orbital of CO_2 see **19**. This charge transfer interaction is supplemented by an important attractive electrostatic interaction between the negatively charged hydride ligand and the positively charged carbon atom and between the positively charged copper atom and the negatively charged oxygen atom. The energy barrier merely results, as shown by the decomposition energy analysis [36], from the linear to bent deformation of CO_2 and from the exchange repulsion between $HCu(PH_3)_n$ and CO_2.

Similar features have also been found in our analysis of the CO_2 insertion into the metal hydrogen bond of $HCr(CO)_5^-$ [37]. This reaction has nevertheless some distinctive characteristics: it apparently involves a dissociative mechanism which in addition might be triggered by the preassociation of CO_2 with $HCr(CO)_5^-$. SCF calculations do indeed point to a structure, see **20**, in which CO_2 is bound through the carbon atom to the hydride ligand. The stability of the adduct has been computed to be 8.4 kcal mol at the SCF level [37], taking into account the basis set superposition error effects. But this value is probably overestimated because it is reduced by about 7 kcal/mol at the CAS-SCF level [37]. The interaction between CO_2 and $HCr(CO)_5^-$ is best described in terms of a 2e charge transfer interaction, see **21**, between the occupied $d_{z^2} + s_H$ orbital and the empty π^* orbital of CO_2 . The doubly occupied $\pi(CO_2)$ can also mix in and bring some repulsion as in the carbon dioxide Rh complex described previously On the other hand one expects some stabilization from an electrostatic interaction between the negatively

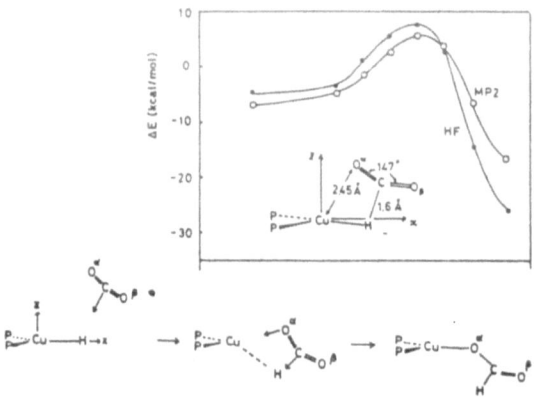

Figure 5. Energy profile and reaction path for the insertion of CO_2 into the Cu-H bond of $HCu(PH_3)_2$ (reproduced with permission of ref. 36a and 36b)

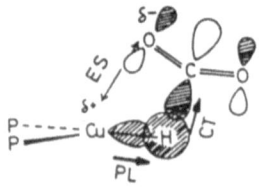

19

(reproduced with permission from ref. 36a)

charged hydride ligand and the positively charged carbon atom of CO_2. The map of the molecular electrostatic potential for $HCr(CO)_5^-$ (Figure 6) does indeed display a potential well (of about -0.170 a.u.)in the vicinity of the hydrogen, slightly off the Cr-H axis. This accounts for the optimum value of 170° computed for the Cr-H-C angle in **20**.

20

21

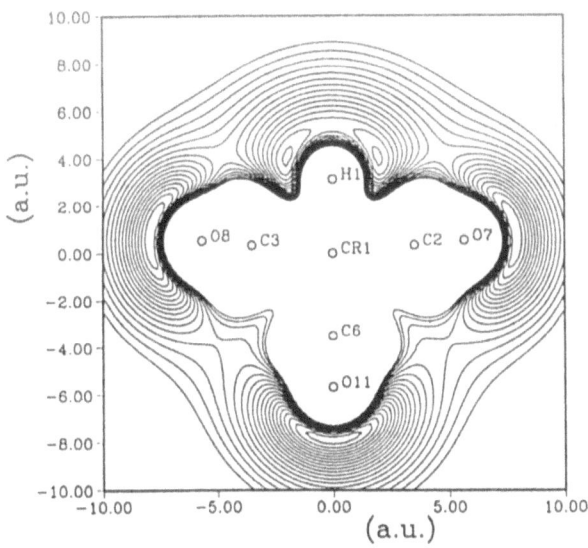

Figure 6. Molecular electrostatic potential map of HCr(CO)$_5^-$. The contour values decrease from -0.105 a.u. (external contour) to -0.195 a.u. by steps of 0.0045 a.u.

The SCF calculations also indicate that in the [Cr(CO)$_5$H...CO$_2$]$^-$ adduct, the dissociation energy of the equatorial CO ligand is lowered (with respect to HCr(CO)$_5^-$) by 5.8 kcal/mol [37] . Thus the CO$_2$ coordination seems to trigger the dissociative mechanism as experimentally hypothesized [38].

The analysis of the respective wavefunctions of the adduct **20** and of the formate product, either **21a** or **21b**, revealed that the electronic distribution was quite similar in both systems. In addition the optimized C-H bond length (1.15Å at the SCF level [37], 1.18Å at the CAS-SCF level [42]) is close to the value expected for the formate ligand. On the other hand the Cr-H bond length (1.88Å at the SCF level , 1.93Å at the CAS-SCF level) has been significantly elongated . Thus the HCO$_2$ unit in **20** has already much of the formate character. We therefore proposed the reaction scheme shown on the next page in which the η^2 - H, O formate intermediate structure plays a critical role [37] This dihapto structure which was found to be only 3.0 kcal/mol above the [Cr(CO)$_5$H...CO$_2$]$^-$ adduct may then rearrange after recombination of CO and some rotational isomerization in **21b** (similar to the isomerization analyzed in the copper system by Sakaki and coworkers) to the thermodynamically most stable isomer **21a.** The O,O η^2 structure **23**, in which the two oxygen atoms are bound to the Cr atom, is found to lie about halfway in energy between the η^1 H, O structure **22b** and the formate structure **21a.** It might therefore be also involved in the insertion process since it would lead directly to **22a.** New experimental studies have now answered this point . The Cr(CO)$_4$(η^2 O$_2$CH)$^-$ system has been synthesized and identified by IR spectroscopy [44]. This finding and new kinetic data have led Darensbourg and coworkers to propose [40] the reaction profile

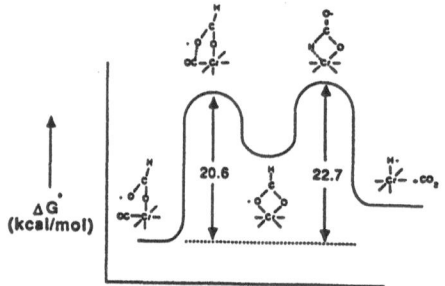

22b

21b

0.

+3.0

−26.4

20

23

−18.1

(SCF relative energies are in kcal/mol)

21a

−30.4

shown on Figure 8 in which the O,H dihapto structure would be the transition state of the rate determining step and the O,O dihapto structure a reaction intermediate.

ΔG°
(kcal/mol)

20.6

22.7

REACTION COORDINATE

(reproduced with permission of ref. 40)

The relatively low destabilization of **22** has been traced [37] to a two electron stabilizing interaction **24** between the empty d_π orbital of the $Cr(CO)_4$ fragment and a doubly occupied valence orbital of the incipient HCO_2^- formate ligand. This latter orbital is characterized by an out of phase combination between the p components on the oxygen atoms and the s component of the hydrogen atom. In the case of the insertion into the metal methyl bond of the $CH_3Cr(CO)_5^-$ analog, the interaction to consider would be **25**. One can easily foresee that it would be much weaker due to an unfavorable overlap between the orbital of $CH_3CO_2^-$ and d_π. In this case the upper lobe of the d_π orbital would point toward the nodal plane of the p component on the carbon atom. This may explain why there would be no decisive gain in going through a dissociative mechanism in the case of the insertion into the metal methyl bond.

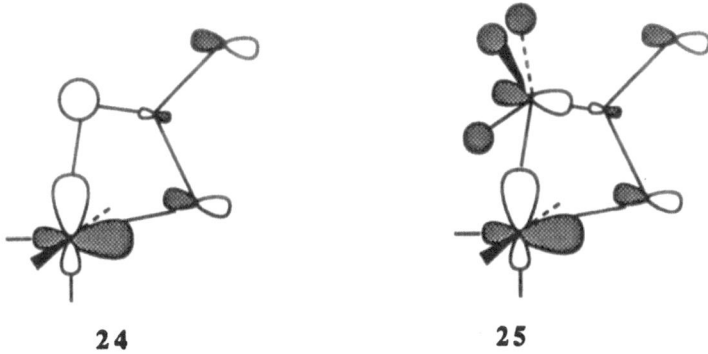

24   25

The same kind of arguments can also explain [37] why in the decarboxylation reaction of square planar d^8 $ML_3(HCO_2)$ formate complexes an ancillary ligand dissociation *does not* seem to be required [45]. In this case a low lying empty orbital of b_2 symmetry is already present on the ML_3 fragment and can interact as shown in **26** in a bonding way with the doubly occupied orbital of HCO_2^- (note the similarity between **26** and **24**). Here again recent kinetic data of the Darensbourg group support this proposal [46].

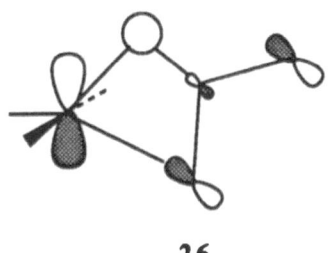

26

5. Conclusions and Perspectives

It is clear from this review that the most essential features which govern the modes of coordination of CO_2 have been unravelled. With this knowledge the theoretical study of the mechanistic details of prototypical carbon dioxide organometallic reactions has been made possible, thus leading to a successful interplay of theory and experiment.

Moreover some aspects of these studies can be extended to other systems and reactions of organometallic or organic chemistry. A first natural extension is from CO_2 to CS_2. Such studies have been carried out, mostly at the extended Hückel level [7, 47], although some comparison has also been provided by a LCAO-HFS calculation [13]. In particular the mechanism found for the coupling of CS_2 with the $Cp_2Mo(\eta^2\text{-}CS_2)$ complex, with an initial cleavage of the metal carbon bond to yield a bent η^1 like CS_2 structure [47c], is quite similar to the one we found for the reaction between $Ni(NH_3)_2(C_2H_4)$ and CO_2. We have also mentionned the analogy between the $Ni(NH_3)_2$ + C_2H_4 reaction and the coupling reaction of C_2H_4 with carbonyl ylids. The same analogy would apply to the coupling of ethylene to the azomethine system $H_2C\text{-}NH\text{-}CH_2$. Interestingly Frühauf and coworkers have advocated the isolobal analogy between azomethine and FeL_3(α-diimine) complexes to support a 1,3 cycloaddition mechanism for the coupling of the FeL_3(α–diimine) complexes with acetylenes [48]. The analogy between this reaction and the $NiL_2(CO_2)$ + C_2H_4 reaction is therefore obvious and one may wonder about the diradical character (and its implications to the reaction mechanism) of these α-diimine complexes. One may also think to the metallacyclotetraazapentadiene analogs. These latter complexes are known to have a smaller HOMO-LUMO gap than the α-diimine complexes[49] and might therefore have a grater diradical character . Work is now in progress along these lines [50].

6. References and Notes

1. Aresta, M. (1990) in M. Aresta and J.V. Schloss (eds.), Enzymatic and Model Carboxylation and Reduction Reactions for Carbon Dioxide Utilization, NATO ASI Series C314, Kluwer Academic, Dordrecht, p.1.
2. a) Darensbourg, D. J.and Kudaroski, R. A. (1983), Adv. Organomet. Chem., **22**, 129.
   b) Palmer, D. A.and Van Eldik, R. (1983), Chem. Rev., **83**, 51.
   c) Ziessel, R. Nouv (1983), J. Chim., **7**, 613.
   d) Walther, D. (1987), Coord. Chem. Rev., 79, 135.
   e) Braunstein, P., Matt, D., Nobel, D. Chem. Rev. 1988, **88**, 747.
   f) Behr, A. Angew. (1988), Chem. Int. Ed. Engl., **27**, 661.
3. Sakaki, S. (1990), in "Stereochemistry of Organometallic and Inorganic Compounds", Elsevier Science Pub., Amsterdam.
4. Huzinaga, S. (1971), Tables of Atomic Fuctions, Technical Report, University of Alberta, Edmonton.
5. d polarization functions of exponent 0.8 were added on both the carbon and oxygen atoms.
6. One has therefore to be careful when considering the results of EH calculations which do not incorporate these effects. This cautionary note has been already given by Hoffmann and coworkers.
7. Mealli, C., Hoffmann, R., Stockis, A. (1984), Inorg. Chem., **23**, 56.
8. Preliminary calculations carried out on the CO_2 insertion into a C-C bond of Li-phosphine enolate systems indicate that differential effects due to the inclusion of d polarization functions on all first row and second row atoms of CO_2 and the phosphine enolate moeity may be as large as 15-20 kcal/mol.
9. a) Gambarotta, S., Arena, S., Floriani., C., Zanazzi, P.F. (1982), J. Am. Chem. Soc., **104**, 5082.
   b) Calabrese, J. C., Herskovitz, T. and Kinney, J. B. (1983) J. Am. Chem. Soc., **105**, 5914.
10. a) Aresta,M., Nobile, C. F., Albano, V. G., Forni, E. and Manassero, M. (1975), J. Chem. Soc., Chem. Comm., 636.
    b) Bristow, G. S., Hitchcock, P. B. and Lappert, M. F. (1981) J. Chem. Soc., Chem. Comm., 1145.
    c) Alvarez, R., Carmona, E., Gutierrez-Puebla, E., Marin, J.M., Monge, A. and Poveda, M.L. (1984), J. Chem. Soc., Chem. Comm., 1326.
    d) Gambarotta, S., Floriani C., Chiesi-Villa, A. and Guastini, C. (1985) J. Am. Chem. Soc., **107**, 2985.
11. Sakaki, S. and Dedieu, A. (1987), Inorg.Chem., **26**, 3278.
12. Sakaki, S., Kitaura, K. and Morokuma, K. (1982), Inorg. Chem., **21**, 760.

13. Ziegler, T. (1986), Inorg.Chem., **25**, 2721.
14. Rosi, M., Sgamelotti, A., Tarentelli, F. and Floriani, C. (1987), Inorg. Chem, **26**, 3805.
15. Branchadell, V. and Dedieu, A. (1987), Inorg. Chem., **26**, 3966.
16. Sakaki, S., Koga, N. and.Morokuma, K. (1990), Inorg. Chem., **29**, 3110.
17. Sakaki, S. (1990), J. Am. Chem. Soc., **112**, 7813.
18. Dedieu, A. (1989), Ingold, F. Angew. Chem. Int. Ed. Engl., **28**, 1694.
19. Dedieu, A., Bo, C. and Ingold, F. (1990), in M. Aresta and J.V. Schloss (eds.), Enzymatic and Model Carboxylation and Reduction Reactions for Carbon Dioxide Utilization, NATO ASI Series C314, Kluwer Academic, Dordrecht, p. 23.
20. a) Aresta, M. and Nobile, C. F. (1977), J. Chem. Soc, Chem. Comm., 708.
    b) Aresta, M., Quaranta, E. and Tommasi, I. (1988), J. Chem. Soc. Chem. Comm., 450.
21. Sirois, S., Salahub, D. (1991), NATO ASI on Metal-Ligand Interactions: from Atoms to MClusters, to Surfaces, Cetraro, Italy, June10-21.
22. see for instance:
    a) Blomberg, M. R. A., Brandemark, V. B., Siegbahn, P. E. M., Mathisen, K. B. and Karlström, G. (1985), J. Phys. Chem., **89**, 2171.
    b) Widmark, P.O., Roos, B.O. and Bäckvall, J. E. (1987), J. Am. Chem. Soc., **109**, 4450.
23. Tsai, J. C., Khan, M. and Nicholas, K. M. (1989), Organomet., **8**, 2967.
24. Sakaki, S., Aizawa, T., Koga, N., Morokuma, K. and Ohkubo, K. (1989), Inorg. Chem., **28**, 103.
25. a) Hoberg, H., Peres, Y., Krüger, C. and Tsay, Y. H. (1987), Angew. Chem. Int. Ed. Engl., **26**, 771.
    b) Hoberg, H., Peres, Y., Milchereit , A. and Gross, S. (1988), J. Organomet. Chem., **345**, C17.
26. The active orbitals of the CAS-SCF wavefunction (d_{z^2}, $d_{x^2-y^2}$ and 4s, $\pi(CO_2)$, $n\pi(CO_2)$, $\pi^*(CO_2)$, $\pi(C_2H_4)$, $\pi^*(C_2H_4)$)account for the main bonding interactions as well as for the sd hybridization features. The geometries were taken from related X-ray crystal structures and calculations except for the O-C-O angle of the CO_2 ligand which was found to be quite sensitive to the correlation effects and optimized at the CAS-SCF level (vide supra). The values of the Table 3 therefore differ from the values reported in reference 19 where the O-C-O angle was optimized at the SCF level.
27. The CI calculations were multireference contracted CI calculation [28] using as references the configurations which were found to have an expansion coefficient greater than 0.05 in the CAS-SCF wavefunction. 10 electrons were correlated (the same as in the CAS-SCF wavefunction) and single and double excitations to all virtual valence orbitals were included.
28. Siegbahn, P. E. M. (1983), Int. J. Quant. Chem., **23**, 1869.
29. A similar results is found in the CAS-SCF wavefunction.
30. Aresta, M., Nobile, C. F. J. (1977), Chem. Soc. Dalton Trans., 708.
31. a) Hoffmann, R. (1982), Angew. Chem. Int. Ed. Engl., 21, 711
    b) Albright, T. A., Burdett, J.K. and Whangbo, M.H. (1985), Orbital Interactions in Chemistry, Wiley, New York, 1985, pp 402-421.
32. a) see for instance Huisgen, R. (1977), Angew. Chem. Int. Ed. Engl., **16**, 172 and references therein.
    b) a theoretical analysis of this attack has been carried out [34] but at the SCF level only. Such (1,3) dipolar cycloaddition should be studied at the MC-SCF

level [35].

33. For theoretical studies on ylids, see for instance:
    a) Volatron, F., Anh, N. T. and Jean, Y. (1983), J. Am. Chem. Soc., **105**, 2359.
    b) Feller, D., Davidson, E. R. and Borden, W. T. (1984), J. Am. Chem. Soc., **106**, 2513.
34. Leroy, G., Ngyen, M. T. and Sana, M. (1976), Tetrahedron, **32**, 1529.
35. Bernardi, F., Bottoni, A., Robb, M. A. and Venturini, A. (1990), J. Am. Chem. Soc., **112**, 2106.
36. a) Sakaki, S. and Ohkubo, K. Inorg. Chem. 1988, **27**, 2020.
    b) Sakaki, S. and Ohkubo, K. Inorg. Chem. 1989, **28**, 2583.
    c) Sakaki, S. and Ohkubo, K. (1989), Organomet, **8**, 2973.
37. Bo, C. and Dedieu, A. (1989), Inorg. Chem., **28**, 304.
38. a) Darensbourg, D. J., Rokicki, A. and Darensbourg, M. Y. (1981), J. Am. Chem. Soc., **103**, 3223.
    b) Darensbourg, D. J. and Rokicki, A. (1982), Organomet., **1**, 1685.
    c) Darensbourg, M. Y., Bau, R., Marks, M. W., Burch, R. R., Jr., Deaton, J.C. and Slater, S. (1982), J. Am. Chem. Soc., **104**, 6961.
    d) Darensbourg, D. J. and Pala, M. (1985), J. Am. Chem. Soc., **107**, 5687.
    e) Darensbourg, D. J., Kudaroski Hanckel, R., Bauch, C. G., Pala, M., Simmons, D. and White, J. N. (1985), J. Am. Chem. Soc., **107**, 7463.
    f) Darensbourg, D. J. and Grotsch, G. S. (1985), J. Am. Chem. Soc., **107**, 7473.
    g) see also Darensbourg, D. J. (1990), in M. Aresta and J.V. Schloss (eds.), Enzymatic and Model Carboxylation and Reduction Reactions for Carbon Dioxide Utilization, NATO ASI Series C314, Kluwer Academic Dordrecht, p.43.

39. a) Beguin, B., Denise, B. and Sneeden, R. P. A. (1981), J. Organomet. Chem., **208**, C18.
    b) Bianchini, C., Ghilardi, C. A., Meli, A., Midollini, S. and Orlandini, A. (1985) Inorg. Chem., **24**, 924.
40. Darensbourg, D. J., Pickner Wiegreffe, H. and Wiegreffe, P. W. (1990), J. Am. Chem. Soc., **112**, 9252.
41. a) Bader, R. F. W. (1985), Acc. Chem. Res., **18**, 9.
    b) Bader, R. F. W. and Nguyen-Dang, T. T. (1981), Adv. Quant. Chem., **14**, 63.
    c) Bader, R. F. W. and Chang, C. (1989), J. Phys. Chem., **93**, 5095.
42. Bo, C. unpublished results
43. Carroll, M. T., Cheeseman, J. R., Osman, R. and Weinstein, H. (1989), J. Phys. Chem., **93**, 5120.
44. Darensbourg, D. J. and Pickner Wiegreffe, H. (1990), Inorg. Chem., **29**, 592.
45. a) Strauss, S. H., Whitmire, K.H. and Shriver, D.F. (1979), J. Organomet. Chem., **174**, C59.
    b) Immirzi, A. and Musco, A. (1977), Inorg. Chim.Acta, **22**, L35.
    c) Paonessa, R.S. and Trogler, W C. (1982), J. Am. Chem. Soc., **104**, 3520.
46. Darensbourg, D. J., Wiegreffe, P. and Riordan, C.G. (1990), J. Am. Chem. Soc., **112**, 5759.
47. a) Bianchini, C., Mealli, C., Meli, A., Sabat, M., Silvestre, J. and Hoffmann, R. (1986), Organomet., **5**, 1733.

b) Li, J., Hoffmann, R., Mealli, C. and Silvestre, J. (1989), Organomet., **8**, 1921.

c) Conan, F., Sala Pala, J., Guerchais, J. E., Li, J., Hoffmann, R., Mealli, C., and Toupet, L. (1989), Organomet., **8**, 1929.

48. Frühauf, H.-W. , Seils, F. and Stam, C. (1989), Organomet., **8**, 2338.

49. Trogler, W. (1990), Acc. Chem. Res., **23**, 426.

50. Liddell, M. and Dedieu, A. to be published.

QUANTUM CHEMICAL MODELS OF CHEMISORPTION ON METAL SURFACES

U. WAHLGREN and P. SIEGBAHN
Institute of Physics
University of Stockholm
Vanadisvägen 9
S-113 46 Stockholm
Sweden

ABSTRACT. The first part of this chapter (sections 1-3) contains a brief review of the independent particle model and the restricted Hartree-Fock equations. The concept of correlation energy is discussed, in particular the division into "static" and "dynamic" correlation. In this context the essential elements in the Multi-Configurational SCF procedure and the direct CI method are described. The second part of the chapter (sections 4 and 5) deals primarily with the cluster model for studying surface-adsorbate interactions. The one-electron ECP method is decribed in some detail, and the Core Polarization Potential (CPP) method is briefly outlined. Issues such as the d-relaxation effects, the importance of valence correlation on the surface-adsorbate bonding, cluster convergency etc. are adressed. The relation between a cluster and an infinite surface is analyzed, and the concept of bond preparation of clusters is introduced.

1. Introduction

The aim of the present chapter is to describe an *ab initio*-type cluster model which can be used to model chemisorption and chemical reactions on metal surfaces with quantitative accuracy. However, in order to discuss the cluster model it is necessary first to describe the molecular orbital concept and the *ab initio* methods on which the cluster model is based.

A molecular orbital describes the motion of one electron in the field generated by the nuclei and some average distribution of the other electrons in the system. This approximation is usually referred to as the independent particle model. The nature of the approximation is easy to see if we compare with the classical case where particles move in trajectories. The total energy of a classical system of charged particles at a given instant is determined by the position of all the particles at that instant, and not by the average charge distribution. We will return to the consequences of the independent particle approximation later.

D. R. Salahub and N. Russo (eds.), Metal-Ligand Interactions: from Atoms, to Clusters, to Surfaces, 199–249.
© 1992 *Kluwer Academic Publishers. Printed in the Netherlands.*

In quantum mechanics similar particles are indistinguishable, and by considering two sequential permutations of two such particles it is easily realized that their wave function must be either symmetric or antisymmetric with respect to particle interchange. Particles with half integral spin are called fermions and are observed to have antisymmetrical wave functions. The electron is a fermion and a many-electron system must thus be described by an antisymmetric wave function. In the independent particle model each electron is described by its own one-particle wave function, which is characterized by a set of quantum numbers. These one-particle wave functions or spin orbitals (SO:s) can be written as a product of a spatial part, which is the molecular orbital, and a spin part. The antisymmetry property of the wave function implies that two electrons cannot carry the same set of quantum numbers, i.e. they cannot occupy the same spin orbital. This is called the Pauli principle. In the independent particle model a many electron wave function thus takes the form of an antisymmetrized product of spin orbitals. It should be pointed out that in the real world two electrons cannot occupy the same point in space simultaneously, since the energy then becomes infinite. However, within the independent particle approximation two electrons (of opposite spin) can, to a certain degree of probability, actually occupy the same point in space simultaneously since each electron moves in the average density of the other electrons. At a more sophisticated level of treatment,when the motion of the electrons is correlated, the proper behaviour of the wave function is restored.

In *ab initio* or first principles methods the MO:s are usually expressed as linear combinations of some predetermined one-electron basis functions. The most common choice of the basis functions are products between Gaussians and some function of the coordinates describing the angular dependence. In practice one uses fixed linear combinations of such functions, where both the exponents and the coefficients are determined from atomic calculations. The problem of calculating the MO:s is thus reduced to determining the coefficients multiplying the known basis functions.

To go beyond the independent particle approximation requires a formalism where the electronic motion can be correlated to describe the effects of the instantaneous interaction between the particles. This is again usually done by expanding the unknown wave function in some predetermined basis functions, which now are many-electron functions. Normally Slater determinants (or linear combinations of Slater determinants) formed from a set of known MO:s are used for the purpose. This approach to the correlation problem, known as Configuration Interaction or CI, is very powerful, but can also be quite time consuming.

For large systems the computational problem soon gets out of hand, both at the independent particle and at the CI levels. Metal clusters containing a large number of atoms (10-100), e.g. metal surface models, cannot be handled even at the independent particle level if no simplifications are used. A very efficient approximation is the Effective Core Potential (ECP) method, where the effects of all core electrons are described by some simplified operators. At least for the first row ferromagnetic metals and for copper it is possible to model clusters using pseudo-one electron atoms, where all electrons except for the outermost s-electrons are described by some effective one electron operators. This cluster model allows calculations on clusters with up to at least a hundred metal atoms.

In the first section below we will give a brief description of the Hartree-Fock equations and the basis set expansion method. This is followed by a section where we discuss the correlation problem, with particular emphasis on dynamic correlation and direct CI methods. In this context we also discuss the important size consistency problem, and some of the more common size-consistent methods used today are briefly reviewed. The next section deals with two approximative methods, the ECP method which serves to approximate the core orbitals in molecular calculations and the Core Polarization Potential method by which the correlation energy due to the instantaneous polarization of the cores (including the d electrons in transition metals) can be estimated to good accuracy without explicit CI calculations. The last section is devoted to the cluster model for surface reactions, with particular emphasis on cluster convergency, i.e. the variation in computed chemisorption energies with cluster size.

A subject of fundamental importance in quantum chemistry is symmetry. In quantum chemical programs symmetry is primarily used to block the problem according to the irreducible representations of the point group of the system. This symmetry blocking is equally important at the independent particle level as in more general applications. Unfortunately it is not possible to cover the subjects of symmetry and group theory in the present chapter.

2. The Hartree-Fock equations

The molecular orbital (MO) concept is probably the most fundamental quantity in applied quantum chemistry. Almost all our models of bonding in molecules and of chemical reactions, as well as band models and surface models used for the condensed phase are based on the MO concept.

The fundamental equations which are used to calculate the molecular orbitals are the Hartree-Fock (HF) equations. These equations are obtained by varying a trial wave function, subject to a normalization condition, to minimize the energy. The form of the trial wave function is an antisymmetrized product, or equivalently a Slater determinant, which fulfills the Pauli principle. In general all MO:s will be different, i.e. we will have different MO:s for different spins, each MO holding one electron only. If the number of α and β spins are the same, the spatial part of SO:s with the same quantum numbers (except of course for the spin) will usually be the same, and in these cases the system will have a closed shell electron configuration. In the most commonly used HF procedures the condition that the spatial part of two SO:s should be the same when they only differ in the spin part is imposed beforehand, in which case only half of the MO:s have to be calculated. The former

202

method is called the Unrestricted Hartree-Fock (UHF) method while the latter, where one MO can hold two electrons with opposite spins is called the Restricted Hartree-Fock (RHF) method. Since a determinant is invariant to transformations among its rows and columns it follows that a wave function which can be written as a single Slater determinant will be invariant under rotations among the occupied orbitals (this is not true for rotations between open and closed orbitals in the RHF method because of the restrictions imposed, however).

Disregarding spin-orbit effects, a closed system of particles must be an eigenfunction of the spin operator, i.e. it has a definite value of the spin. It is easy to show that a closed shell system is a singlet, but the situation becomes more complicated for open shell systems. If the number of α and β spins is not equal in the UHF scheme we have what in the RHF picture would be called an open shell system. However, the resulting UHF wave function is not an eigenfunction to the spin operator, and properties such as multiplet structures cannot be studied with this method (the UHF method is quite appropriate for certain other problems, e.g. for calculating spin densities, and it has the advantage that it often gives correct dissociation products, but these subjects fall outside the scope of the present lecture series). The extension of the RHF formalism to include open shells is somewhat more complicated, but the resulting state has the advantage of being a true eigenfunction to the spin operator. The complication arises because the wave function is no longer invariant to rotations among the occupied orbitals (it is invariant to rotations among the closed shells, however), and a correct multiplet must in general be described by several determinants.

The form of the Hartree-Fock equations for a closed shell system is

$$\hat{F} = \hat{h} + \sum_{i=1}^{N}(2\hat{J}_i - \hat{K}_i) \tag{2.1}$$

where $\hat{h}$ is a one electron operator which includes the kinetic energy of the electrons and all external fields, e.g. the nuclear attraction, $\hat{J}_i$ are the coulomb operators describing the electrostatic repulsion between the electrons and $\hat{K}_i$ are the exchange operators which have no classical analog. The exchange terms occur because the form of the wave function is an antisymmetrized product, and they do not correspond to any operator in the Hamiltonian.

The Hartree-Fock equations are normally solved by expanding the unknown molecular orbitals in some preselected basis set. In matrix form the HF equation for the closed shell case reads:

$$F_{pq} = h_{pq} + \sum_{rs} D_{rs} P_{pq,rs} \tag{2.2}$$

where h_{pq} are the matrix elements $\int \chi_p(1)(\hat{T} + \hat{V})\chi_q(1)dr_1$, D_{rs} are density matrix elements given by $D_{rs} = 2\sum_i c_r^i c_s^i$ (the c:s are molecular orbital coefficients for MO number i), and $P_{pq,rs}$ are *super matrix* elements defined as:

$$Ppq, rs = (pq \mid rs) - \frac{1}{4}((pr \mid qs) + (ps \mid qr)) \qquad (2.3)$$

$$(pq \mid rs) = \int \chi_p(1)\chi_q(2)\frac{1}{r_{12}}\chi_r(1)\chi_s(2)dr_1dr_2 \qquad (2.4)$$

The most common type of basis sets used in molecular calculations are the Cartesian Contracted Gaussian Type Orbital (CGTO) basis sets, where the basis functions have the form:

$$f_{l,jkn}(x, y, z) = \sum d_i x^j y^k z^n e^{-\alpha_i r^2} \qquad (2.5)$$

where the azimuthal quantum number l is given by $j+k+n$. It should be noted that for e.g. a d-function there are six cartesian components (ijk = $200(x^2)$, $020(y^2)$, $002(z^2)$, $110(xy)$, $101(xz)$ and $011(yz)$). In most programs a transformation to the spherical harmonics representation is carried out at some point during the calculations, and the totally symmetric component $x^2+y^2+z^2$ (which is a 3s-type function) is deleted from the basis set. Similar transformations are also carried out for the higher azimuthal quantum numbers.

The contraction coefficients d_i and the exponents α_i are determined from atomic calculations. There are two different types of contraction schemes used today: the segmented contraction scheme where a given primitive Gaussian (usually) only appears once in the contracted functions and the general contraction scheme where all the contracted basis functions of a given type on a given centre may contain all the primitive functions of that type on that centre. There are several advantages with the general contraction scheme. Nearly optimal inner shell orbitals are easily obtained from the atomic calculations, which both increases the flexibility in the basis set choice (there are only a few segmented contracted basis sets available which allow a complete contraction of the inner shells) and minimizes the risk for basis set superposition errors (this problem is discussed below). It is also possible to construct Atomic Natural Orbital (ANO) basis sets which are particularly well suited for CI calculations. The general contraction scheme is intimately connected with modern vectorized computors.

Clearly a calculation will never be better than the basis set permits, and care has to be exercised when the basis set is selected. In general the basis set should be balanced, i.e. all atoms should be equally well described. If the SCF calculation is to be followed by a CI (see below) higher angular momentum functions must be included in the basis set. A serious pitfall is if too small basis sets are used, such that the basis set on one atom may improve the purely atomic character of a neighbour at short distances. A typical situation when this may occur is for segmented contracted minimal basis sets with a poor description of the inner shells. A lot of energy can then be gained from improving i.e. a 1s orbital, and this is what the basis set on a neighbouring atom will try to do. There are many published calculations on binding energies where most of the binding energy comes from this so called basis set superposition error (BSSE). This type of error is even more serious at the CI level of approximation; we will return to this point in the next section. The size of the BSSE can be estimated using the Counter-Poise method, which simply amounts to carrying out a calculation on a part of the system with only the basis set (and no nuclei) for the other part of the system present for some short distance geometry.

3. Beyond the independent particle model

The independent particle model is an approximation, and the question arises how large the error is and when the method is applicable. In the independent particle model an electron is only experiencing the average field of the remaining electrons in the system, while the instantaneous interaction is neglected. The motion of the electrons is thus not correlated in the classical sense, where charged particles can avoid each other at all instants. The difference between the "correct" energy (not including relativistic corrections) of the system and the energy obtained at the independent particle level is called the correlation energy. The question is thus how large the correlation energy is and to what extent it is necessary to account for it in e.g. chemical reactions. The answer is that it can sometimes be amazingly large. Take the example of the H_2 molecule. The molecular orbitals in H_2 which can be formed from two H 1s atomic orbitals are σ_g and σ_u. In the ground state of the H_2 molecule the σ_g orbital is doubly occupied while the σ_u orbital is empty. If we dissociate the molecule we arrive at a binding energy which is about 7 eV too large (incidentally a very reasonable binding energy for H_2 is obtained from a UHF wave function).

Suppose we use two determinants to describe the wave function for H_2:

$$\Psi = C_1\Phi_1 + C_2\Phi_2 \tag{3.1}$$

where Φ_1 and Φ_2 are Slater determinants with a doubly occupied σ_g and a σ_u MO respectively. At short internuclear separation C_1 is close to 1 while C_2 is small, while at large internuclear separation they are equal. Using this wave function, where C_1 and C_2 are variational parameters, we obtain a binding energy which is very reasonable (about 1 eV too low if we do not include p-functions in the calculation). The correlation energy is in this case as large as about 6 eV! The main reason is that the H_2 molecule does not dissociate correctly in the RHF picture, and at large internuclear separation the system is not described as two hydrogen atoms but as an equal mixture of two neutral atoms, a proton and a hydronium ion. This type of correlation effect is called near degeneracy or static correlation. If we instead consider the computed bond length, which is 0.735 Å at our present RHF level while the experimental result is 0.746 Å, the result is quite satisfactory. At short distances the HF approximation thus works quite well for H_2. The same type of situation occurs for many systems where a covalent bond is broken during the dissociation process.

A quite different correlation effect occurs e.g. if we want to compute an electron affinity. The difference between the energies of neutral oxygen and the oxygen anion is -0.54 eV at the HF level, i.e. the neutral oxygen is more stable than the anion, while the experimental result is 1.46 eV (this result has been reproduced in large CI calculations). The reason for the large correlation energy contribution to the electron affinity is that the charge density in the 2p shell is high, which makes the instantaneous interaction between the p-electrons important. The dominating effect is that the electrons try to stay on opposite sides of the nucleus (a type of correlation often referred to as left-right correlation). The correlation has a purely dynamic character (the electrons simply try to avoid each other at all instants). This type of correlation is consequently called dynamical correlation.

A reasonable definition of static and dynamical correlation would seem to be that static correlation is picked up at the Multi-Configuration SCF (MCSCF) level and dynamic correlation at the CI level. However, the difference between static and dynamic correlation is actually rather fuzzy. Since an MC expansion in principle be very long, it is necessary also to specify that the static correlation energy is obtained from short MC expansions $\Psi = \sum_i C_i \Phi_i$ and only from configurations with a C_i larger than a certain threshold (0.05-0.1).

In H_2 there are no appreciable near degeneracy effects at the equilibrium geometry, but the binding energy obtained with a two component MC is 4.13 eV. The improvement over the SCF result (using atomic calculations at large internuclear separations, which gives a dissociation energy of 3.64 eV) is about 0.5 eV, which is substantial (the coefficient in front of the σ_u^2-configuration is about 0.1). The experimental dissociation energy is 4.75 eV, and using the above definition of static correlation, we obtain a dynamical correlation contribution to the H_2 bond of 0.6 eV. The contribution from dynamical correlation to the binding energy when one covalent bond is broken is normally around 0.5 eV.

The H_2 example is a very simple one since only one covalent bond is broken, and the configurations which give correct dissociation products are easily identified (only the σ_g^2 and the σ_u^2 configurations have to be included in an MCSCF wave function). The structure of the bond is also very simple. A more complicated example of diatomic bonding is presented by the F_2 molecule, which also is bound by one sigma bond but where there are three lone pairs on each fluorine. In this case the binding energy obtained at the one configuration SCF level is -1.37 eV, i.e. the molecule is unbound, while the experimental binding energy is 1.68 eV. This rather small experimental value is a consequence of the repulsion from the non-bonding 2p electrons on fluorine. A two component MCSCF calculation gives a binding energy of 0.54 eV, which is a substantial improvement but still rather far from the experimental value (the error at the MCSCF level is 1.1 eV which is almost a factor of two larger than the corresponding error in H_2). The reason for the large correlation effect is the lone pairs on each fluorine. The lone pair electrons in the p-shells try to avoid the electrons in the bond at the same time as the electron cloud contracts, which gives rise to large dynamical correlation effects. In the most accurate CI calculations performed so far for F_2 eight reference states and large basis sets including g-functions were used. The error for D_e at this level is 0.02 eV, which is chemical accuracy (less than 1 kcal/mol). It should be noted that a CI calculation based on two reference configurations still gives an error of 0.5 eV for the binding energy. The reason is that 3p orbitals must be included in the reference space to allow for radial correlation.

Among the diatomic molecule formed from the first row atoms N_2, which is bound by a triple bond, is the most difficult to treat quantum chemically. The RHF value for the dissociation energy is about 5.3 eV, compared to the experimental D_e of 9.90 eV. Using all configurations needed for proper dissociation and a basis set including g-functions a dissociation energy of 9.8 eV is obtained, which is still 0.1 eV off the experimental value.

We have up to now only considered dissociation energies. If we instead look at bond distances and vibrational frequencies some interesting properties of the approximations used emerge. Consider the first row homonuclear diatomic molecules H_2, F_2 and N_2. The computed bond distances at the one configurational SCF level are 1.39 a_0, 2.50 a_0 and 2.013 a_0 respectively. At the MCSCF level the corresponding numbers are 1.42 a_0, 2.74 a_0, and 2.077 a_0, while the (multireference) CI results are 1.41 a_0, 2.672 a_0, and 2.081 a_0. The experimental values are 1.41 a_0, 2.668 a_0 and 2.074 a_0. The trend is thus the one configuration SCF calculations tend to underestimate the bond distances, MCSCF tend to overestimate them and MR-CI gives values close to experiment. The reason for these trends is systematic; one configurational closed shell wave functions dissociate incorrectly, which increases the energy too much when the bond is stretched from its equilibrium value. On the other hand the inclusion of only the dissociative configurations in the MCSCF calculations evidently favour longer bond distances. Exactly the same trends, with few exceptions, are found also for the vibrational frequencies (overestimated at the one-configuration SCF level and underestimated at the MCSCF level).

MC effects at equilibrium geometries are often very important for transition metal complexes. The reason is that for metal atoms the bonding frequently involves both the $3d$ and the $4s/4p$ orbitals, and the spatial extent of these orbitals are quite different. For nickel the $< r >$ values for the $3d$ and the $4s$ orbitals are about 1 a_0 and 3 a_0 respectively, which means that for at least one of these orbitals the bond to a ligand will have to be formed with a non optimal overlap.

Quantum chemical calculations on transition metals are in general quite difficult to perform. One reason is the one given above, but another severe problem is the number of possible states for an atom with an open d-shell, and the difficulty to describe these states satisfactorily. Take the Ni atom as an example. The experimental value for the splitting between the d^9 and the d^{10} states of the Ni atom is 1.74 eV. At the non-relativistic Hartree-Fock limit the splitting is as poor as 4.20 eV, even though there are no significant near degeneracy effects in any of these states. Adding relativistic effects (by perturbation theory) leads to an even worse result of 4.41 eV. In the following numbers this relativistic effect is added. The (essentially dynamic) correlation effect on the splitting is thus 2.67 eV. With one reference state, size consistency corrections and very large basis sets including h-functions the splitting is improved to 2.27 eV, which is still 0.53 eV away from the experimental value. This was actually the best computational result obtained until a few years ago. At this time a multireference treatment was performed where the most important $3d$ to $4d$ excitations were included in the reference space. Above a threshold of 0.01 there are about 20 such configurations for both states, but these configurations turn out to be much more important for the d^{10} state. At this MR-CI level the splitting is 1.87 eV using a basis set including g-functions, which is a quite acceptable value with an error of only 0.13 eV.

We will illustrate the problems encountered at the molecular level with the examples NiH and NiCO. For NiH we will concentrate our interest on the bond distance R_e and the frequency ω_e. The experimental values are 2.75 a_0 and 1993 cm^{-1}, respectively. At the Hartree-Fock level the results are 2.88 a_0 and 1726 cm^{-1}, which is contrary to the trends for molecules containing first row atoms. The main reason for these trends is the high polarizability of the $3d$ orbitals, which in a sense belong to the core orbitals of the Ni atom since they are far inside the $4s$ orbital. The instantaneous influence of this polarizability leads to a reduced repulsion for the between the $3d$ electrons and the valence electrons, which will shorten the bond. This is a typical dynamical correlation effect and is thus missing at the Hartree-Fock level but appears at the MR-CI level through configurations in which one electron is excited from the valence orbitals and the other is excited from the $3d$ (core) orbitals. These excitations are therefore normally called core-valence excitations. In a multi-reference CI treatment with a threshold selection of 0.05 the bond distance is decreased to 2.69 a_0 and the frequency is increased to 1971 cm^{-1}, which are acceptable values. To obtain high quantitative accuracy a lower selection threshold for the reference configurations of 0.02 is needed, just as for the splitting of the Ni atom. For NiH this leads to 29 reference configurations and the results are 2.76 a_0 and 1997 cm^{-1} with a basis set including several f-functions.

For both the nickel atom and for NiH it was necessary to go to very long CI expansions to reach high accuracy. The linear molecule NiCO will instead serve to illustrate the necessity to correlate many electrons in these systems. The D_e value for this molecule for dissociation into Ni and CO is experimentally uncertain but should be in the interval 30-40 kcal/mol. At the Hartree-Fock level a value of -67 kcal/mol is obtained. We have earlier seen an example where the dissociation energy is negative at the Hartree-Fock level, namely for F_2, but NiCO is thus much worse. Since there is no actual standard bond formation in this molecule one might suspect that the main reason for the poor value lies in the treatment of the $3d$ electrons. This turns out to be the case and an MR-CI calculation (reference selection threshold 0.05) correlating only the nickel $3d$ and $4s$ electrons leads to a positive value for D_e of 16 kcal/mol, which is still less than half of the actual dissociation energy. The only electrons on CO which are close to the nickel atom are the carbon lone pair electrons. Correlating also these electrons improves D_e to 24 kcal/mol. Continuing to account for the correlation effects of the remaining CO valence electrons gradually improves D_e to a value of 33 kcal/mol, with even the carbon and oxygen $2s$ electrons contributing a few kcal/mol to the dissociation energy. If high quantitative accuracy should be reached it is thus necessary to correlate all valence electrons in a rather large surrounding of the transition metal atoms.

We have mentioned two extensions to the independent particle model: the MCSCF method where the orbitals are optimized at the same time as the coefficients in a (usually relatively) short CI expansion, and the general CI methods where the Hamiltonian matrix is diagonalized in a given many-electron basis set (Slater determinants or spin adapted Configuration State Functions which are linear combinations of Slater determinants). We will be very brief in the description of MCSCF since this method is less central in cluster applications. We will only mention one of the current methods, the CASSCF method, where the CI expansion includes all configurations which can be formed when a number of valence electrons are distributed among a set of valence orbitals (i.e. a full CI in that space). However, before describing the MCSCF and the CI methods it is necessary to give a brief introduction to the method of second quantization, which forms the basis for all modern CI methods.

3.1 SECOND QUANTIZATION

Suppose that we have at our disposal an ordered set of orthonormalized spin orbitals:

$$\{\phi_i; i = 1, 2n\} \tag{3.2}$$

where each SO is a product of a space and a spin part (we use the RHF picture, i.e. the number of space orbitals is n). With these SO:s we can construct an N-electron basis consisting of $2n!/(N!(2n-N)!)$ determinants. These determinants can be written in the *occupation number* representation as:

$$| 1, 1, 0, 0, 0, 1, 0, 1, 1 > \tag{3.3}$$

which signify a Slater Determinant with the occupied SO:s 1,2,6,8,9. In general we write a Slater determinant as

$$| m_1, m_2, m_3, \ldots\ldots, m_{2n} > \tag{3.4}$$

where m_i is the occupation number (0 or 1) for SO number i. The SO:s must be ordered, since the wave function is antisymmetric with respect to (occupied) orbital interchange. The occupation number representation of the Slater determinants define a vector space, the *Fock space*, comprising determinants with all possible occupation numbers including the vacuum state where all occupations are zero:

$$| vac >=| 0_1, 0_2, 0_3, ...0_{2n} > \tag{3.5}$$

An annihilation operator is defined as:

$$\hat{a}_i | m_1, m_2, ..., 1_i, ..., m_{2n} >= m_i (-1)^{p_i} | m_1, m_2, ..., 0_i, ..., m_{2n} > \tag{3.6}$$

where p_i is the number of transpositions needed to move orbital i to the first position in the ket. The corresponding creation operator is defined as:

$$\hat{a}_i^+ | m_1, m_2, ..., 0_i, ..., m_{2n} >= (1 - m_i)(-1)^{p_i} | m_1, m_2, ..., 1_i, ..., m_{2n} > \tag{3.7}$$

Note that the effect of the operators is zero if the orbital occupation should become different from 0 or 1 as a result of the operation.

It may be shown that the annihilation and the creation operators are adjoint to each other and that they fulfill the following anti-commutator relations:

$$\hat{a}_i \hat{a}_j + \hat{a}_j \hat{a}_i = 0 \tag{3.8}$$

$$\hat{a}_i^+ \hat{a}_j^+ + \hat{a}_j^+ \hat{a}_i^+ = 0 \tag{3.9}$$

$$\hat{a}_i^+ \hat{a}_j + \hat{a}_j \hat{a}_i^+ = \delta_{ij} \tag{3.10}$$

From the annihilation and creation operators we can define operators which generate single, double etc. excitations from a Slater determinant. The single excitation operator, which excites an electron from SO j to SO i is:

$$\hat{a}_i^+ \hat{a}_j | m_1, m_2, .., 0_i, .., 1_j, .., m_{2n} >= (-1)^{p_j - p_i} | m_1, m_2, .., 1_i, .., 0_j, .., m_{2n} > \tag{3.11}$$

where we assume $j \geq i$. Since we are using the same spatial orbitals for both α and β spin, we can define a spin summed excitation operator defined as:

$$\hat{E}_{ij} = (\hat{a}_{i\alpha}^+ \hat{a}_{i\alpha} + \hat{a}_{i\beta}^+ \hat{a}_{i\beta}) \tag{3.12}$$

The indices i and j now refer to the n spatial molecular orbitals. The $\hat{E}_{ij}$ are called the generators of the unitary group since they fulfill the appropriate commutation relations

$$[\hat{E}_{ij}, \hat{E}_{kl}] = \hat{E}_{il} \delta_{jk} - \hat{E}_{kj} \delta_{il} \tag{3.13}$$

210

which follow from (3.8)-(3.10). The following relations follow immediately from the definition of the generators:

$$\hat{E}_{ij}^+ = \hat{E}_{ji} \tag{3.14}$$

$$\hat{E}_{ii} \mid m > = n_i \mid m > \tag{3.15}$$

where n_i is the occupation number for molecular orbital i (0,1 or 2).

All operators can now be written in the second quantization language. For a one electron operator it is easy to show that

$$\hat{F} = \sum_{i,j}^{SO} F_{ij} \hat{a}_i^+ \hat{a}_j \tag{3.16}$$

with

$$F_{ij} = \int \phi_i^*(x) \hat{F}(x) \phi_j(x) dx \tag{3.17}$$

where x denotes both space and spin coordinates. Expressed in the generators rather than in the creation and annihilation operators the matrix elements become:

$$\hat{F} = \sum_{i,j}^{MO} F_{ij} \hat{E}_{ij} \tag{3.18}$$

where i and j are molecular orbital indices. A matrix element between two Slater determinants can be written:

$$< m \mid \hat{F} \mid n > = \sum_{i,j} F_{ij} < m \mid \hat{E}_{ij} \mid n > = \sum_{i,j} F_{ij} D_{ij}^{mn} \tag{3.19}$$

D_{ij}^{mn} are called one-electron coupling elements. If we have a wave function Ψ on the form:

$$\Psi = \sum_m c_m \mid m > \tag{3.20}$$

we have for the first order density matrix:

$$D_{ij} = < \Psi \mid \hat{E}_{ij} \mid \Psi > = \sum_{mn} c_m^* c_n D_{ij}^{mn} \tag{3.21}$$

For the two-electron operator $\frac{1}{r_{12}}$ we have similarly:

$$\hat{G} = \frac{1}{2} \sum_{ijkl} (ij \mid kl)(\hat{E}_{ij}\hat{E}_{kl} - \delta_{jk}\hat{E}_{il}) \tag{3.22}$$

where $(ij \mid kl)$ are two-electron integrals. The matrix elements of $\hat{G}$ can be written:

$$< m \mid \hat{G} \mid n > = \sum_{ijkl} (ij \mid kl) P_{ijkl}^{mn} \qquad (3.23)$$

where the two-electron coupling coefficients P_{ijkl}^{mn} are defined as

$$P_{ijkl}^{mn} = \frac{1}{2} < m \mid \hat{E}_{ij}\hat{E}_{kl} - \delta_{jk}\hat{E}_{il} \mid n > \qquad (3.24)$$

The Hamiltonian in the second quantization language becomes

$$\hat{H} = \sum_{ij} h_{ij}\hat{E}_{ij} + \frac{1}{2}\sum_{ijkl} (ij \mid kl)(\hat{E}_{ij}\hat{E}_{kl} - \delta_{kj}\hat{E}_{il}) \qquad (3.25)$$

where h_{ij} are the one electron integrals including the kinetic energy and the nuclear attraction terms, and g_{ijkl} are the two electron repulsion integrals. The total energy in second quantized form reads:

$$E = \sum_{ij} h_{ij}D_{ij} + \sum_{ijkl} (ij \mid kl)P_{ijkl} \qquad (3.26)$$

where D_{ij} is the first order density matrix and P_{ijkl} is the second order density matrix defined as:

$$P_{ijkl} = \sum_{mn} c_m^* c_n P_{ijkl}^{mn} \qquad (3.27)$$

The problem of calculating the matrix elements over the Hamiltonian is in the second quantization formalism reduced to calculating the one and two-electron coupling coefficients and a two-electron integrals list over the molecular orbitals. This approach has generated something of a revolution in CI calculations.

3.2 THE CASSCF METHOD

In the CASSCF method each symmetry block of MO:s is divided into the following subsets:

1. Inactive orbitals
2. Active orbitals
3. External orbitals

The inactive and active orbitals are occupied in the wave function, while the external orbitals span the rest of the orbital space. The total wave function is formed as a linear combination of all the configurations, in the N-electron space, that fulfill the given space and spin symmetry requirements. The wave function is thus "complete" in the configuration space spanned by the active orbitals. The inactive orbitals are always doubly occupied, while the active orbitals have fractional occupation numbers since they are occupied only in some reference configurations.

The eigenvalue problem is solved either by the Newton-Raphson method or by the Super CI method.

The expansion coefficients (for the moment we only consider the MO:s) are determined by minimizing the expectation value for the total energy. Assume that the energy is a function of a set of parameters p_i, which we arrange as a column vector $\mathbf{p}$. A Taylor expansion of the energy $E = E(\mathbf{p})$ around a point $\mathbf{p_0}$, which can be arbitrarily set to zero gives:

$$E(\mathbf{p}) = E(0) + \sum_i \frac{\partial E}{\partial p_i} + \frac{1}{2} \sum_{i,j} p_i \frac{\partial^2 E}{\partial p_i \partial p_j} p_j + .. \tag{3.28}$$

or in matrix notation

$$E(\mathbf{p}) = E(0) + \mathbf{g}^+ \mathbf{p} + \frac{1}{2} \mathbf{p}^+ \mathbf{H} \mathbf{p}.. \tag{3.29}$$

$$g_i = \frac{\partial E}{\partial p_i} \tag{3.30}$$

$$H_{ij} = \frac{\partial^2 E}{\partial p_i \partial p_j} \tag{3.31}$$

The vector $\mathbf{g}$ and the matrix $\mathbf{H}$ are called the energy gradient and the Hessian matrix. The stationary points on the energy hypersurface are obtained by putting $\frac{\partial E}{\partial \mathbf{p}}$ equal to zero. In the Newton-Raphson procedure terms up to second order are retained in the expansion of the energy, and we obtain the equations:

$$\mathbf{g} + \mathbf{H}\mathbf{p} = 0 \quad \text{or} \quad \mathbf{p} = -\mathbf{H}^{-1}\mathbf{g} \tag{3.32}$$

which gives a new $\mathbf{p_0}$. The Newton-Raphson procedure is quadratically convergent close to the true minimum.

The orbital optimization involves a unitary rotation of the orbitals between the three spaces (the energy is invariant to rotations within each subspace). A unitary transformation can always be written as

$$\hat{U} = e^{-\hat{T}} \tag{3.33}$$

where $\hat{T}$ is an antihermitian operator. $\hat{T}$ can be expressed in terms of the generators of the unitary group as

$$\hat{T} = \sum_{i,j} T_{ij} \hat{E}_{ij} = \sum_{i>j} T_{ij} (\hat{E}_{ij} - \hat{E}_{ji}) \tag{3.34}$$

where the antisymmetric property of $\hat{T}$ has been used in the last step. When the unitary operation $\hat{U}$ is applied to the orbital set the Slater determinants will evidently also be transformed. It can be shown that

$$| m' >= e^{-\hat{T}} | m >$$ (3.35)

where $| m >$ is a Slater determinant. The variational parameters in the (orbital) optimization step are T_{ij}.

The Newton-Raphson procedure converges very fast provided that the trial MO:s are good. However, since the full Hessian matrix has to be calculated the method is quite time consuming, and a cheaper alternative is provided by the Super CI method. In the Super CI method the calculations start by solving the CI problem using a set of trial MO:s. A function space is constructed from all singly excited states which can be obtained from the reference state, and the Super CI state is constructed from these so called Brillouin states:

$$| pq >= (\hat{E}_{pq} - \hat{E}_{qp}) | 0 >$$ (3.36)

$$| SCI >=| 0 > + \sum_{p<q} t_{pq} | pq >$$ (3.37)

where either p or q is occupied. The Super CI method implies solving the corresponding secular problem using t_{pq} as the exponential parameters for the orbital rotation. At convergency all t_{pq} will vanish. The Super CI method is in in this form a linearly convergent method.

A normal computational procedure is to start with the Super CI method until the MO:s are reasonably converged, and then switch to the Newton-Raphson method.

An attractive alternative to a full Newton-Raphson calculation is to use approximative Hessian matrices combined with a line search procedure. An approximative Hessian can be generated iteratively during the SCF cycles by an update procedure. This method has been shown to be very efficient for transition metal clusters, which normally are very difficult to converge with linear methods.

It should be mentioned that an expression similar to (3.35) may be set up also for the CI coefficient optimization. By combining these expression one obtains a formula for a full one step optimization procedure.

3.3 THE CI METHOD

The configuration interaction equations are derived using the variational method. Here, one starts out by writing the energy as a functional F of the approximate wavefunction ψ.

$$F = \frac{< \psi | \hat{H} | \psi >}{< \psi | \psi >}$$ (3.38)

The value of the functional has to be larger than or equal to the lowest exact eigenvalue of the Hamiltonian. This is trivial to see if the approximate wavefunction ψ is expanded in terms of the complete set of exact eigenfunctions to the Hamiltonian operator, as obtained from the Schrödinger equation. The variational method is said to yield an "upper bound" to the true energy. In the derivation of the CI equations the wavefunction is written as

$$\psi = \sum_i C_i \Phi_i \tag{3.39}$$

where Φ_i are (orthogonal) configurations which are chosen as described below. Assuming the configurations to be orthogonal, minimization of the energy functional with respect to the parameters C_i leads to the CI equations

$$\sum_j (H_{ij} - E\delta_{ij})C_j = 0 \tag{3.40}$$

where E is the energy and H_{ij} are the Hamiltonian matrix elements between the configurations. Unlike the MCSCF method the form of the configurations, i.e. the orbitals, are not varied in the CI method.

Most CI methods today take the spin symmetry explicitly into account, i.e the many-electron basis functions are eigenfunctions of the spin operator (Configuration State Functions or CSF:s). The effect of spin adapting the configurations in the CI expansion is that the individual configurations no longer are single but linear combinations of Slater determinants with fixed spin adaption coefficients. The brute force evaluation of matrix elements between configurations will therefore be to first evaluate the matrix elements between the individual Slater determinants building up the configurations followed by a summation of these matrix elements using the spin adaptation coefficients. This is an inefficient procedure, and a large part of the development of modern CI methodologies has been devoted to the problem of obtaining matrix elements directly over spin adapted configurations. Here the language of the unitary group has been particularly useful and we will return to how this formalism has been used for the evaluation of matrix elements between spin adapted configurations later on in this chapter.

There are some very useful formulas for the spin operators, which we will give here for reference. From the definition of $\hat{S}^2$ and by using the step up and step down spin operators it is easy to show that

$$\hat{S}^2 = \hat{S}_z(\hat{S}_z + 1) + \hat{S}_-\hat{S}_+ \tag{3.41}$$

where for a many-electron system

$$\hat{S}_z = \sum_i \hat{s}_{zi} \tag{3.42}$$

and similarly for $\hat{S}_+$ and $\hat{S}_-$. The sum is over the electrons in the system. From the above expression for $\hat{S}^2$ it is reasonably simple to show that when $\hat{S}^2$ acts on a determinant the result can be written as

$$\hat{S}^2\Phi = \{\sum_P \hat{P}_{\alpha\beta} + \frac{1}{4}[(n_\alpha - n_\beta)^2 + 2n_\alpha + 2n_\beta]\}\Phi \qquad (3.43)$$

where $\hat{P}_{\alpha\beta}$ is an operator which exchanges α and β spins in Φ. The sum is over all possible such interchanges. It is easily realized from the expression for $\hat{S}^2$ that the contribution from doubly occupied orbitals to the spin expectation value will be zero, which means that these orbitals can be disregarded when the electrons of a configuration are spin coupled.

In order to obtain a spin adapted CSF directly from a Slater determinant the projection operator technique can be used. The operator

$$\hat{O}_K = \frac{\hat{S}^2 - K(K+1)}{S(S+1) - K(K+1)} \qquad (3.44)$$

will annihilate the spin component with S=K in the slater determinant. In order to obtain a spin adapted CSF all undesired spin-components are annihilated until a proper spin state is found (or 0 in which case one has to start out with another trial determinant). This method will not automatically lead to orthogonal spin-adapted configurations, but has to be followed by an orthogonalization procedure if this is desired.

3.3.1. *The MR-CI configuration expansion.* Suppose that the Hartree-Fock configuration is a qualitatively good approximation to the true wave-function. We can then select the Hartree-Fock configuration as a single reference state. Since the Hamiltonian operator does not contain more than two-electron operators only determinants which differ by at most two orbitals will have non-zero matrix elements with each other. From perturbation theory we know that the configurations which have non-zero matrix elements with the zeroth order wave-function should be the dominant corrections to the wave-function. Therefore, the simplest choice of CI expansion, which is actually often a very good approximation to the true wave-function, can be written as

$$\Psi_{SD} = C_0\Phi_0 + \sum_{ia} C_i^a \Phi_i^a + \sum_{ijab} C_{ij}^{ab} \Phi_{ij}^{ab} \qquad (3.45)$$

where Φ_0 is the Hartree-Fock configuration and Φ_i^a and Φ_{ij}^{ab} are single and double replacements (excitations) out of Φ_0. The occupied (internal) orbitals i,j are replaced by (excited to) the unoccupied (external or virtual) orbitals a,b. This wave-function is usually called the SD-CI (singles and doubles CI) wave-function. Including size-consistency corrections, which will be described later in this chapter, this type of wave-function has been shown to yield results of very high quantitative accuracy for cases where the Hartree-Fock wave-function is a good zeroth order wave-function, i.e. normally for molecules around their equilibrium geometries.

In situations where the Hartree-Fock wave function is a less adequate zeroth order starting point the SD-CI formalism must be extended. We have already discussed how the qualitative problems with the single-configuration Hartree-Fock approximation can normally be corrected by adding a few other configurations. The simple idea to extend the SD-CI wavefunction is then to add also the single and double replacements from these other important configurations, which in the CI formalism will be termed reference states. The wave-function can then be written

$$\Psi_{MR-CI} = \sum_I \Psi_{SD}(I) = \sum_I \{ C(I)\Phi(I) + \sum_{ix} C_i^x(I)\Phi_i^x(I) + \sum_{ijxy} C_{ij}^{xy}(I)\Phi_{ij}^{xy}(I) \}$$

(3.46)

where we have used the letters x and y, rather than a and b for the orbitals to which electrons are excited, to emphasize that an orbital which is unoccupied (external) for one reference state may be occupied (internal) for another reference state. The sum over I runs over all selected reference states. This is the MR-CI (multi-reference CI) wave-function. It should be noted that since a single or double replacement from one reference state can also be a single or double replacement from another reference state, a unique list of configurations has to be constructed by deleting doubly counted configurations. It turned out that for somewhat technical reasons the extension of the direct CI method to the MR-CI expansion became very difficult. Perhaps surprisingly, a major step towards the generalization of the direct CI method was taken by rewriting the MR-CI wavefunction in the following way,

$$\Psi_{MR-CI} = \sum_\mu C_\mu \Phi_\mu + \sum_{\mu a} C_\mu^a \Phi_\mu^a + \sum_{\mu ab} C_\mu^{ab} \Phi_\mu^{ab}$$

(3.47)

where the letters a and b have been used for orbitals which are unoccupied in all reference states and μ is used as a compound index for the occupied orbitals. This can be exactly the same wave-function as (3.46) but the number of external orbitals rather than the excitation level for the configurations has been emphasized. This reformulation turns out to reduce the difficulties mentioned above drastically. The three groups of configurations in (3.47) will in the following be denoted as valence, singly external and doubly external configurations. The relation between the groups of configurations in (3.46) and (3.47) is the following. The configurations in the first group in (3.46), the reference configurations, is included in the valence configurations in (3.47). The single excitations in (3.46) can be either valence or singly external configurations, whereas the double excitations in (3.46) can belong to any of the three groups in (3.47). One should also note that all the doubly external configurations in (3.47) are double excitations from the reference states but this group does not contain the full list of double excitations.

3.3.2 *The occupation graph and the branching diagram.*   One of the first things that is done in an MR-CI program is to generate a unique set of configurations. For this purpose the occupation graph and the branching diagram are very useful. An example of an occupation graph is given in Fig.1.

The idea is that all possible occupations in the MR-CI expansion can be generated by following a path in the diagram from the bottom to the top of the graph. The simplest organization is to leave out the external orbitals and instead have three different graphs for the internal orbitals, one for each group of configurations in (3.47). The graph consists of points called vertices and lines called arcs. The vertices are arranged along horizontal lines, one horizontal line for every orbital. The inclination of the arcs gives the orbital occupation. The arc with the smallest angle with the horizontal lines denotes double occupation, the one with a somewhat larger angle denotes single occupation and the vertical arc denotes zero occupation. At each vertex in the graph a number is given which is equal to the total number of electrons at that vertex summed along any path to that vertex from the bottom of the graph. This number must consequently not exceed the total number of electrons that should be distributed in the internal orbitals, which means that there can be no vertices to the left of the top vertex. In the construction of the graph one must also exclude vertices which can not be reached from the top vertex even if all the top orbitals are doubly occupied. By having the orbitals which are doubly occupied in all reference states at the bottom of the graph it is furthermore easy to exclude vertices with more than two holes in these orbitals. Any further reduction of the vertices and arcs in the graph is not generally possible for a normal MR-CI case. Therefore the occupations which are generated by following all possible paths from the bottom to the top of the graph will normally exceed the number which is actually going to be used in the calculation. These unwanted occupations have to be deleted by a comparison to each individual reference state.

Having constructed all the desired occupations in the MR-CI expansion, there remains to construct all possible spin-couplings for each occupation, before a full list of configurations is obtained. The total number of spin-couplings for a given number of singly occupied orbitals is conveniently given by the branching diagram which is shown in Fig.2.

The x-axis in this diagram gives the number of electrons N in singly occupied orbitals and the y-axis gives the total spin quantum number S. The number given at each vertex in the graph is equal to the number of spin-couplings for given values of N and S. The construction of this diagram starts from the bottom left part. The number at this vertex gives the number of spin-couplings when there are no singly occupied orbitals. This number clearly has to be one. Arcs are then

218

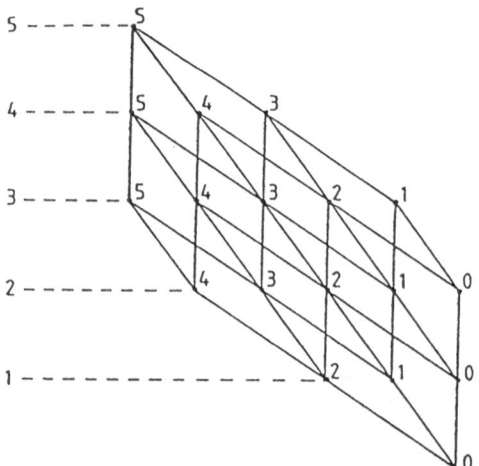

Figure 1. Occupation graph with 5 electrons in 5 orbitals

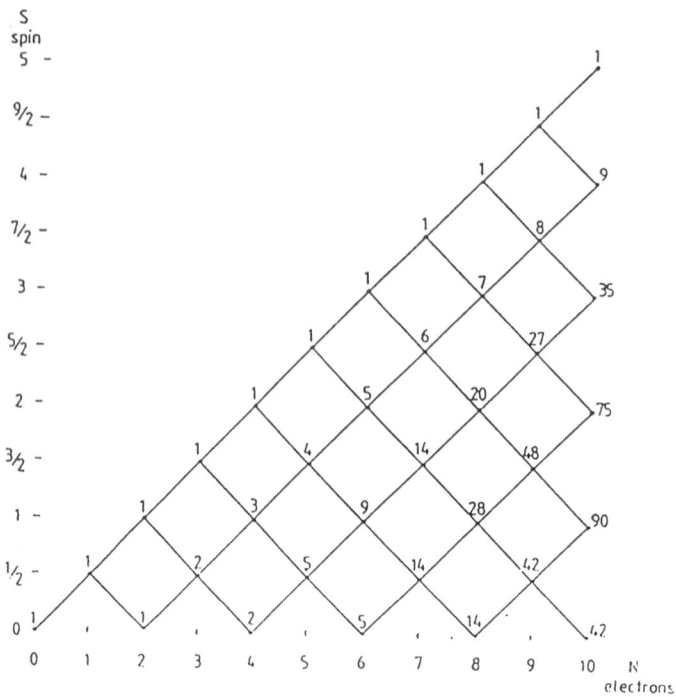

Figure 2. Branching diagram

drawn to connect the vertices with a certain number of singly occupied orbitals to the vertices with one more singly occupied orbital. The numbers at the vertices are obtained by adding the numbers of the vertices which are connected to that vertex by the arcs coming from the left, i.e. in most cases two numbers. The fact that there are two connecting vertices is a simple consequence of the fact that the spin of an electron can couple in two ways, up or down, to generate $S=S+1/2$ and $S=S-1/2$ respectively. One notable fact in this diagram is that the number of possible spin couplings grows rapidly with the number of singly occupied orbitals. Since for many methods, prototype matrix elements between different spin-couplings have to be stored in core-storage, a calculation using these methods can be limited by the maximum allowed number of singly occupied orbitals in the program. This organisation using separate graphs for the occupations and of the spin-couplings is referred to as the "symmetric group" approach to CI.

We will end this section by pointing out that there also exists a graphical representation where both the occupations and spin-couplings are given for each configuration. This graph, which was actually prior in time to the occupation graph described above, was first used by Shavitt. An organisation where occupation and spin-coupling are treated on an equal footing is usually referred to as the "unitary group" approach to CI.

3.3.3 *Diagonalization and the direct CI equations.*

The direct CI equations are obtained by combining the normal CI equations (3.40) with an iterative diagonalization procedure. (The same direct CI equations can also be obtained within a perturbation theory approach). The simplest iterative procedure is obtained by moving everything but the diagonal terms in the CI equations over to the right hand side and assume that this side of the equations can be obtained from the CI vector of the previous iteration $\mathbf{C}^k$. An improved CI vector $\mathbf{C}^{k+1}$ is then obtained as,

$$C_\mu^{k+1} = C_\mu^k + \frac{1}{E^k - H_{\mu\mu}} (\sum_\nu H_{\mu\nu} C_\nu^k - E^k C_\mu^k) \qquad (3.48)$$

where E^k is the variational energy evaluated using the vector $\mathbf{C}^k$. The start, $\mathbf{C}^0$, of this iterative procedure is normally obtained by a diagonalization of the small reference space, which can usually be done using the Jacobi method. A straightforward application of this iterative procedure will not be very efficient and actually diverge in many cases. A very efficient scheme is instead obtained by setting up a small CI equation system where the basis vectors are the $\mathbf{C}^k$ vectors. The Hamiltonian matrix elements over these vectors are very easily obtained as a byproduct in constructing $\mathbf{C}^{k+1}$. The dimension of this secular equation system is thus equal to the number of CI iterations and the equation system is therefore always so small that it can easily be solved. This is the Davidson diagonalization method.

The most important point in the above outlined procedure is that essentially all the work in a CI iteration is performed in generating the vector σ, which can be written as

$$\sigma_\mu = \sum_\nu H_{\mu\nu} C_\nu^k \tag{3.49}$$

In order to be able to write out all the terms of the direct CI equations explicitly, the Hamiltonian operator is needed in a form where the integrals appear. This is done using the language of second quantization. The Hamiltonian written in terms of the generators of the unitary group is:

$$\hat{H} = \sum_{pq} h_{pq} \hat{E}_{pq} + \frac{1}{2} \sum_{pqrs} (pq \mid rs)(\hat{E}_{pq}\hat{E}_{rs} - \delta_{qr}\hat{E}_{ps}) \tag{3.50}$$

which is the same equation as (3.25). Combining this expression for the Hamiltonian with expression the expression for σ, the final direct CI equations for the σ vector can be written as

$$\sigma_\mu = \sum_\nu \{\sum_{pq} h_{pq} A_{pq}^{\mu\nu} + \frac{1}{2} \sum_{pqrs} (pq \mid rs) A_{pqrs}^{\mu\nu}\} C_\nu \tag{3.51}$$

where

$$A_{pq}^{\mu\nu} = <\mu \mid \hat{E}_{pq} \mid \nu> \tag{3.52}$$

and

$$A_{pqrs}^{\mu\nu} = <\mu \mid \hat{E}_{pq}\hat{E}_{rs} - \delta_{qr}\hat{E}_{ps} \mid \nu> \tag{3.53}$$

We will in the following call these coefficients "the direct CI coupling coefficients" or simply the coupling coefficients.

The main point in this reformulation is that the explicit reference to the Hamiltonian matrix is avoided. The so called "residual vector" σ is obtained directly from the one- and two-electron integrals, the previous CI vector and the coupling coefficients. It is clear from the expression for σ_μ that the number of coupling coefficients in the direct CI equations is as large as the number of Hamiltonian matrix elements. Since these coefficients do not depend on the actual values of the integrals but only on the structure of the interacting configurations they can in principle be calculated as they are needed given the proper algorithms. Of key importance for the efficiency of the direct CI method is thus that extremely efficient such algorithms for the direct CI coupling coefficients can be obtained.

3.3.4 *Evaluation of the direct CI coupling coefficients.* A key step in the development of the multi reference direct CI method was the realization that the full list of coupling coefficients $A_{pqrs}^{\mu\nu}$ can be generated from a small set of prototype coupling coefficients $B_{pqrs}^{\mu\nu}$, initially termed "internal" coupling coefficients. The relation between these coefficients can simply be written as

$$A_{pqrs}^{\mu\nu} = B_{pqrs}^{\mu\nu} C_{pqrs}^{\mu\nu} \tag{3.54}$$

where $C_{pqrs}^{\mu\nu}$ were initially termed "external" coupling coefficients. The idea is to determine the internal coupling coefficients within an orbital space containing the full list of internal orbitals but replacing the long list of external orbitals by a few prototype external orbitals. The function of the external coupling coefficients $C_{pqrs}^{\mu\nu}$ would then be to relate a particular prototype internal coupling coefficient $B_{pqrs}^{\mu\nu}$ to a general coupling coefficient $A_{pqrs}^{\mu\nu}$ with the same internal orbitals involved as in $B_{pqrs}^{\mu\nu}$, but where the external orbitals involved will be completely general. The key to the efficiency of this reformulation is first that the number of internal coupling coefficients is so small that they can be stored on a peripheral device and also that the external coupling coefficients are extremely easy to determine from the structure of the interaction. Due to the simple structure of the external coupling coefficients we will in the following make no distinction between the full coupling coefficients $A_{pqrs}^{\mu\nu}$, and the internal coupling coefficients $B_{pqrs}^{\mu\nu}$.

The presently most efficient method for evaluating the direct CI coupling coefficients is based on a factorization algorithm. The steps in this factorization are easy and are based on the use of projection operators. In an infinite configuration basis the following identity holds

$$\sum_{\kappa} | \kappa >< \kappa | = 1 \tag{3.55}$$

This expression is often called "the resolution of the identity". The expression (3.3) for the two-electron coupling coefficient can now be rewritten by introducing the projection operator between the generators of the unitary group. The result is

$$A_{pqrs}^{\mu\nu} = \sum_{\kappa} A_{pq}^{\mu\kappa} A_{rs}^{\kappa\nu} - \delta_{qr} A_{ps}^{\mu\nu} \tag{3.56}$$

At first sight this reformulation only seems to complicate matters since an infinite sum over configurations is introduced. It should, however, be noted that the sum is only infinite in principle. First, it is clear that since the generators of the unitary group replace one of the orbitals in the selected orbital set by another orbital in the same set, there will be no matrix elements for these operators outside the chosen orbital space (Hilbert space). This restriction still makes the summation very long, in principle over the complete CI list in the selected orbital basis. If we now look at each individual matrix element, with fixed values of μ,ν,p,q,r and s, we find that the summation in (3.56) is cut down to a quite tractable length. It is clear that for this coupling coefficient to be different from zero the occupation of κ must simultaneously be a single excitation from p to q in μ and s to r in ν.

The in principle infinite sum over configurations in (3.56) is thus for a particular matrix element reduced to a sum where κ has at most one particular orbital occupation. In fact, most matrix elements over the product $E_{pq}E_{rs}$ for different p,q,r,s and between fixed μ and ν will be zero, which is of course a consequence of the normal selection rules between determinants. The reduction of the sum to a single occupation for a particular matrix element does not mean that there is only one term in the sum. For a particular occupation of κ there are still normally several different possible spin couplings which should enter into the summation. The final form of the two-electron coupling coefficient can consequently be written

$$A_{pqrs}^{\mu\nu} = \sum_{\kappa_s} A_{pq}^{\mu\kappa_s} A_{rs}^{\kappa_s\nu} - \delta_{qr} A_{ps}^{\mu\nu} \tag{3.57}$$

where κ_s only runs over the spin couplings of the fixed occupation in κ. The expression for a two-electron coupling coefficient is thus reduced to a rather simple expression of products of one-electron coupling coefficients.

3.3.5. Approximate CI methods.

Since the solution of the CI problem with a long CI expansion is time consuming there is often a need for faster approximate methods. In this section we will describe two such methods both based on contractions of the CI expansion.

In the so called "externally contracted CI" (CCI) method the general MR-CI expansion (3.47) is rewritten as

$$\Psi_{MR-CI} = \sum_{\mu(N)} C_\mu \Phi_\mu + \sum_{\mu(N-1)} C_\mu \sum_a \tilde{C}_\mu^a \Phi_\mu^a + \sum_{\mu(N-2)} C_\mu \sum_{ab} \tilde{C}_\mu^{ab} \Phi_\mu^{ab} =$$

$$= \sum_\mu C_\mu \varphi_\mu \tag{3.58}$$

where the summations have been written to emphasize that the internal part of the configurations contain different number of electrons in each sum. The C_μ are obtained variationally but the contraction coefficients $\tilde{C}_\mu^a$ and $\tilde{C}_\mu^{ab}$ are obtained by perturbation theory as

$$\tilde{C}_\mu^{ab} = \frac{<O \mid \hat{H} \mid \Phi_\mu^{ab}>}{E_0 - <\Phi_\mu^{ab} \mid \hat{H} \mid \Phi_\mu^{ab}>} \tag{3.59}$$

The number of variational parameters in (3.58) is therefore much smaller than it is in (3.47) to the price that some accuracy is lost and that each function φ_μ is more complicated than the original configurations Φ_μ^{ab}. The loss of correlation energy by the contraction is usually on the order of 1-3 %. It should be added that the CCI approximation becomes better the larger the reference space is, and it is not to be recommended if configurations from a CASSCF calculation with coefficients larger than 0.10 are not included in the reference space.

The second approximate scheme we will discuss here is the internally contracted CI (ICCI) method. In this method correlating configurations are formed by applying excitation operators (the generators of the unitary group) directly on the full reference CI vector. The four types of configurations thus formed can be written as

$$\Psi_0 = |\ O\ > = \sum_\mu C_\mu \Phi_\mu \tag{3.60}$$

$$\Psi_i^a = \hat{E}_{ai} |\ O\ > = \sum_\mu d_\mu \Phi_\mu^a \tag{3.61}$$

$$\Psi_{ij}^{ak} = \hat{E}_{ai}\hat{E}_{kj} |\ O\ > = \sum_\mu d_\mu \Phi_\mu^a \tag{3.62}$$

$$\Psi_{ij}^{ab} = (\hat{E}_{ai}\hat{E}_{bj} + p\hat{E}_{aj}\hat{E}_{bi}) |\ O\ > = \sum_\mu d_\mu \Phi_\mu^{ab} \tag{3.63}$$

where p is +1 for singlet coupling and -1 for triplet coupling of the external orbitals a and b. The configuration given in (3.62) has not been explicitly discussed in this chapter before. It is usually called a "semi-internal" configuration. The contraction coefficients d_μ will be linear combinations of the coefficients C_μ which define the reference CI vector of interest. The total ICCI wavefunction is then written as

$$\Psi = C_0\Psi_0 + \sum_{ia} C_i^a \Psi_i^a + \sum_{ijka} C_{ij}^{ak} \Psi_{ij}^{ak} + \sum_{ijab} C_{ij}^{ab} \Psi_{ij}^{ab} \tag{3.64}$$

In the single-reference case the internally contracted CI method is identical to the uncontracted case, but it is easy to see that in the multi-reference case the number of variational parameters can be drastically reduced. In fact, the number of variational coefficients C in (3.64) is almost independent of the number of reference states used. This method has therefore its main strength in cases with very many reference states. The largest multi-reference CI calculation performed to this date has been done using this method for the molecule Cr_2 for which 3088 reference configurations were used. Since the contraction error in the ICCI method is usually quite small, on the order of 0.1-0.2 % of relative correlation energy contributions, this method is very promising. At present, however, there is no existing program where the full internal contraction according to (3.60)-(3.63) is implemented. The calculation on the Cr_2 molecule was done with the singly excited (3.61) and semi-internal configurations (3.62) left uncontracted. This calculation was therefore quite expensive. To construct a program where the internal contraction is fully utilized therefore still remains a challenge. It is likely that such a program will be made the coming 1-2 years.

3.3.6 *Size consistency.* A method is said to be size-consistent if the computed energy of the composite system A + B, with A and B at infinite distance from each other, yields the same energy as if the method is applied to A and B separately and the energies are added, i.e. $E(A+B)=E(A)+E(B)$. Some of the methods we have discussed are automatically size-consistent. This is true, for example, for the Hartree-Fock method and the complete CI method, and it is also true for perturbation methods such as the coupled cluster method. It is, however, not true for the SD-CI or the MR-CI method. This is easy to see if we consider the case of N He atoms at large internuclear separation. The correlation energy for one He atom is

$$E_c = \frac{< \psi_0 + \psi_c \mid \hat{H} - E_0 \mid \psi_0 + \psi_c >}{< \psi_0 \mid \psi_0 > + < \psi_c \mid \psi_c >} \qquad (3.65)$$

where $\psi = \psi_0 + \psi_c$. An SD-CI calculation on N He atoms at large separation gives

$$E^N = \frac{N < \psi_0 + \psi_c \mid \hat{H} \mid \psi_0 + \psi_c >}{< \psi_0 \mid \psi_0 > + N < \psi_c \mid \psi_c >} \qquad (3.66)$$

and thus $N E_c \neq E^N$.

We will in this section show that it is possible, by a slight modification of the formalism, to correct these CI methods to be approximately size-consistent. The experience gathered over the past two decades on size-consistency corrections indicates that the calculated results are much improved at the SD-CI level, whereas relative energies are improved at the MR-CI level but the situation for geometries is less clear at this level.

The reformulation of the CI method which leads to approximate size-consistency starts out by writing the correlation energy functional for an SD-CI wave function as

$$F_c = \frac{< \psi_0 + \psi_c \mid \hat{H} - E_0 \mid \psi_0 + \psi_c >}{< \psi_0 \mid \psi_0 > + g < \psi_c \mid \psi_c >} \equiv \frac{Nu}{De} \qquad (3.67)$$

where E_0 is the reference energy, ψ_0 is the reference part of the CI vector and ψ_c is the remaining part of the vector. With $g=1$, F_c in (3.67) reduces to the normal expression for the correlation energy as obtained from the functional F in (3.38). By inspection of (3.67) it can be realized that the problem of size-inconsistency appears in the expression for the denominator De. This is clear since for infinitely separated systems A and B the numerator must be $Nu(A+B)=Nu(A)+Nu(B)$, which is easily realized since $< A \mid \hat{H} \mid B >$ must be zero. With a constant denominator independent of the size of the system, F_c must also be size-consistent. The simplest way to achieve this property for the denominator is to put $g=0$. When this is done we have a size-consistent method which has many names, simply because it has been rediscovered several times. Some of these names are CEPA-0 (coupled electron pair approximation 0), L-CPMET (linearized coupled pair many-electron theory) and DE-MBPT (doubly excited many-body perturbation theory). A related entity is the so called "Davidson correction" which is frequently used to correct a CI calculation for size-inconsistency. The expression for this correction to the CI energy is simply obtained by taking the difference between the size-consistent functional with g=0 and the normal CI functional with g=1. The result for the single-reference SD-CI case is

$$E_{Dav} = \frac{1 - C_0^2}{C_0^2} E_c \qquad (3.68)$$

where C_0 is the coefficient for the reference configuration and E_c is the correlation energy. A completely analogous result is obtained in the multi-reference case. It is worth noting that adding the Davidson correction to the CI energy will not make the energy exactly size-consistent even though the use of the energy functional with g=0 is exactly size-consistent. This is so since the Davidson correction is evaluated using CI coefficients which have been optimized using a functional which is not size-consistent.

The above description is one way to realize that the SD-CI method is not size-consistent. Another way is to look in detail at what happens when this method is used on the composite and on the separated systems. It is clear that if the energy of A and B should be additive the corresponding wavefunction for (A+B) should be equal to the wavefunction of A times the wavefunction of B. This means that since in the calculations on the separated systems there are local double excitations on both A and B, the product wavefunction will contain certain quadruple excitations. In the SD-CI calculation on the composite system these quadruple excitations are clearly missing and this is the reason for the size-inconsistency. It is also clear that for the SD-CI method E(A)+E(B) must be lower in energy than E(A+B).

Another point worth making is that since the SD-CI method is exact within the chosen basis set for a two-electron system, it must be size-consistent in this particular case. Nevertheless, when Davidson's correction is applied to an SD-CI wave-function for a two-electron system it will give a non-zero contribution, which is thus an artefact of this correction. (The same error appears also when the functional (3.67) is used with g=0.) This artefact can be simply removed and this is done in the Averaged Coupled Pair Functional (ACPF) method. In this method the factor g is considered to be a function of the number of electrons N, g=g(N), and one considers the special case of n separated He atoms. If the denominator De in (3.67) for one He atom is

$$1 + g(2) < \psi_c^0 \mid \psi_c^0 > \qquad (3.69)$$

where it has been assumed that the wavefunction is normalized with $< \psi_0 \mid \psi_0 >= 1$, then the denominator for n He atoms will be

$$1 + g(2n) < \psi_c \mid \psi_c >= 1 + g(2n) * n < \psi_c^0 \mid \psi_c^0 > \qquad (3.70)$$

where a corresponding normalization has been used. If we now use the information that for a two-electron system as in (3.69) the normal CI expression is exact we

226

will require that g(2) should be equal to one. Since the requirement for size-consistency is that the denominator should be constant, independent of the number of electrons, then g(2n)*n should also be equal to one. With $n = \frac{N}{2}$ these two conditions can be fulfilled by requiring that

$$g = \frac{2}{N} \tag{3.71}$$

Since in a normal CI calculation g is equal to one, the effect of using the ACPF g-value will be particularly important when there are many electrons. It should also be noted that the ACPF method is not exactly size-consistent in the general case. However, in test calculations the energy E(A+B) has been calculated for a variety of systems and been found to be very nearly equal to E(A)+E(B). Since the ACPF method is easily generalized to the MR-CI case, this method appears very promising for MR-CI calculations on systems with a large number of electrons.

3.4 RECOMMENDED FURTHER READING

For the MCSCF method:

B.O.Roos, The Multiconfiguratonal (MC) SCF Method, in Methods in Computational Physics (G.H.F. Diercksen and S. Wilson, eds.), D.Riedel Publishing Company, Dordrecht (1983).

Three articles in: *Ab Initio Methods in Quantum Chemistry, Part II*, (K.P. Lawley, ed.), John Wiley & Sons Ltd., Chichester (1987):

B.O.Roos, The Complete Active Space Self-Consintent Field Method and its Application in Electronic Structure Calculations.

R.Shepard, The Multiconfigurational SCF method.

H.J.Werner, Matrix-Formulated Direct Multiconfigurational Self Consistent Field and Multiconfigurational Reference Configuration-Interaction Methods.

For the CI method:

I. Shavitt, The Method of Configuration Interaction, in Methods of Electronic Structure Theory (H.F. Schaefer III, ed.), Plenum Press, New York (1977).

B.O. Roos and P.E.M. Siegbahn, The Direct Configuration Interaction Method from Molecular Integrals, in Methods of Electronic Structure Theory (H.F. Schaefer III, ed.), Plenum Press, New York (1977).

H.-J. Werner, Matrix-Formulated Direct Multiconfiguration Self-Consistent Field and Multiconfiguration Reference Configuration-Interaction Methods, in Ab Initio Methods in Quantum Chemistry - II (K.P. Lawley, ed.), John Wiley & Sons Ltd, Chichester (1987).

P.E.M. Siegbahn, The Direct CI Method, in Methods in Computational Molecular Physics, (G.H.F. Diercksen and S. Wilson, eds.) D. Reidel Publishing Company, Dordrecht (1983).

P.E.M. Siegbahn, The Direct Configuration Interaction Method with a Contracted Configuration Expansion, Chem. Phys. 25, 197 (1977).

P.E.M. Siegbahn, Multiple Substitution Effects in Configuration Interaction Calculations, Chem. Phys. Letters 55, 386 (1978).

V.R. Saunders and J.H. van Lenthe, The Direct CI Method- A Detailed Analysis, Mol. Phys. 48, 923 (1983).

P.E.M. Siegbahn, A New Direct CI Method for Large CI Expansions in a Small Orbital Space, Chem. Phys. Letters 109, 417 (1984).

P.E.M. Siegbahn, Current Status of the Multiconfiguration-Configuration Interaction (MC-CI) Method as Applied to Molecules Containing Transition-metal Atoms, Faraday Symp. Chem. Soc. 19, 97 (1984).

R.J. Gdanitz and R. Ahlrichs, The Averaged Coupled-Pair Functional (ACPF): A Size-Extensive Modification of MR CI(SD), Chem. Phys. Letters 143, 413 (1988).

H.-J. Werner, P.J. Knowles, An Efficient Internally Contracted Multi-Configuration-Reference Configuration Interaction Method, J. Chem. Phys. 89, 5803 (1988).

4. Approximative methods

In systems with a large number of electrons approximations to the full *ab initio* methods described in the previous sections quickly become a necessity. The problems occur at several levels. The storage of the raw integrals becomes a problem for more than say 200-300 basis functions; the transformation to MO:s which must be done prior to CASSCF or CI calculations generates a storage problem of the same order, the convergency in the SCF procedures may become overwhelming for metal clusters with open d-shells; to correlate large number of electrons becomes extremely time consuming; and finally the size consistency problem may become prohibitive for a very large number of electrons.

In the present section we will describe the Effective Core Potential (ECP) model, where all effects from the core electrons are parametrized. The one-electron version of this approximation is basic in the cluster model, where the metal atoms are described as pseudo one-electron atoms (i.e. all remaining electrons including the d electrons are regarded as core). We will also briefly describe the Core Polarization Potential (CPP) model, which is semi-classical model describing the effects of the most important part of the d-correlation.

4.1 THE ECP METHOD

The ECP method dates back to 1960, when Phillips and Kleinman suggested an approximation scheme for discarding core orbitals in band calculations [1]. The idea was to modify the Fock operator such as to make the core and the valence orbitals degenerate:

$$\hat{F} \rightarrow \hat{F} + \sum_c (\epsilon_v - \epsilon_c) \mid c > < c \mid \qquad (4.1)$$

The solutions corresponding to the core and the valence orbitals could then be rotated in such a way as to remove the nodes in the valence orbital while keeping its valence character in the outer atomic regions. This method was called the Pseudopotential method by Phillips and Kleinman. Only one single valence electron was considered in the original pseudopotential method.

The generalization of the pseudopotential method to molecules was done by Bonifacic and Huzinaga[2] and by Goddard, Melius and Kahn[3] some ten years after Phillips and Kleinman's original proposal. In the molecular pseudopotential or Effective Core Potential (ECP) method all core-valence interactions are approximated with l dependent projection operators and a totally symmetric screening type potential. The new operators, which are parameterized such that the ECP operator should reproduce atomic all-electron results, are added to the Hamiltonian and the ECP equations are obtained variationally in the same way as the usual Hartree Fock equations. The total energy is calculated with respect to this approximate Hamiltonian.

There are essentially two types of ECP's in general use, one which uses explicit core orbitals in the projection operators and one which uses projection operators on the orbital angular momentum with a parameterized local radial part. Using the symbol $M(r)$ for the totally symmetric screening potential the two types of ECP operators can be written:

$$\hat{F} = \hat{F}^{val} + \hat{M}(r) + \sum_c^{core} \mid \phi_c > B_c < \phi_c \mid \qquad (4.2)$$

$$\hat{F} = \hat{F}^{val} + \hat{M}(r) + \sum_l^{valence} \mid l > f_l(r) < l \mid \qquad (4.3)$$

The former approach is used by Bonifacic and Huzinaga[2] while orbital angular momentum projection operators are used by Goddard, Kahn and Melius[3], by Barthelat and Durand[4] and others. Explicit core orbital projection operators can, in the full basis set, be viewed as shift operators which ensure that the first root in the Fock matrix really corresponds to a valence orbital. The by far most important feature of the ECP method is to reduce the size of the basis set, however, and when the basis set is modified the role of the core orbital projection operators partly changes.

Experience shows that the two types of ECP's are of a comparable accuracy. This is not so surprising since the projection operators are l dependent in both cases and the operators are parameterized such as to reproduce selected atomic properties, usually valence orbital energies and shapes. There are, however, some practical differences between the methods. Nodeless valence orbitals are used in all ECP methods which makes use of local r-dependent operators, while valence orbitals with a nodal structure are usually employed in explicit core orbital projection operator type ECP's. While the ECP' become reasonably basis set independent if only local r-dependent operators are used, it is fairly laborious to determine the parameters, while the full nodal structure of the valence orbitals and the particular form of the Bonifacic-Huzinaga type ECP makes this relatively easy. A full nodal structure in the valence orbitals is clearly an advantage in CI calculations, since a full nodal structure in the valence orbitals will minimize the error in the interaction integrals.

The ECP method which will be discussed henceforth is derived from Huzinaga and Bonifacic, and the full nodal structure of the valence orbitals is always kept. In the early ECP application on first row transition metals the only orbitals which were included were $3d$ and $4s$[5-7]. However, experience showed that in certain cases it was important also to include (frozen) $3s$ and the $3p$ orbitals in the valence space[6,8], and ECP's with these characteristics were accordingly developed[9].

The equations used in the frozen ECP formalism are as follows:

$$\hat{F} = \hat{h}^{eff} + \sum_{c}^{val}(2\hat{J}_i - \hat{K}_i) \tag{4.4}$$

where

$$\hat{h}^{eff} = \hat{T} + \hat{V}^{eff} + \hat{P} + \sum_{j}^{frozen}(2\hat{J}_j - \hat{K}_j) \tag{4.5}$$

$$\hat{V}^{eff} = (\frac{-Z^{eff}}{r})(1 + M_1 + M_2) \tag{4.6}$$

$$M_1 = \sum_{p} A_p exp(-\alpha_p r^2) \tag{4.7}$$

$$M_2 = \sum_{q} r C_q exp(-\gamma_p r^2) \tag{4.8}$$

$$\hat{P} = \sum_{k} |\phi_k> B_k <\phi_k| \tag{4.9}$$

The parameters which enter these equations are B_k, A_p, α_p, C_q and γ_q. The B_k values are usually chosen as the absolute value of the corresponding core orbital energies. A_p, α_p, C_q and γ_q are calibrated in atomic calculations such as to reproduce the orbital energies and shapes of the valence orbitals. The outer core orbitals are expressed in the valence basis set by a least squares fitting procedure and are kept frozen during both the parameter fitting and the molecular calculations.

As a consequence of the development of the new integral programs for general contraction schemes the traditional ECP method became rather obsolete for the first row transition metals. The first obvious case where the method still is important is for heavier elements where relativistic effects can be parametrized into the ECP:s. This method has been successfully applied by many workers for the second and third transition metal series.

A different use of the ECP formalism was developed with the emerging interest in reactions on metal surfaces. With present day computers it is not realistic to carry out quantitative *ab initio* calculations on more than 5-10 metal atoms. However, if the troublesome d shells can be approximated, much larger clusters can be handled. Melius *et al.*[10-12] were the first to suggest a computational scheme in which the d electrons are also included in an ECP. This type of ECP, in which only the $4s$ electrons are explicitly accounted for in the calculations, was used by Goddard *et al.*[3] in calculations of oxygen and hydrogen chemisorption on nickel surfaces. These and similar one-electron ECP's have also been used successfully by Bagus *et al.*, by Bauschlicher and by Siegbahn *et al.* in applications where one metal atom is described at the all-electron level and the one-electron ECP atoms are used to mimic a metallic surrounding[13-15].

There are several prerequisites which have to be fulfilled for the one electron ECP approach to be applicable. In the case of metal clusters the atomic configuration must be known, i.e. one must safely be able to assume a $d^n s^2$ or a $d^{n+1} s^1$ configuration on the atoms in the cluster. The d orbitals should not form covalent bonds neither within the cluster nor between the cluster and the adsorbate. Ferromagnetic metals and copper are likely to have these properties, but for other metals, particularly to the left in the periodic table and in the second and third transition series, this is not so clear.

Since the d orbitals are not allowed to relax in a one-electron ECP it may appear that a third prerequisite is that the frozen d approximation should be valid, i.e. the relaxation of the d orbitals should not influence the bonding appreciably. In fact the effect of d orbital relaxation on chemisorption energies is appreciable, and this assumption is thus not fulfilled. We will adress the problem concerning the role of the d-orbitals and the development of the one-electron ECP model in separate sub-sections below.

4.1.1 *The role of the d-orbitals* The d orbital relaxation energy is defined as the difference at the SCF level between the chemisorption energy obtained at the all-electron SCF level when the d orbitals are frozen in their atomic shapes and when they are allowed to relax in the molecular surrounding. The metals selected for the investigation of the role of the d orbitals were nickel and copper. The ferromagnetism of nickel makes it likely that the d orbitals have an essentially atomic character in the metal, and for both nickel and copper it is reasonable to assume that the d orbitals are largely inert during surface reactions. The cluster used was a five-atom cluster, arranged in a square pyramid with an oxygen atom approaching the base of the pyramid. This system serves as a model for oxygen chemisorption at a fourfold hollow site on a (100) surface of the metal.

The results of the all-electron investigations were quite surprising, particularly for the nickel system. The d shell relaxation energy, i.e. the difference obtained for the oxygen binding energy with frozen and relaxed d orbitals, was 44 kcal/mol for Ni_5O and 17 kcal/mol for Cu_5O[16,17]. The total oxygen binding energy in Ni_5O is 42 kcal/mol at the SCF level, which means in a sense that all the binding at this level comes from the d orbital relaxation. As will become clear below, the qualitative aspects of the bond formation are already correctly described at the frozen $3d$ level, however.

There are two possible explanations of the large effect of the d orbital relaxation on the oxygen binding energy. The system may be trying to minimize the repulsion between the adsorbate and the d orbitals, in which case the origin of the relaxation is a polarization or hybridization of the d orbitals away from the adsorbate. This picture is consistent with the larger relaxation obtained for Ni than for Cu, since Ni_5 has the possibility to minimize the repulsion by rotating the open and the closed d orbitals. The d orbitals may of course also bind covalently to the adsorbate. This possibility seemed initially not very likely since a localization of the d orbitals in the relaxed case yielded quite pure d orbitals. However, orbital coefficients may not be a certain measure of covalency effects, and in order to quantify the covalent contribution of the d orbitals to the cluster adsorbate bond a CSOV[18] analysis was carried out on Cu_5O[19] (the presence of open d orbitals on Ni_5O makes a corresponding analysis more difficult).

TABLE 1. CSOV analysis of the bonding in Cu_5O. ΔE is the difference from a relaxed $3d$ calculation. The chemisorption energy D_e is calculated relative to $Cu_5(^4A_2) + O(^3P)$. All energies in kcal/mol.

Step	ΔE	D_e	$3d$-pop	Q(O)
Step 1	31.8	7.6	50.00	−0.85
Step 2	1.6	25.4	49.78	−0.88
Step 3	0.0	27.0	49.76	−0.84

The Constrained Space Orbital Variation, or CSOV technique is a method for analysing in detail different contributions to a chemical bond[18]. In the CSOV method, subsets of the occupied orbitals are frozen or otherwise constrained at different stages of the calculation, sometimes in combination with constraints on the virtual space. Our CSOV analysis consisted essentially of three steps; a first step where the binding energy was calculated with all d orbitals frozen in their atomic shapes, a second step in which the d orbitals were relaxed but where the rest of the valence space (as obtained from step 1) was frozen , and finally a fully relaxed calculation. Since the bonds between the valence orbitals of the cluster and the adsorbate are formed in the first step and not allowed to change in the second step the energy difference between the second and the third steps gives a measure of the importance of the d orbitals for the covalent bonds. The result of the CSOV analysis are shown in Table 1. The energy difference obtained between the second and third steps was only 2 kcal/mol, and the effect of the d orbitals on the binding is thus caused almost exclusively by polarization of the d orbitals in the cluster. The repulsion between the adsorbate and the d orbitals at the frozen orbital level can be rationalized in terms of Pauli exclusion effects, i.e. the orthogonalization of the adsorbate orbitals against the d orbitals in the cluster. We should mention here that in the second step described above we actually carried out several calculations using different subsets of the virtual space as well but, this gave no conflicting information.

A similar investigation on the Ni_5O[20] cluster gave a somewhat different result. In this case the increase in the oxygen binding energy obtained between the calculation where the d orbitals were relaxed in a frozen valence space and the fully relaxed calculation was 18 kcal/mol. However, the CSOV analysis becomes more complicated and actually somewhat ambiguous on metals with open d shells, largely because of Pauli repulsion effects. Our experience shows that the orbital configuration selected for the frozen orbital calculation does not change when the d orbitals are relaxed in the field of the frozen valence space chosen for the open d shells. The computed relaxation will thus depend on the d occupation used at the frozen level, and Pauli repulsion terms may become severely overestimated since even minute deformations in the outer regions of the d orbitals can involve rather large energy effects. In fact, the high localizability of the $3d$ orbitals makes a $3d$ covalency effect of 18 kcal/mol appear rather high. Considering the ambiguities in the CSOV analysis in the Ni_5O case we believe that 18 kcal/mol represents an overestimate of the $3d$ covalency effects in Ni_5O by maybe 8-10 kcal/mol . However, the results do indicate that the separability of the valence and the d orbitals is better for copper than for nickel which may be interpreted as a higher degree of d covalency in the adsorbate substrate bonding in Ni_5O than in Cu_5O. The total effect is at any rate rather small compared to the total binding energy of 115-130 kcal/mol.

All the above calculations were carried out using a fully contracted d orbital on the metal atoms. If a radial polarization of the d shell or if covalency is important one would expect the results to be sensitive to the basis set used to describe the d orbital. In order to investigate this effect a larger calculation using a triply split d shell was carried out on Ni_5O[21]. The chemisorption energy obtained with the two different basis sets was remarkably stable, 41.5 kcal/mol for the small basis and 45 kcal/mol with the large basis, a result which confirms the conclusion that d covalency is not important in the bond between a nickel surface and oxygen.

4.1.2 *The one-electron ECP:s*   Actually, a small d orbital relaxation is not a necessary prerequisite for the development of a one-electron ECP. The essential point is that the relaxation effect is dominated by the cluster and not by the adsorbate, and this is likely to be the case since the relaxation is not dominated by covalency effects.

In the one-electron ECP the effects from all electrons except for the 4s are parametrized into the ECP operators. One evident consequence is that the atomic ground state can no longer be used to find all the parameters needed. However, the most important consequence of the approximation is that the d orbital projection operator will not play the role of a simple level shift operator any longer. On the contrary, it will contribute actively to the bonding by raising the energy of the system when an approaching orbital is starting to overlap with the d orbital, which of course is precisely the effect of a "real" d orbital as well. The only way to determine the parameters of the d projection operators at the atomic level would be to consider excited states of the atom with an occupied $4d$ orbital. The sensitivity of the $4d$ orbital to the parameters of the d projection operator is, however, very small[22]. The only remaining alternative is to optimize the d orbital projection operator by comparing with molecular or cluster calculations at the all-electron level. In our one-electron ECP approach we have decided to make the parameterization of the d projection operator by comparing with all electron calculations on oxygen chemisorption at the hollow position on a five atom metal cluster[22]. The parameterization of a one-electron ECP describing chemisorption on a cluster with d orbitals frozen in their atomic shapes yielded a B_k value about twice as large as the orbital energy. In the relaxed d orbital calculations it turned out to be impossible to obtain both the chemisorption energy and the distance above the surface with the original shape of the d orbital from the atomic calculation. This result is not so surprising, since a large fraction of the d relaxation energy is caused by a polarization of the d orbitals which of course affects their shape. The solution to this problem was to modify the form of the d projection operator to "soften" it in the outer regions by introducing a second diffuse and slightly attractive d projection operator.

Of course the projection operators of s and p symmetry also play a role in interatomic interactions. At short internuclear separation the electron in the bonds will start to penetrate the core region. When this happens the orthogonality restriction between the valence and the core orbitals will cause an exponential energy increase. If an ECP does not contain any frozen outer core orbitals this repulsion must be described entirely by the projection operators. Often, but not always, do such ECP's work in the vicinity of the equilibrium geometry, but they always break down at short internuclear separations. In the case of the one-electron ECP's it turned out that for values of B_{3d} close to the 3d orbital energy oxygen collapsed through the surface in the absence of a frozen outer core. The inclusion of a frozen 3s orbital in the ECP prevented collapse for all values of B_{3d}, however. Although oxygen did not collapse through the surface for the B_{3d} values actually used in the cluster ECP's, we still decided to keep a frozen 3s orbital on all centers close to the adsorbate in the cluster models mimicking adsorption at hollow positions. All centers further away were described by smaller ECP's without a frozen $3s$. A further improvement of the one-electron is to also include a frozen $3p$ orbital in the ECP. By using general contracted basis set this extension requires the use of three s functions and two p functions (including one diffuse p function) in the ECP basis set.

The exponents and the expansion coefficients in the basis set used with the ECP were determined by a least-squares fit to the all-electron orbitals. All the parameters entering the screening potential were determined by fitting the shape and the energy of the $4s$ orbital to the corresponding all-electron results. The B_k values entering all but the $3d$ type projection operators were chosen in the standard way, i.e. as the negative orbital energies of the corresponding core orbitals. In the following the basis set used with the one-electron ECP only had a frozen $3s$ orbital, while the $3p$ orbital was described by a projection operator.

The final calibration of the B_{3d} parameters were carried out on a five atom metal cluster with an adsorbed oxygen atom at the fourfold hollow position. The binding energies and the bond distances were reproduced to within 1 kcal/mol and 0.05 bohr for Ni_5O[22] and Cu_5O[17](see table 2). The vibrational frequencies computed using the one electron ECP's are somewhat too low, by 40 cm^{-1} for Ni_5O and by 10 cm^{-1} for Cu_5O, indicating a somewhat underestimated repulsion at shorter bond distances.However, the geometrical configuration, with an oxygen atom sort of sliding in between four nickel atoms, will make the vibrational frequencies very sensitive even to minor errors in the interatomic potentials.

It is not evident that one-electron ECP's which have been optimized for oxygen can be used for other adsorbates. If e.g. the d orbitals bind covalently to oxygen to any appreciable extent, we would expect the model to describe adsorption of other adsorbates less satisfactorily. Calculations on Ni_5H and Cu_5H yielded, however, completely satisfactory results(see table 2). The difference between the ECP and the all-electron binding energies were 3 kcal/mol for Ni_5H and 1 kcal/mol for Cu_5H, and the distance above the surface was overestimated by 0.1 a.u. (for Ni_5H) . This is not much considering that the potential energy surface is very flat.

Adsorption of oxygen and hydrogen at hollow sites on Ni(100) and Cu(100) is described satisfactorily by one-electron ECP clusters. When adsorption at on-top positions are considered, effects directly involving the metal d shells become important. An example of this is the dissociation of molecular hydrogen at the on-top position on a Ni(100) surface where $s - d$ hybridization is very important both for intermediate states and for the barrier to dissociation. In such cases it is necessary to include one all-electron atom at the on-top site in the calculations. For reactions which take place at bridge sites it is desirable to use two all-electron atoms, but in this case the results from all ECP calculations can be corrected by comparing with calculations with an adsorbate and only two all-electron metal atoms.

TABLE 2. SCF results for hydrogen and oxygen on Ni_5O and Cu_5O. The all-electron r_e value for CU_5H is not geometry optimized

System	Case	r_e	D_e	ω_e	O−pop
Ni_5O	All-electron	1.92	41.6	381	−1.04
Ni_5O	ECP	1.95	42.9	347	−1.04
Cu_5O	All-electron	1.71	27.3	328	−0.85
Cu_5O	ECP	1.66	27.6	316	−0.88
Ni_5H	All-electron	2.38	36.8	835	−0.63
Ni_5H	ECP	2.49	33.4	847	−0.51
Cu_5H	All-electron	(2.50)	25.3	−	−0.64
Cu_5H	ECP	2.51	24.5	764	−0.42

4.2 THE CORE POLARIZATION POTENTIAL METHOD

The Core Polarization Potential (CPP) method[23] is based on a classical effective field operator where the core-valence correlation is viewed as resulting from the interaction between the field generated by the valence electrons and an induced core-dipole moment proportional to the core-polarizability α_c.

$$\hat{V}_{CPP} = -\frac{1}{2}\sum_c(\alpha_c\mathbf{f}_c^2 - 2\cdot\mathbf{f}_c\cdot\mu_c^0 + \mathbf{f}_c^0\cdot\mu_c^0) \qquad (4.10)$$

The operator $\hat{V}_{CPP}$ acts on the valence electrons only and must thus be projected onto the valence space. The static polarization of the cores is included in the operator through the term $-\frac{1}{2}\sum_c\alpha_c\mathbf{f}_c^2$ and the terms containing the expectation value, μ_c^0, of the core dipole moment thus correct for the static core polarization actually obtained with the basis set used[23]. The fields, $\mathbf{f}_c$, contain contributions from both the surrounding cores and the valence charge density. In order to avoid unphysical contributions from penetration into the core region a cut-off function, $C(\rho_c, \mathbf{r}_{ci})$, is applied to the fields generated by the valence electrons

$$C(\rho_c, \mathbf{r}_{ci}) = (1 - exp(-(\frac{\mathbf{r}_{ci}}{\rho_c})^2))^2 \qquad (4.11)$$

where ρ_c is a cut-off parameter which is determined so that the experimental or core-valence CI ionization potential for metal atom is reproduced. The CPP operator thus contains two basic parameters, the core dipole polarizability (α_c) and a cut-off radius (ρ_c) inside which the field generated by the valence electrons is made to disappear exponentially.

The core polarization potential (CPP) method has been shown to give very accurate results for the alkali metals[23,24]. Pettersson and Åkeby have modified and extended the method for the more difficult copper compounds, and investigated the importance of core-valence polarization on the oxygen chemisorption energy on a Cu_5 cluster[25]. The calculations have been carried out at the all electron level using both a completely contracted and a triply split d shell on copper[25,26]. Size-consistent CI calculations were also carried out using the correlated pair function (CPF) method[27] on Cu_5O correlating, the d electrons on the upper four copper atoms and the $2s$ and $2p$ shells on oxygen. The results of these calculations are shown in table 3.

TABLE 3. Computed chemisorption energy for Cu_5O with and without explicit correlation of the $3d$ orbitals. The 3_d basis set has a triply split and 1_d basis set a totally contracted $3d$ orbital. The numbers within parenthesis are the number of correlated electrons. The 3_d results have been corrected for the basis set superposition error.

Basis	Method	r_e	D_e
3_d	SCF	1.75	27.1
3_d	CPF(11)	1.94	88.8
3_d	CPF(51)	1.71	102.0
1_d	CPF(11)	1.92	86.5
1_d	CPF(11)+CPP	1.53	99

5. The cluster model.

The cluster model referred to in the present chapter is largely based on the one electron ECP approximation. It was originally designed as a model for chemisorption and reactions on metal surfaces, but lately a new application has become reactivities of small real metal clusters. The model is used as follows:

- At hollow positions on a metal surface all metal atoms are described by one electron ECP:s

- At on top positions the metal atom immediately underneath the adsorbate is described at the all-electron level. All remaining metal atoms are described by one electron ECP:s

- The valence correlation contribution to the substrate-surface bond is calculated using normal CI procedures.

- The core(3d)-valence correlation contribution to the bonding is, when included, obtained using the CPP method.

One of the central features of the model is that it is designed to mimick all electron calculations on small clusters, not primarily experimental results. This aspect is important, since all results can then be understood in terms of the basic computational models (independent particle model, correlation etc). Clearly a finite cluster is not equivalent to a full surface, and in order to analyze the model it is necessary to have all approximations used under control.

A central issue in the development of a surface cluster model is to understand the differences between a finite cluster and an infinite surface. In the early applications chemisorption energies showed erratic variations with the size of the cluster. This behavious of the cluster model could be understood from the electronic properties of the cluster. In a finite cluster there are a discrete set of excited levels, while for an infinite surface this discrete set of states is replaced by a continuous band. For a finite cluster there is thus a problem in just how to select the proper states to use for calculating e.g. chemisorption energies, since states which are excited relative to the ground state by perhaps 1-2 eV in the cluster becomes degenerate in the limit of an infinite surface. The concept of bond preparation makes it possible to select the relevant states for a small cluster such that the cluster model can be used as quantitative model for describing reactions on an infinite surface.

This section is divided into two subsections. The first of these is concerned with the concept of bond-preparation and examples are given from hydrogen, fluorine, oxygen and methylene chemisorption in the four-fold hollow site of Ni(100). In the second subsection calculations on the adsorbtion of CO on Cu_n^+ clusters with n=1,10 will be compared to corresponding experimental results.

5.1 BOND-PREPARATION FOR HYDROGEN AND OTHER ADSORBATES

The simplest possible adsorbate for studying the cluster size dependence on the chemisorption energy is the hydrogen atom. In this case there is only one bonding mechanism, hydrogen is bound to the substrate with one covalent bond, and there are no other interactions that influence the chemisorption energy such as repulsion from closed shells on the adsorbate. Nevertheless, even for the hydrogen atom it is found that if the chemisorption energy is calculated as the difference between ground states at short and long distances between the adsorbate and the cluster (E_{Gr}), strongly oscillating results as a function of cluster size is obtained[28]. Of seven different clusters modelling the Ni(100) surface (Ni_5-Ni_{50}), it was found that in particular $Ni_{21}(12,9)$ gave a very poor chemisorption energy of 43.0 kcal/mol compared to the experimental value of 63 kcal/mol[29]. All the other six clusters gave chemisorption energies in the range 50-65 kcal/mol (these results include correlation effects). When the Ni(111) surface was modelled by ten different clusters (Ni_4-Ni_{40}) even more strongly oscillating results were obtained. This surface has the same experimental chemisorption energy of 63 kcal/mol, but the $Ni_{19}(12,7)$ cluster, for example, gave only a value of 27.5 kcal/mol and $Ni_{22}(12,7,3)$ gave 37.8 kcal/mol. It turns out that all these results can be understood in a rather simple way. The clusters that gave poor chemisorption energies all had only closed shells in symmetry A_1 (the same symmetry as the hydrogen 1s orbital), and also had large excitation energies to the first excited state which had a singly occupied a_1 orbital. However, if the chemisorption energies instead were calculated with respect to this excited state quite reasonable chemisorption energies were obtained for every cluster. This procedure was termed bond-preparation. The bond-prepared results for the seven clusters modelling the Ni(100) surface gave an average value of 60.3 kcal/mol with a standard deviation of only 3.0 kcal/mol, whereas for the ten clusters modelling the Ni(111) surface the corresponding results are 61.7 kcal/mol and 4.5 kcal/mol, respectively. The simplest way to understand what bond-preparation means is to interpret the chemisorption bond as covalent, thus requiring one singly occupied orbital both on the adsorbate and also locally on the substrate. For a real surface (unlike the situation for a finite cluster) this type of bond-prepared state should be accessible without any energy cost due to the band nature of the electronic states of an infinite surface. As will be seen below there is also another way to interpret the results for hydrogen, which turns out to be more applicable for the case of atomic oxygen chemisorption in particular.

The first adsorbate studied after the hydrogen atom was the methyl radical[30], and as already mentioned it behaved in essentially the same way as hydrogen. If the cluster was prepared with one singly occupied orbital of A_1 symmetry, stable chemisorption energies were obtained. Calculations of this type led to the presently probably best estimate for the chemisorption energy of methyl on Ni(111) of 50-55 kcal/mol. Previous estimates ranged from 30 kcal/mol to 70 kcal/mol. It should be added that accurate experimental determinations of the chemisorption energies for short lived adsorbates like methyl are extremely difficult to obtain.

TABLE 4. Chemisorption of atomic fluorine in the 4-fold hollow site of Ni(100), energies in kcal/mol. E_{Gr} are the chemisorption energies calculated with respect to the ground states at short and long distances, whereas E_{Bp} are calculated relative bond-prepared states.

Cluster	E_{Gr}	E_{Bp}
$Ni_5(4,1)$	121.7	121.7
$Ni_9(4,5)$	111.7	121.9
$Ni_{17}(12,5)$	99.3	117.9
$Ni_{21}(12,5,4)$	125.1	125.1
$Ni_{25}(12,9,4)$	120.4	120.4
$Ni_{29}(16,9,4)$	113.5	120.4
$Ni_{33}(12,9,12)$	128.1	128.1
$Ni_{37}(12,13,12)$	120.1	127.9
$Ni_{41}(16,9,16)$	119.2	122.8
$\overline{D}$	118.4	122.9
σ	7.7	3.3

The first more critical test case for the generality of bond-preparation was recently made for the case of atomic fluorine chemisorption on Ni(100). Since the concept of bond-preparation might appear to require the formation of a covalent bond, it is interesting to see what happens for an extremely ionic adsorbate. For the related system of F on Cu(100), the bonding has recently been analyzed in detail and it was concluded that the bonding is entirely ionic, and F on the Ni(100) surface should be similar[31]. One complication expected for an ionic adsorbate is that the ionization energy of the cluster could enter into the estimate of the chemisorption energy, and the ionization energy is known to converge very slowly with cluster size (as $1/R$, where R is the size of the cluster). This expectation turns out to be wrong, however. The results for the fluorine chemisorption energy for nine different clusters, calculated both with respect to ground states and with respect to bond-prepared states, are given in Table 4. Bond-preparation is done in exactly the same way as for the hydrogen atom. The most striking effect of bond-preparation is obtained for the $Ni_{17}(12,5)$ cluster, which has the poorest ground state chemisorption energy (E_{Gr}) of 99.3 kcal/mol but a bond-prepared chemisorption energy (E_{Bp}) of 117.9 kcal/mol in line with the results for the other clusters. Overall, the standard deviation is reduced from 7.7 kcal/mol down to 3.3 kcal/mol using bond-preparation.

As mentioned above, with a totally ionic bond it might be expected that the chemisorption energy should be sensitive to the ionization energy (the Fermi level) of the cluster. It is therefore rather surprising that no such sensitivity is found in the calculations. If one should try to explain the variation of the chemisorption energies in Table 4, which are successfully explained by the bond-preparation concept, by variation in IP's (ionization potentials) instead, there will be a total failure. For example, Ni_{17} has the smallest chemisorption energy at the SCF level of 75.3 kcal/mol but does not at all have the largest ionization energy. Using Koopmans' theorem an IP of 3.92 eV is obtained for ionizing an e-electron. Ni_{33} has the largest chemisorption energy of the clusters at the SCF level with 97.8 kcal/mol, but still has an IP of 4.69 eV which is thus as much as 18 kcal/mol larger than the one for Ni_{17}. The situation does not change much, even if only bond-prepared clusters and only ionization of an a_1 electron is considered. This is perhaps even more surprising since an a_1 electron has definitely moved from the cluster to fluorine. For example, the energy for ionizing an a_1 electron from Ni_{29} is 4.36 eV, while the corresponding IP of Ni_{37} is 5.06 eV, which is thus 16.1 kcal/mol larger. With the larger IP for Ni_{37} a smaller chemisorption energy would thus be expected for this cluster compared to Ni_{29}. The bond-prepared results show that the opposite is actually true, with a larger chemisorption energy for Ni_{37} (at the SCF level the difference is 9.9 kcal/mol). This contradiction is not a result of using Koopmans' theorem. Separate ΔSCF calculations for Ni_{29} and Ni_{37} lead to IP's of 4.07 eV and 4.47 eV, respectively, which still shows a large energy difference in the same direction as the Koopmans' values. The conclusion for fluorine chemisorption must be the same as the conclusion drawn earlier for hydrogen chemisorption[32], that the position of the Fermi level (lowest IP) is nearly irrelevant for describing chemisorption on clusters of these sizes.

In order to understand the results for fluorine, which is so ionic, an alternative formulation, which does not require a covalent bond formation, of how bond-preparation should be viewed is preferable. In this new picture, a reliable chemisorption energy is obtained if the electrons on the adsorbate fit into the electronic structure of the bare cluster. The practical consequence of this picture for hydrogen and fluorine is exactly the same as the bond-preparation described above. In the chemisorbed situation, essentially two electrons are located on hydrogen. With bond-preparation, these two electrons will occupy part of the space of a previously singly occupied cluster orbital. From the cluster point of view, no new orbitals are thus required to describe the chemisorption and a reliable chemisorption energy is obtained. The electrons on the adsorbate fit into the electronic structure of the bare cluster. If, on the other hand, the outermost cluster orbital with the same symmetry as the hydrogen orbital is doubly occupied, the addition of the hydrogen electron would mean that an additional, previously unoccupied, cluster orbital would have to be utilized for this electron, and a poor chemisorption energy is generally obtained. For fluorine the key orbital is the lone pair orbital pointing towards the cluster. For this orbital to fit into the electronic structure of the cluster, the outermost cluster orbital of the same symmetry has to be singly occupied, exactly as for hydrogen. With this new picture of the bonding, the insensitivity of the results on the IP of the cluster is expected. It is clearly in this case a poor picture, even for an ionic adsorbate, to view the bond formation in two steps with first the ionization of an electron from the cluster and then an addition of an electron on the adsorbate, since the electron basically stays in the same region of space when the bond is formed.

A few more points are worth mentioning concerning the results for fluorine. First, fluorine has one additional bonding mechanism compared to hydrogen, and this is with bond formation parallel with the surface in the E symmetry. This bonding mechanism was used by only one of the clusters, $Ni_{21}(12,9)$, and led to a chemisorption energy of 116.6 kcal/mol. With only one such example, it is at present not clear how this type of bonding should be compared to the bonding in the other clusters and this result was therefore not included in Table 4. Secondly, to reach a best prediction for the fluorine chemisorption energy a few corrections to the average value of 122.9 kcal/mol in Table 4 has to be made. These correction are obtained from all-electron calculations on Ni_5F, from all-ECP calculations using a large basis set on fluorine (including up to two f-functions) and from counterpoise calculations for the basis set superposition error. Finally the $3d$ correlation effects were estimated based on CPP calculations and on all-electron correlated calculations on Cu_5F. The best estimate for the chemisorption energy is then obtained as 118.9 kcal/mol, which should be read as about 120 kcal/mol. There are no available experimental results to compare to. For further details of these calculations, see ref [33].

The description of the chemisorption bond, and thereby the concept of bond-preparation, is considerably more complicated for oxygen than for the adsorbates discussed above. The main reason for this is that not only one but two bonds are formed to the surface. To analyze the bonding for each cluster, the ground state occupation of the bare cluster should be subtracted from the ground state occupation for the cluster with oxygen adsorbed. When this is done three different bonding mechanisms can be identified. The first two of these are the obvious ones for the formation of two covalent bonds, with either two bonds parallel with the surface or with one bond parallel and one perpendicular to the surface. The third bonding mechanism is less obvious but turns out to be the one adopted by most clusters including the three largest clusters studied. For these clusters the occupations at short distance differ from the optimal bare cluster occupation only in the number of closed shells. The open shell occupations are the same. The closed shell occupations differ by two orbitals in symmetry A_1 and one orbital of E symmetry (occupied by four electrons). There is essentially only one way in which this bonding mechanism can be understood. The starting point is an oxygen atom which approaches the cluster with two electrons in the $2p_z$ lone pair (pointing down towards the surface) and with singly occupied $2p_x$ and $2p_y$ orbitals. As the chemisorption bond starts to form the $2p_z$ lone pair electrons will replace and thus kick out two electrons of the same symmetry (A_1) for the cluster. These two cluster electrons are moved over to the $2p_{x,y}$ orbitals of oxygen which are parallel to the surface and which originally have two holes. In this process the oxygen $2p$ electrons are partly delocalized over on the cluster, particularly for the $2p_{x,y}$ orbitals.

The above description of the main bonding mechanism for oxygen may seem quite different from the one found for fluorine, but it is in fact possible to describe also this latter bonding in a similar way. The starting point for fluorine is then with a doubly occupied $2p_z$ orbital pointing down towards the surface and a set of $2p_{x,y}$ orbitals with one hole. If the cluster has one singly occupied orbital of A_1 symmetry, which is the requirement for bond-preparation, the bonding proceeds with the $2p_z$ orbital replacing the singly occupied a_1 orbital and thus kicking out the electron from this orbital. This electron is then moved over to the hole in the $2p_{x,y}$ orbitals of fluorine, and the bonding is completed by delocalization of the $2p$ orbitals over on the cluster in complete analogy with the main bonding mechanism for oxygen (although the delocalization is less pronounced for fluorine than for oxygen). We note that the requirement of one singly occupied cluster a_1 orbital in the case of fluorine chemisorption follows from the fact that there is only one hole in the $2p_{x,y}$ orbitals on fluorine. With two holes in the $2p_{x,y}$ orbitals as on oxygen, there will be no such requirement on the cluster occupation in that case. It might further be added that with the above ionic description of the bonding mechanism, where electrons jump from the cluster to the adsorbate, the fact that electrons move between symmetries is not a significant aspect of the mechanism.

In particular, for fluorine the orbital occupancy per symmetry can be kept by simply thinking of the fluorine $2p_z$ orbital as initially singly occupied.

TABLE 5. Chemisorption of atomic oxygen in the 4-fold hollow site of Ni(100), energies in kcal/mol. E_{Gr} are the chemisorption energies calculated with respect to the ground states at short and long distances, whereas E_{Bp} are calculated relative bond-prepared states.

Cluster	E_{Gr}	E_{Bp}
$Ni_5(4,1)$	110.6	110.6
$Ni_9(4,5)$	104.7	104.7
$Ni_{17}(12,5)$	106.3	106.3
$Ni_{21a}(12,9)$	101.8	101.3
$Ni_{21b}(12,5,4)$	112.6	112.6
$Ni_{25}(12,9,4)$	112.9	112.9
$Ni_{29}(16,9,4)$	98.6	105.4
$Ni_{33}(12,9,12)$	113.2	113.2
$Ni_{37}(12,13,12)$	114.1	114.1
$Ni_{41}(16,9,16)$	111.1	111.1
$\overline{D}$	108.6	109.2
σ	5.1	4.2
Exp.		115-130

With three possible bonding mechanisms available for oxygen and a main bonding scheme which does not lead to any requirements on the occupation of the cluster, it is perhaps not too surprising that already the ground state chemisorption energies E_{Gr} are reasonably stable from cluster to cluster, see Table 5. There are, however, still two clusters modelling the Ni(100) surface, Ni_{21a} and Ni_{29}, for which no reasonable bonding scheme can be attributed. It is gratifying to note that precisely these clusters are the ones which give the poorest chemisorption energies. The assignment of bonding schemes has thus allowed us to identify the clusters for which the results should be most questionable. Another consequence of the large flexibility in selecting bonding scheme is that bond-preparation by going to an excited state of the cluster is less successful for oxygen than it has been for hydrogen and fluorine. For a cluster which is not bond-prepared, this flexibility often leads to an adjustment of the wave-function which tends to give quite reasonable chemisorption energies also in these cases. Bond-preparation is thus barely capable of moving the result for Ni_{29} into the range covered by the other clusters and the result for Ni_{21a} hardly changes.

TABLE 6. Chemisorption of methylene in the 4-fold hollow site of Ni(100), energies in kcal/mol. E_{Gr} are the chemisorption energies calculated with respect to the ground states at short and long distances, whereas E_{Bp} are calculated relative bond-prepared states.

Cluster	E_{Gr}	E_{Bp}
$Ni_5(4,1)$	79.3	79.3
$Ni_9(4,5)$	66.7	66.7
$Ni_{17}(12,5)$	72.7	75.6
$Ni_{21a}(12,9)$	70.9	72.0
$Ni_{21b}(12,5,4)$	80.3	80.3
$Ni_{25}(12,9,4)$	79.9	79.9
$Ni_{29}(16,9,4)$	73.4	79.2
$Ni_{33}(12,9,12)$	78.3	78.3
$Ni_{37}(12,13,12)$	80.6	80.6
$Ni_{41}(16,9,16)$	80.3	80.3
$\overline{D}$	76.2	77.2
σ	4.7	4.3
Exp.		70-100

Exactly as discussed for fluorine above, an improved prediction of the chemisorption energy for oxygen on Ni(100) can be obtained by doing additional calculations. When this is done a value of 130.0 kcal/mol is obtained, which should be read as about 130 kcal/mol. This is in perfect agreement with the most reliable experimental value [34] but differs from earlier calorimetric measurements giving 115 kcal/mol [35]. More than half of the increase from the average value 109.2 kcal/mol in Table 5 to the final estimate of 130 kcal/mol comes from $3d$ correlation effects (13 kcal/mol) and most of the other half from increasing the basis set on oxygen (8.0 kcal/mol) to include more d- and also f-functions.

Similar calculations as described above for the Ni(100) surface have also been performed for the Ni(111) surface. Just as for hydrogen chemisorption, modelling the Ni(111) surface by different clusters leads to somewhat larger oscillations of the chemisorption energies than for the Ni(100) surface. This appears to be related to a more complicated nodal pattern of the orbitals on the Ni(111) than on the Ni(100) surface, which can be seen by studying the orbital contour plots. The final best prediction of the oxygen chemisorption energy in the three-fold hollow site of Ni(111) is 115 kcal/mol. There are no available experimental measurements for this energy but the result agrees well with expectations based on other similar results[36]. More details of the oxygen calculations can be found in ref[37].

The final adsorbate discussed in this subsection is methylene (CH_2). The results for the ground state and the bond prepared chemisorption energies of 10 different clusters modelling the Ni(100) surface are given in Table 6. Methylene was in all these calculations oriented in a plane perpendicular to the metal surface. The main difference in the chemisorption of oxygen and methylene is that oxygen has three bonding mechanisms but methylene only two. The bonding mechanism missing for methylene is obviously the one with two bonds parallel to the surface which is not possible with the present orientation of the molecule. Just as for oxygen, the E_{Gr} results are already quite stable for methylene, and in the cases where bond preparation is required the effect is only minor. The major reason for this is that the main bonding mechanism (the same as for oxygen) does not set any requirements on the occupation of the bare cluster in contrast to the case for hydrogen and fluorine. The final best estimate for the chemisorption energy of methylene is 99.2 kcal/mol, which should be read as about 100 kcal/mol. No direct measurements of the chemisorption energy of this short-lived radical exist but previous estimates range from 70 to 100 kcal/mol. The main corrections to the average value of 77.2 kcal/mol from Table 6 comes from the inclusion of $3d$ correlation effects (9.7 kcal/mol) and the increase of the adsorbate bassis set (6.5 kcal/mol), similar to oxygen, but for methylene there are also some minor (4.7 kcal/mol) multi-reference CI effects.

5.2 CO CHEMISORPTION ON COPPER CLUSTERS

The above analysis shows that for strongly bound adsorbates it is possible to reach an understanding, whereby the cluster oscillations of the chemisorption energies can be brought down to a tolerable level. The situation is much less satisfactory for a seemingly ordinary adsorbate like CO. For on top CO chemisorption on five different clusters Hermann et al[38] found a strongly oscillating behaviour of the chemisorption energies on cluster size, with SCF results of -0.55 eV for Cu_1, +0.45 eV for Cu_5, -0.40 eV for Cu_{10}, +0.28 eV for Cu_{14} and -0.55 eV for Cu_{34}. The experimental surface result is +0.58 eV[39]. It should first be noted that with the picture of the chemisorption bond, where the electrons on the adsorbate should fit into the electronic structure of the bare cluster, it is not totally unexpected that the bonding of CO is difficult to describe. The key interaction when CO is adsorbed should be the same as when oxygen and fluorine is adsorbed, namely the repulsion between the cluster and the lone pair pointing down towards the surface. In all these cases the lone pair of the adsorbate should kick out the electron(s) in the outermost orbital of the cluster with the same symmetry as the lone pair. The question is where these electrons should go. In the case of oxygen there is no problem, the $2p_{x,y}$ orbitals can host these electrons since there are originally two holes in these orbitals. For fluorine, with only one hole in the $2p_{x,y}$ orbitals a requirement is that the outermost cluster a_1 orbital should be singly occupied. The corresponding requirement for CO would be that there should initially be an empty a_1 cluster orbital. This requirement may not be easy to fulfil without either charging the cluster or placing electrons in very diffuse orbitals, in both cases with expected undesirable effects on the chemisorption. However, before these types of bond-preparations are tried, a systematic study of CO chemisorption on different clusters is required, based on more examples than those studied in ref[38]. The preliminary results of such a study is presented in this subsection.

In a recent experiment Leuchtner et al[40] measured the association rate constant of CO in collisions with Cu_n^+ clusters, with n=1-14. With the above theoretical results[38] in mind a surprisingly fast convergence of the rate constant to infinite surface results were obtained. Almost the same value was obtained for the rate constant for clusters with 7 atoms and more. A full quantitative understanding of these results were also obtained based on a simple model for the bonding. With the assumption that there is a covalent contribution which is essentially constant for all clusters and apart from that only simple electrostatic effects due to the charge of the cluster, of ion-dipole and ion-polarizability type, a formula for the bond strength as a function of cluster size is obtained

$$E_n = E_{bulk} + \frac{(r_{Cu} + r_{CO})^4}{(r_{Cu}n^{1/3} + r_{CO})^4}\Delta E \qquad (5.1)$$

ΔE is the difference between E_{bulk} (0.58 eV) and E_1 which was taken from a calculation to be 1.30 eV.

It is interesting to speculate over possible origins of the differences between the experimental results and the previous theoretical results. One obvious difference between the two studies is that the theoretical study was concerned with neutral clusters whereas the experimental clusters were cations. Another difference is that the experimental clusters have optimal geometries whereas the theoretical clusters were selected as models (of the same type as in Tables 4-6) of the Cu(100) surface. A similar type of difference is that in the experiments CO will find an optimal position to adsorb on the cluster, whereas in the theoretical study CO was always adsorbed at the central on-top position. The final two differences we will list here are more trivial. First, an accurate calculation of CO adsorbed on a cluster is quite difficult and the cluster oscillations could be due to inaccuracies in the calculations. Secondly, the experiment, where a convergence of rate constants was actually observed, could be misinterpreted in terms of a convergence of the chemisorption energies.

The first point which was reexamined in the present study was the accuracy of the previous calculations. At the SCF level about the same results as in ref[38] were obtained using the present basis sets and ECP's. Since the previous calculations and the present ones differ significantly in these latter aspects, the cluster oscillations are not likely to be simply due to inaccuracies in the calculations. On the other hand, rather large correlation effects of 0.6-0.7 eV on the chemisorption energies were also found in the present calculations. These effects are much larger than the effect of 0.25-0.30 eV which were estimated in ref[38]. A preliminary study indicates that the basis set superposition error with the present basis set (about 0.2 eV) will be roughly cancelled by the effect of adding more basis functions, such as f-functions on the on-top copper atom. It is clear that the size of the correlation effect does not change the oscillating behaviour of the chemisorption energies but it does give a different perspective on the results for the different clusters. For example, the largest cluster studied (Cu_{34}) gave a chemisorption energy for CO of at the SCF level of -0.55 eV, but CO will be bound when correlation effects are added. Also, our calculations indicate another ground state for this cluster than what was found in ref[38], and this also seems to add a few tenths of an eV to the binding energy. Altogether, it can not be ruled out that Cu_{34} actually gives a rather reasonable chemisorption energy. It must, however, be clearly emphasized that our calculations are not accurate enough either to give a definite value for the CO chemisorption energy on this large cluster. It is therefore important to go back and analyze smaller clusters.

TABLE 7. Chemisorption of CO on singly ionized and neutral copper clusters, energies in eV. The numbers quoted from the experimental study by Leuchtner et al[40] come from an assumed formula and concerns the cationic clusters (equation (5.1) in the text).

Cluster	Neutral	Cation	Ref.40
Cu_1	0.10	1.18	1.30
Cu_2	0.60	1.06	0.96
Cu_3	0.93	1.00	0.83
Cu_4	1.03	1.04	0.77
Cu_5	0.46	1.19	0.73
Cu_6	1.06	1.23	0.70
Cu_7	0.59	1.42	0.69
Cu_8	0.44	0.93	0.67
Cu_9	0.72	0.88	0.66
Cu_{10}	0.31	0.66	0.65
Exp.	0.58	0.58	0.58

The calculated chemisorption energies for CO on optimal neutral and cationic Cu_n clusters with n=1,10 are shown in Table 7. The results include correlation effects from all the valence electrons including the $3d$ electrons of the on-top atom and CPP effects on the other copper atoms. The results using equation (5.1) are also shown in the table. The correlation effect for the three largest clusters is taken from Cu_7. To properly understand the trends in the results, a few points need to be noted. First, the smallest clusters, with n up to 3, have a quite different type of binding than the larger clusters and should be looked at separately. Second, there are notable shell closing effects, particularly for 8 electrons as usual. Since CO contributes two electrons (the carbon lone pair) to the Cu_n cluster electronic structure, the shell closing should appear at $n=6$ for neutral clusters and at $n=7$ for cationic clusters. These clusters are therefore expected to give values which are larger than the general trend. There are no major differences in the convergence pattern for neutral and cationic clusters. Even if shell closing effects are subtracted from the results in the table and the smaller clusters with n up to 3 are disregarded, the chemisorption energies can not be considered converged at $n=10$, in contrast to the experimental conclusion. However, the trend of the energies in the table is still rather promising and it seems as if converged values might appear already at about 15-20 atoms, but this remains to be shown. The shell closing effects should at least be rather small for clusters of that size.

Even though the presently calculated results show the general trend predicted by the experiments, there are some notable differences. First, the calculated results show typical shell closing effects but such effect were excluded in the analysis of the experiments. Second, the calculated results still show oscillations even for the largest clusters. The simplest interpretation of the experimentally observed convergence of the association rate constants, is that this convergence is totally dominated by dynamical effects. With the rather large chemisorption energies for the cationic clusters in Table 7, and since there should not be any barrier for the association, it is not unlikely that every collision will lead to a possible formation of a complex, and this possibility is more or less independent of cluster size. For this complex to stay bound a quick dissipation of the kinetic energy is necessary but this requires clusters with many degrees of freedom. It appears that about 7 atoms are necessary for this latter step, and that more atoms will not increase this dissipation significantly.

Based on the results in Table 7, we tentatively draw the conclusion that geometry optimized clusters lead to a faster convergence of the CO chemisorption energies than clusters which are selected to be representations of a particular surface like the Cu(100) surface. The detailed reason for this has to be further investigated. One tempting explanation is that the rough edges in the latter type of clusters will have a tendency to lead to charge buildup which by a charge wave will transmit to the center of the cluster. This hypothesis was investigated by studying CO chemisorption on one-dimensional ring clusters, where edges are missing, but oscillations of the chemisorption energies were still obtained. Our main hypothesis at the moment is instead that non-optimal clusters will have sites where the binding capacity is not fully used, leading to too large CO binding energies, and other sites where the binding capacity is exhausted, leading to too small CO binding energies. The geometry optimization of the cluster will tend to even these differences out, leading to more stable CO chemisorption energies. Connected with these effects, should be a smoother convergence of the orbital energy pattern for optimal clusters than for non-optimal clusters. Clearly, more research on this problem is needed. Also, larger optimal clusters should be studied but a full geometry optimization of these clusters will be tedious. Therefore, models to construct near optimal structures of clusters without an explicit optimization, as for example suggested in ref[41], is at present of large interest in this context.

6. Acknowledgement

We are grateful to prof. Björn Roos for letting us use his teaching material on MCSCF methods from the European Summerschool on Quantum Chemistry 1989 (ESQC-89).

References

1. Phillips,J.C. and Kleinman,L. (1959), Phys.Rev. **116**, 287
2. Bonifacic,V. and Huzinaga,S. (1974), J.Chem.Phys., J.Chem.Phys. **60**, 2779
3. Melius,C.F. and Goddard,W.A. (1974), Phys.Rev. **A10**, 1528
4. Barthelat,J.C., Durand,Ph. and Serafini,A. (1977), Mol.Phys. **33**, 159
5. Wahlgren,U. (1978), Chem.Phys. **32**, 215
6. Gropen,O., Wahlgren,U. and Pettersson,L.G.M. (1982), Chem.Phys. **66**, 459
7. Gropen,O., Wahlgren,U. and Pettersson,L.G.M. (1982), Chem.Phys. **66**, 453
8. Pettersson,L.G.M. and Strömberg,A. (1983), Chem.Phys.Lett. **99**, 122
9. Pettersson,L.G.M., Wahlgren,U. and Gropen,O. (1983), Chem.Phys. **80**, 7
10. Melius,C.F., Upton,T.H. and Goddard,W.A. (1978), Solid State Comm. **28**, 501
11. Upton,T.H. and Goddard,W.A. (1981), in CRC critical reviews, solid state and materials sciences, CRC press, Boca Raton, page 261
12. Upton,T.H. and Goddard,W.A. (1979), Phys.Rev.Lett, **42**, 472
13. Bagus,P.S., Bauschlicher,C.W., Nelin,C.J. and Laskowski,B.C. (1984), J.Chem.Phys. **81**, 3594
14. Bauschlicher,C.W. (1986), J.Chem.Phys. **84**, 250
15. Siegbahn,P.E.M., Blomberg,M.R.A. and Bauschlicher,C.W. (1984), J.Chem.Phys.**81**, 2103
16. Panas,I., Siegbahn,P. and Wahlgren,U. (1988), Theor.Chim.Acta **74**, 167
17. Mattsson,A., Panas,I., Siegbahn,P., Wahlgren,U. and Åkeby,H. (1987), Phys.Rev. **36**, 7389
18. Bagus,P.S., Hermann,K. and Bauschlicher,C.W. (1984), J.Chem.Phys. **80**, 4378
19. Wahlgren,U., Pettersson,L.G.M. and Siegbahn,P. (1989), J.Chem.Phys **90**, 4613
20. Wahlgren,U., Pettersson,L.G.M. and Siegbahn,P., Unpublished results
21. Wahlgren,U., Almlöf,J. and Siegbahn,P.E.M. (1990), Theor.Chim.Acta, in press

22. Panas,I., Siegbahn,P. and Wahlgren,U. (1987), Chem.Phys. **112**, 325

23. Müller,W., Flesch,J. and Meyer,W. (1984), J.Chem.Phys **80**, 3297

24. Partridge,H., Bauschlicher,C.W., Pettersson,L.G.M., McLean,A.D., Liu,B., Yoshimine,M. and Komornicki,A. (1990), J.Chem.Phys, in press

25. Pettersson,L.G.M. and Åkeby,H. (1990), J.Chem.Phys. **94**, 2968

26. Pettersson,L.G.M., Åkeby,H., Siegbahn,P. and Wahlgren,U. (1990), J.Chem.Phys. **93**, 4954

27. Ahlrichs,R., Scharf,P. and Ehrhardt,C. (1985), J.Chem.Phys. **82**, 890

28. I. Panas, J. Schüle, P. Siegbahn and U. Wahlgren (1988), Chem. Phys. Letters **149**, 265

29. G. Ertl, ch. 5. in *The Nature of the Surface Chemical Bond*, T.N. Rhodin and G. Ertl (Eds.)(1979), North-Holland, Amsterdam.

30. J. Schüle, P. Siegbahn and U. Wahlgren (1988), J. Chem. Phys. **89**, 6982

31. L.G.M. Pettersson and P.S. Bagus (1986) Phys. Rev. Letters **56**, 500

32. I. Panas and P.E.M. Siegbahn (1990), J. Chem. Phys. **92**, 4625

33. P.E.M. Siegbahn, L.G.M. Pettersson and U. Wahlgren (1991) J. Chem. Phys. **94**, 4024

34. W.F. Egelhoff, Jr. (1984), Phys. Rev. **B29**, 3681

35. D. Brennan, D.O. Hayward and B.M.W. Trapnel (1960), Proc. Roy. Soc. **A256**, 81

36. E. Shustorovich (1986,1989), Surf. Sci. Rep. **6**, 1; Adv. Catalys. **37**, 1

37. P.E.M. Siegbahn and U. Wahlgren, Intern. J. Quantum Chem. in press.

38. K. Hermann, P.S. Bagus and C.J. Nelin (1987), Phys. Rev. **B35**, 9467

39. J.C. Tracy (1972), J. Chem. Phys. **56**, 2748; C.F. McConville, D.P. Woodruff, N.C. Prince, G. Paolucci, V. Chab, M. Surman and A.M. Bradshaw (1986), Surf. Sci. **166**, 221.

40. R.E. Leuchtner, A.C. Harms and A.W Castleman, Jr (1990), J. Chem. Phys. **92**, 6527

41. M.H. McAdon and W.A. Goddard, III (1985), Phys. Rev. Letters **55**, 2563

CATALYTIC REACTIONS OF TRANSITION METAL CLUSTERS AND SURFACES FROM AB-INITIO THEORY
–Cluster and Dipped Adcluster Model Studies Combined with the SAC/SAC-CI Method–

H. Nakatsuji, H. Nakai, and M. Hada

Department of Synthetic Chemistry
Faculty of Engineering
Kyoto University, Kyoto, Japan

ABSTRACT. Reactions of transition-metal clusters are of considerable interests by themselves and from an analogy to surface catalytic reactions. We first review here our ab-initio theoretical studies on the reactions of small metal clusters with a hydrogen molecule. The clusters investigated are palladium, platinum, and ZnO. We also investgate the hydrogenation reaction of acetylene catalyzed by palladium. We study the energetics, mechanism, cluster-size dependence, surface stability, etc. and propose a molecular beam study for confirming the results. We next consider the case in which the electron transfer between an admolecule and a surface is important. We propose dipped adcluster model (DAM) in which adcluster (admolecule + cluster) is dipped onto the electron bath of the solid metal and an equilibrium is established for the electron exchanges. The role of the electrostatic image force is investigated. Electron correlations, electron transfers, and participations of lower excited states which are important for surface electronic processes are described by the SAC/SAC-CI method. We apply these methods to oxygen chemisorptions on palladium and silver surfaces.

1. INTRODUCTION

For theoretically studying catalytic reactions on a metal surface, the first question is "how do we describe molecule-surface interaction". It involves the interaction between finite and infinite systems so that a modelling is necessary for an adequate description of the system. In this review article, we consider two models; cluster model and dipped adcluster model.[1]

Cluster model is based on the locality of the interaction between an admolecule and a surface: namely the admolecule can interact directly with only a few atoms of a surface. We apply this model to hydrogen chemisorptions on Pd, Pt, and ZnO surfaces.[2-4] We then study the hydrogenation reaction of acetylene on a Pd surface.[5] Strictly speaking, these studies are actually for the reactions of the small metal clusters with hydrogen and acetylene.

When electron transfer between admolecule and surface is important, the clus-

D. R. Salahub and N. Russo (eds.), Metal-Ligand Interactions: from Atoms, to Clusters, to Surfaces, 251–285.
© 1992 *Kluwer Academic Publishers. Printed in the Netherlands.*

ter model may be inadequate. In such a case we propose a model[1] in which the adcluster, a combined system of the cluster and admolecule, is dipped onto the free electron bath of a bulk metal and an equilibrium is established for electron exchanges between them. This model is called dipped adcluster model (DAM) and applied to the study of oxygen chemisorptions on Pd and Ag surfaces.[1,6]

Electron correlations are very important for metal cluster and surface electronic processes. We very often deal with transition metals. Metal clusters and surfaces have many dangling bonds, so that they have many lower excited states which are sometimes involved in catalytic processes. Electron transfer is often important in such processes. Therefore, a reliable and efficient method for dealing with such electronic processes is necessary. We here use the SAC (symmetry adapted cluster)/SAC-CI method for groud, excited, ionized, and electron attached states[7-10] for describing electron correlations in surface lower states and electron transferred states. Since the review article on the SAC/SAC-CI method is written separately,[10] we do not explain it here.

We review in this article our ab-initio theoretical studies for the reactions of hydrogen molecule with small Pd and Pt clusters[2,3] and the hydrogenation reaction of acetylene on palladium.[5] The reaction modes are considerably different between these two metals. We then investigate the hydrogen chemisorption on a semi-conductor surface, ZnO.[4] The electronic mechanism is very different from that on a metal surface. We next review an idea of the dipped adcluster model for chemisorptions and catalytic reactions involving electron transfer between admolecules and surfaces,[1] and apply it to oxygen chemisorptions on palladium[1] and silver surfaces.[6]

The basis sets of calculations are as follows. For Pd, Pt, and Ag atoms, the contracted [3s2p2d] sets are used and the Kr and Xe cores are replaced by the relativistic effective core potential (ECP).[11] For Zn, the [2s2p2d] CGTO's are used and the Ar core is replaced by the ECP.[11] For hydrogen, we use the [2s] set of Huzinaga-Dunning,[12] and for carbon the 4-31G set. We have added the first derivative bases for H and C so that the Hellmann-Feynman theorem is approximately satisfied for the forces acting on the hydrogens and carbons.[13] For oxygen, the [4s2p] CGTO plus anion s,p bases[14] and d-polarization functions are used for the calculations with silver and in ZnO, and 4-31 G set for those with palladium. Most Hartree-Fock calculations are done with the use of the program GAMESS.[15]

2. HYDROGEN CHEMISORPTION ON PALLADIUM

We first study the interaction of a hydrogen molecule with a Pd atom. The ground $^1S(d^{10})$ state of the Pd atom shows an affinity to the H_2 molecule, but does not cleavage the H-H bond. The equilibrium geometry is an equilateral triangle with the H-H distance of 0.768 Å, very close to that of the free molecule 0.741 Å and the Pd-H distance of 1.898 Å. The excited states, $^{1,3}D(d^9s^1)$ are repulsive.

We next consider an approach of H_2 to Pd_2 as illustrated in Figure 1. This side-on orientation of H_2 was shown to be most stable. The Pd-Pd distance is fixed to 2.7511 Å, the bulk fcc crystal structure.[16] In Figure 1, we show the potential curves of H_2 at several Pd_2-H_2 separations, R. They were calculated by the CAS-SCF method.[17] When R is larger than 2.5 Å, the potential of H_2 is essentially the

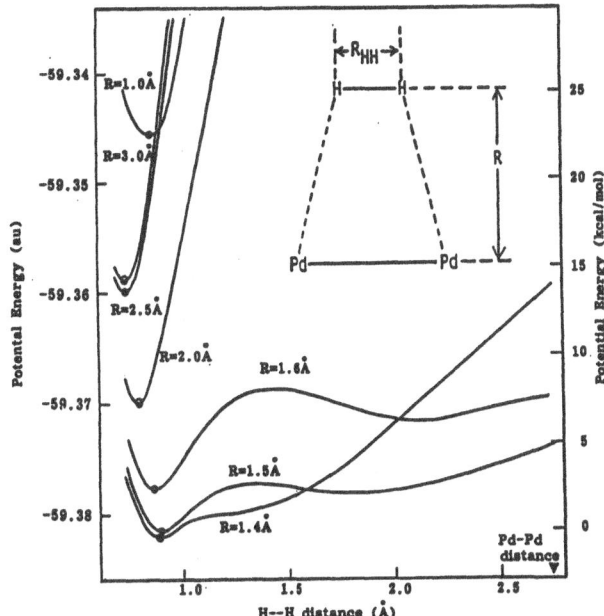

Figure 1. Potential curves for the H-H stretching of the Pd_2-H_2 system at different Pd_2-H_2 separations (CAS-SCF method).

same as that of the free molecule. When H_2 approaches Pd_2 up to $R = 2.0$ Å, the H-H distance becomes longer but the potential is still very sharp. However, at $R = 1.4 \sim 1.6$ Å, the potential curve suddenly (catastrophically) becomes very flat for an elongation of the H-H distance. At $R = 1.6$ Å, a double-well potential appears, and at $R = 1.5$ Å, the system becomes considerably more stable than that at $R = 1.6$ Å. Here, the second minimum appears at $R(\text{H-H}) = 1.75$ Å, besides the first minimum at $R(\text{H-H}) = 0.847$ Å. At $R = 1.4$ Å, the first minimum is more stabilized than that at $R = 1.5$ Å, but the second minimum disappears. When the H_2 molecule further approaches Pd_2 up to $R = 1.0$ Å, the system becomes very much unstable. Thus, a stable adsorption of the H_2 molecule seems to occur at about 1.5 Å, from the Pd surface. The calculated heat of adsorption is about 15 kcal/mol which is smaller than the experimental value, $20.8 \sim 24.4$ kcal/mol, for the bulk Pd surface.[18]

In order to obtain more reliable potential curves of H_2 interacting with Pd_2 at $R = 1.5$ Å, we calculated the potential curve of the ground state by the SAC method, and those of the singlet and triplet excited states by the SAC-CI method. Figure 2 shows the results. In the ground-state curve, we clearly see two potential minima. The minimum at $R(\text{H-H}) = \sim 0.89$ Å, corresponds to the molecular adsorption form, and that at $R(\text{H-H}) = \sim 2.1$ Å, corresponds to the dissociative form. The dissociative form is more stable than the molecular form by 2.2 kcal/mol and the barrier height is 5.6 kcal/mol. However, since the motion along the metal surface was not energetically optimized, the actual barrier could be smaller. We note that at the second minimum, the Pd-H distance is ~ 1.5 Å, which is close to the experimental internuclear distance of a free PdH molecule, 1.529 Å.[19] Thus, the H_2 molecule with a binding energy of

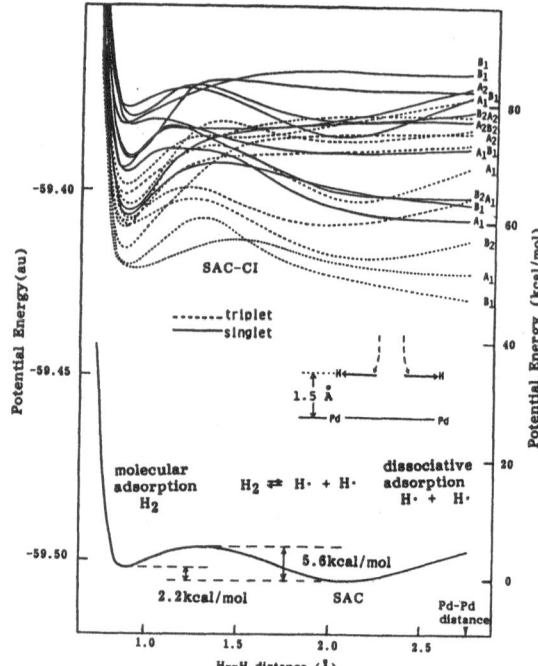

Figure 2. Potential energy curves of the ground and excited states of the Pd_2-H_2 system as a function of the H-H distance of the H_2 molecule 1.5 Å from the Pd_2 fragment (SAC and SAC-CI methods).

about 104 kcal/mol is dissociated, with almost no barrier, into two atomic hydrogens on the Pd_2 "surface", like on an extended surface.

From Figure 2, we see that the excited states of the Pd_2-H_2 system are well separated from the ground state, throughout the process, by more than 50 kcal/mol. There is almost no chance for the excited states to participate in the dissociative process. Therefore, the mechanism of the dissociative adsorption on the Pd surface is different from that proposed for a Ni surface by Melius et al.[20]

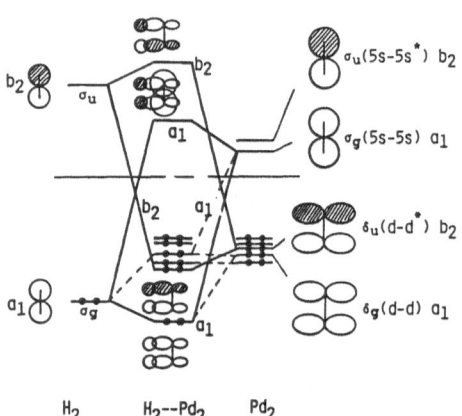

Figure 3. Schematic orbital correlation diagram for the interaction of H_2 and Pd_2.

Then, by what mechanism does the Pd_2 show such a catalytic ability ? Figure 3 shows a schematic orbital correlation diagram of the Pd_2-H_2 system. The left-hand side is the MO's of H_2, the right-hand side is the valence MO's of Pd_2, and the center is for the Pd_2-H_2 system. Two interactions are important. One is the electron transfer from the $\delta_u(d$-$d^*)$MO of Pd_2 to the antibonding MO of H_2. This transfer works to weaken the H-H bond. The other is the electron back-transfer from the bonding MO of H_2 to the bonding σ_g (5s-5s)MO of Pd_2. This back-transfer also works to weaken the H-H bond. These interactions increase as the H_2 approaches the Pd_2, and finally lead to a cleavage of the H-H bond. Other implications of this diagram are that the d electrons are important in the newly formed Pd-H bond and that the Pd-Pd bond is not weakened (rather strengthened) by the adsorption of H_2.

The last point is because, on the Pd_2 side, the electron flows out from the antibonding δ_u MO and flows into the bonding σ_g MO, resulting in a net increase in the Pd-Pd bond order. This aspect seems to be important in relation to the stability of the catalyst, implying that the Pd atom is not exfoliated as a PdH molecule from the metal surface. We note that these $4d_\delta$ and 5s AO's constitute the so-called "dangling" bonds of the metal surface. This mechanism may be simplified as the bond alternation mechanism shown below.

We have obtained a density profile which confirms such a bond alternation.

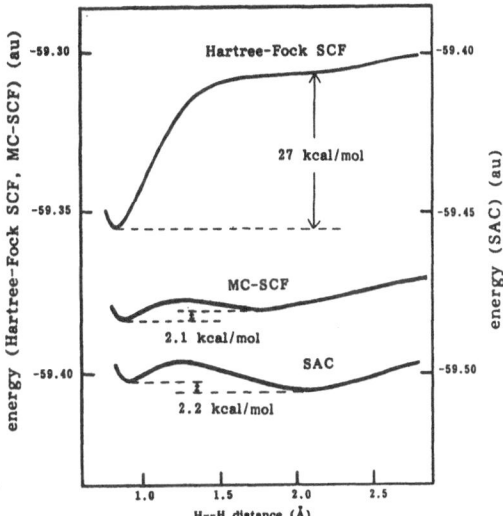

Figure 4. Potential energy curves for the H-H separation of the Pd_2-H_2 system with the Pd_2-H_2 distance at 1.5 Å calculated by the Hartree-Fock, CAS-SCF, and SAC methods.

We must say that the electron correlation is an origin of the dissociative adsorption, since it can not be explained even qualitatively without including the electron correlation. Figure 4 shows the potential curves of H_2, as that shown in Figure 2, calculated by the Hartree-Fock, CAS-SCF, and SAC/SAC-CI methods. We see that the dissociative state is more stable than the molecular state only when the SAC method is used. The Hartree-Fock result fails to predict the existence of the dissociative adsorption state.

We have further shown that the existence of molecular and dissociative states of H_2 on a Pd surface explains the irreversibility in the observation of the photoemission spectra of the H-exposed Pd surface.[21-23] The discussions are found in ref. 2.

3. HYDROGENATION OF ACETYLENE ON PALLADIUM

As we have thus obtained dissociatively adsorbed hydrogens on the Pd_2 cluster on a purely theoretical ground, we investigate here the activity of this system for the hydrogenation reaction. We have chosen acetylene as a reactant because the catalytic reaction

$$C_2H_2 + H_2 \rightarrow C_2H_4 \tag{1}$$

is useful in chemical industry for converting acetylene included as impurities in ethylene gas. Palladium is a good catalyst of this reaction and shows the selectivity which is practically very important.[24] Namely, as far as acetylene exists in the mixture, it is hydrogenated selectively to ethylene, and ethane is not formed. Another reason we have chosen this reaction is that it is typically a symmetry-forbidden reaction.[25,26] Without an existence of the catalyst the barrier of this reaction is too high to occur smoothly. Then, an actual occurrence of this reaction on a palladium surface should be due to the catalytic activity of palladium. We want to know the electronic origin of this catalytic activity.

We consider here two modes of the reaction. One is that acetylene in a gas phase or in a Van der Waals layer of the catalyst reacts with the hydrogen molecule dissociatively adsorbed on palladium. This pathway is called Eley-Rideal (ER) mode. This mode is suitable for investigating the reactivity of the hydrogens dissociatively adsorbed on Pd_2. Experimentally this mode is not necessarily realistic, since acetylene is more easily adsorbed on palladium than hydrogen.[27] However, this reaction mode would become realistic when molecular beam experiment is undertaken. It would give valuable information on the reactivity of hydrogens adsorbed on a metal cluster.

Another mode we investigate is the surface reaction in which hydrogen dissociatively adsorbed on palladium attacks acetylene also adsorbed on the surface. This mode is called Langmuir-Hinshelwood (LH) mode. This mode is experimentally natural,[27] but theoretically more difficult than the ER mode, because of the larger freedom in the reaction pathway.

3.1. Eley-Rideal Mode

The assumed pathway of the ER mode and the forces acting on the carbons and hydrogens of the system during the reaction are shown in Figure 5. This force is calculated by using the Hellmann-Feynman theorem, which is satisfied since the basis sets for C and H include derivative bases.[13]

As seen from Figure 5, the system is repulsive in the beginning of the reaction, showing an existence of potential barrier. This is seen for the positions 1 and 2 of Figure 5. The hydrogens adsorbed on Pd_2 and acetylene are repulsive to each other.

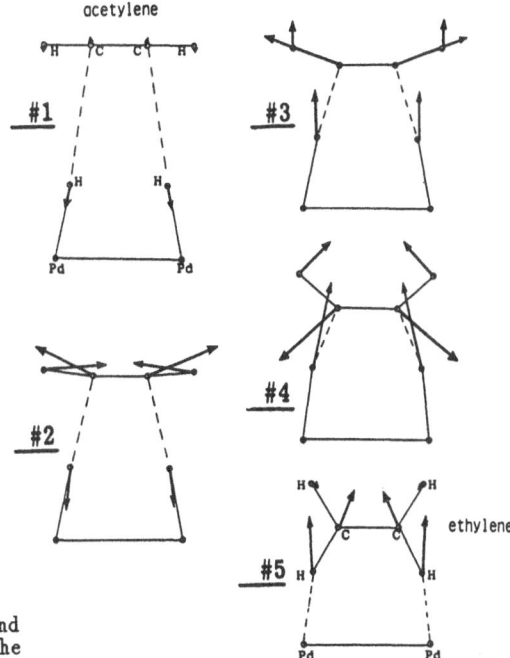

Figure 5. Force acting on the carbons and hydrogens during the Eley-Rideal mode of the reaction.

In the form 3, the forces acting on carbons are still repulsive, but the forces acting on the hydrogens on Pd$_2$ are attractive and the terminal hydrogens of acetylene feel the bending force. In the form 4, all of the forces acting on hydrogens and carbons work to accelerate the reaction. In the form 5, ethylene is formed and the force works to push out the product ethylene out from the Pd$_2$. Therefore, the ethylene is released automatically from the catalytic 'surface' of Pd$_2$. The active site of Pd$_2$ thus generated again adsorbs H$_2$ and enter into the catalytic cycle of the reaction. We should note that though ethylene is attractive to palladium when its π-orbital attacks the surface, it is repulsive in the form shown in 5 of Figure 5.

The corresponding potential energy curve for the reaction is shown in Figure 6. It also shows the potential curves for the same reaction without Pd$_2$. The curve starting from the level, C$_2$H$_2$ + H$_2$(2.1 Å), is the potential curve of the system entirely same as that shown in Figure 5 except for the non-existence of the Pd$_2$. Another sharp curve is also without Pd$_2$ starting from C$_2$H$_2$ and molecular hydrogen (R(H-H) = 0.74 Å).

The barrier of the reaction with an existence of Pd$_2$ is about 32 kcal/mol with respect to the free system. It is much smaller than that of the same reaction without Pd$_2$, which is as large as 138 kcal/mol due to the symmetry forbidden nature of the reaction. When ethylene is formed on Pd$_2$, it is repelled out automatically from the surface, since there, C$_2$H$_4$ and Pd$_2$ are coplanar so that the system is more unstable than the free system by 22 kcal/mol. The Pd$_2$ fragment thus generated again adsorbs H$_2$ and enters again into the reaction cycle. This is the catalytic cycle of the hydrogenation reaction in the Eley-Rideal mode involving Pd$_2$ as a catalyst.

258

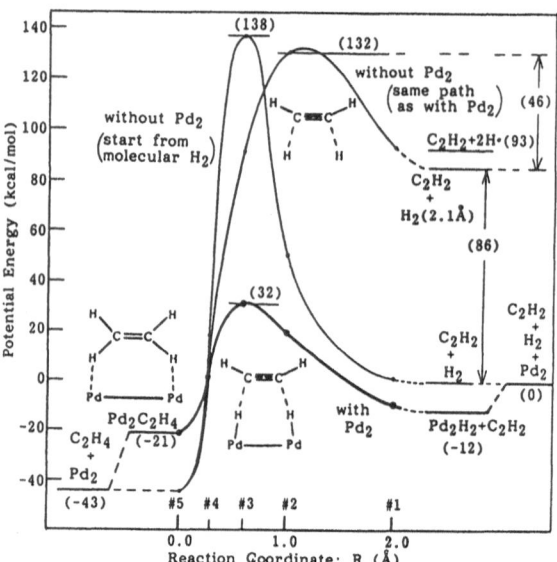

Figure 6. Potential energy curves for the hydrogenation reaction of acetylene with and without Pd$_2$. For the reaction without Pd$_2$, the steep curve on the left-hand side corresponds to the path starting from C$_2$H$_2$ and molecular hydrogen, and the curve on the right-hand side corresponds to the same pathway as that in Figure 5 except for the non-existence of the Pd$_2$. The numbers in the parentheses show relative energies in kcal/mol obtained by the CAS-SCF method.

When we start from C$_2$H$_2$ and two almost atomic hydrogens with the H-H distance of 2.1 Å, the H-H distance of the dissociatively adsorbed hydrogen on Pd$_2$, the barrier is about 46 kcal/mol, which is almost the same as the barrier 44 kcal/mol for the curve with Pd$_2$. The slopes of the two curves up to the transition states are also similar. This result implies two facts. First, the most important step, energetically, in this catalytic process is the dissociative adsorption of H$_2$ on a palladium surface. Second, the hydrogen dissociatively adsorbed on Pd$_2$ is essentially as reactive as a free atomic hydrogen, despite of the existence of the Pd-H bonds on the surface. This is indeed surprising and shows the catalytic activity of palladium in the second hydrogenation step.

3.2. Langmuir-Hinshelwood Mode

We next consider the Langmuir-Hinshelwood (LH) mode in which the olefin adsorbed on a metal surface is attacked by hydrogen also adsorbed on the surface.

Figure 7 shows the assumed pathway for the two step LH mode involving vinyl radical as a surface intermediate. Figure 8 shows the potential energy diagram for this reaction mode. The barrier is only 7 kcal/mol relative to the separated system. Afterwards, the reaction proceeds smoothly to form vinyl radical adsorbed on Pd$_2$, which is more stable than the separated system, Pd$_2$ + C$_2$H$_3$, by 7 kcal/mol. Therefore, the vinyl radical remains on the surface and receives second attack of the surface hydrogen to form ethylene. In this final product form, ethylene is coplanar with Pd$_2$ so that it is repelled automatically from the surface. This is the completion of the catalytic cycle. The naked site of palladium thus released adsorbs acetylene and hydrogen and enters again into the cycle. Thus, the two step LH mode involving vinyl radical as a surface intermediate is favorable as the mode of the hydrogenation

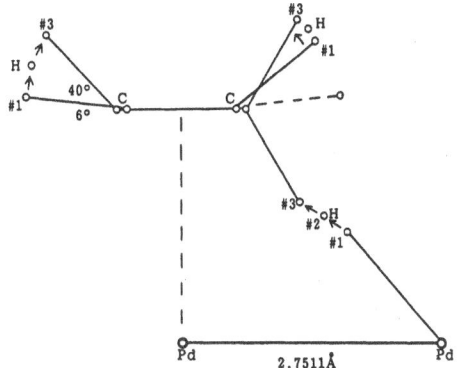

Figure 7. Assumed reaction pathways in the Langmuir-Hinshelwood mode. The Pd-Pd and Pd-C_2 distances are fixed throughout the reaction.

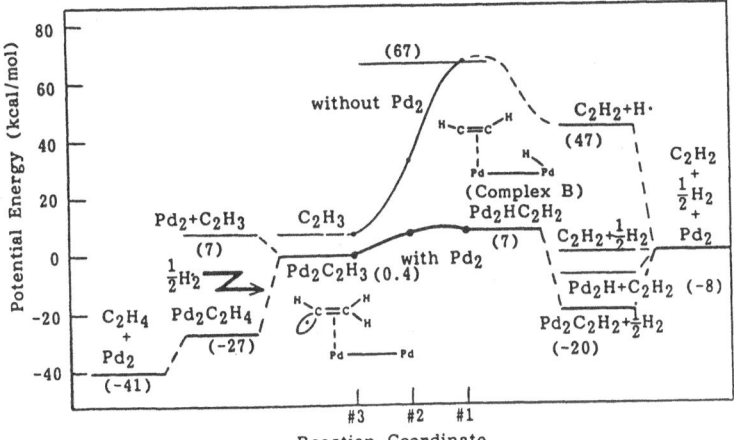

Figure 8. Energetics for the hydrogenation reaction of acetylene in the Langmuir-Hinshelwood mode. The coordinates 1 ~ 3 are shown in Figure 7. Energetics for the same pathway without Pd_2 is also shown. The numbers in parentheses show relative energies in kcal/mol.

reaction. This result agrees with the experimental observations.[27]

3.3. Selectivity in the Hydrogenation Reaction

Activity and selectivity are two major roles of catalyst. We explain here the selectivity of the palladium catalyst. It is summarized as follows. (1) Hydrogenation of ethylene does not occur until all acetylene impurities are converted to ethylene. (2) Ethane is generated only scarcely from acetylene. This selectivity occurs even though palladium is a better catalyst for the hydrogenation of ethylene.

A possible explanation based on the present calculation is as follows. For the first selectivity, the origin is that the heat of adsorption is larger for acetylene than for ethylene as observed experimentally.[27] Further, the sticking probability of acetylene

260

is larger than that of ethylene because acetylene has active π orbitals in all angles around the C-C axis but ethylene has the π orbital only in the plane perpendicular to the molecular plane. The second selectivity is due to the fact that the ethylene produced by the hydrogenation reaction of acetylene is coplanar with the active palladium atoms, so that, ethylene is repelled from the surface and released out automatically from the reaction cycle.

3.4. Proposed Molecular Beam Experiment

We propose here molecular beam experiments of the palladium clusters Pt_n $(n \geq 2)$ as shown in Figure 9. When the beam of the Pd_n cluster is introduced through H_2 gas, it will chemisorb hydrogen molecule and the product beam $Pt_n(H_2)_m$ will come out, where H_2 is dissociatively adsorbed on the Pt_n cluster. When this beam is further introduced through acetylene gas at some temperature, it will be hydrogenated by the Eley-Rideal mechanism and ethylene will be generated. The acetylene gas will then become a mixture of ethylene and acetylene gases. The bare palladium sites produced may adsorb acetylene, but if the speed of the beam or the length of the acetylene tank is well regulated, we may get a bare palladium beam, which may be put again into the hydrogen gas. This is a cycle of the catalytic reaction with the use of the molecular beam as a catalyst.

Crossed molecular beam experiment shown below would give more detailed information on the dynamics of the reaction. The difference from the experiment shown above is that basically only one H_2 molecule will be chemisorbed on the Pd_n cluster and also only one acetylene molecule will attack this $Pd_n H_2$ cluster. Therefore when n is large, both ER and LH modes would occur on the cluster surface, giving rise to different behaviors in the product beams.

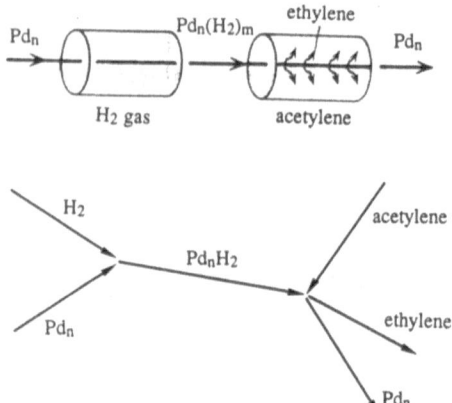

Figure 9. Proposed molecular beam experiments.

4. REACTIONS OF A HYDROGEN MOLECULE WITH SMALL PLATINUM CLUSTERS

We study here reactions of a hydrogen molecule with small platinum clusters Pt_n $(n = 1,2,3)$ by ab initio methods. This provides a cluster model study for hydrogen chemisorption on a Pt surface.

4.1. Pt-H₂ System

We first examine side-on, on-top approach of a hydrogen molecule to a Pt atom. The definition of the reaction path is shown in Figure 10 and the accounts are found in ref. 3. In the Pt-H$_2$ system, only the central Pt atom (denoted as Pt$_a$) is considered.

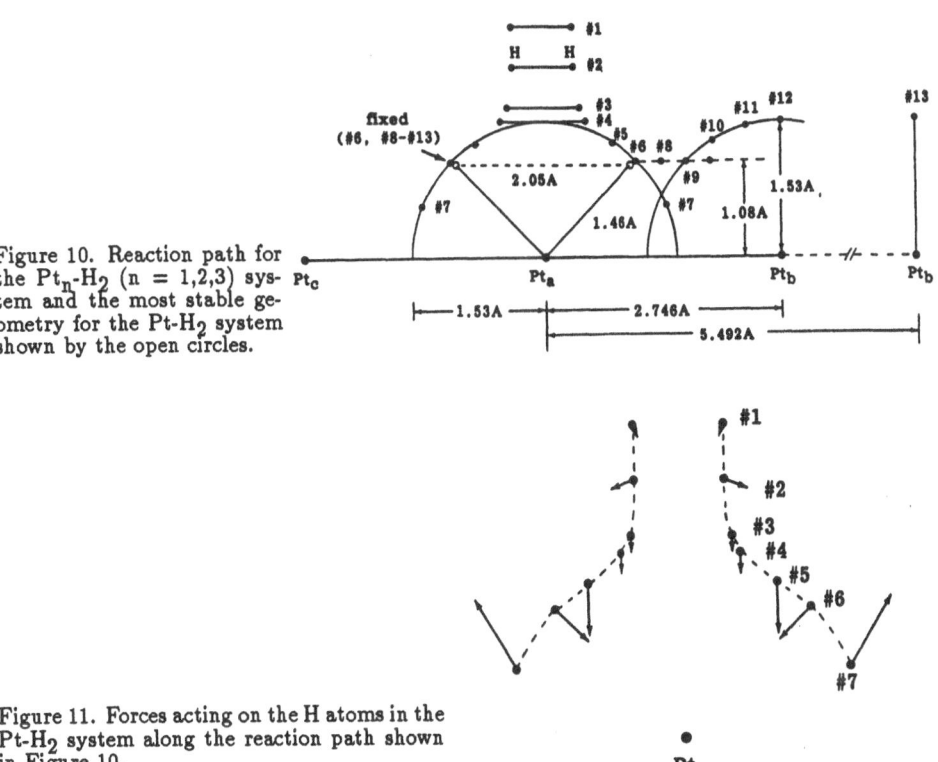

Figure 10. Reaction path for the Pt$_n$-H$_2$ (n = 1,2,3) system and the most stable geometry for the Pt-H$_2$ system shown by the open circles.

Figure 11. Forces acting on the H atoms in the Pt-H$_2$ system along the reaction path shown in Figure 10.

Figure 11 is a display of the forces acting on the H atom of the Pt-H$_2$ system calculted by the Hartree-Fock method for the $^1A_1$ state. At all the points, #1 ~ #6, the H atoms are attracted by the Pt atom. At the point #2, the H$_2$ molecule experiences the forces which act to elongate the H-H distance. At the points #3 ~ #5, that force is zero because the H-H distance has been optimized. At the point #7, the H atoms are repelled by the Pt atom. The optimized position of H$_2$ (open circles, Figure 10) is a little bit inside of the position #6, as indicated from the force.

The potential energy curves of the Pt-H$_2$ system calculated by the SAC/SAC-CI method are shown in Figure 12. The left-hand side (Figure 12a) is without the spin-orbit coupling effect and the right-hand side (Figure 12b) is with the spin-orbit coupling. All the triplet states arising from the ground state (^3D) of the Pt atom show large energy barriers of more than 16 kcal/mol (B$_1$, A$_2$, and B$_2$ states) with the stabilization energies less than 10 kcal/mol (B$_2$ state). Only the $^1A_1$ state, which originates from the excited ^1S state of the Pt atom, is very attractive. The calculated

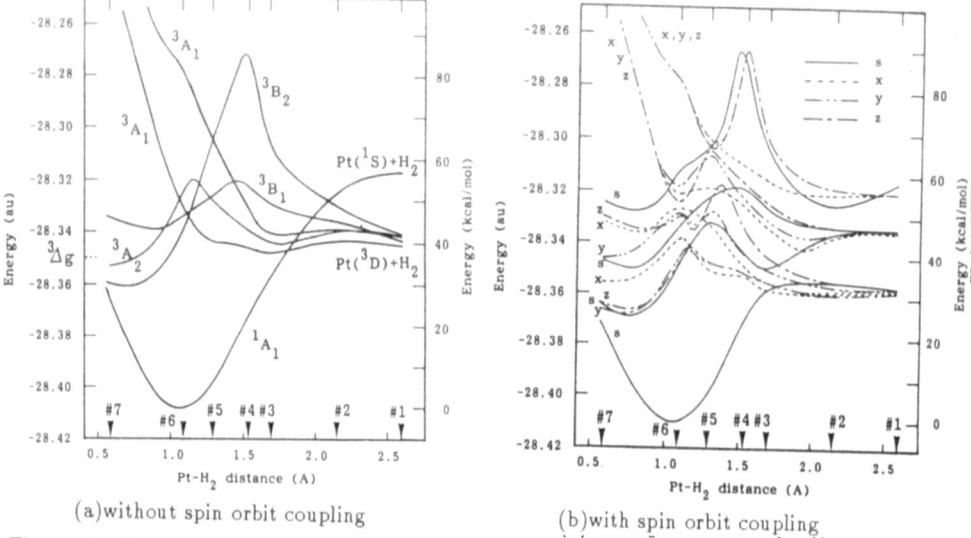

Figure 12. Potential energy curves of the ground and excited states of the Pt-H$_2$ system in the side-on, on-top approach calculated by the SAC/SAC-CI method.

stabilization energy of the $^1A_1$ state, relative to the dissociation limit of Pt(^{3}D) + H$_2$, is about 40 kcal/mol, which is larger than the experimental heat of adsorption, 26 kcal/mol.[28] Thus, even a single Pt atom leads to a dissociative adsorption of a H$_2$ molecule. The catalytically active state for this adsorption is not the ground state, but the excited ^{1}S state of the Pt atom. When we consider spin-orbit interactions, we get the potential curves shown in Figure 12b. The potential well of the ground state at #5 and #6 geometries dissociates smoothly to the essentially ^{3}D state of Pt and a H$_2$ molecule. The heat of adsorption is calculated as 32.0 kcal/mol which is closer to the experimental value of 26 kcal/mol than that calculated from Figure 12a. On the basis of the present calculations, only the dissociative adsorption state is realized. The molecular adsorption state seems not to exist in contrast to the palladium case.

Next we analyze the electronic mechanism of the reaction. There are two important orbital interactions between H$_2$ and Pt as shown in Figure 13. At an early stage of the interaction, the electron transfer from the σ_g orbital of H$_2$ to the 6s orbital of the Pt atom is important. To accelerate the cleavage of the H-H bond, electron back-transfer from the d_{yz} orbital of the Pt atom to the σ_u orbital of H$_2$ is important. As the reaction proceeds, the electronic configuration of the Pt atom changes from d^{10} to a mixture of d^9s^1 and d^8s^2. As the Pt atom prefers the d^9s^1 and d^8s^2 configurations by 17.5 kcal/mol and 15.2 kcal/mol, respectively, to the d^{10} configuration,[29] this reaction proceeds very smoothly to dissociate completely the H$_2$ molecule. By contrast, as the Pd atom prefers the d^{10} configuration by 17.5 kcal/mol to the d^9s^1 configuration,[29] the dissociative adsorption by a single Pd atom does not occur, though it occurs smoothly on the Pd$_2$ cluster, as shown in the previous

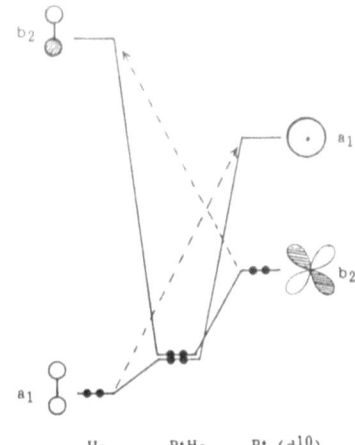

Figure 13. Orbital correlation diagram for the Pt-H_2 system.

H₂   PtH₂   Pt (d^{10})

section.

4.2. Pt_2-H_2 System

Here we consider the adsorption process of a hydrogen molecule on Pt_2 cluster. New aspects here are the migration process of a hydrogen atom on a Pt surface and the stability of the Pt-Pt bond during the catalytic process. The reaction path is shown in Figure 10. The metal atoms considered here are Pt_a and Pt_b, with the Pt-Pt distance being fixed at 2.746 Å.[16] The points #1 ~ #6 are the same as those described for the Pt-H_2 system. After reaching the position #6, which is very close to the optimized geometry of the Pt-H_2 system, the surface migration of the dissociated H atom is considered. The H atom on the left-hand side is fixed at this geometry and only the H atom on the right-hand side is moved from #6 to #12. This is the migration process of a hydrogen atom from Pt_a to Pt_b.

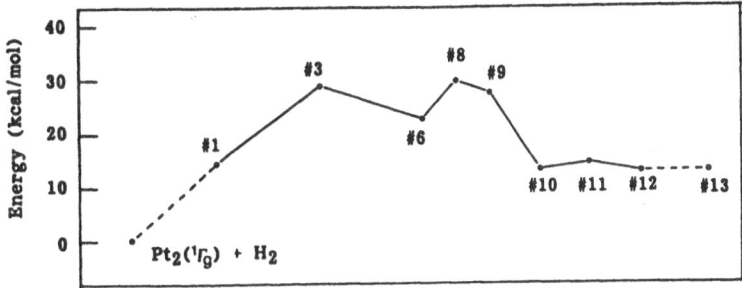

Figure 14. Potential energy diagram of the singlet ground state of the Pt_2-H_2 system calculated by the CAS-SCF method.

Figure 14 shows the energetics for the adsorption and migration processes of H_2 on Pt_2. The H_2 molecule must overcome a high barrier of 27.7 kcal/mol on going from the free system to point #3. Even after reaching point #6, the system is not stabilized relative to the free system. This is very different from the Pt-H_2 system. Further, as the right H migrates on the Pt_2 surface from point #6 to #10, the energy of the system is lowered by about 15 kcal/mol, with the barrier of about 8 kcal/mol. Points #10 ~ #12 have almost the same energy. The adsorption energy calculated for this system is -12 kcal/mol which is the energy difference between #10 and the free system, $Pt_2(^1\Gamma_g) + H_2$. Thus all of the processes considered for this system prove to be a game at an energy level higher than the free $Pt_2 + H_2$ system. This implies that the dissociative adsorption reaction of H_2 does not occur on the Pt_2 cluster, at least in the side-on, on-top form and also in the side-on bridge form as we investigated.[3,30]

Another important aspect of the model study of chemisorption on a cluster surface is the stability of the cluster during catalytic processes. In order to examine it, we calculate the energy of point #13 in Figure 10, where the Pt_a-Pt_b distance is 5.492 Å, twice the original length of 2.746 Å, with the right H atom just above Pt_b. If any bonding remains between Pt_a and Pt_b after the migration of the H atom (#12), the energy of #13 would become higher than that of #12. However, the calculated results shown in Figure 14 indicate that the energy of #13 is almost the same as that of #12, implying that the Pt_a-Pt_b bond is completely broken after the hydrogen migration.

We thus conclude that the Pt_2 cluster does not react with the H_2 molecule and hence it is not a good model for chemisorption of a hydrogen molecule.

4.3. Pt3-H2 System

Here we examine, using a larger cluster Pt_3, the dissociative adsorption of an H_2 molecule and the migration of an H atom on a platinum surface. The reaction path is again the side-on, on-top approach shown in Figure 10 and all of the three Pt atoms Pt_a, Pt_b and Pt_c are involved. The geometries of the approach #1 - #12 are the same as those defined and used previously for the Pt-H_2 and Pt_2-H_2 systems. The results are shown in Figure 15.

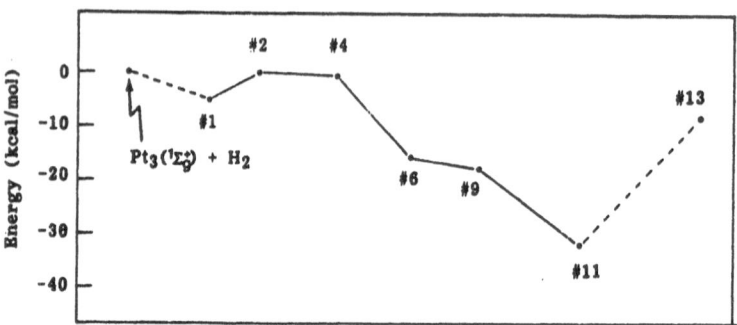

Figure 15. Potential energy diagram of the singlet A_1 state of the Pt_3-H_2 system calculated by the MC-SCF method.

The energy of #1 is lower than that of the free system, $Pt_3(^1\Sigma_g) + H_2$ by 5.0 kcal/mol, probably indicating the physisorption state. At point #2, the system is unstabilized by 4.7 kcal/mol in comparison with #1, but this is almost at the same level as the free system. The energy barrier which has to be overcome is considerably smaller in this system than in the previous Pt_2-H_2 system. At point #6, which is close to the most stable geometry of the Pt-H_2 system, the energy lowering from the free system is 15.7 kcal/mol. Although this is smaller than that of the Pt-H_2 system (about 40 kcal/mol), it is larger than that of the Pt_2-H_2 system (-21.9 kcal/mol). As the right H atom migrates on the Pt_3 surface from #6 to #11, the energy of the system is lowered again by 15.6 kcal/mol, with no energy barrier. The system prefers one H atom on each Pt atom rather than two H atoms on one Pt atom. The adsorption energy finally calculated for this system is 31.3 kcal/mol. This is obtained by substracting the energy of #11 from that of the free system $Pt_3(^1\Sigma_g)$ + H_2. This energy may be compared with the experimental adsorption energy, 24 kcal/mol of H_2 on a real Pt surface.[28]

We next examine the strength of the Pt-Pt bond during the migration process, as in the case of the Pt_2-H_2 system. At #13, the Pt_a-Pt_b distance is elongated to 5.492 Å, twice of the original distance, with the right H atom kept on Pt_b. The energy at #13 is higher than that at #11 by 23.3 kcal/mol, ensuring that even after the migration a reasonably strong bond exists between Pt_a and Pt_b. In contrast to the Pt_2 case, the stability of the Pt_3 cluster is related to the existence of the non-bonding σ MO which is considerably stabilized by the participation of the p_σ AO of the central Pt atom.

We thus conclude that the Pt_3 cluster reacts with the H_2 molecule as

$$\text{Pt-Pt-Pt} + H_2 \quad \rightarrow \quad \overset{\displaystyle H \quad H}{\underset{\displaystyle \text{Pt-Pt-Pt}}{\mid \quad \mid}}$$

essentially without an energy barrier. This reaction gives a good model for actual surface reactions both in reproducing the adsorption energy and in showing the stability of the Pt-Pt bond during the dissociative adsorption process.

4.4. Short Summary

Thus, we have studied the reactions of a hydrogen molecule with small platinum clusters, Pt_n (n = 1,2,3). For the Pt_3 cluster, a linear geometry was assumed. We predict that the Pt atom and the Pt_3 cluster would react smoothly with H_2, but the Pt_2 cluster would not. The side-on on-top approach seems to be preferable. The catalytically active state of the Pt atom is not the ground $^3D(d^9s^1)$ state, but the excited $^1S(d^{10})$ state. In particular, the reaction with the Pt_3 cluster is suitable as a model of the hydrogen chemisorption on an actual platinum surface. The calculated heat of reaction was 40 kcal/mol (32 kcal/mol with including the spin orbit coupling) for the Pt-H_2 system, -12 kcal/mol for the Pt_2-H_2 system, and 32 kcal/mol for the Pt_3-H_2 system, in comparison with the experimental value, 26 kcal/mol for an extended surface.[28] The migration of H after dissociative adsorption occurs very smoothly without an energy barrier. The Pt-Pt bond of the Pt_3 cluster is stable during these processes. Experiments must now be undertaken on the reactivity of the small Pt clusters.

We summarize here the gross charges of the Pd_2H_2, PtH_2 and Pt_3H_2 systems

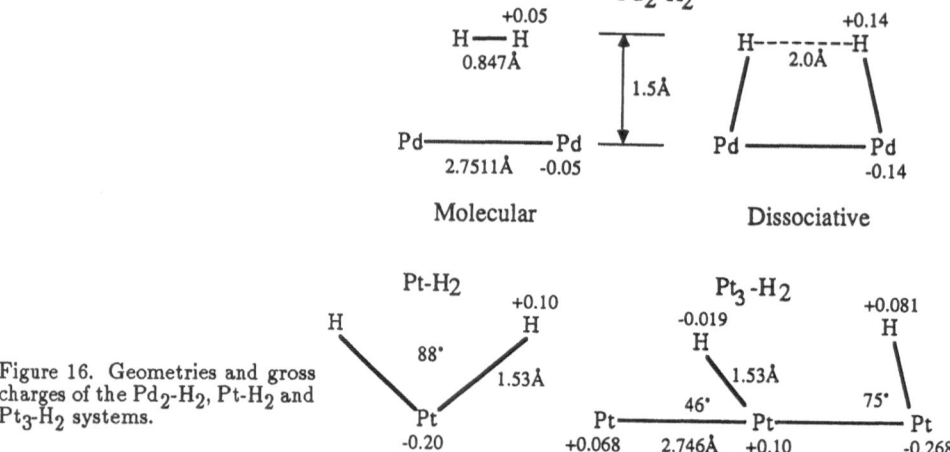

Figure 16. Geometries and gross charges of the Pd_2-H_2, Pt-H_2 and Pt_3-H_2 systems.

in Figure 16. The charges on the hydrogens are $+0.02 \sim +0.14$ which are small. Thus the electron transfers between the admolecules and the metal surfaces are small for hydrogen chemisorptions. This is why the cluster model is successful for describing the hydrogen chemisorptions on the Pd and Pt surfaces.

5. HYDROGEN CHEMISORPTION ON A ZnO SURFACE

Zinc oxide is an n-type semiconductor and has a catalytic activity for hydrogenations of olefins. It is a wultzite-type crystal and has many stable surfaces. It dissociatively adsorbs hydrogen molecule and the existence of some adsorbed hydrogen species is known.[31] Type I hydrogen shows a rapid and reversible adsorption and is responsible for the O-H and Zn-H IR peaks observed at 1710 and 3510 cm^{-1}, respectively.[32] This species is the principal source of hydrogens for the hydrogenation reaction of ethylene.[32] Type II hydrogen, on the other hand, contributes little to the hydrogenation of ethylene and does not give the Zn-H and O-H bands, but it promotes the rate of the catalytic reaction. Type III hydrogen exists at the temperature near 78 K. This is molecularly adsorbed on the same site as the type I species.[33]

On the theoretical side, some relevant papers have been published. In particular, Witko and Koutecky[34] studied the potential curves of $ZnO + C_2H_4$ and $(ZnO + C_2H_4)^+$ systems using the pseudo-potential MRD-CI and all electron MRD-CI methods. Attractive interactions have been found for several excited states of the $ZnO + C_2H_4$ system.

We have studied the hydrogen chemisorption on a ZnO surface.[4] The reaction path is calculated for the H_2 chemisorption on a ZnO surface. For simulating the ZnO(1010) surface, one ZnO molecule embeded in a Madelung potential is used. The ZnO distance is fixed at 1.95 Å, an experimental value for the crystal. The Madelung potential is expressed by the 32 point chargres of ± 0.5 situated on the first and second layers. The electrostatic potential due to the ionic layer decreases exponentially, so that the electrostatic potential made by the 32 point charges located around the ZnO molecule reach to 92 % of the one due to 6886 point charges. The Madelung potential is proportional to the ionic charge, q, in $Zn^{+q}O^{-q}$. The Mulliken's atomic charge of ZnO calculated by the Hartree-Fock method is ± 0.6. However the smaller value should be used for q because of the electron spacial distribution. We then choose the point charge of ± 0.5.

5.1. ZnO

The potential curves of the low-lying states of an isolated molecule ZnO are calculated by the SAC/SAC-CI method. The results are shown in Figure 17. The ground state is $^1\Sigma^+$ for the Zn-O distance shorter than 1.98 Å, but at a larger distance the $^3\Pi$ state becomes the ground state. The equilibrium bond lenght of the $^1\Sigma^+$ state is 1.76 Å, which is shorter than the distance in the crystal, 1.95 Å. The bonding orbital is made of the 4s orbital of Zn and the $2p\sigma$ orbital of O. The binding energy is 20 kcal/mol and the dipole moment at the point of equilibrium is 5.82 Debye.

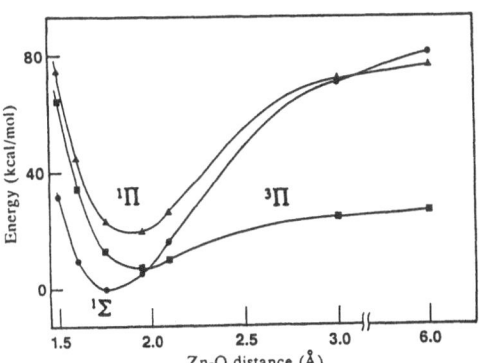

Figure 17. Potential energy curves for the lower singlet and triplet states of ZnO calculated by the SAC/SAC-CI method.

5.2. ZnO + H$_2$ System within the Madelung Potential

Figure 18 shows the reaction path for the hydrogen chemisorption on ZnO surrounded by the Madelung potential and the forces acting on the H atoms. The transition state exists between the points 3 and 4. The point 6 is the most stable geometry of this system. There, the H-H distance is 3.73 Å, which is 5 times as large as the one of a free H$_2$: namely, the H-H bond is completely broken. The H-Zn-O angle is 146°, in contrast to 180° obtained without the Madelung potential. The reason is the electrostatic repulsion between the adsorbed hydrogen and the surrounding Madelung potential. The H-O-Zn angle, on the other hand, is 111° which is slightly larger than 95° obtained without the Madelung potential.

The potential energies along the path calculated by the SAC/SAC-CI method are shown in Figure 19. We see that only the ground state is active for the hydrogen chemisorption. The reaction is exothermic by 73.5 kcal/mol and the reaction barrier is 11.5 kcal/mol. The Madelung potential lowers the barrier by 2.5 kcal/mol. The excited states are all repulsive. The calculated vibrational frequencies ν_{O-H} and

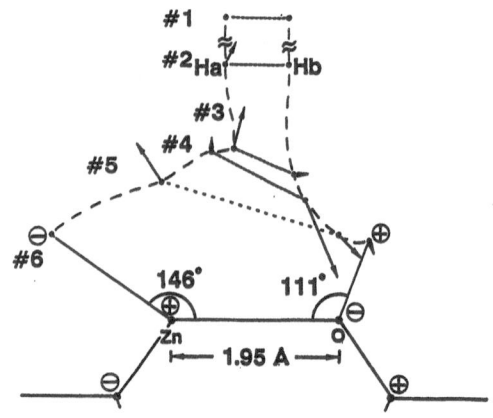

Figure 18. Forces acting on the H atoms in the ZnO + H$_2$ system with the Madelung potential.

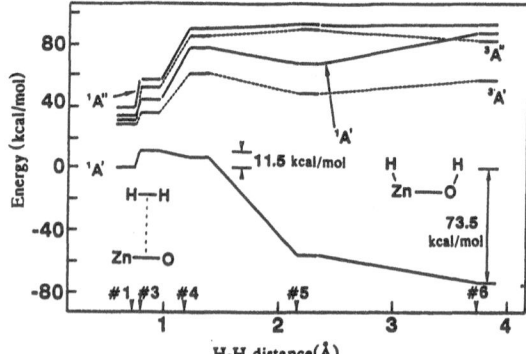

Figure 19. Potential energy diagram of the ground and several singlet and triplet excited states of the ZnO + H$_2$ system with the Madelung potential calculated by the SAC/SAC-CI method.

νZn-H at the point 6 are 4090 and 1730 cm^{-1}, which are to be compared with the experimental values of 3510 and 1710 cm^{-1}, respectively.

Figure 20 shows the contour maps of the density difference defined by

$$\Delta\rho = \rho(ZnO - H_2) - \rho(ZnO) - \rho(H) - \rho(H) \tag{2}$$

At point 2, the density of H$_2$ is polarized by the long-range electrostatic dipole field of Zn$^+$O$^-$, so that the right-hand-side hydrogen becoms protonic. There is a large difference between the densities at 3 and 4, though there is a little difference in geometry between 3 and 4. It indicates that the transition state exists between 3 and 4. At 5, the H-H bond is completely broken, and the Zn-H and O-H bonds are formed. Along the Zn-H bond, the density on the left of the Zn-H bond increases and induces to the left the force acting on the hydrogen. Then this hydrogen moves and at 6, the final Zn-H and O-H bonds are formed. Throughout the reaction, the density in the ZnO region does not decrease, indicating that the Zn-O bond is kept stable. This is related to the stability of the catalytic surface.

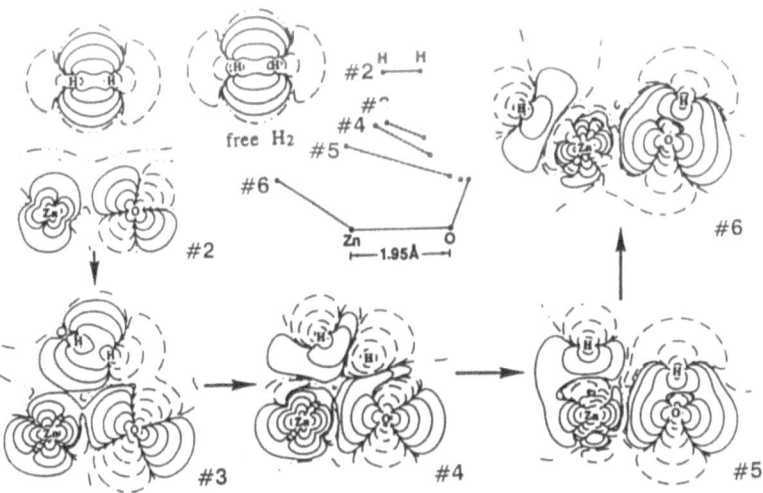

Figure 20. Reorganization of the electron density of the ZnO + H_2 system along the reaction path. The density difference is defined in the text.

5.3. Two Point Charges Plus H_2 System

Here, we study the role of the electrostatic polarization of ZnO for the reaction. To estimate the effect of the charges of ZnO on H_2, we replace Zn^+O^- by the two point charges (± 0.5) placed at the positions of Zn and O, and let H_2 approach along the reaction path shown in Figure 18. The energy is calculated by the full-CI method and given in Table I along with the atomic charges on H_a and H_b. We see that the electrostaic potential works to stabilize the system especially before reaching the barrier, namely the #3 geometry. It also considerably decreases the instability of the #4 and #5 geometries. The electrostaic polarization on ZnO induces a polarizatrion of the bonding σ_g MO of H_2 on the side of H_a (Figure 18) and the antibonding σ_u MO on the other side. This results in an increase of the overlaps between the σ_g orbital of H_2 and the LUMO of ZnO and between the σ_u orbital of H_2 and the $2p_\pi$ orbital of O. Thus, the electrostaic potential due to the charge polarization of Zn^+O^- makes the cleavage of H_2 easier.

Table I. Energies and atomic charges of H_2 along the reaction path where ZnO is replaced by the two point charges of ± 0.5. The energy of H_2 at the equilibrium bond length and without point charges is taken as a standard.

| | | with point charges | | without point charges |
| | | | atomic charge | |
point	R_{H-H}(Å)	ΔE(kcal/mol)	H_a/H_b	ΔE(kcal/mol)
2	0.7417	-0.19	-0.03/+0.03	0
3	0.7846	-10.23	0.00/0.00	0.69
4	1.2191	18.14	-0.05/+0.05	39.91
5	2.1674	43.36	-0.45/+0.45	97.70

5.4. Mechanism of the Reaction

The mechanism of this reaction is qualitatively explained as follows. The electron donation from the $2p_\pi$ orbital of O to the antibonding σ_u MO of H_2 and the backdonation from the bonding σ_g MO of H_2 to the LUMO of ZnO are important. In the initial stage of the reaction, the charge polarization of ZnO induces a polarization of the HOMO of H_2 on the side of Zn and LUMO on the side of O, which makes the electron transfer and back-transfer interaction with ZnO easier, because such a deformation of MO's increases the overlaps betwen the active MO's of H_2 and ZnO. Among the lower-lying states of ZnO, only the $^1\Sigma^+$ state (ground state) is catalytically active for the H_2 chemisorption. All the other low-lying states are repulsive. The Madelung potential enhaces the polarization of ZnO, namely Zn^+-O^-, and the reactivity with H_2 as a result. It also affect the geometry of the dissociatively adsorbed H_2 on a ZnO surface, and the energy gap between the ground and excited states. Throughout the reaction, another type of the $2p\pi$ orbital of O, which is parallel to the surface, and the bonding HOMO of ZnO is inactive and works to keep the ZnO bond stable during the catalytic process.

6. DIPPED ADCLUSTER MODEL (DAM) FOR CHEMISORPTIONS AND CATALYTIC REACTIONS ON A METAL SURFACE

In the previous sections we have used the cluster model for studying hydrogen chemisorptions and hydrogenation reactions on metal and metal oxide surfaces. A reason of the success is that the electron transfers between the admolecules and surfaces are small. Actually the gross charges of the admolecules in the hydrogen chemisorptions are shown in Figure 16. There, the charges on the admolecules are essentially neutral. However, when electron transfers between admolecules and surfaces are large, the cluster model may be inadequate, because the cluster itself has to supply or absorb electrons by much affecting the bondings within the cluster when the size of the cluster is small. In actual metal surfaces, a sufficient number of electrons are involved in the extended orbitals so that the transfer of electrons to or from the admolecule does not much affect the local bonding nature of the metal atoms of the cluster directly interacting with the admolecule. Thus, the cluster model would be incomplete for the surface reactions which accompany large electron (or spin) transfers between admolecules and bulk metals.

6.1. Dipped Adcluster Model (DAM)

We here propose a model for chemisorptions and surface reactions in which the electron transfers between admolecules and surfaces are large. We define 'adcluster' as a combined system of the admolecule and a cluster. We dip it onto the electron 'bath' of the solid metal and let an equilibrium be established for the electron and/or spin transfer between them. The equilibrium condition is described with the use of the chemical potentials of the adcluster and the solid surface. Namely, at equilibrium, the adcluster is at the $\min[E(n)]$ in the range,

$$- \frac{\partial E(n)}{\partial n} \geq \mu \tag{3}$$

where $E(n)$ is the energy of the adcluster with n being the number of electrons transferred from the bulk metal to the adcluster and μ the chemical potential of the electrons of the metal surface. Since the adcluster is a partial system, the number of the transferred electrons, n, is not necessarily an integer. In this model, the external effects such as those of promoters, cocatalysts, supports, temperature, electric potential, light, etc., are included through the variations of the chemical potential μ.

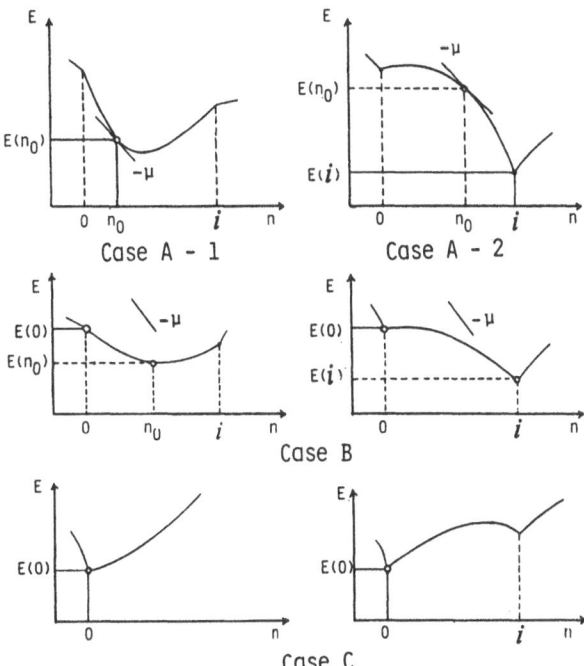

Figure 21. Some typical behaviors of the $E(n)$ curve as a function of n, the number of electrons transferred into the adcluster. On the horizontal axis, i denotes an integer number of electrons. The gradient μ is the chemical potential of the metal surface.

Some typical behaviors of the $E(n)$ curve are illustrated in Figure 21. In case A, the gradient becomes $-\mu$ at $n = n_0$. The $E(n)$ curve is lower and upper convexes in cases A-1 and A-2, respectively. The electrons flow into the adcluster up to n_0 and i in cases A-1 and A-2, respectively. We note that n_0 may be a non-integer but i is an integer. In case B, there is a region of n where $E(n)$ is lower than $E(0)$, but the gradient is smaller than μ in that region, so that the electron flow does not occur. A device for lowering μ is necessary for realizing the electron transfer. In case C, $E(0)$ is most stable in some wide region of n. In this case, the electron transfer can not be expected, so that the cluster model plus image force correction described below would be appropriate.

6.2. Molecular Orbital Model of the Dipped Adcluster

We give here molecular orbital model of the dipped adcluster and calculate the energy $E(n)$ and the electronic structure. We assume that the adcluster exchanges electrons and spins with the solid through its HOMO (highest occupied mo), LUMO (lowest unoccupied mo), SOMO (singly occupied mo), or some other active MO, with the other MO's being doubly occupied or completely unoccupied. Such active MO is denoted by m. Two types of spin coupling are assumed for the electrons occupying the m-th MO. One is called highest spin coupling, in which the m-th MO is first occupied by α spin electron and after its occupation becomes equal to unity, it is then occupied by β spin electron. In this case, the adcluster is paramagnetic. The other is paired spin coupling in which the same amounts of α and β spin electrons occupy the m-th MO. Here, the adcluster is diamagnetic. The energy of the adcluster with x electrons occupying the m-th MO, which is assumed to be non-degenerate, is given by

$$E^{(0)} = 2 \sum_k H_k + \sum_{k,l}(2J_{kl} - K_{kl}) + x \sum_k (2J_{km} - K_{km}) + x \, H_m + Q \quad (4)$$

where k, l run over the doubly occupied MO's. H_k, J_{kl} and K_{kl} denote core-hamiltonian integral, coulomb repulsion integral, and exchange repulsion integral, respectively. The meaning of the superscript (0) will be self-evident later. The quantity Q represents the electron repulsion within the active orbital m and depends on the nature of the spin coupling as described above. It is given by

$$Q = \| x - 1 \| \, J_{mm} \quad (5a)$$

for the highest spin coupling and

$$Q = (\frac{x}{2})^2 \, J_{mm} \quad (5b)$$

for the paired spin coupling. Here, $\|a\| = 0$ if $a < 0$, $\|a\| = a$ if $0 \leq a \leq 1$, and $\|a\| = 1$ if $a > 1$. It is easy to prove that Q is minimum for the highest spin coupling and maximum for the paired spin coupling. Therefore, the energy of the adcluster itself is lowest in the highest spin coupling. The actual preference of the way of the spin coupling would also depend on the nature of the solid and of the interaction between the adcluster and solid. For example, when some amount of α spin is transferred from the solid metal to the adcluster, the system is spin polarized and paramagnetic near the adcluster. For cases in which the orbital m is degenerate, see ref. 1.

The energy of the open-shell restricted Hartree-Fock (RHF) method is written as

$$E = \sum_k \lambda_k H_k + \frac{1}{2} \sum_k \sum_l \lambda_k \lambda_l (\alpha_{kl} J_{kl} - \beta_{kl} K_{kl}). \quad (6)$$

By a comparison between Eqs.(6) and (4), the occupation parameter λ_k and the spin coupling parameters α_{kl} and β_{kl} in Eq.(6) are fixed, so that performing the RHF-MO SCF calculation involving the non-integral occupation number x, we obtain the molecular orbitals and the energy $E(x)$.

6.3. Electrostatic Image Force

When an adatom A at the position a has a charge q, it induces an opposite charge on a metal surface. At point $x(x, y, 0)$ of the surface, the induced charge density is given by,[35]

$$\sigma(\mathbf{x}) = -\frac{q|\mathbf{a} - \mathbf{a}'|}{4\pi|\mathbf{a} - \mathbf{x}|^3} \tag{7}$$

where $\mathbf{a}'$ is the positional vector of the mirror image of the adatom A. The electrostatic interaction between the charge q and the hole $\sigma(\mathbf{x})$ sums up to the well-known image force given by

$$\mathbf{F}_{if} = \int\int_{xy}^{\text{surface}} \frac{\sigma(\mathbf{x})q}{|\mathbf{a} - \mathbf{x}|^3}(\mathbf{a} - \mathbf{x})dxdy = \frac{q^2(\mathbf{a}' - \mathbf{a})}{|\mathbf{a} - \mathbf{a}'|^3}, \tag{8}$$

and the stabilization energy is given by

$$E_{if} = \int\int_{xy}^{\text{surface}} \frac{\sigma(\mathbf{x})q}{2|\mathbf{a} - \mathbf{x}|}dxdy = -\frac{q^2}{2|\mathbf{a}' - \mathbf{a}|} \tag{9}$$

where the factor 2 in the denominator is due to the integration over the half space $(z \geq 0)$. For polyatomic systems, see ref. 1.

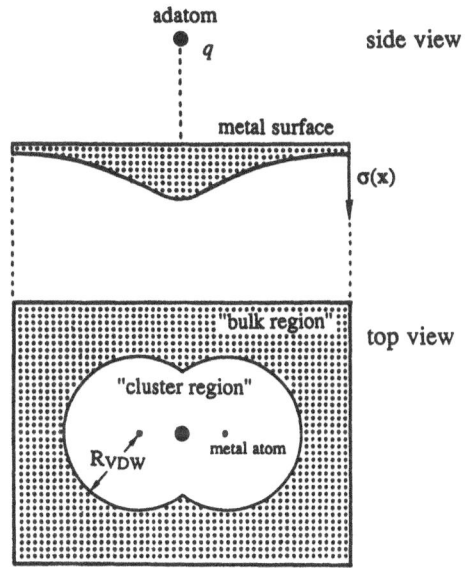

Figure 22. Schematic representation of the adatom and the induced charge density on the surface. The cluster region is estimated from the Van der Waals radius of the metal atom.

When the energy $E^{(0)}$ of the adcluster alone is calculated by an ab initio method, the electrostatic interaction within the adcluster is already included. Therefore, the electrostatic term $E^{(1)}$ is estimated, as illustrated in Figure 22, by integrating the electrostatic interaction between the charge q and the hole $\sigma(\mathbf{x})$ at the point $\mathbf{x}$ outside the cluster region of the surface. In practice, the energy $E^{(1)}$ is calculated by subtracting from E_{if} the electrostatic energy (E_{in}) for $\sigma(\mathbf{x})$ inside the cluster region, that is,

$$E^{(1)} = E_{if} - E_{in}. \tag{10}$$

As sketched in Figure 22, the cluster region of the surface is estimated by the Van der Waals radius of the metal atom, and the Mulliken's atomic charge is used for q.

The energy of the system thus becomes,

$$E = E^{(0)} + E^{(1)} \tag{11}$$

where $E^{(0)}$ is the energy of the adcluster alone and $E^{(1)}$ the energy of the electrostatic interaction between the adcluster and the bulk metal.

The estimation of the electrostatic energy may be done before or after MO calculations. In principle, the effect should be included before MO calculations because, then, we can include the relaxation of the electron cloud of the adcluster in the electrostatic field of the surface. Since q and $\sigma(\mathbf{x})$ depends on each other, the calculations are done iteratively. We have shown in the previous paper[1] that these two methods give very similar results. Therefore, we here calculate the image force correction after MO calculations.

6.4. Palladium-O_2 System

We apply the dipped adcluster model to palladium-O_2 system, representing a palladium surface by a single Pd atom. The O_2 molecule is put at end-on, on-top position of the Pd atom, so that the adcluster is a linear Pd-O_a-O_b system. At an infinite separation, the O_2 molecule is in the $^3\Sigma_g$ state and the Pd atom is in the $^1S(d^{10})$ state. We first apply the highest spin coupling model and then the paired spin coupling model. Though the former gives a continuous picture leading to the correct separation limit, the latter does not. The electrons are transferred into the degenerate π^* MO's of the PdO_2 system from the Pd solid.

6.4.1. Highset Spin Coupling

Figure 23 is a display of the $E(n)$ curve, the energy of the adcluster calculated as a function of n, the number of the electrons transferred into the adcluster. This figure is for the Pd-O_a distance fixed at 2.0 Å, which is almost the most stable distance. The O-O distance is changed from 1.20752 Å, which is an equilibrium length R_{eq} of O_2,[19] to 1.35 Å, which is an equilibrium length of O_2^-,[19] and further to 1.5 Å. The chemical potential of the solid palladium metal is 5.12 eV,[36] which is shown by the gradient in Figure 23. Clearly, the behavior of the curves corresponds to Case A-2. Therefore, after some barrier, one electron flows from the bulk metal into the adcluster, so that the adcluster becomes $(Pd-O_2)^-$. The energy of this charged state is lower than that of the neutral one.

Figure 24 shows the potential energy curve for the end-on approach of the O_2 molecule to the palladium surface. The broken lines are for $E^{(0)}$ alone and the solid lines for $E^{(0)} + E^{(1)}$. The broken and solid curves for $n = 0.0$ nearly overlap each other, since the electrostatic term is very small. The broken curve for $n = 0.0$ corresponds to the cluster model, which results in that the O_2 molecule is not adsorbed onto Pd in contrary to the experiment. This result also shows that a linear PdO_2 molecule does not exist. On the other hand, when the electron transfer from the bulk metal to the adcluster is admitted, the system becomes stable as O_2 approaches Pd, which is shown by the broken curve of $n = 1.0$. Furthermore, the image force term is also important for stabilizing the system as shown by the solid curve of $n = 1.0$.

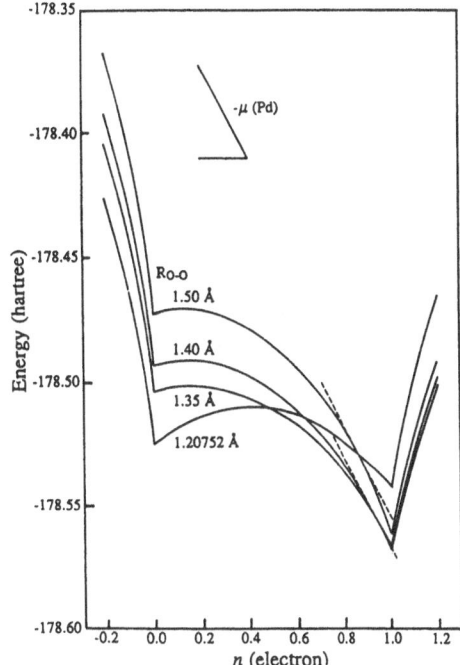

Figure 23. $E(n)$ curves for the Pd-O_a-O_b system in the highest spin coupling model with the O-O distances of 1.20752, 1.35, 1.40, and 1.50 Å and the Pd-O_a distance fixed at 2.00 Å.

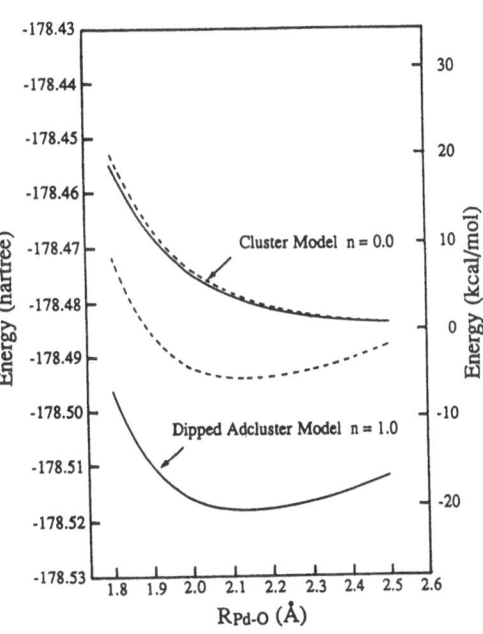

Figure 24. Potential energy curves for the end-on approach of the O_2 molecule onto palladium calculated by the highest spin coupling. The O-O length is fixed at 1.20752 Å (equilibrium length of the free O_2) for the upper one and at 1.35 Å for the lower one. Broken and solid lines are calculated, respectively, without and with the electrostatic energies $E^{(1)}$. The upper broken line with $n = 0.0$ corresponds to the cluster model. The energy scale on the right-hand side is in kcal/mol relative to the free Pd plus O_2 system.

Table II. The adsorption energy, geometry, and vibrational frequency of the PdO_2 adcluster in the highest spin coupling model.

	adsorption energy [a]	bond length (Å)		vibrational frequency (cm^{-1})	
	(kcal/mol)	R_{Pd-O}	R_{O-O}	ω_{Pd-O} [b]	ω_{O-O}
$E^{(0)}$ alone	6.4	2.15	1.39	329	1229
$E^{(0)} + E^{(1)}$	21.9	2.15	1.40	338	1250
experiment	7.6 ~12.3		1.32±0.05 [c]	485 [c]	1035

[a] Relative to the HF energy of Pd (^1S) + O_2 ($^3\Sigma_g^-$); -178.48500 hartree.

[b] O_2 is assumed to vibrate as a unit.

[c] For O_2 on a Pt(111) surface.

Table II shows a summary of the adsorption energy and some geometrical and vibrational parameters. It also shows the effect of the image force. For the adsorption energy, the effect of $E^{(1)}$ is large but for the geometrical and vibrational parameters, it is very small, since the image force is a long-range force. The agreement between theory and experiment is reasonable, considering the simplicity of the theoretical model. Probably the effect of $E^{(1)}$ is overestimated here since the adcluster is too small in this calculation.

The Mulliken population at the optimal geometry is Pd(-0.186)-O_a(-0.584)-O_b(-0.236). The inner O_a atom is more negatively charged than the outer O_b atom. However, the frontier orbital of this adcluster has largest amplitude on the outer O_b atom, so that the O_b atom is expected to be more reactive than the O_a atom.

6.4.2. Paired Spin Coupling

The $E(n)$ curves calculated for the paired spin coupling model are displayed in Figure 25. The curves are lower convexes in contrast to the upper ones of Figure 23. At n = 0.25, the tangent of the curve for R(O-O) = 1.35 Å coincides with the chemical potential $-\mu$ of the solid palladium metal 5.12 eV,[36] and the adcluster is the most stable there in the range defined by Eq.(3). Therefore, about 0.25 electron flows into the adcluster from the bulk metal; thus, the adcluster has a non-integer number of electrons. When the chemical potential μ of the metal is regulated, for example, by an external potential, the number of electrons n transferred into the adcluster are regulated. At the limit of $\mu = 0.0$, about 1.15 electrons flow into the adcluster, which corresponds to the minimum of the $E(n)$ curve for R(O-O) = 1.40 Å.

Figure 26 shows the potential energy curves for the Pd-O distance for the Pd-O_2 adcluster with fixed n of 0.25, though of course, the number of electrons n should be optimized as the functions of these distances. Broken and solid lines are calculated respectively without and with the image force term $E^{(1)}$. The geometries and the vibrational frequencies of the adsorbed system are calculated from these curves and shown in Table III. They are similar to the results of the highest spin coupling model shown in Table II, though the bond lengths of the Pd-O and O-O bonds calculated by the paired spin coupling model are a little shorter than those by the highest spin coupling model, and the force constants are a little larger. A reasoning for these

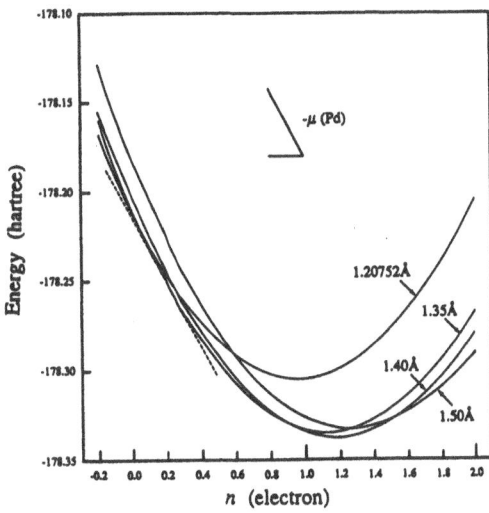

Figure 25. $E(n)$ curves for the Pd-O_a-O_b system in the paired spin coupling model with the O-O distances of 1.20752, 1.35, 1.40, and 1.50 Å and the Pd-O_a distance fixed at 2.00 Å.

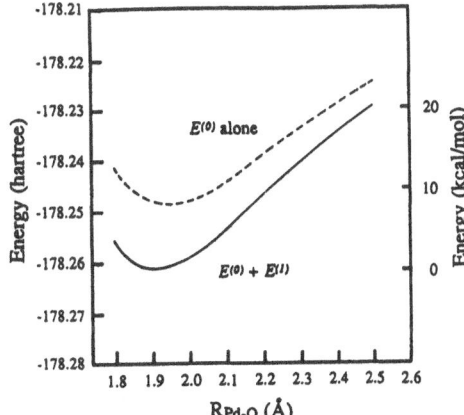

Figure 26. Potential energy curves for Pd-O vibration calculated by the paired spin coupling. The O-O length is fixed at 1.35 Å. Broken and solid lines are calculated, respectively, without and with the electrostatic energies $E^{(1)}$.

small differences is given in ref. 1.

The Mulliken's atomic charge of the adcluster is Pd(+0.301)-O_a(-0.349)-O_b(-0.202). The charge on Pd is positive, which we think more reasonable than the previous result of the highest spin coupling. The inner oxygen O_a is again more negative than the outer O_b atom.

6.5. Brief Remarks

Here, we have explained the dipped adcluster model for the study of chemisorptions and catalytic reactions on a metal surface which involve large electron transfer between an admolecule and a metal surface. The size of the cluster necessary for investigating such electronic processes would be reduced if the dipped adcluster model is adopted instead of the conventional cluster model. This merit is practically very

Table III. Geometry and vibrational frequency of the PdO_2 adcluster calculated by the paired spin coupling model.

	bond length (Å)		vibrational frequency (cm⁻¹)	
	R_{Pd-O}	R_{O-O}	ω_{Pd-O} a)	ω_{O-O}
$E^{(0)}$ alone	1.96	1.30	389	1259
$E^{(0)} + E^{(1)}$	1.93	1.31	399	1279
experiment		1.32±0.05 b)	485 b)	1035

a) O_2 is assumed to vibrate as a unit.
b) For O_2 on a Pt(111) surface.

important since electron correlations are often very important for describing the model reactions on cluster surfaces and further since sometimes the catalytically active state is not necessarilly the ground state but an excited state of the surface cluster. We therefore have to make the size of the cluster as small as possible. Although we have restricted the present formulation only within the molecular orbital model, an inclusion of electron correlation is necessary. Further, it is interesting to investigate the effect of the cluster size on the dipped adcluster model. If the size is large enough to be able to describe the real surface, the dipped adcluster model would, in principle, become unnecessary. But, how large is such size ?

7. MOLECULAR AND DISSOCIATIVE CHEMISORPTIONS OF AN O_2 MOLECULE ON AN Ag SURFACE

Partial oxidation of ethylene on a silver surface is an important catalytic reaction, for which no catalysts except for silver have been found effective.[37] However, the mechanism of this catalytic reaction is not yet completely elucidated. We do not yet understand why only a silver surface has such a reactivity and selectivity.

Experimentally, four different species are known for the adsorbed oxygens on a silver surface;[38-44] namely, physisorbed species (O_2),[38] molecularly adsorbed species, which are superoxide (O_2^-)[39,40] and peroxide (O_2^{2-}),[40-43] and dissociatively adsorbed species (O⁻ or/and O^{2-}).[41,44]

Some theoretical papers have been published on the oxygen chemisorptions.[45-50] The GVB-CI study by Upton et al.[47] gave geometric and spectroscopic parameters of O_2 on an Ag surface in good agreement with the experimental data. However, the molecular adsorption energies relative to the ground state were not reproduced: they were negative. Carter et al.[48] reported an extensive study on the mechanism of the partial oxidation of ethylene on a silver surface. All of these studies used the cluster model, so that the effects of the bulk metal are only insufficiently included in the calculations. So far, no ab initio studies have been able to describe the dissociative adsorption of an O_2 molecule on an Ag surface, though recently Panas et al.[49] have successfully described the O-O dissociation on a nickel surface.

We have applied the DAM for studying the mechanism of the chemisorption of an oxygen molecule on a silver surface. Figure 27 shows the $E(n)$ curve for the

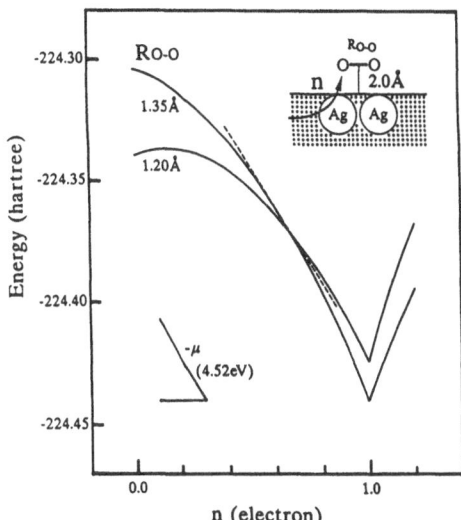

Figure 27. $E(n)$ curves for the Ag_2-O_2 adcluster in the highest spin coupling model with the Ag_2-O_2 distance of 2.00 Å and the Ag-Ag distance of 2.8894 Å.

Ag_2O_2 adcluster in the highest spin coupling model. The geometry is shown in the figure. The Ag-Ag distance is fixed to 2.8894 Å, the lattice distance in a crystal. The curve is A-2 type, so that one electron transfer from the bulk metal into the adcluster is expected. For surface electronic processes involving transition metals, electron correlations are often quite essential. Since we are going to study several different adsorption states of O_2, we have to calculate several lower states of the system. Sometimes, the catalytically active state is not necessarily the ground state of the system. For this purpose, we use the SAC/SAC-CI method which is proved to be very accurate and useful for studying ground, excited, ionized, and electron-attached states of a molecule.[7-10] Since Figure 27 shows only one-electron transfer is important in this case, we can calculate the correlated wave functions of this system by the SAC/SAC-CI method.

7.1. Approach of O_2 onto a Silver Surface

We first show the energetics for the approach of an O_2 molecule onto an Ag surface. The active site of an Ag surface is represented by Ag_2, and an O_2 molecule approaches the surface in a side-on bridge form keeping C_{2v} symmetry. This geometry was suggested by Backx et al. for O_2 adsorbed on an Ag(110) surface.[42] Figure 28 shows the potential energy curves of the Ag_2O_2 adcluster calculated by the SAC/SAC-CI method as a function of the Ag_2-O_2 distance. The curves are calculated for the O-O distance fixed at 1.35 Å, which is an equilibrium distance of an O_2 anion.[47] The asterisks show the energies at the optimized O-O distances.

The $^3B_2$ state, which does not involve an electron transfer from the bulk metal to the adcluster, is the ground state of the separated system; namely, the $^1\Sigma_g$ state of Ag_2 and the $^3\Sigma_g$ state of O_2. The energy of the dissociation limit is estimated by optimizing the O-O distance at the Ag_2-O_2 distance of 5.0 Å, and is shown by the

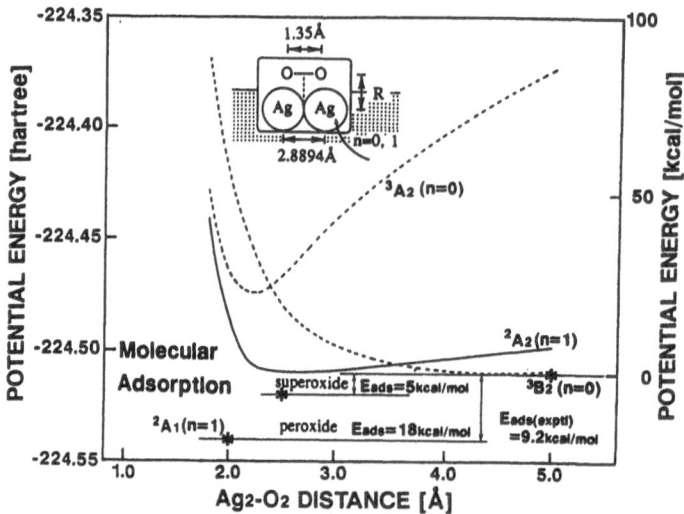

Figure 28. Potential energy curves for the approach of O_2 onto Ag_2 in the Ag_2-O_2 adcluster. n denotes the number of the electrons transferred from the bulk metal to the adcluster.

asterisk in Figure 28. There, the optimized O-O length is 1.29 Å, in comparison with 1.27 Å, the optimized length for the free O_2 molecule (the experimental value is 1.207 Å[50]). The potential curve for the $^3B_2$ state rises monotoniously as O_2 approaches Ag_2, showing that no chemisorption occurs along this state.

When one electron transfer is admitted from the bulk metal to the adcluster, namely $n = 1$, the potential of the $^2A_2$ state becomes attractive and a minimum is found at $R(Ag_2$-$O_2) = 2.6$ Å. When the O-O distance is further optimized at this minimum, we get $R(O$-$O) = 1.50$ Å and the system is stabilized down to the asterisk shown in Figure 28. This $^2A_2$ state corresponds to the superoxide species, O_2^-. There is another state, $^2A_1$ state, which also results from the one-electron transfer from the bulk metal to the adcluster. This $^2A_1$ state corresponds to the peroxide species, O_2^{2-}, and has a potential minimum at $R(Ag_2$-$O_2) = 2.0$ Å and $R(O$-$O) = 1.66$ Å. The corresponding energy is shown by the asterisk in Figure 28. The calculated adsorption energies of the superoxide and peroxide species are 5.5 and 17.8 kcal/mol, respectively, in comparison with the experimental molecular adsorption energy of 9.2 kcal/mol.[43]

The $^3A_2$ state shown in Figure 28 is an electron transferred state from Ag_2 to O_2, but no electron is supplied from the bulk metal ($n = 0$). The image force term is included and works to stabilize the system. Though the system is stabilized as O_2 approaches the surface, the energy is always higher than the free molecule limit. This state corresponds to the molecularly adsorbed species obtained by the conventional cluster model and the adsorption energy is negative as in the previous studies.[47] This failure is mainly due to a limitation of the cluster model. In the cluster model, all the electrons transferred to O_2 must be supplied from Ag_2, but in DAM some of the electrons are supplied from the bulk metal. We see that the electron transfer from the bulk metal to the adcluster is essential for the occurrence of the chemisorption of an O_2 molecule on an Ag surface. We have confirmed that the stabilization of the charged O_2 admolecule by the electrostatic image force of the Ag metal is also important.

7.2. Potential Curve for O-O Stretching and Dissociation on an Ag Surface

Figure 29 shows the potential energy curves for the O-O elongation on the Ag_2 site. These potentials are calculated for the Ag_2O_2 adcluster with $n = 1$ with fixing the Ag_2-O_2 distance at 2.0 Å. As before, the $^2A_2$ and $^2B_1$ states correspond to the superoxide species and the $^2A_1$ state to the peroxide species. These potentials are calculated by the SAC/SAC-CI method.

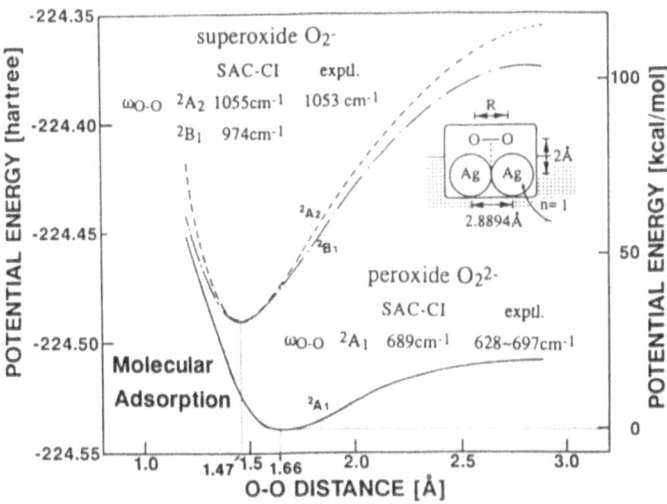

Figure 29. Potential energy curves for the O-O elongation in the Ag_2-O_2 adcluster.

The equilibrium adsorption geometries and the vibrational frequencies of the superoxide and peroxide are calculated and shown in Table IV. The calculated vibrational frequencies agree well with the experimental values. The calculated O-O distance of the peroxide is larger than the experimental one. The gross charge of superoxide is calculated to be $O_2^{-0.5} \sim {-0.6}$ and that of peroxide as $O_2^{-1.4}$, which are smaller than the formal charges in the notations O_2^- and O_2^{2-}, respectively. Backx et al. estimated the charge of -1.7 for peroxide from the consideration on the vibrational frequency as a function of the number of electrons in the π^* antibonding orbitals.[42]

It is expected that the dissociative adsorption is led from the peroxide species ($^2A_1$), because the curves of the superoxide species ($^2A_2$, $^2B_1$) rise more rapidly than that of the peroxide as the O-O distance is elongated. However, the potential curve of the peroxide rises monotonically up to R(O-O) = 2.8894 Å, which is twice

282

Table IV. The adsorption geometries and vibrational frequency of the molecular adsorption species of O_2 on an Ag surface.

Species	State	Bond Length (Å)		O-O Vibrational
		$R(Ag_2\text{-}O_2)$	$R(O\text{-}O)$	Frequency (cm^{-1})
superoxide	$^2A_2$	2.6	1.47	1055
	$^2B_1$	—	1.47	974
	exptl.	—	—	1053
peroxide	$^2A_1$	2.0	1.66	689
	exptl.	—	1.47 ± 0.05	$628 \sim 697$

as large as the O-O distance of the free O_2 molecule. We could not obtain the second minimum corresponding to the dissociatively adsorbed state from the calculations for the Ag_2O_2 adcluster. A reason is attributed to the electrostatic repulsion between the negative charges on oxygens. It is estimated as large as 60 kcal/mol from the gross charges on the oxygens (-0.72) separated by 2.8894 Å. For realizing the stabilization of the dissociative state, two oxygen atoms must be separated further on the surface. We, therefore, need a larger surface of the Ag atoms, so that we next consider the dissociation of O_2 on the linear Ag_4.

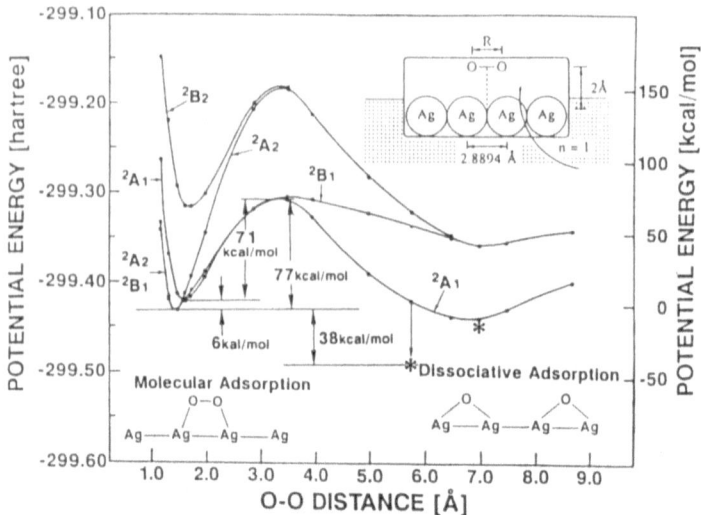

Figure 30. Potential energy curves against the O-O distance for the Ag_4-O_2 adcluster.

Figure 30 shows the potential energy curves for the ground and excited states of the Ag_4O_2 adcluster with $n = 1$. They are again due to the SAC/SAC-CI calculations. In contrast to the calculations for the Ag_2O_2 adcluster, we here have frozen the electron correlations of the d-orbitals of the silvers. In Figure 30, we get the potential minima of not only the molecular adsorption states ($^2A_1$, $^2A_2$, and $^2B_1$) near $R(\text{O-O}) = 1.5$ Å, but also the dissociative adsorption state ($^2A_1$) at $R(\text{O-O}) = 7.0$ Å. When we further optimize the $Ag_4\text{-}O_2$ distance at $R(\text{O-O}) = 5.7788$ and 7.0 Å, it becomes shorter up to 1.60 and 1.90 Å, respectively, and the corresponding energies are shown by the asterisks in Figure 30. The system is most stable at $R(\text{O-O}) = 5.7788$ Å, and $R(Ag_4\text{-}O_2) = 1.60$ Å. The dissociated oxygens are adsorbed at the two-fold bridge site of the Ag surface, as illustrated in Figure 30, and the Ag-O length is calculated to be 2.16 Å, which agrees with the observed distance of 2.06 ~ 2.17 Å.[46] The gross charge of oxygen at the optimized geometry is -0.98 and so the dissociated oxygen is essentially O^-.

Table V. Adsorption energies of O_2 on a silver surface. (kcal/mol)

Adsorption State		Theoretical	Experimental	
			Ag(110)	Ag(111)
molecular	superoxide	5.5	9.3	9.2
	peroxide	17.8		
dissociative		44.0 61.4	42.5	39.9

The dissociative adsorption state is lower by 38.5 and 43.6 kcal/mol than the molecular superoxide and peroxide species, respectively. By the thermal desorption spectra of O_2 adsorbed on Ag(111) and Ag(110),[45] the dissociative adsorption state is observed to be lower by 31.6 and 34.7 kcal/mol, respectively, than the molecular adsorption state. We summarize the calculated and experimental adsorption energies in Table V. The energy barrier between the molecular and dissociative adsorptions lies at $R(\text{O-O}) = 3.5$ Å, with the height of 77.2 kcal/mol from the superoxide species and 71.2 kcal/mol from the peroxide species. This barrier may be too large, in comparison with that in Figure 29, and this may be attributed to the neglect of the correlations of the d electrons for the Ag_4O_2 adcluster. The reorganization of the Ag surface would also work to reduce the barrier.

ACKNOWLEDGEMENT

The authors thank Prof. T. Yonezawa, Mr. Y. Matsuzaki, and Y. Fukunishi for collaborations in some of the studies included in this review. The calculations have been carried out at the computer centers of the Institute for Molecular Science and of Kyoto University. Parts of these studies have been supported by the Grant-in-Aids for Scientific Research from the Ministry of Education, Science, and Culture of Japan.

REFERENCES

[1] H. Nakatsuji, J. Chem. Phys., **87**, 4995 (1987); H. Nakatsuji, H. Nakai, and Y. Fukunishi, J. Chem. Phys., **95** , 640 (1991).

[2] H. Nakatsuji, M. Hada, and T. Yonezawa, J. Am. Chem. Soc., **107**, 8264 (1985); **109**, 1902 (1987).

[3] H. Nakatsuji, Y. Matsuzaki, and T. Yonezawa, J. Chem.Phys., **88**, 5769 (1988).

[4] H. Nakatsuji and Y. Fukunishi, Intern. J. Quantum Chem., in press.

[5] H. Nakatsuji, M. Hada, and T. Yonezawa, Surface Sci., **185**, 319 (1987).

[6] H. Nakatsuji, H. Nakai, Chem. Phys. Letters, **174**, 283 (1990).

[7] H. Nakatsuji and K. Hirao, J. Chem. Phys., **68**, 2035 (1978).

[8] H. Nakatsuji, Chem. Phys. Letters, **59**, 362 (1978); **67**, 329, 334 (1979); Chem. Phys. **75**, 425 (1983).

[9] H. Nakatsuji, Program System for SAC and SAC-CI calculations, Program Library No. 146 (Y4/SAC), Data Processing Center of Kyoto University (1985); Program Library SAC85, No. 1396, Computer Center of Institute for Molecular Science (1986).

[10] H. Nakatsuji, Report in Molecular Theory, CRC Press, in press.

[11] (a) P. J. Hay, J. Am. Chem. Soc., **103**, 1390 (1981); (b) J. O. Noell and P. J. Hay, Inorg. Chem., **21**, 14 (1982). (c) P. J. Hay and W. R. Wadt, J. Chem. Phys., **82**, 270 (1985).

[12] (a) S. Huzinaza, J. Chem. Phys., **42**, 1293 (1965); (b) T. H. Dunning Jr., ibid **53**, 2823 (1970).

[13] (a) H. Nakatsuji, K. Kanda, and T. Yonezawa, Chem. Phys. Letters, **75**, 340 (1980); (b) H. Nakatsuji, T. Hayakawa, and M. Hada, Chem. Phys. Letters, **80**, 94 (1981); (c) H. Nakatsuji, K. Kanda, H. Hada, and T. Yonezawa, J. Chem. Phys., **77**, 3109 (1981).

[14] T. H. Dunning, Jr. and P. J. Hay, "Modern Theoretical Chemistry", edited by H. F. Schaeffer III, Plenum, New York, 1977, Vol.3.

[15] B. R. Brooks, P. Saxe, W. D. Laidig and M. Dupuis, Program System GAMESS; Program Library No.481, Computer Center of the Institute for Molecular Science, (1981).

[16] R. C. Weast, Ed. "Handbook of Chemistry and Physics", CRC Press, Cleveland, 1984–1985, F-167.

[17] (a) B. Roos, P. Taylor, and P Siegbahn, Chem. Phys., **48**, 157 (1980); (b) P. Siegbahn, A. Heiberg, B. Roos, and B. Levy, Phys. Scr., **21**, 323 (1980)

[18] (a) H. Conard, G. Ertl, and E. E. Latta, Surface Sci., **41**, 435 (1974); (b) R. J. Behm, K. Christmann, and G. Ertl, Surface Sci., **99**, 320 (1980); (c) C. Nyberg and G. C. Tengstal, Surface Sci., **126**, 163 (1983).

[19] K. P. Huber and G. Herzberg, "Molecular Spectra and Molecular Structure. IV. Constants of Diatomic Molecules" Van Nostrand Reinhold Co., New York, 1979.

[20] (a) C. F. Melius, Chem. Phys. Letters, **39**, 287 (1976); (b) C. F. Melius, J. W. Moskowitz, A. P. Mortola, M. B. Baillie, and M. A. Ratner, Surface Sci., **59**, 279 (1976).

[21] W. Eberhardt, S. G. Louie, and E. W. Plummer, Phys. Rev., **B28** 465 (1983).

[22] J. P. Muscat, Surface Sci., **148**, 237 (1984).

[23] N. A. Baykara, J. Andzelm, S. Z. Baykara, and D. R. Salahub, Intern. J. Quantum Chem., **29**, 1025 (1986).

[24] (a) G. C. Bond, D. A. Dowden, and N. Mackenzie, Trans. Faraday Soc. **54** 1537 (1958). (b) G. C. Bond and P. B. Wells, J. Catal., **4**, 211 (1965); **5** 65, (1965).

[25] R. B. Woodward and R. B. Hoffmann, "The Conservation of Orbital Symmetry", Academic Press, New York, 1970.

[26] G. Henrici-Olive and S. Olive, "Coordination and Catalysis", Verlag Chemie: Weinheim, 1977.

[27] G. C. Bond, "Catalysis by Metals" Academic Press, New York, 1962.

[28] J. R. Anderson, "Structure of Metallic Catalysts", Academic, New York, 1975.

[29] C. E. Moore, "Atomic Energy Levels", National Bureau of Standards, Washington, D. C., 1971, vol. 3

[30] H. Nakatsuji and M. Hada, Croat. Chem. Acta, **57**, 1371 (1984).

[31] R. P. Eischens, W. A. Pliskin, and M. J. D. Low, J. Catal., 1, 80 (1962); R. Ugo, Catal. Rev., 11, 225 (1975).

[32] A. L. Dent and R. J. Kokes, J. Phys. Chem., 73, 3781 (1969).

[33] W. C. Conner, Jr. and R. J. Kokes, J. Catal., 36, (1975).

[34] M. Witko and V. B. Koutecky, Intern. J. Quantum Chem., 24, 1535 (1986).

[35] J. H. Jeans, "The Mathematical Theory of Electricity and Magnetism", chapter VIII, Cambridge University, New York (1966).

[36] H. B. Michaelson, J. Appl. Phys., 48, 4729 (1977).

[37] A. Ayame and H. Kanoh, Shokubai 20 381 (1978) in Japanese.

[38] D. Schmeisser, J. E. Demuth and Ph. Avouris, Phys. Rev., B26 4857 (1982).

[39] K. C. Prince and A. M. Bradshow, Surface Sci., 126 49 (1983).

[40] C. Pettenkofer, I. Pockrand and A. Otto, Surface Sci., 135, 52 (1983); C. Pettenkofer, J. Eickmans, U. Erturk and A. Otto, Surface Sci., 151, 9 (1985).

[41] (a) A. Sexton and R.J. Madix, Chem. Phys. Letters 76, 294 (1980). (b) M. A. Bartreau and J. Madix, Chem. Phys. Letters, 97, 85 (1983).

[42] C. Backx, C. P. M. deGroot and P. Biloen, Surface Sci., 104, 300 (1981); C. Backx, C. P. M. deGroot, and P. Biloen, Appl. Surface Sci., 6, 256 (1980).

[43] D. A. Outka, J. Stohr, W. Jark, P. Stevens, J. Solomon and R. J. Madix, Phys. Rev. B35, 4119 (1987); J. Stohr and D. A. Outka, Phys. Rev., B36, 7891 (1987).

[44] C. T. Campbell, Surface Sci., 157, 43 (1985).

[45] J.-H. Lin and B. J. Garrison, J. Chem. Phys., 80, 2904 (1984).

[46] A. Selmani, J. Andzelm, and D. R. Salahub, Intern. J. Quantum Chem., 29, 829 (1986).

[47] T. H. Upton, P. Stevens, and R. J. Madix, J. Chem. Phys., 88, 3988 (1988).

[48] E. A. Carter and W. A. Goddard III, Surface Sci., 209 243 (1989).

[49] I. Panas, P. Siegbahn and U. Wahlgren, J. Chem. Phys., 90, 6791 (1989).

[50] P. H. Krupenie, J. Phys. Chem. Ref. Data 1 423 (1972).

REACTIVITY AND ELECTRONIC STRUCTURE OF ORGANOMETALLIC RADICALS

W. C. TROGLER
Department of Chemistry
University of California, San Diego
La Jolla, California 92093-0506
USA

ABSTRACT. Reactions of 18-electron organometallic complexes often proceed by ligand dissociation, reductive elimination, or insertion to yield 16-electron coordinatively unsaturated intermediates. Another class of reactions proceeds by associative attack of a ligand or nucleophile at the metal in an 18-electron complex, when the complex contains a ligand that can undergo a structural rearrangement to accept an electron pair to avoid a 20-electron intermediate. Organometallic radicals with a 17-electron configuration exhibit perhaps the greatest tendency to engage in associative reaction mechanisms. When compared to eighteen electron mononuclear complexes we have seen rate accelerations of 10^9 to 10^{10}. These reactions all appear to involve 19-electron intermediates or transition states. The previous observation of catalysis of CO insertion by single electron oxidation has also been shown to arise from an electron-rich 19-electron species, which can form by nucleophilic addition of solvent to a 17-electron complex. These and other reactions of 17-electron complexes correlate with the presence of a sterically accessible odd electron localized on the metal. This orbital can σ bond to a nucleophile to form a 19-electron species stabilized by a 2-center 3-electron bond. Parallels can also be drawn between the reactivity of closed shell cluster complexes and those oxidized by one electron.

1. Reactivity and Electronic Structure of 18-Electron Compounds

1.1. THE 18-ELECTRON RULE

The 18-electron rule, which holds for a wide number of organotransition metal complexes, derives from the simple observation that 18 electrons complete the valence ns $(n-1)d$ np shell of electrons around these elements. In the context of ligand field or molecular orbital theory this number represents the maximum occupation of the metal-ligand bonding and non-bonding molecular orbitals. An 18-electron configuration represents an upper limit to the electron count at the metal, because the removal of electrons from non-bonding or weakly bonding orbitals can lead to 17-, 16-, 15-, or 14-electron systems. Isolable compounds are known for all these electron counts [1]. By contrast 19- and 20-electron complexes must contain electrons in highly antibonding metal-ligand orbitals. The few examples of such complexes usually fall into a category where the excess electrons either occupy a special non-bonding orbital derived from the ligands present (e.g. $W(CO)(PhC\equiv CPh)_3$ - 20-electrons), or else consist of compounds

287

D. R. Salahub and N. Russo (eds.), Metal-Ligand Interactions: from Atoms, to Clusters, to Surfaces, 287–310.
© 1992 *Kluwer Academic Publishers. Printed in the Netherlands.*

1

where an extra electron occupies a low-lying, ligand-centered orbital, as in the 19-electron complex **1** [2,3].

In **1** the extra electron resides in an orbital delocalized over the π-system of the bipyridine ligand. In this historical context it is understandable that for reaction mechanisms of 18-electron complexes mechanistic chemists usually invoke 16-electron transition states and intermediates.

1.2. DISSOCIATIVE MECHANISMS IN 18-ELECTRON COMPLEXES

1.2.1. *Dissociative Ligand Substitution.*
Many 18-electron complexes, for example all the mononuclear metal carbonyl complexes, exchange ligands at the metal center predominantly by a rate-limiting dissociation of a bound ligand to generate a 16-electron coordinatively unsaturated species. Numerous examples of these reactions can be found in various textbooks and review articles [4]. The dissociation of carbon monoxide from chromium hexacarbonyl, eq. 1, generates the 16-electron species $Cr(CO)_5$ as an

intermediate in the net reaction. In this mechanism k_1 represents the slow rate-limiting step, and k_2 and k_{-1} are fast. Under the usual conditions, where [CO] is low, the back reaction (k_{-1}) can be neglected, so first order kinetics obtains. These reactions are not restricted to octahedral complexes and carbon monoxide ligands. For example, $Ni[P(OCH_3)_3]_4$ and $Re_2(CO)_{10}$ both react by rate limiting dissociation of a ligand to generate a 16-electron species at the metal center [5,6]. A key step in the mechanism of action of Wilkinson's catalyst, $RhCl(PPh_3)_3$, for the hydrogenation of alkenes is the dissociation of PPh_3 from a $RhH_2Cl(PPh_3)_3$ intermediate species to generate the 16-electron complex, $RhH_2Cl(PPh_3)_3$, which can bind alkene substrate [7]. Indeed, as pointed out by Tolman, many homogeneous catalyst systems (e.g. hydroformylation and hydrogenation) function by cycling between 16- and 18-electron complexes [8].

1.2.2. *16-Electron Mechanisms for Other Reactions.*
Oxidative addition reactions usually add 2 electrons to the coordination sphere of the metal and require a 16-electron complex, or intermediate, for the addition reaction to proceed. The majority of mechanistic studies

of oxidative addition have been performed with stable 16-electron square planar complexes of Rh(I), Ir(I), Pd(II), and Pt(II) [4]. When 18-electron complexes are precursors to oxidative addition they often must dissociate a ligand before oxidative addition can occur. One striking example is the use of photochemical dissociation of a CO ligand from Ir(Cp*)(CO)$_2$ to generate a 16-electron species capable of reacting with C-H bonds in a variety of hydrocarbons [9].

Besides the straightforward dissociation of a 2-electron donor ligand, a variety of other reactions of 18-electron complexes proceed through 16-electron intermediates. For example, reductive elimination of hydrogen, alkanes, alkyl halides, and other substrates from metals usually leads to 16-electron intermediates [1]. Insertion reactions, such as carbonylations of 18-electron systems, often involve a rate limiting insertion step, followed by trapping of the 16-electron species by an entering ligand, eq. 2 [10].

$$(2)$$

Although this mechanism may lead to second-order kinetics, an electron-rich intermediate is not involved. Whether the 16-electron intermediate derives any stabilization from weak coordination to the oxygen of the acyl ligand, remains unresolved. From these examples it is clear that the 18- to 16-electron conversion occurs frequently in organometallic reaction mechanisms.

1.3. ASSOCIATIVE MECHANISMS IN 18-ELECTRON COMPLEXES

1.3.1. *Associative Ligand Substitution.* One unusual aspect, observed early in the study of organometallic reaction mechanisms, was the tendency for complexes that contained certain ligands to react according to associative processes. Complexes that contain π-unsaturated organic ligands, such as cyclopentadienyl, indenyl, allyl, benzene, naphthalene, and other aromatic ligands react associatively [11]. Even simple mononuclear carbonyl complexes, such as V(CO)$_5$NO and Co(CO)$_3$NO, have been observed to react associatively when a nitrosyl ligand is present [12]. These reactions have been reviewed by Basolo [13]. A key feature in these systems is the presence of a ligand π-system, which can accept an electron pair from the metal, undergo a structural change, and thereby maintain an 18-electron configuration in the transition state. For example, with π-unsaturated organic ligands, such as cyclopentadiene, ring slippage can account for the associative reaction mechanism shown by numerous metal complexes, eq. 3. Authentic ring-slipped species have been isolated and structurally characterized in several cases, which provides evidence that such intermediates are at least feasible [14].

For nitrosyl complexes it is necessary to invoke population of the π^*-orbitals of the NO group, and a structural reorganization from a linear to a bent nitrosyl ligand, as shown in eq. 4 [12]. For electron bookkeeping the NO ligand can be viewed as isomerizing from a linear NO$^+$ to a bent NO$^-$ group as the reaction progresses.

Dissociative ligand substitution has been observed for unsaturated metallacyclopentadiene complexes, such as 2 [15]. However, in complexes that contain lower energy π^* metallacycles, such as in 3 or 4, carbon monoxide is replaced by

$$(3)$$

$$(4)$$

associative processes [16]. Here the possibility of delocalizing an electron pair into the ligand π^* system has been suggested as a way to avoid a 20-electron transition state or intermediate.

Relative rates for associative attack at **3** and **4** seem to parallel the availability of the lowest empty π^* orbital. Complex **4**, with four highly electronegative ring nitrogens reacts 10^5 more rapidly with PMe$_3$ nucleophile than does **3** [16].

Small second order terms have even been observed in the rate laws for substitution of carbon monoxide in the 18-electron hexacarbonyls of Cr, Mo, and W [17]. The origin of the small associative, k_2, contribution to the overall rate law, $k_1[M(CO)_6] + k_2[M(CO)_6][L]$, has been attributed to a dissociative interchange mechanism where a precursor outer-sphere complex $\{M(CO)_6,L\}$ forms, followed by a dissociative interchange of a CO ligand and L in this weakly bound complex. The motivation for this explanation derives in part from the strong aversion organometallic chemists have toward 20-electron intermediates.

2   **3**   **4**

1.3.2. *Associative Mechanisms for Other Reactions of 18-Electron Complexes.* The observation of a ligand-dependant rate law does not necessarily indicate associative attack at a metal center. For example, a complex that undergoes rapid reversible dissociation of a ligand, L, and trapping of the coordinatively unsaturated intermediate by a different ligand, L', will show a first-order dependence on [L']. When associative mechanisms have been seen for other kinds of reactions of 18-electron complexes they usually can be understood in the context of the principles outlined in section 1.3.1 above. Thus, $Mo(Cp)(CO)_3(CH_3)$ undergoes insertion in the presence of $PMePh_2$ ligand by both first- and second-order paths, with a catalytic effect of donor solvents [18]. One possible explanation for direct nucleophilic attack would invoke ring slippage of the cyclopentadienyl ligand. An alternative rationale, suggested for catalysis of CO insertion in $Mn(CO)_5(CH_2C_6H_4OCH_3)$ by $O=PPh_3$ nucleophile, requires that the nucleophile catalyzes insertion, without direct attack at the metal [19].

1.3.3. *The Special Case of Metal Cluster Complexes.* It would be fair to say that the level of understanding of reactivity in metal cluster complexes lies well below that for mononuclear complexes. By analogy to mononuclear metal carbonyl complexes, compounds such as $Mn_2(CO)_{10}$ and $Re_2(CO)_{10}$, undergo substitution of carbon monoxide ligands by simple dissociation of carbon monoxide to form a coordinatively unsaturated intermediate [6]. Flash photolysis and low temperature matrix photolysis experiments studies suggest that the intermediate species in these reactions can be stabilized by bridging of a carbon monoxide ligand between two metal centers [20]. Thus the mechanism can be described as in eq. 5.

$$(5)$$

Similar CO-bridged intermediates have been observed during photodissociation of carbon monoxide from $Ru_3(CO)_{12}$ and probably occur during thermal CO-dissociative reactions of these cluster compounds [21].

The odd aspect about the thermal reactivity of trinuclear and higher cluster systems is the availability of facile associative pathways for CO substitution. This even occurs in simple binary carbonyls, such as $M_3(CO)_{12}$ (M = Ru and Os) or $M_4(CO)_{12}$ (M = Co, Rh, and Ir) [22]. Phosphine substituted clusters exhibit similar trends, and the second row transition metals, Ru and Rh, can show dramatic (10^6) increases in associative reactivity as compared to first or second row transition metals [23]. Often two-term rate laws are obtained. For example, in $Rh_4(CO)_9[HC(PPh_2)_3]$ the reaction can occur predominantly through a dissociative path with a weak nucleophile, such as PPh_3. For strong nucleophiles, such as PBu_3, the associative term in the mechanism dominates. Orbital origins of the associative mechanism in cluster systems remain to be established. From $X\alpha$ studies of these compounds, we suggested that low-lying σ^* orbitals derived from edge bonding in a triangular array of metals could act as an acceptor orbital for nucleophilic attack, since these would be unique to tri- and tetranuclear clusters in comparison to the mononuclear complexes [24]. We speculate that the corresponding metal-metal antibonding orbital in the dinuclear $M_2(CO)_{10}$ complexes would not be sterically accessible for attack by a nucleophile, because of the axial carbonyl groups. Our understanding about the reactivity of metal cluster compounds would clearly benefit from some quantitative theoretical studies of the reaction mechanism.

2. Reactivity and Electronic Structure of 17-Electron Compounds

For years the 18-electron rule was thought to limit the possible reaction intermediates in organometallic mechanisms to an 18 or less electron count. Indeed, when the high ligand substitution lability of the Re(CO)$_5$ radical was discovered by Byers and Brown, they first attributed this to facile loss of a CO ligand [25]. We became involved in a study of the origin of this high kinetic lability of 17-electron complexes as an outgrowth of studies about the electronic structure of V(CO)$_6$ [25].

2.1. ASSOCIATIVE MECHANISMS IN 17-ELECTRON COMPLEXES

2.1.1. *Associative Ligand Substitution.* Vanadium hexacarbonyl is isostructural with chromium hexacarbonyl, and differs only by the removal of a single valence t_{2g} electron (and of course a proton). In contrast to the CO substitution reactions of Cr(CO)$_6$, which occur primarily by a dissociative mechanism, those of V(CO)$_6$ proceed entirely by an associative process. Thus, strict second-order rate laws, k [V(CO)$_6$] [Nuc], were observed for the substitution of carbon monoxide by a variety of phosphorus, nitrogen, oxygen, and arsenic donor nucleophiles, Nuc, according to eq. 6 [26]. Under comparable conditions the rate of CO substitution in V(CO)$_6$ increases by a factor of 10^{10} over that in Cr(CO)$_6$. This tremendous enhancement in kinetic lability explains how small amounts of 17-electron complexes, generated by oxidation of 18-electron systems, can participate as the reactive species in electron transfer and radical chain catalysis mechanisms [27, 28].

$$V(CO)_6 + Nuc \longrightarrow V(CO)_5(Nuc) + CO \qquad (6)$$

The kinetics studies of octahedral V(CO)$_6$, as well as V(CO)$_5$(PR$_3$), show several general features. First the rates of reaction with a variety of phosphorus donor ligands depend strongly on the electron-donor properties of the entering ligand [26]. In a comparison of the sensitivity of various associative reactions of organometallic systems with phosphorus donor nucleophiles, those of V(CO)$_6$ show the greatest variation in rate, as the electron-donor properties of the ligand are changed [29]. This suggests a strong bonding interaction occurs between the nucleophile and V(CO)$_6$ to stabilize the associative reaction path. Activation energy parameters are consistent with this conclusion. For V(CO)$_6$ the $\Delta H^{\ddagger}$ are relatively low (7 to 10 kcal/mole) and $\Delta S^{\ddagger}$ are highly unfavorable (-23 to -28 cal/mole K). This might be expected for a highly organized 19-electron transition state stabilized by significant bonding between the metal and incoming nucleophile [26].

We find similar reactivity patterns in 5-coordinate 17-electron complexes [30,31]. These systems, which are more sterically accessible to nucleophilic attack, react much faster than 6-coordinate radicals. Most are unstable and can not be isolated, so one needs to first generate the radical for characterization by transient techniques. Electrochemistry offers a simple means of generating 17-electron compounds by oxidation of 18-electron complexes. We have used this technique extensively to study reaction mechanisms of organometallic radicals. The application of this method to organometallic systems poses a particular problem because of the low dielectric constant solvents that need to be used. While cyclic voltammetry provides useful qualitative information about reaction chemistry, it suffers from the distortion of the wave-shape by poorly conducting solvent media. Also, the simulation of cyclic voltammetry wave shapes to obtain kinetic information about solution processes becomes complicated by the contribution of heterogeneous electron transfer rates to the wave shape [32]. Instead we employ the double potential step chronocoulometry technique. In this experiment one applies a voltage pulse to generate the radical under diffusion-limited conditions, allows the radical to react for a certain length of time, and then reverses the voltage to recover unreacted radical. By monitoring the ratio of the forward and reverse charges passed, as a function of the delay time and nucleophile concentration, one obtains kinetic information about the radical [33]. Because this is a ratio technique the effects of solution resistance tend to cancel, and since the experiment is performed under diffusion limited conditions the rate of heterogeneous charge transfer does not effect the measurement. This latter fact is very important if one desires to obtain reliable activation energies, as has been pointed out in the electrochemical literature [34]. Artifacts due to adsorption and cell capacitance can be readily detected from Q vs $t^{1/2}$ plots [35].

These techniques enabled us to generate the series of radical cations $Fe(CO)_3L_2^+$, where L = PMe_3, PBu_3, PPh_3, and PCy_3 [30]. These species undergo disproportionation in the presence of nitrogen nucleophiles and halide ions. We observed a similar tendency for $V(CO)_6$ to undergo nucleophile-induced net disproportionation with oxygen and nitrogen donor ligands. With phosphines only simple substitution occurs. The stoichiometry of these disproportionation reactions, shown in eqs. 7 and 8, always results in an 18-electron complex as one product.

$$3\,V(CO)_6 + 6\,py \longrightarrow 2\,V(CO)_6^- + V(py)_6^{2+} + 6\,CO \qquad (7)$$

$$2\,Fe(CO)_3L_2^+ + 6\,py \longrightarrow Fe(CO)_3L_2 + Fe(py)_6^{2+} + 3\,CO + 2\,L \quad (8)$$

The role of the nucleophile in promoting these reactions is believed to involve formation of a $V(CO)_5(py)$ or $Fe(CO)_2(py)L_2^{2+}$ intermediate. When pyridine replaces the strong π-acceptor CO ligand this should destabilize the highest occupied metal-localized d orbital in the substituted radical, so there now exists a driving force to transfer the electron to the more stable d orbital in the parent complex $V(CO)_6$ or $Fe(CO)_3L_2^+$. Because electron transfer occurs much more rapidly than nucleophilic attack, the kinetics still is determined by associative attack of the nucleophile at the metal center.

The kinetics for the reaction of eq. 8 follow a second-order rate law and the activation parameters ($\Delta H^{\ddagger} = 10$ kcal/mol, $\Delta S^{\ddagger} = -21$ cal/mol K for L = PPh_3) are quite similar to those observed for associative substitution in $V(CO)_6$ [26,30]. This series of complexes illustrates the importance of steric effects at the metal center, as shown in Table 1. The rate of nucleophilic attack shows a dramatic decrease as the steric bulk of the phosphine bound to the metal increases. This is expected if nucleophile-metal bonding occurs in the transition state.

Further evidence for bond formation between iron and the entering nucleophile derives from the large value of the Hammett ρ (-3.25) determined for a series of substituted pyridines [30]. This resembles the Hammett ρ of -3.3 determined for the S_N2 displacement reaction between $ArNMe_2$ and methyl iodide, where Ar is the aryl ring whose substituent effect is being examined [36].

TABLE 1. Rate of Nucleophilic Attack by
Pyridine at $Fe(CO)_3L_2^+$ at 25°C in CH_2Cl_2.

L	k $(M^{-1} s^{-1})$
$P(CH_3)_3$	85.
$P(n\text{-}C_4H_9)_3$	8.1
$P(c\text{-}C_6H_{11})_3$	0.0001

More recently our attention has turned to determining periodic effects on the reactivity of 17-electron radicals. The reactivity of 18-electron complexes generally maximizes for second row-transition metals [4]. For example, in the iron triad the relative rates of CO substitution in the $M(CO)_5$ complexes follow the order: Fe (1) < Os (800) < Ru (5 × 10^7) [37]. Similar trends are seen for phosphine substituted compounds [38]. We have recently shown that Ru and Os analogues of $Fe(CO)_3L_2^+$ can be generated and kinetically characterized [39]. These radicals decay in the presence of [NBu₄]Cl and [NBu₄]Br by a process first order in the halide species when the concentration of [NBu₄][PF₆] supporting electrolyte exceeds that of the halide. However, when the halide concentration becomes large relative to the supporting electrolyte concentration then the reaction becomes independent of the concentration, of [NBu₄]Cl or [NBu₄]Br. This unusual behavior arises from the importance of ion pairing in the low polarity CH_2Cl_2 solvent.

From conductivity measurements the K_d equilibrium constants for dissociation of the ion pairs range from 2.6 × 10^{-5} for [NBu₄]Cl to 1.7 × 10^{-4} for $[Fe(CO)_3L_2][PF_6]$. Thus, the concentration of free ions is extremely low. The unusual concentration dependence for the reaction rate fits well to the kinetic model shown in eqs. 9 and 10.

$$[M(CO)_3L_2,PF_6] + [NBu_4, Cl] \overset{K}{\rightleftharpoons} [M(CO)_3L_2,Cl] + [NBu_4,PF_6] \quad (9)$$

$$[M(CO)_3L_2,Cl] \overset{k}{\longrightarrow} \text{products} \quad (10)$$

The derived K for this mechanism are near unity, as expected for a relatively nonspecific ion-pairing equilibrium. This suggests the 19-electron species represents a transition state, rather than a true intermediate in this system. In this scheme the first-order rate process, k, can be regarded as an associative interchange process where bond formation between M and Cl⁻ occurs in the transition state as an M-CO bond breaks. On changing M the rate varies as Fe (15 s^{-1}), Ru (56 s^{-1}), and Os (5 s^{-1}) at 25° C. For Br⁻ as the nucleophile the variation is even less Fe (3.3 s^{-1}), Ru (5.5 s^{-1}), and Os (4.4 s^{-1}). This change on varying the metal is greatly reduced from that in analogous 18-electron carbonyls. This may reflect the intrinsic mechanistic difference between the dissociative path obeyed by the 18-electron systems and the associative character to the radical reactions.

An interesting question arose when we considered 17-electron radicals that contain a ligand such as cyclopentadiene, known to lead to associative reactions in 18-electron complexes. Would the reaction proceed through a 19-electron intermediate, as for other 17-electron radicals, or would ring slippage occur? Both possibilities are shown in eq. 11 for reactions of carbonylbis(cyclopentadienyl)vanadium.

One way to answer this question was to compare the reactivity of the cyclopentadienyl complexes to those that contain the open pentadienyl ring (C_5H_7 = pd). The latter ligand

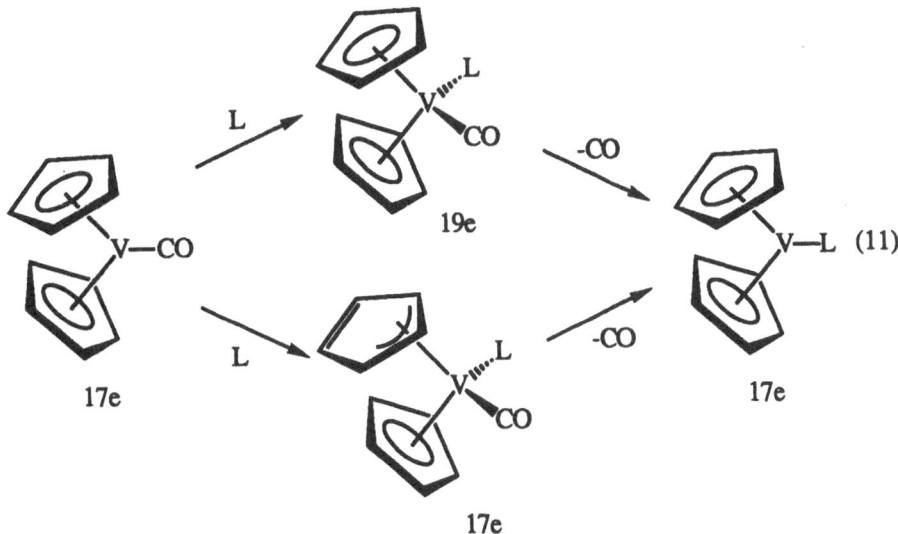

facilitates associative reaction paths in 18-electron complexes more effectively than cyclopentadienyl, because ring slippage occurs more readily [40]. Therefore we measured the rate of eq. 11, which obeys second-order kinetics, $k[VCp_2CO][L]$, for $L = {}^{13}CO$ and $MeOOCC \equiv COOMe$. At $60°$ C the rate for ^{13}CO nucleophile was about 800 $M^{-1}s^{-1}$. For $V(\eta^5\text{-}C_5H_7)_2CO$ the corresponding rate was 0.0038 $M^{-1}s^{-1}$[41]. Although the low reactivity of the pentadienyl complex ruled out ring slippage, it raised the new question as to why the pentadienyl complex reacted so much slower! This question will be considered later in the section on electronic structure.

An alternative mechanism also had to be considered, as shown in eq. 12. This seemed possible because the 15-electron vanadocene complex can be isolated [42]. Also, for the pentamethylcyclopentadienyl (Cp^*) complex the equilibrium constant ($K = k_1/k_{-1}$) can be measured as $\sim 10^{-5}$ [41]. Fortunately the k_{-1} rate could be measured directly by flash photolysis of VCp_2CO and VCp^*_2CO. To our surprise the rate of CO addition to the 15-electron metallocenes was slower than the rate of addition to the 17-electron carbonyl adducts! For example at $25°$ C the rate of CO addition to VCp^*_2 is 1.4 $M^{-1}s^{-1}$, but for CO exchange in VCp^*_2CO the rate is 48 $M^{-1}s^{-1}$ [41]. At first we thought the low reactivity of the 15-electron complexes might arise from a spin-selection rule. Both VCp_2 and VCp^*_2 exist as ground-state quartets, while the reaction products, VCp_2CO and VCp^*_2CO, are ground-state doublets. If this were the explanation then Vpd_2, which is a

$$15e \qquad\qquad 17e \qquad\qquad 17e \qquad (13)$$

ground-state doublet, might show an enhanced reactivity. When the latter rate was measured, however, it was found to be slightly (a factor of 1/2) slower than for VCp$_2$. Therefore, we instead attribute the low reactivity of the 15-electron metallocenes to the large structural reorganization that has to take place - from a parallel ring metallocene to a bent metallocene - on addition of carbon monoxide.

2.1.2. Associative Ring Slippage.

The preceding evidence suggests that nucleophilic attack at the metal in a 17-electron metallocene complex does not require ring slippage, but it does not mean that the latter process cannot occur. This became evident when complexes that contain the indenyl ligand, C$_9$H$_7$, were examined. This ligand is well known to enhance the susceptibility of 18-electron complexes toward nucleophilic attack by as much as 10^6 to 10^9, as compared to analogous cyclopentadienyl compounds [43]. Thus when CO was added to V(η^5-C$_9$H$_7$)$_2$ the unusual slipped ring complex V(η^5-C$_9$H$_7$)(η^3-C$_9$H$_7$)(CO)$_2$ was isolated and structurally characterized [44]. Presumably this reaction occurred by the sequential addition of CO shown in eq. 13.

The same slipped ring species can be made by a different route, which involves single electron transfer [45]. One-electron reduction of the stable 18-electron complex [V(η^5-C$_9$H$_7$)$_2$(CO)$_2$]PF$_6$ produces V(η^5-C$_9$H$_7$)(η^3-C$_9$H$_7$)(CO)$_2$, as evidenced by a clean reversible reduction wave in the cyclic voltammogram. This provides an interesting case where one-electron reduction of an 18-electron complex produces a 17-electron complex! In contrast to the indenyl complex [VCp$_2$(CO)$_2$]PF$_6$ and [VCp*$_2$(CO)$_2$]PF$_6$ produce free CO and VCp$_2$CO or VCp*$_2$CO, within a tenth of a second at -40°C, when they are reduced by one-electron [45]. A ring-slipped species, if it exists at all, must be very short lived for the cyclopentadienyl complex.

2.1.3. Associative Carbon Monoxide Insertion.

For years it was known that oxidation of certain 18-electron complexes that contained carbonyl and alkyl ligands led to migratory insertion at tremendously accelerated rates [46]. This seemed somewhat odd to us because the insertion reaction, eq. 2, creates coordinative unsaturation and less electron density at the metal. Therefore it wasn't clear why oxidation would accelerate such a process. Indeed this anomaly had been noted by previous investigators [47]. Because all previous mechanistic electrochemical studies of this reaction had been performed in coordinating solvents, we wondered whether the solvent could be playing a noninnocent role and coordinate to the 17-electron species to drive the insertion reaction, as shown in eq. 14 for an iron carbonyl system. The first evidence we obtained to support this mechanism, was the stability of the FeCp(PPh$_3$)(CO)(CH$_3$)$^+$ radical in the non-coordinating solvent dichloromethane [48]. Insertion only occurred when ligands, such

$$(14)$$

as acetonitrile or pyridine were added to this solution. The kinetics of the insertion reaction at room temperature follows a second-order rate law, $k[FeCp(PPh_3)(CO)(CH_3)+]$ [L]. For a series of para-substituted pyridine nucleophiles the Hammett ρ obtained (-4.7) was reminiscent of that observed for nucleophilic attack at the 5-coordinate iron(I) radicals discussed earlier.

At low temperature the second step in eq. 14 becomes rate limiting and this permits the 19-electron intermediate (or a species with a slipped or partly-slipped cyclopentadienyl ring) to be observed directly. The IR spectra for L = pyridine-d_5 showed carbonyl stretching frequencies of 1966 cm^{-1}, 1916 cm^{-1}, and 1645 cm^{-1}, respectively, for the three species of eq. 14. At 0°C the second step in eq. 14 becomes rate limiting for certain pyridine nucleophiles and the rate constants obtained for the insertion step show an enhanced reactivity as the pyridine nucleophile became more basic. Thus, oxidatively catalyzed CO insertion works in this case because the 17-electron oxidized species can undergo nucleophilic attack to form an electron rich 19-electron species, which undergoes insertion.

2.1.4. *Associative Reactions of Cluster Radicals.*

From our work described above and related studies of Kochi [49], Poë [50], Tyler [51], and Brown [28] it has become evident that 17-electron organometallic radicals are generally susceptible to nucleophilic attack. Here we describe one example of how one-electron oxidation of a metal-metal bonded dimer can lead to enhanced reactivity toward associative attack. The earliest such example for clusters was described by Rieger [52].

The synthesis, redox chemistry, and structures of several 33- and 34-electron phosphido-bridged metal complexes has been developed by Baker [53]. Qualitative observations about the reactivity of these systems can be summarized as in Fig. 1. The 34-electron diamagnetic species are kinetically inert, while the 33-electron species substitute a carbon monoxide ligand readily at room temperature. This system is ideal for a kinetic investigation, because one can oxidize species A at the electrode surface to form B, which can then be allowed to react to form C, whose reduction wave to form D can be observed at a different potential. Potential step experiments between the redox couples for A/B and C/D yield reliable kinetics data for the rate of the B —> C conversion. These experiments show that the rate law follows second order behavior, $k_1[Fe_2(\mu\text{-}PPh_2)(CO)_7]$ [L], where L = PCy$_3$, PPh$_3$, PPh$_2$Me, PPhMe$_2$, PMe$_3$, PPh(OMe)$_2$, and P(OMe)$_3$. We estimate the rate acceleration, compared to the diamagnetic Fe$_2$(μ-PPh$_2$)(CO)$_7^-$ or FeCo(μ-PPh$_2$)(CO)$_7$ undergoing the same reaction, as 10^5 to 10^6 [53]. The dependence of the rate constant, k_1, on steric and electronic properties of the phosphine ligands closely resembles the trends observed for mononuclear radicals. Thus, singly oxidized metal dimer radicals exhibit a high lability toward associative substitution just as for 17-electron mononuclear transition metal complexes. The rate accelerations appear to be somewhat less than in the mononuclear complexes, which may reflect delocalization of the odd electron over more atoms.

298

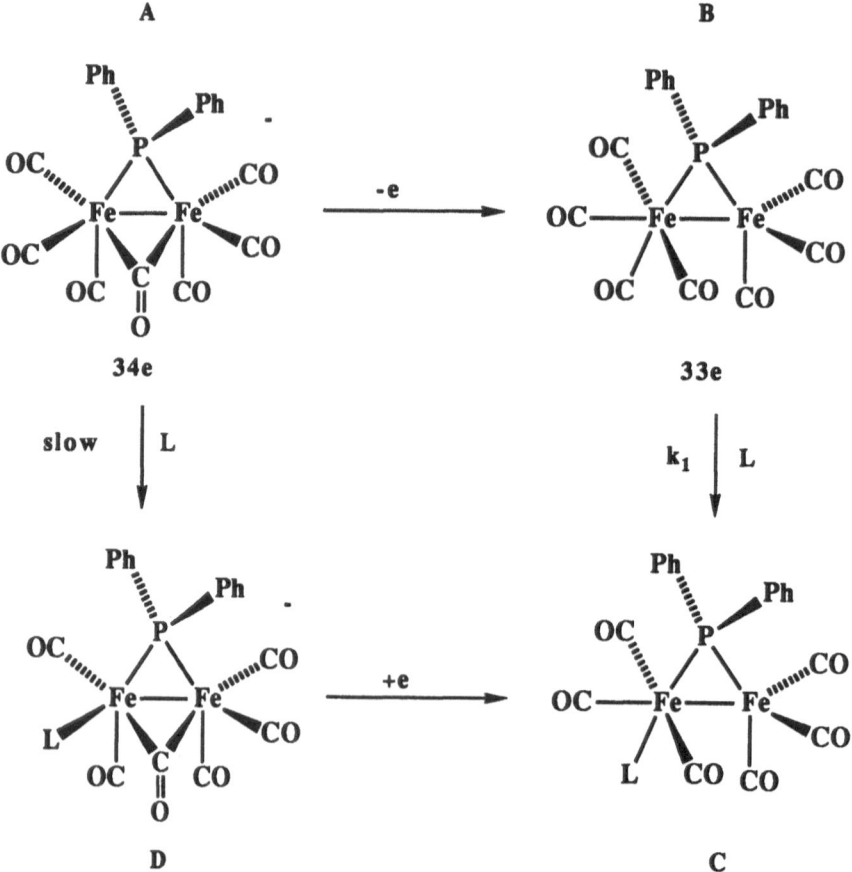

Figure 1. Redox and ligand substitution chemistry of phosphido-bridged iron clusters.

2.1.5. *Oxidative Cleavage of the Metal-Metal Bond in Radical Dimers.* Dimetal decacarbonyls, $M_2(CO)_{10}$, have been prototypical systems for the study of metal-metal bonding. All attempts to oxidize them by one-electron lead instead to a net two-electron process with the production of 2 $M(CO)_5(Sol)^+$ species, where Sol is a solvent or ligand and M = Mn or Re. This is somewhat surprising, because estimates for the metal-metal bond strength in these compounds range between 35 kcal/mol for Mn to 60 kcal/mol for Re. One might expect that the radical cation with a half-order metal-metal bond would persist for some length of time. It should require a higher oxidation potential to remove the second electron. Thus, the origin of this behavior is not clear.

Given our preceding experience with the high preference for associative attack at electron deficient radicals we thought that the solvent or the counterion, present in all electrochemical experiments, might be good enough nucleophiles to promote metal-metal bond cleavage. This would yield a mononuclear radical, which would be immediately oxidized at the same potential. Indeed this sort of ECE mechanism had been suggested by earlier studies of Kadish [54]. One way to reduce the susceptibility of the putative

$M_2(CO)_{10}^+$ radicals to nucleophilic attack would be to look at isoelectronic anions $M_2(CO)_{10}^-$, where M = Cr, Mo, or W. As shown in Fig. 2, the cyclic voltammograms of the parent diamagnetic complexes $[PPN^+]_2[M_2(CO)_{10}^{2-}]$ all show chemically reversible one-electron waves at fast scan rates [55]. For M = W this transforms to an irreversible two-electron wave at slower scan rates, and the molybdenum complex does so to a lesser extent. The $Cr_2(CO)_{10}^-$ radical is even stable for seconds at room temperature. The stability order W < Mo < Cr was unexpected, because the third row metals generally form the strongest metal-metal bonds and the most inert complexes. This, however, might be consistent with nucleophilic attack at the metal. We saw above that the third row metals would be most sterically accessible to nucleophilic attack and gain the most stabilization by bonding to the nucleophile.

When the potential step experiments were performed and the data fit to a working curve for nucleophilic attack by solvent, it clearly did not fit the ECE mechanism. As shown in Fig. 3, the data closely follows the working curve for an EC homogeneous disproportionation mechanism. In the EC homogeneous disproportionation mechanism the rate limiting step would be outer-sphere electron transfer between two radical intermediates, as shown in eq. 15.

$$2\ M_2(CO)_{10}^{2-} \xrightarrow{-2e} 2\ M_2(CO)_{10}^- \xrightarrow{k} M_2(CO)_{10}^{2-} +\ 2\ M(CO)_5(THF) \quad (15)$$

For $Mo_2(CO)_{10}^-$ and $W_2(CO)_{10}^-$ the values of k at 20° C were 590 $M^{-1}s^{-1}$ and 2,500 $M^{-1}s^{-1}$, respectively. This mechanism suggests that the inability to detect isoelectronic $Mn_2(CO)_{10}^+$ and $Re_2(CO)_{10}^+$ radicals may not arise from the instability of the half-order metal-metal bond or from rapid nucleophilic attack, but instead because of rapid electron transfer as in eq. 15.

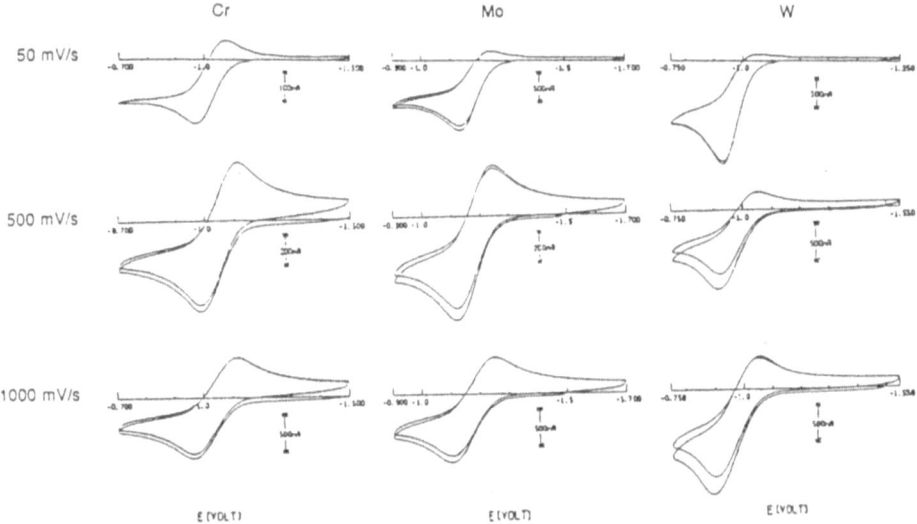

Figure 2. Cyclic voltammograms of the $M_2(CO)_{10}^{2-}$ complexes in THF solvent at 23°C and at several scan rates.

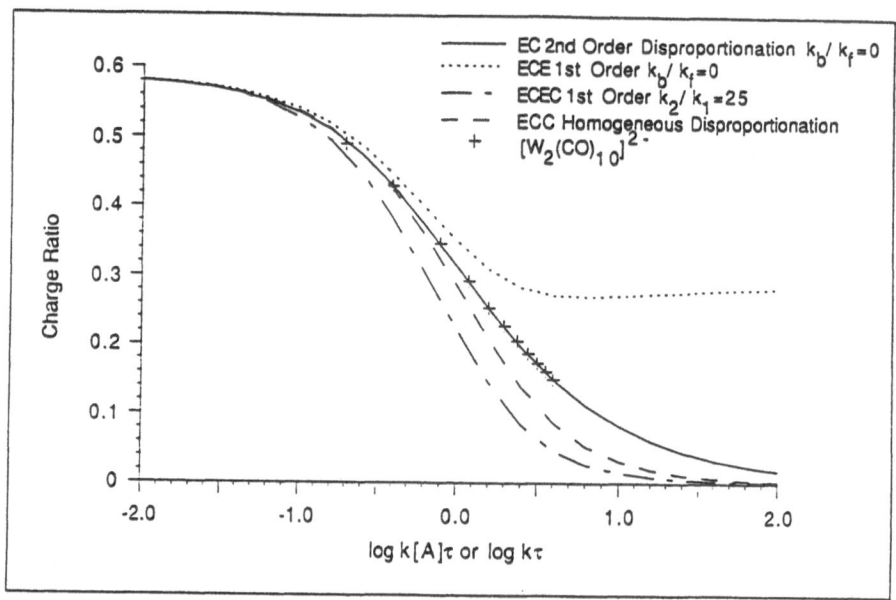

Figure 3. Experimental data from potential step experiments (+) and the response functions expected for various mechanisms.

2.1.6. *Orbital Origins of Associative Reactivity in 17-Electron Complexes.*

The observations about the reactivity of 17-electron radicals suggest that just as even-electron organometallic complexes tend to cycle between 16- and 18-electron configurations in their reactions, so to do 17-electron radicals cycle between 17- and 19-electron configurations. Associative reactions appear to be the rule rather than the exception for the reactions of 17-electron systems. Such paths are generally much more rapid than associative reactions in 18-electron systems where, for example, ring slippage is possible. Comparisons of the sensitivity of reaction rates to steric and electronic properties of the incoming nucleophile show the rates of 17-electron radicals depend strongly on the electronic properties of the nucleophile. All this points to significant metal-nucleophile bonding in the 19-electron intermediates or transition states involved in these reactions. One of the simplest models advanced to explain this behavior centers on the bonding interaction possible between the nucleophile and the odd electron on the metal-localized radical. It is important to emphasize the difference between the 19-electron complexes involved in these reaction mechanisms and the 19-electron complexes one might generate by reduction of an 18-electron complex.

First consider a 19-electron complex one might generate by reduction of an 18-electron complex, such as $Cr(CO)_6$, as shown in Fig. 4. Because the odd electron resides in e_g, a metal-CO antibonding orbital, or one of the CO π^* orbitals, it is highly unfavorable energetically. The $Cr(CO)_6^-$ species is extremely short lived and dissociates CO to form 17-electron $Cr(CO)_5^-$. In the presence of H-atom donors, such as $HSnBu_3$, it can form a formyl complex $Cr(CO)_5(CHO)^-$ [56]. Pulsed radiolysis experiments to generate species, such as $Fe(CO)_5^-$ in rigid matrices show the odd electron in these 19-electron complexes can localize on a carbonyl ligand and cause it to adopt a bent geometry [57].

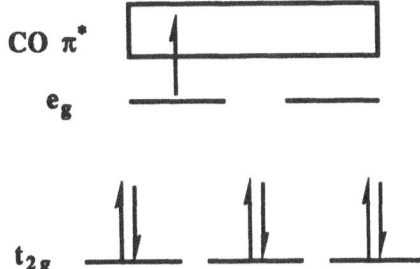

Figure 4. 19-electron species formed by reduction of 18-electron $Cr(CO)_6$.

The preceding situation can be contrasted with the interaction possible when a nucleophile bonds to a 17-electron complex to generate a 19-electron species quite different from that of Fig. 4. We show this situation in Fig. 5 for an electron pair donor interacting with a half-filled d-t_{2g} orbital in $V(CO)_6$. Here the odd electron occupies an antibonding σ^* orbital of the metal-nucleophile bond.

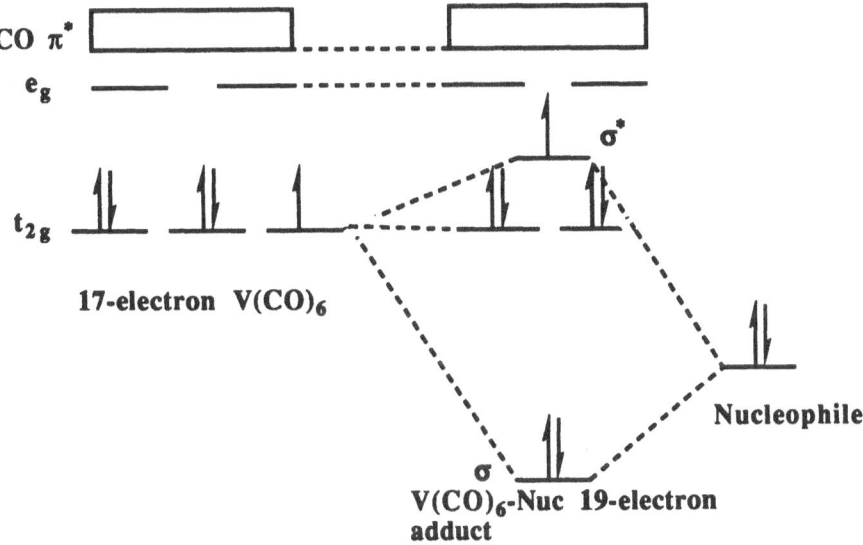

Figure 5. 19-electron species formed by interaction between a 17-electron complex and a nucleophile.

In some exploratory SCF-Xα–DV calculations, where a model PH_3 nucleophile was used to approach six-, five-, and four-coordinate metal carbonyl radicals along faces and edges of their polyhedra, we found that certain geometries of approach led to well developed σ-bonding interactions between PH_3 and the half-occupied orbital on the metal carbonyl radical [58]. For example, Fig. 6 shows the interaction between the phosphorus lone pair and one of the t_{2g}-d orbitals of $V(CO)_6$. The PH_3 ligand

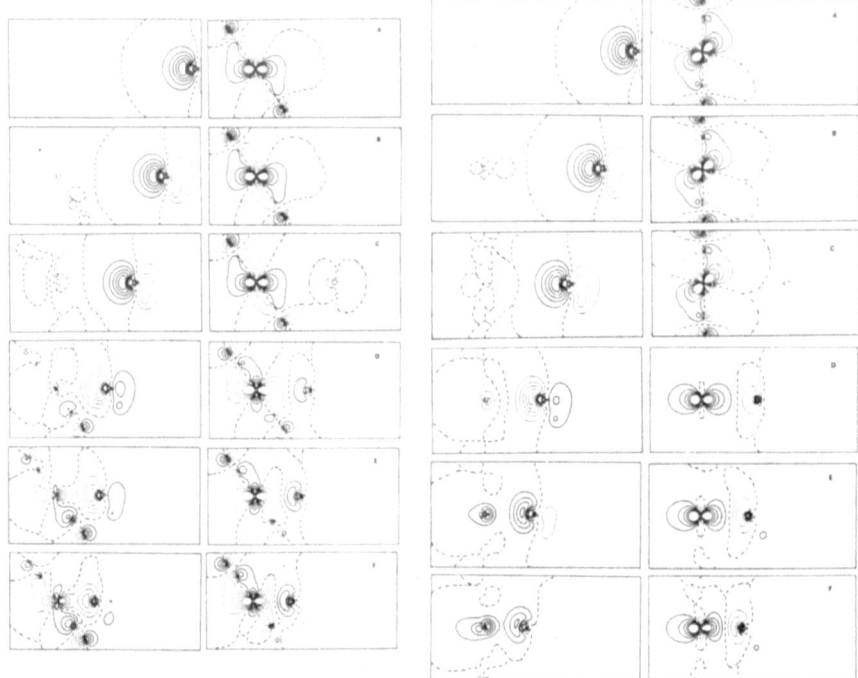

Figure 6. Overlap of phosphorus lone pair orbital (right most species in each box) with a t_{2g} orbital of $V(CO)_6$ (left most orbital in each box) to generate the bonding and antibonding σ-orbital pair. The two left columns represent attack of the PH_3 ligand on an octahedral face of $V(CO)_6$, while the two right columns represent edge attack. The V-P separation varies from 9Å at the top (A) to 2.6Å at the bottom (F) of the figure. Reproduced with permission from ref. 58.

approaches V along an octahedral face (panels A-F left). The σ-bond which develops clearly exhibits better overlap than for the edge approach geometry (panels A-F right).

Similar studies for square pyramidal, trigonal bipyramidal, and tetrahedral complexes suggest that attack will be preferred at an open face of a square pyramid, the equatorial edge of a trigonal bipyramid, and at an open face of a tetrahedral complex. Recent *ab initio* studies reported by Hall, support this model of transition state stabilization by nucleophile bonding to the half-occupied orbital in 17-electron complexes [59].

Recall one of the unexpected results in our study of the reactivity of VCp_2CO and Vpd_2CO was the unusual low reactivity of the pentadienyl complex toward associative attack. Initially we performed Xα calculations (spin polarized) for these two compounds, but were disappointed because their electronic structure appeared to be very similar, as shown in Fig. 7 [60]. In both cases the highest occupied orbital, which contains the odd electron, has a_1 symmetry.

We examined the EPR spectra of these complexes, and their ^{13}CO adducts, both at room temperature and at 77K. From the isotropic EPR g values and vanadium hyperfine splitting constants shown in Table 2, it seemed that the complexes with one or more

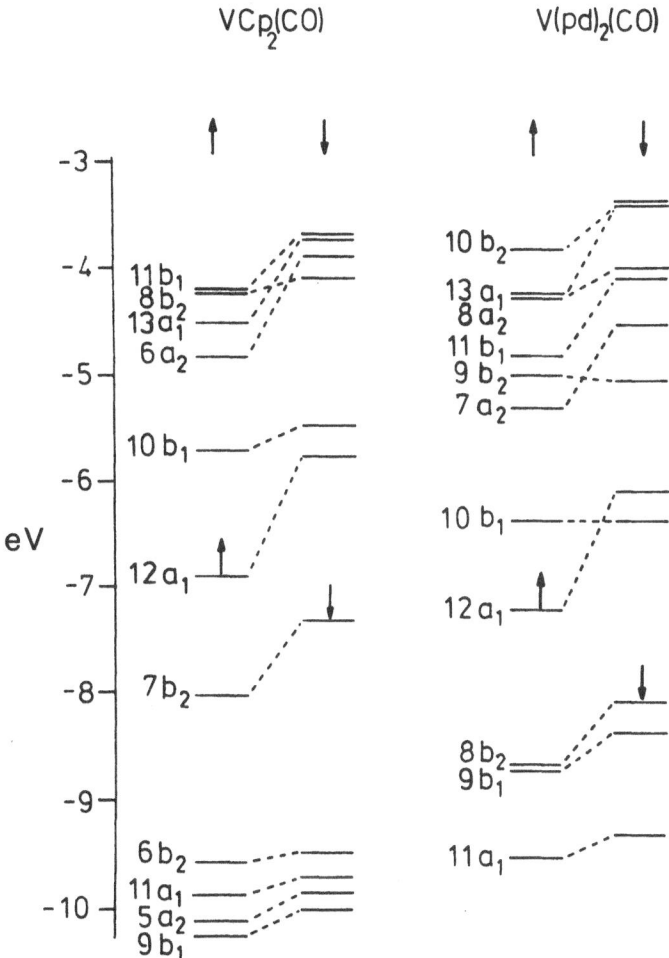

Figure 7. Spin polarized SCF-Xα-DV results for VCp$_2$CO and Vpd$_2$CO. Reproduced with permission from ref. 60.

pentadienyl ligands behaved differently from the cyclopentadienyl complexes. All the compounds that contain a pentadienyl ligand have g values less than the free electron value, and exhibit large vanadium hyperfine splittings. The superhyperfine splitting observed in the ^{13}CO derivatives shows only a small spin density (~5%) delocalized on the carbonyl ligand. A large anisotropy in the rhombic components of the vanadium hyperfine tensor could be detected by simulation of the low temperature EPR spectra, and this can only occur if more than one d orbital contributes to the orbital containing the odd electron. In C$_{2v}$ symmetry this is only allowed for the a$_1$ representation, which supports the prediction of the a$_1$ symmetry HOMO by the calculations. These facts suggested the difference between the complexes arose in t he differential contribution of the various

TABLE 2. EPR Parameters and CO Stretching Frequencies for Cyclopentadienyl and Pentadienyl Complexes of V(II), and the Rates of ^{13}CO Exchange
at 60°C.

Complex	A_{iso} (10^{-3} cm^{-1})	g_{iso}	v_{CO} (cm^{-1})	Rate (M^{-1}s^{-1})
Vpd$_2$CO	-7.36	1.991	1959	0.004
V(pd)CpCO	-5.86	1.993	1938	0.006
VCp$_2$CO	-2.68	2.005	1881	800
VCp$^*_2$CO	-1.73	2.004	1842	260

d orbitals to the a_1 HOMO, and an orbital plot shown in Fig. 8 supports this view. The
12 a_1 orbital plots for VCp$_2$CO and Vpd$_2$CO appear to be similar, except they are
hybridized 90° relative to one another, as shown in the Figure. The xz plane contains the
Cp and pd rings. For Vpd$_2$CO the 12 a_1 HOMO directs up toward the center of the
pentadienyl ring. This makes it less sterically accessible to bind to a nucleophile. In
contrast, the VCp$_2$CO 12 a_1 orbital lies mainly in the yz plane and points out the edge of
the complex between the two cyclopentadienyl rings. This should facilitate its binding to
an entering nucleophile. It is also noteworthy that the mixed ligand V(pd)CpCO exhibits
spectroscopic properties much more like Vpd$_2$CO than VCp$_2$CO, and its reactivity also
more closely resembles Vpd$_2$CO than VCp$_2$CO.

Another possible case for the importance of the frontier orbital character of the odd
electron may be the differing reactivities of the one-electron oxidized clusters Fe(μ-
PPh$_2$)(CO)$_7$ and M$_2$(CO)$_{10}^-$ for M = Cr, Mo, or W. In the iron cluster both MO
calculations and EPR results indicate the odd electron lies in a nonbonding d-orbital
contained in the trigonal plane of one of the two pseudotrigonal bipyramidal iron atoms,
as shown in 5 [61]. Thus nucleophilic attack on the exposed edge of thetrigonal
bipyramid should be possible, as suggested by our model Xα calculations for such
5-coordinate radicals [58]. As already noted in section 2.1.4, the radical Fe$_2$(μ-
PPh$_2$)(CO)$_7$ undergoes rapid CO substitution reactions.

For the M$_2$(CO)$_{10}^-$ radical the half-filled orbital may be the metal-metal σ-bonding
level, which would be localized along the metal-metal axis. Nucleophilic attack along this
axis is blocked by the axial CO groups present. This concept of steric protection of a
radical orbital also may explain the anomalous low reactivity of 17-electron Mn(η^4-
butadiene)$_2$CO. Extended Hückel calculations suggest the odd electron occupies a d_{z^2}-
type orbital directed toward the center of each butadiene ligand in this compound [62].

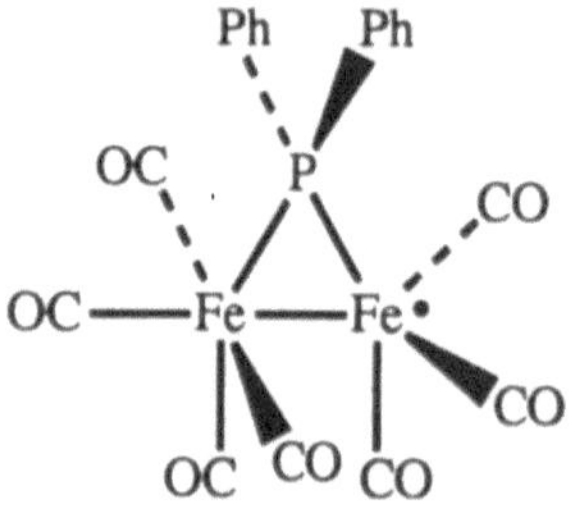

5

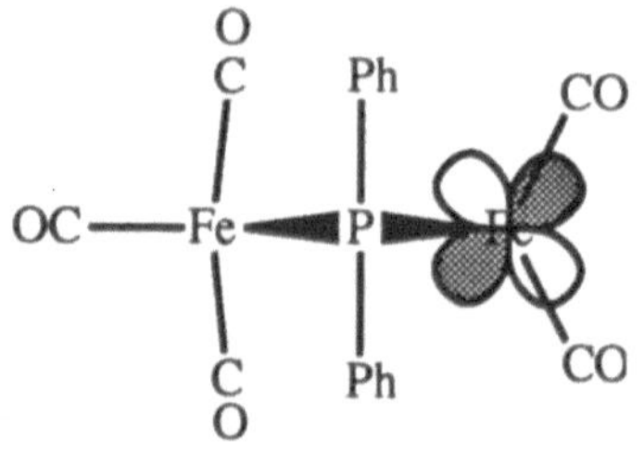

Top view of HOMO in 5

12 a₁

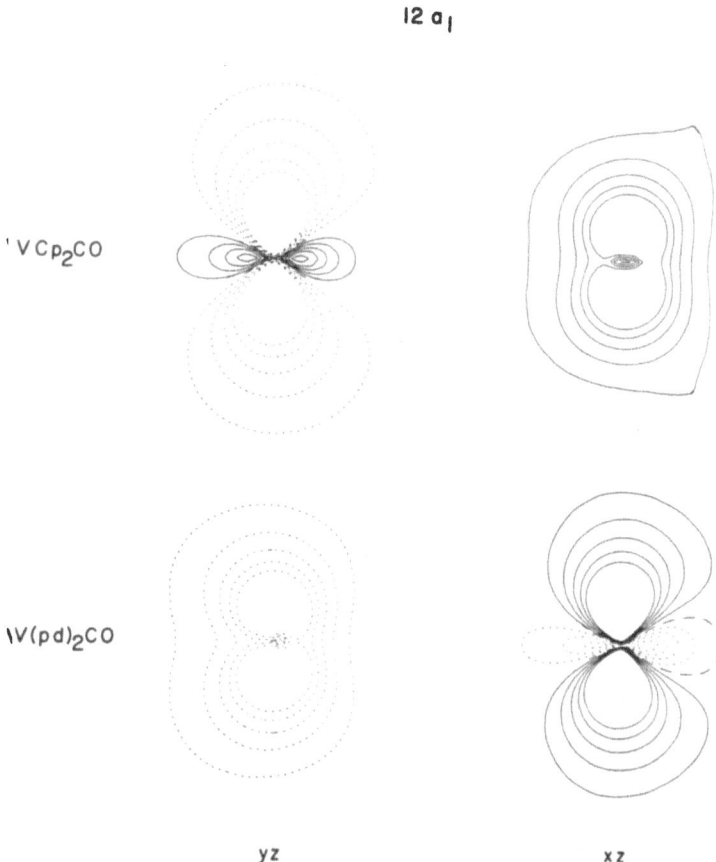

'VCp₂CO

\V(pd)₂CO

yz   xz

Figure 8. Plots of the highest occupied a₁ orbitals in VCp₂CO and Vpd₂CO. The (y,z)-plane lies between the plane of the Cp or pd ligands. Reproduced with permission from ref. 59.

3. Frontier Orbital Interactions in Electron Transfer Reactions

In the course of our studies to chemically generate radicals, or study their redox processes, we have found several examples where an inner sphere path for electron transfer may be operative. For example, in studies of the disproportionation of $V(CO)_6$ IR spectral studies revealed the presence of isocarbonyl-bridged intermediates and for THF nucleophile the species **6**, which had previously been characterized by X-ray diffraction, formed as the stable product [26,63]. Later we showed that VCp^*_2, which could function as a one-electron reductant, readily reacts with the one-electron oxidant $V(CO)_6$ to form the stable isocarbonyl product shown in Figure 9 [64]. These observations suggest that electron transfer reactions between $V(CO)_6$ and other metals may be facilitated by an inner-sphere mechanism that involves an isocarbonyl bridge.

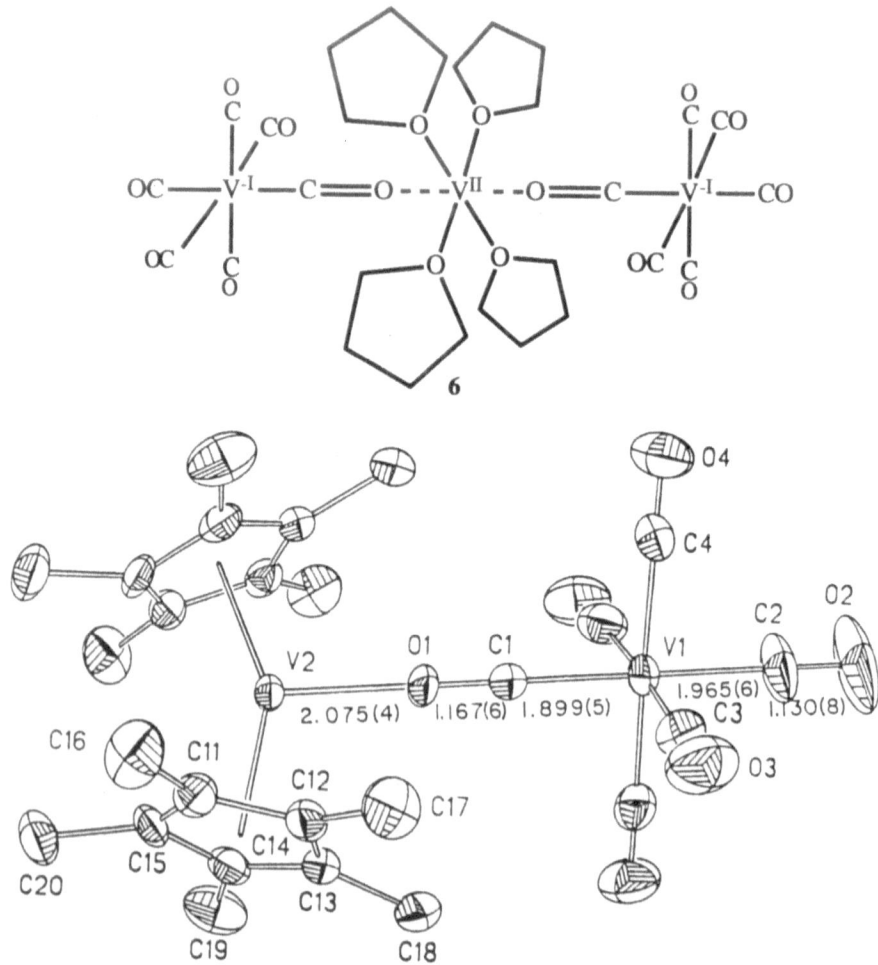

Figure 9. Thermal ellipsoid plot (50%) of [VCp*+2][V(CO)6−] reproduced with permission from ref. 64.

In closing we further illustrate the importance of frontier orbital interactions in determining localization of the odd electron. While attempting to isolate the $Os(CO)_3(PPh_3)_2^+$ radical by oxidation of the neutral complex with $Ag(O_2CCF_3)$ we instead isolated a precursor complex to electron transfer [65]. From a thermodynamic standpoint silver(I) should completely oxidize $Os(CO)_3(PPh_3)_2$. However, some crystals form if benzene is layered onto the reaction solution and the structure, shown in Fig. 10, is that of an $Os(CO)_3(PPh_3)_2$-$Ag(O_2CCF_3)$ adduct. The osmium maintains an approximate trigonal bipyramidal geometry and the Ag(I) oxidant binds to an edge of the trigonal bipyramid. We view this structure as derived from a trigonal bipyramid, rather than an octahedron, because of the acute Ag-Os-C angles of 69° and 73.5° in the trigonal

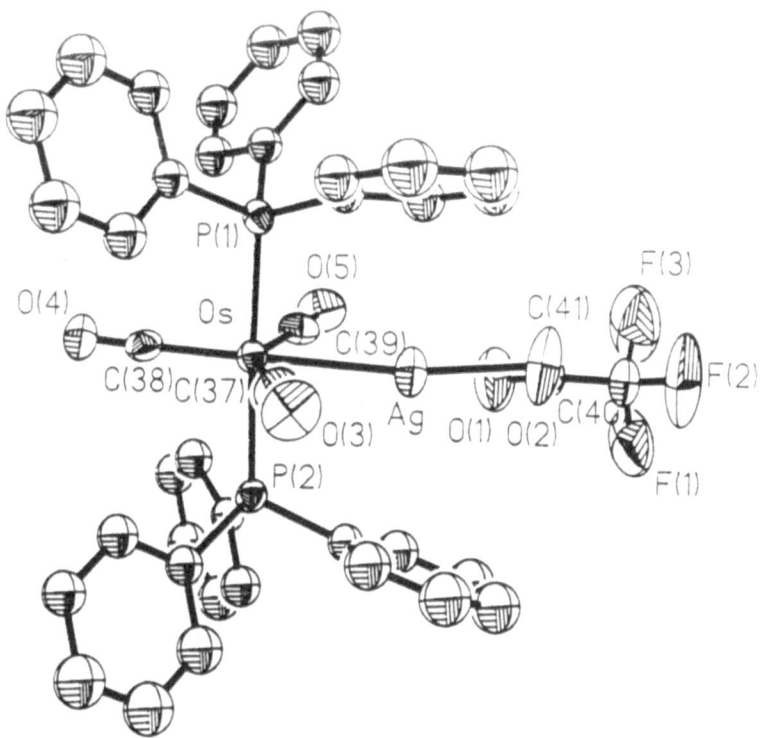

Figure 10. Thermal ellipsoid plot (50%) of Os(CO)₃(PPh₃)₂-Ag(O₂CCF₃).

plane. In the context of theoretical studies presented by Hall elsewhere in this volume, the linear geometry for the semibridging carbonyls suggests a repulsive interaction between them and the Ag^+ cation [59]. Indeed, part of the activation barrier responsible for the isolation of this metastable complex may be that further opening of the Ag-Os-C angle to 90° needs to occur before electron transfer can commence. In our theoretical studies of the $Fe(CO)_3(PH_3)_2^+$ radical we found the highest occupied orbital to be an e-type orbital composed of Fe d orbitals in the equatorial plane of the trigonal bipyramid (*e.g.*, **7**) [66]. Thus, the precursor complex represents the Ag^+ oxidant binding to the frontier orbital density in the HOMO on iron prior to the oxidation step. When these crystals are dissolved, or on irradiation, the EPR signal for the radical appears. Thus, this complex represents a snapshot of the reaction coordinate for electron transfer.

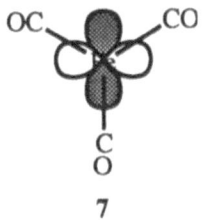

7

ACKNOWLEDGMENTS

The contributions of the student, faculty, and postdoctoral colleagues to this research are gratefully acknowledged, as well as the support of the U. S. National Science Foundation by Grant (CHE-8815958).

REFERENCES

1. Collman, J. P., Hegedus, L. S., Norton, J. R. and Finke, R. G. (1987), Principles and Applications of Organotransition Metal Chemistry, University Science Books, Mill Valley, California.
2. Laine, R. M., Moriarty, R. E. and Bau, R. (1973), J. Am. Chem. Soc., **94**, 1402.
3. Dessey, R. E. and Bares, L. A. (1972), Acc. Chem. Res., **1**, 95; Tyler, D. R. and Mao, F. (1990), Coord. Chem. Rev., **97**, 119.
4. Basolo, F. and Pearson, R. G. (1985) in Wiley & Sons (eds), Mechanisms of Inorganic Reactions, 2nd ed., New York, 1967; Atwood, J. D. Inorganic and Organometallic Reaction Mechanisms, Brooks/Cole Publ., Monterey, California.
5. Meier, M., Basolo, F. and Pearson, R. G. (1969), Inorg. Chem., **8**, 795.
6. Atwood, J. D. (1981), Inorg. Chem., **20**, 4031; Schmidt, S. P., Trogler, W. C. and Basolo, F. (1982), Inorg. Chem., **21**, 1699; Muetterties, E. L., Burch, R. R. and Stolzenberg, A. M. (1982), Annu. Rev. Phys. Chem., **33**, 89.
7. Halpern, J. (1982), Science, **217**, 401.
8. Tolman, C. A. (1972), Chem. Soc. Rev., **1**, 337.
9. Hoyano, J. K., McMaster, A. D. and Graham, W. A. G. (1983), J. Am. Chem. Soc., **105**, 7190.
10. Noack, K. and Calderazzo, F. J. (1967), Organomet. Chem., **10**, 101.
11. Basolo, F. (1990), Polyhedron, **9**, 1503.
12. Shi, Q.-Z., Richmond, T. G., Trogler, W. C. and Basolo, F. (1984), Inorg. Chem., **23**, 957.
13. Basolo, F. (1985), Inorg. Chim. Acta, **100**, 33.
14. O'Connor, J. M. and Casey, C. P. (1987), Chem. Rev., **87**, 307.
15. Yamazaki, H. and Wakatsuki, Y. J. (1978), Organomet. Chem., **149**, 377.
16. Trogler, W. C. (1990), Acc. Chem. Res., **23**, 426.
17. Schneider, K. J., VanEldik, R. (1990), Organometallic, **9**, 1235.
18. Wax, M. J. and Bergman, R. G. (1981), J. Am. Chem. Soc., **103**, 7028; Mawby, R. J., Basolo, F. and Pearson, R. G. (1964), J. Am. Chem. Soc., **86**, 3994.
19. Webb, S. L., Giandomenico, C. M. and Halpern, J. (1986), J. Am. Chem. Soc., **108**, 345.
20. Hepp, A. F. and Wrighton, M. S. (1983), J. Am. Chem. Soc., **105**, 5934; Firth, S., Hodges, P. M., Poliakoff, M., Turner, J. J. and Therien, M. J. (1987), J. Organomet. Chem., **331**, 347.
21. Bentsen, J. G. and Wrighton, M. S. (1987), J. Am. Chem. Soc., **109**, 4518, 4530; Ford, P. C. (1990), J. Organomet. Chem., **383**, 339.
22. a) Shojaie, A. and Atwood, J. D. (1985), Organometallics, **4**, 187.
    b) Poë, A. J. and Sekhar, V. C. (1985), Inorg. Chem. , **24**, 4376.
    c) Kennedy, J. R., Basolo, F. and Trogler, W. C. (1988), Inorg. Chim. Acta, **146**, 75.

    d) Norton, J. R. and Collman, J. P. (1973), Inorg. Chem., **12**, 476.

    e) Darensbourg, D. J. and Baldwin-Zuschke, B. J. (1982), J. Am. Chem. Soc., **104**, 3906.

    f) Albano, V., Bellow, P. L. and Scatturin, V.(1967), Chem. Commun., 730.

23. Kennedy, J. R., Selz, P., Rheingold, A. L., Trogler, W. C. and Basolo, F. (1989), J. Am. Chem. Soc., **111**, 3615; Brodie, N. M. J. and Poë, A. J. (1990), J. Organomet. Chem., **383**, 531.

24. Trogler, W. C. (1990), Acc. Chem. Res., **23**, 239.

25. Holland, G. F., Manning, M. C., Ellis, D. E. and Trogler, W. C. (1983), J. Am. Chem. Soc., **105**, 2308.

26. Shi, Q.-Z., Richmond, T. G., Trogler, W. C. and Basolo, F. (1984), J. Am. Chem. Soc., **106**, 71; Richmond, T. G., Shi, Q.-Z., Trogler, W. C. and Basolo, F. (1984), J. Am. Chem. Soc., **106**, 76.

27. Kochi, J. K. (1990), in Trogler, W. C. (ed.), Organometallic Radical Processes, Elsevier, Amsterdam.

28. Brown, T. L. (1991) in Trogler, W. C. (ed.), Organometallic Radical Processes, Elsevier, Amsterdam.

29. Golovin, M. N., Rahman, M. M., Belmonte, J. E. and Giering, W. P. (1985), Organometallics, **4**, 1981; Poë, A. J. (1988), Pure Appl. Chem., **60**, 1209.

30. Therien, M. H., Ni, C.-L., Anson, F. C., Osteryoung, J. G. and Trogler, W. C. (1986), J. Am. Chem. Soc., **108**, 4037.

31. Song, L. and Trogler, W. C. (1991), submitted.

32. Amatoré, C. (1990) in Trogler, W. C. (ed.), Organometallic Radical Processes, Elsevier, Amsterdam.

33. Hanafey, M. K., Scott, R. L., Ridgway, T. H. and Reilley, C. N. (1978), Anal. Chem., **50**, 116.

34. Van Duyne, R. P. and Reilley, C. N. (1972), Anal. Chem., 44, 142,153,158.

35. Christie, J. H., Osteryoung, R. A. and Anson, F. C. (1967), J. Electroanal. Chem., **13**, 236.

36. Hammett, L. P. (1970), Physical Organic Chemistry; 2nd Ed., McGraw Hill, New York.

37. Shen, J. K., Gao, Y. C., Shi, Q.-Z. and Basolo, F. (1989), Inorg. Chem., **28**, 4304.

38. Siefert, E. E. and Angelici, R. J. (1967), J. Organomet. Chem., **8**, 374; Johnson, B. F. G., Lewis, J. and Twigg, M. V. (1975), J. Chem. Soc., Dalton Trans., 1876.

39. Song, L. and Trogler, W. C., (1991), submitted.

40. Ernst, R. D. (1988), Chem. Rev., **88**, 1255.

41. Kowaleski, R. M., Basolo, F., Trogler, W. C., Gedridge, R. W., Newbound, T. D. and Ernst, R. D. (1987), J. Am. Chem. Soc., **109**, 4860.

42. Fischer, E. O. and Hafner, W. (1954), Z. Naturforsch., Teil B., **9**, 503.

43. Kowaleski, R. M., Trogler, W. C. and Basolo, F. (1986), Gazz. Chim. Ital., **116**, 105.

44. Kowaleski, R. M., Rheingold, A. L., Trogler, W. C. and Basolo, F. (1986), J. Am. Chem. Soc., **108**, 2460.

45. Miller, G. L., Therien, M. H. and Trogler, W. C. (1990), J. Organomet. Chem., **383**, 271.

46. a) Magnuson, R. H., Zulu, S., T'sai, W.-M. and Giering, W. P. (1980), J. Am. Chem. Soc., **102**, 6887.

b) Magnuson, R. H., Meirowitz, R., Zulu, S. and Giering, W. P. (1982), Ibid., **104**, 5790.

c) Magnuson, R. H., Meirowitz, R., Zulu, S. J., Giering, W. P. (1983), Organometallics, **2**, 460.

47. Doxsee, K. M., Grubbs, R. H. and Anson, F. C. (1984), J. Am. Chem. Soc., **106**, 7819.

48. Therien, M. H. and Trogler, W. C. (1987), J. Am. Chem. Soc., **109**, 5127.

49. Kochi, J. K. (1986), J. Organomet. Chem., **300**, 139.

50. Fox, A., Malito, J. and Poë, A. J. (1981), J. Chem. Soc., Chem. Commun., 1052.

51. Tyler, D. R. (1990), in Trogler, W. C. (ed.), Organometallic Radical Processes, Elsevier, Amsterdam.

52. Bezems, G. J., Rieger, P. H. and Visco, S. (1981), J. Chem. Soc., Chem. Commun., 265.

53. Baker, R. T., Calabrese, J. C., Krusci, P. J., Therien, M. H. and Trogler, W. C. (1988), J. Am. Chem. Soc., **110**, 8392.

54. Lacombe, D. A., Anderson, J. E. and Kadish, K. M. (1986), Inorg. Chem., **25**, 2074.

55. Phillips, J. R. and Trogler, W. C., (1991), submitted.

56. Harayanan, B. A., Amatoré, C. and Kochi, J. K. (1986), Organometallics, **5**, 926.

57. Fairhurst, S. A., Morton, J. R. and Preston, K. F. (1982), J. Chem. Phys., **77**, 5872.

58. Therien, M. H. and Trogler, W. C. (1988), J. Am. Chem. Soc., **110**, 4942.

59. Hall, M. B. (1992), in Salahub, D. R. and Russo, N. (eds.), Metal-Ligand Interactions, NATO ASI Volume.

60. Kowaleski, R. M., Basolo, F., Osborne, J. H. and Trogler, W. C. (1988), Organometallics, **7**, 1425.

61. Krusic, P. J., Baker, R. T., Calabrese, J. C., Morton, J. R., Preston, K. F. and LePage, Y. (1989), J. Am. Chem. Soc., **111**, 1262.

62. Harlow, R. L., McKinney, R. J. and Whitney, J. F. (1983), Organometallics, **2**, 1839.

63. Schneider, M. and Weiss, E. (1976), J. Organomet. Chem., **121**, 365.

64. Osborne, J. H., Rheingold, A. L. and Trogler, W. C. (1985), J. Am. Chem. Soc., **107**, 6292.

65. Song, L. and Trogler, W. C. (1991), submitted.

66. Therien, M. H. and Trogler, W. C. (1986), J. Am. Chem. Soc.6, **108**, 3697.

DENSITY FUNCTIONAL THEORY - PRINCIPLES AND APPLICATIONS TO METAL-LIGAND INTERACTIONS

D. R. SALAHUB
Département de chimie, Université de Montréal
C.P. 6128, Succursale A
Montréal, Québec H3C 3J7
Canada

ABSTRACT. A historical and conceptual overview of Density Functional Theory is given, along with sketches of the proofs for the fundamental theorems, to make them plausible. The Kohn-Sham approach is compared and contrasted with wave function-based methodologies. Practical implementations using Gaussian basis sets have been developed over the last few years(including the program deMon) and the validation of these for transition metal-ligand interactions is discussed, as is their use in revealing the properties of transition-metal clusters interacting with ligands or adsorbates.

1. Introduction - Quantum Molecular Modeling in the Nineties

Computer molecular modeling is having a profound effect on the practice of chemistry. The appearance on the scene of empirical force fields with sufficient accuracy to predict a surprising variety of chemical, biochemical, and materials properties, and of sufficient rapidity on modern workstations to allow complex "real" systems to be treated, is changing the way chemists "do" chemistry. Much of the current activity involves classical approaches (molecular dynamics or molecular mechanics) or, for the least complex of the complex systems, semi-empirical quantum chemical methods. The success of these approaches, both scientific and commercial, is starting to have an influence on the more traditional areas of *ab initio* quantum chemistry. The "pie-in-the-sky" dream of quantum chemistry is to predict the properties and reactivities of just such complex systems "from first principles". Although some statements of this dream may still seem quixotic, recent advances in methodology and in computer power make them ring more and more true.

Looking ahead to probable developments over the next decade or so, it seems clear that the trend to more accurate treatments of more complex systems will continue at an ever-increasing rate. Quantum Chemistry will complete a metamorphosis into what one might call Quantum Molecular Modeling. The usual mainstays of Quantum Chemistry, geometry optimisations, vibrational frequency calculations, reaction path following, and the like, will be extended to systems with more atoms and more complex atoms, including the transition metals. Moreover, such studies will include the effects of the surrounding environment, treated at appropriate levels of approximation. The future state-of-the-art will then involve a quantum mechanical "core", ranging from semiempirical to very precise and expensive correlated *ab initio* methods, surrounded by "layers". Rick Fine [1] has called this an onion and I have attempted to show this in Figure 1. (Clementi's

D. R. Salahub and N. Russo (eds.), Metal-Ligand Interactions: from Atoms, to Clusters, to Surfaces, 311–340.
© 1992 *Kluwer Academic Publishers. Printed in the Netherlands.*

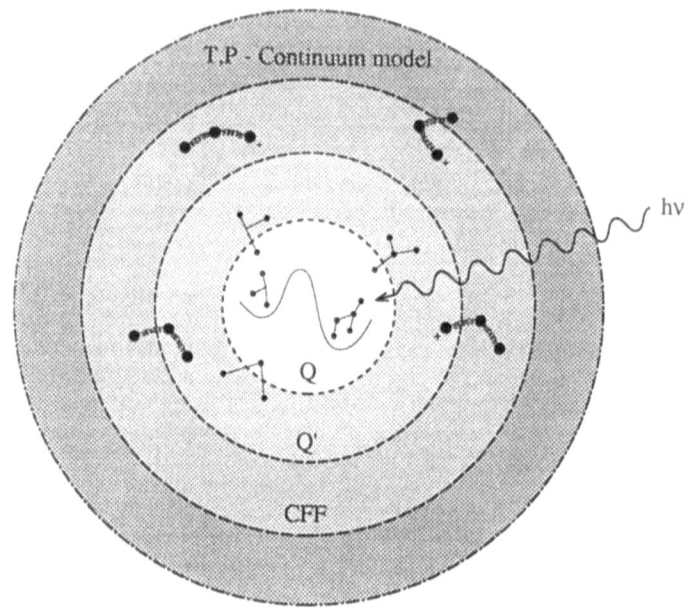

Figure 1. Schematic representation of Quantum Molecular Modeling. Q represents quantum methods, CFF classical force fields. (With a nod to Rick Fine for mentioning the "onion" and to Enrico Clementi for "global simulation".)

"global simulation" [2] is also clearly related to what follows.) I like the onion analogy particularly because of the presence in an onion of permeable membranes between the layers. In the first layer after the quantum mechanical core one might find a set of surrounding atoms treated at a lower level of theory, but still within quantum chemistry. These atoms or molecules have to interact with the quantum object and help to define its properties. Further out, but still communicating with the quantum object one can find empirical force-fields, discrete models for solvation, pseudopotentials for "embedding" atoms, and finally continuum models for a solvent. All of this can be penetrated by external electric or magnetic fields, or by light. Right now, much work is going on within each layer and some first attempts have been made towards accomplishing the very difficult task of joining methods with different conceptual, methodological, and computational frameworks across the boundaries [3].

A particularly interesting and challenging area for Quantum Molecular Modeling is represented by the present ASI on Metal-Ligand Interactions. These interactions are often studied in the "real world" of complex structural, electronic, magnetic, or optical materials, or in the "real world" of homogeneous or heterogeneous catalysis, or in the "real world" of metallo-enzymes performing biochemical functions, to name only a few "real worlds" that come to mind. There is a clear need for quantum methodologies that have the potential to treat metal-ligand interactions not only for the case of an isolated molecule but also are able to plug into the environmental framework that will be provided by the Quantum Molecular Modeling of the near future. There is also a clear advantage if such methods can span the levels of aggregation presented in the subtitle to the ASI "from atoms, to clusters, to surfaces". The rest of this chapter will present the essence of a theory that has already taken us a large step towards this goal, the so-called Density Functional Theory (DFT). A sketch of some of the methodology of DFT will be presented, followed by some examples of "validation" for transition-metal ligand interactions and then some examples involving clusters that I hope will illustrate the present power of this methodology and give an inkling of future prospects. Further applications of DFT may be found in the chapters of Russo and Ziegler.

2. Foundations of Density Functional Theory

The most demanding quantum chemical problems, in general, and the problem of transition-metal bonding and reactivity, in particular, require a treatment of electronic correlation at a high level. This can be achieved in several ways. Most straightforwardly, for small systems, a wave function can be constructed as a superposition of Slater determinants [4]

$$\Psi = \sum_i c_i \Phi_i$$

(1)

This method of Configuration Interaction and its variants (see, e.g., [4,5] and the chapter by Zerner in this book) can lead to the exact wave function in the limit of an infinite expansion. Of course, this is not practical for large systems so one often has to restrict the questions asked to those that require smaller models, or lesser accuracy. The computational effort involved with these methods increases rapidly with the size of the system (scaling as N^5 or usually higher, where N is the number of electrons). In practice this means that only the smallest transition-metal systems (diatomics, complexes involving one or, rarely, two "real" transition-metal centers) can, in general, be treated at a meaningful level of accuracy *via* CI. The semi-empirical methods (see Zerner's chapter) may also be viewed as including correlation. Here "Nature's correlation" is introduced by replacing some of the elements of an *ab initio* approach which are difficult or too time consuming to evaluate by parameters adjusted to reproduce experimental data.

Density Functional Theory(DFT) also deals with correlation (and exchange). The methods of DFT are also "*ab initio* " in that no semi-empirical parameters are introduced. They do, of course, involve approximations, some of which allow the rapid treatment of larger systems on which their practical success is based(methods can be formulated that scale as N^3 or even better). But before getting to the more practical details, I would like to spend some time with the basic structure of the theory. Although the final working equations will be reminiscent of those involved in Hartree-Fock (or, really, Hartree) theory, there are some fundamental differences which I think are worth understanding, particularly if even more powerful approximations are eventually to be formulated.

Some of the constituent ideas of modern DFT calcualtions can be traced back to pre-quantum mechanical days. We will see that the model of a homogeneous (constant density) electron gas [6], which goes back to the Drude model for the conduction electrons of metals, plays a central role. Sommerfeld incorporated Fermi-Dirac statistics into this model and the kinetic energy of this gas was folded into the Thomas-Fermi model for the electronic structure of atoms. The first *bona fide* density functional of which I am aware is the Thomas-Fermi [7, 8] expression for the kinetic energy:

$$\langle T_{TF} \rangle = \int \rho \, t_{TF} \, d\bar{r} \; \propto \; \int \rho^{5/3} \, d\bar{r} \tag{2}$$

Equation (2) also represents a so-called "local" approximation in that the kinetic energy density, $\rho^{2/3}$, which is multiplied by ρ and integrated over all of space to yield the total kinetic energy, depends only on the density at a given point. An example of a "nonlocal" (gradient-corrected) functional is provided by the Von Weiszacker [11] correction:

$$t_W = \frac{1}{8} \frac{\bar{\nabla}\rho \bullet \bar{\nabla}\rho}{\rho} \tag{3}$$

This depends on spatial derivatives of the density as well as on the density itself (We will meet other nonlocal functionals later on in the discussion of exchange and correlation).

The Thomas-Fermi-Dirac [10] model for atoms also includes a local density approximation for exchange, which is based on the homogeneous electron gas. Dimensional analysis shows that, if a simple power law is assumed for the exchange energy density, then this must scale as $\rho^{1/3}$ so the exchange energy is given by:

$$\varepsilon_x = -\frac{3}{8}\left(\frac{3}{\pi}\right)^{1/3} \rho^{1/3} \qquad \text{(hartree)} \tag{4}$$

In a classic paper, Slater [11] showed that the Hartree-Fock exchange terms could be averaged in such a way that simplified working equations involving the $\rho^{1/3}$ potential resulted. Introduction of a parameter α, often adjusted to reproduce atomic Hartree-Fock energies [12], yields the so-called $X\alpha$ equations:

$$(-\frac{1}{2}\nabla^2 + V_N + V_{COUL} + C_x \, \alpha \rho^{1/3}) \phi_i^{X\alpha} = \varepsilon_i^{X\alpha} \, \phi_i^{X\alpha} \tag{5}$$

where V_N is the nuclear potential, V_{COUL} is the classical coulomb potential due to the electrons, including the self-interaction, and the last term in parentheses is the local $X\alpha$ approximation for exchange.

These rather simple self-consistent field equations were extensively used by Slater's school, laying the ground work for much of modern band theory. Some inroads into chemistry were made using this approximation, either using a "Muffin-Tin" potential and multiple scattering equations [13] or using various LCAO approximations [14-17]. In the sixties and seventies, the band structure work and molecular applications appeared to be doing well and it appeared that they could continue to thrive independently of the development of the structure of an underlying formal theory. However, such a development took place in 1964, with the publication of two remarkable papers by Hohenberg and Kohn [18] and by Kohn and Sham [19]. The importance of these papers has become more and more apparent over the years. They have given rise to a school of thought that has allowed a sequence of approximate methodologies of increasing accuracy and utility to be formulated. It is doubtful that such techniques could have developed in the absence of the solid conceptual basis provided by the HK and KS theorems and their derivatives.

The 1964 paper proved two theorems. First, it was shown that the (nondegenerate) ground state of an N-electron system is fully determined by the electron density. This is a powerful observation; there is enough information in the electron density, a measureable quantity that depends on only the three spatial coordinates, to entirely fix the wave function, a much more complex function of the 3N electronic coordinates. The second theorem provided a variational principle. It showed the existence of an energy functional of the density which takes its minimum value for the exact density. The HK theorems are of the nature of existence proofs and in themselves are not of practical utility. The 1965 KS paper, however, took the HK concepts and through some truly ingenious arguments distilled them into a set of, still exact, self-consistent equations for which approximations could be formulated. Most of the practical DF calculations being performed today are within the KS framework. The proofs of the basic theorems are simple(although not without subtlety) and we will run through them now. A good reference for further details and background is the book by Parr and Yang [20].

2.1. THE HOHENBERG-KOHN THEOREMS

An N-electron system is represented by a Hamiltonian:

$$\hat{H} = \hat{T} + \hat{V} + \hat{U}. \tag{6}$$

where $\hat{T}$ is the kinetic energy, $\hat{V}$ is the so-called external potential (external to the electron system), which in the case that interests us is just the coulomb field of the nuclei, and $\hat{U}$ is the electron-electron repulsion. The usual, wave function, approach is to insert this Hamiltonian into Schrödinger's equation:

$$\hat{H}\Psi = E\Psi \tag{7}$$

If this can be solved, then all of the properties can be calculated from the wave function, including the electron density:

$$\rho(\bar{r}) = \int \Psi^*(1,...,N)\Psi(1,...,N)dv_2...dv_N \tag{8}$$

One might say that ρ is a functional of the wave function, or of the Hamiltonian. HK proved that the converse is also true. The external potential, and hence the Hamiltonian, is completely fixed once the density is known. Of course, once the Hamiltonian is fixed one can, in principle, calculate the wave function with Schrödinger's equation and thereafter derive all of the properties of the system by "normal" quantum mechanics.

The proof is by *reductio ad absurdum*. We assume that another potential, V', exists with an associated wave function Ψ', that gives rise to the same density ρ as the system of interest. Consider the normal variational principle and use Ψ as a trial function for the Hamiltonian, H', that contains V':

$$E' = \left\langle \Psi' \middle| \hat{H}' \middle| \Psi' \right\rangle$$
$$\leq \left\langle \Psi \middle| \hat{H}' \middle| \Psi \right\rangle = \left\langle \Psi \middle| \hat{H} + \hat{V}' - \hat{V} \middle| \Psi \right\rangle \tag{9}$$

$$E' \leq E + \int \left[\hat{V}' - \hat{V} \right] \rho(\bar{r}) d\bar{r} \tag{10}$$

The important thing to note is that the only difference between the two Hamiltonians, since we are dealing with an N-electron system, arises from the assumed difference in the external potential and that this difference gives rise to a term in the energy calculated by a simple integral over the density. Now, the whole argument can be repeated, interchanging the primed and unprimed variables

$$E \leq E' + \int \left[\hat{V} - \hat{V}' \right] \rho(\bar{r}) d\bar{r} \tag{11}$$

Adding equations (10) and (11) we obtain the contradiction:

$$E + E' \leq E + E' \tag{12}$$

Hence our original assumption is false and we conclude that **V is a unique functional of ρ.** (A "poor man's proof" of the idea that there is enough information in the density to fix the Hamiltonian goes as follows (it has been attributed to E.B. Wilson, but I do not know of a formal reference to it). If one has a density map (determined from X-ray measurements, or otherwise) then the positions of the nuclei may be determined by the presence of cusps in the density. The quantum mechanical cusp condition [21] then allows the nuclear charges to be determined. Finally, one can integrate the density to find out the number of electrons present and this fixes all of the terms in the Hamiltonian. Of course, the HK framework is more powerful than this, as we now proceed to show.)

The proof of the HK variational principle goes as follows. Since Ψ is a functional of ρ, so also are the kinetic and electron-electron interaction energies so we can define a universal functional, $F[\rho]$, which contains everything except the external potential term:

$$F[\rho(\bar{r})] = \left\langle \Psi \middle| \hat{T} + \hat{U} \middle| \Psi \right\rangle \tag{13}$$

For a given external potential:

$$E_V[\rho] = \int \hat{V}(\bar{r}) \rho(\bar{r}) d\bar{r} + F[\rho] \tag{14}$$

$E_V[\rho]$ assumes its minimum value for the correct ρ, if the admissible functions are restricted by normalisation:

$$N[\rho] = \int \rho(\bar{r}) d\bar{r} = N \tag{15}$$

Now the usual energy expression, involving the wave function:

$$\varepsilon_V[\Psi'] = \left\langle \Psi' \middle| \hat{V} \middle| \Psi' \right\rangle + \left\langle \Psi' \middle| \hat{T} + \hat{U} \middle| \Psi' \right\rangle \tag{16}$$

has its minimum for $\Psi' = \Psi$ and is stable to arbitrary small variations of the wave function for which the particle number is kept fixed. Now if Ψ' is the wave function associated with an external potential V', we can write the (usual) energy in terms of the functional $F[\rho']$:

$$\varepsilon_V[\Psi'] = \int V(\bar{r})\rho(\bar{r})d\bar{r} + F[\rho'] \tag{17}$$

and this must lie above the true ground state:

$$\varepsilon_V[\Psi] = \int V(\bar{r})\rho(\bar{r})d\bar{r} + F[\rho] \tag{18}$$

But the right hand sides of equations (17) and (18) are just the expressions for $E_V[\rho']$ and $E_V[\rho]$, so, by the usual variational theorem, we can write:

$$\varepsilon_V[\rho'] > E_V[\rho] \tag{19}$$

The above follows the original treatment of HK which, in fact, gave rise to much discussion over two "problems". First, the restriction to nondegenerate states (required by the use of the usual variational principle) resulted in a less-than-general theory. Second, there is an implicit assumption in the HK variational theorem that all of the trial densities do, in fact, correspond to physically realisable external potentials. If this is not so (and counterexamples have been constructed) then such a "pathological" density could yield a value for the energy below the true value. Fortunately, Levy [22] has found a formulation that addresses these issues and confirms the validity of the theorems, and extends them to nondegenerate states. The central idea is to search through wave functions that correspond to the density, rather than the density itself. Assume that ρ' is the density corresponding to V' and that ρ_0 is the exact ground state density. Ψ_0 is the exact ground state wave function. Now there are many well behaved functions that can yield ρ_0. The wave function Ψ_0 is special in that it yields ρ_0 **and** at the same time minimises the "internal" terms of the energy:

$$\langle\hat{T}+\hat{U}\rangle \tag{20}$$

($V_{ext} = V_N$ is already fixed once the density is fixed.)

If several functions minimise $<\hat{T} + \hat{U}>$ then one has a degenerate state so, with this approach, degeneracy does not present any special difficulty. Now, perform a "constrained search". Simply define the universal variational functional:

$$Q[\rho] = Min\langle\Psi_\rho|\hat{T}+\hat{U}|\Psi_\rho\rangle = \langle\Psi_\rho^{min}|\hat{T}+\hat{U}|\Psi_\rho^{min}\rangle \tag{21}$$

$Q[\rho]$ searches each and every antisymmetric function , Ψ_ρ, which yields the trial ρ, and delivers the minimum in $<\hat{T} + \hat{U}>$. The energy is then found by adding the external potential term:

$$\int V(\bar{r})(\bar{r})d\bar{r} + Q[\rho] = \langle\Psi_\rho^{min}|\Sigma V(\bar{r}_i) + \hat{T}+\hat{U}|\Psi_\rho^{min}\rangle \tag{22}$$

318

The equality holds when $\rho = \rho_0$.

These are powerful and elegant theorems but it should not be forgotten that all of the difficulties of the N-body problem are still present. The HK theorems serve to reveal a truth of Nature, that there is a route from the density to the energy that doesn't necessarily require the wave function. But the functional is unknown. For some, fortunately, this has represented a call to arms behind the "density banner", to seek density-based approximations that can be applied to real systems. The hope, which has already been realised to a remarkable extent, is that the approximations will be significantly different from the usual Hartree-Fock + Configuration Interaction approach so as to provide both new computational tools and a new conceptual framework. A big step in this direction was taken by Kohn and Sham in 1965.

2.2. THE KOHN-SHAM FORMULATION OF DENSITY FUNCTIONAL THEORY

Kohn and Sham [19] found a route to a set of equations that have a very simple, orbital, structure yet are formally exact. Their work showed that pre-DFT work on the Slater or $X\alpha$ Hamiltonian should, in fact, be properly viewed as belonging to the DFT family. This not only removed some confusion and unfruitful comparisons from the area but, more importantly, has provided a framework for testing proposed new functionals. The final equations are not only simpler to solve than the Hartree-Fock equations, they can also incorporate an approximate treatment of correlation which is often of similar or better accuracy than far more expensive correlated methods. This is not to say that KS-DFT is, at present, the solution to all problems. For example, much work remains to see whether fundamental questions of applicability to problems involving multiplet structure, or other multideterminantal situations can be solved. However, even ignoring such areas, the utility of the KS approach and its applicability to a wide range of systems makes it probable that it will play an increasingly important role in quantum chemistry. So it is, I believe, worthwhile to understand this approach at the deepest possible level.

The first step in the KS development is to separate the classical coulomb energy from the rest of the HK functional, F:

$$F[\rho(\bar{r})] = \frac{1}{2} \int \frac{\rho(\bar{r})\rho(\bar{r}')}{|\bar{r}-\bar{r}'|} d\bar{r}' d\bar{r} + G[\rho(\bar{r})] \tag{23}$$

This defines a new universal functional, G, which contains the kinetic energy and all terms due to exchange and correlation (it is perhaps worth mentioning at this stage, that the separation of exhange and correlation energy, or even of some parts of the kinetic energy is not unambiguously defined in DFT). Then (one of the clever constructs I like most, in my scientific readings), KS separated in G, a kinetic energy term, but not the kinetic energy of the real system. They, rather, defined T_s as the kinetic energy of a system of fictitious noninteracting particles that is constrained to have the same density as the real system. The reason for doing this will soon become clear. So:

$$G[\rho(\bar{r})] \equiv T_s[\rho(\bar{r})] + E_{xc}[\rho(\bar{r})] \tag{24}$$

E_{xc} contains all the exchange and correlation terms and the "residual" kinetic energy. The total energy is:

$$E = \int V(\bar{r})\rho(\bar{r})d\bar{r} + F[\rho] = \int V(\bar{r})\rho(\bar{r})d\bar{r}$$

$$+ \frac{1}{2}\int \frac{\rho(\bar{r})\,\rho(\bar{r}')}{|\bar{r}-\bar{r}'|}\,d\bar{r}'d\bar{r} + T_S[\rho] + E_{xc}[\rho] \qquad (25)$$

Now, if we use the second HK theorem and calculate the energy variation for an arbitrary variation of the density which maintains normalisation, then the following Euler equations result:

$$\frac{\delta T_S[\delta(\bar{r})]}{\delta\rho(\bar{r})} = V(\bar{r}) + \int \frac{\rho(\bar{r})d\bar{r}'}{|\bar{r}-\bar{r}'|} + \frac{\delta E_{xc}}{\delta\rho} = 0 \qquad (26)$$

However, if one instead applies the HK theorem to a system of N noninteracting electrons, but moving in a fictitious "external" potential given by the last three terms of eqn.(26).

$$V_{\text{"ext"}} = V(\bar{r}) + \frac{\delta E_{xc}}{\delta\rho} + \int \frac{\rho(\bar{r}')}{|\bar{r}-\bar{r}'|}\,d\bar{r}' \qquad (27)$$

Then **exactly the same Euler equation results.** However, we know how to solve the problem of independent, noninteracting particles. Schrödinger's equation is separable and we can simply write

$$\left(-\frac{1}{2}\nabla^2 + V(\bar{r}) + \int \frac{\rho(\bar{r}')}{|\bar{r}-\bar{r}'|}\,d\bar{r}' + \frac{\delta E_{xc}}{\delta\rho}\right)\phi_i = \varepsilon_i\phi_i \qquad (28)$$

The density is given simply by:

$$\rho(\bar{r}) = \sum_i n_i\,\phi_i^*(\bar{r})\phi_i(\bar{r}) \qquad (29)$$

where n_i are occupation numbers. Occupation numbers for levels degenerate at the Fermi energy may be fractional.

These are the Kohn-Sham equations. They are still exact; all of the difficulties are contained in the term E_{xc}. However, they have a simple structure which is amenable to approximations. But before discussing the most common approximations, let us briefly compare the basic structures of Kohn-Sham-DFT with the conventional wave function, CI, approach. These are shown schematically in Figure 2. In the CI world, the "operator part", the Hamiltonian, is easy to write down. The difficult part, the part where many quantum chemists spend their lives working, is to determine the wave function, a multiconfigurational expansion. In KS-DFT, on the contrary, the "wave function part" is relatively easy. One has orbitals and some self-consistent equations to obtain them and hence the density. The KS approach is inherently "single-configurational"; one has only the KS orbitals and their occupation numbers, in order to synthesize the density. The difficulty is in the operator and progress in DFT will depend, more than on anything else, on our ability to find new, more accurate exchange-correlation functionals. My own feeling is that quantum chemists, with their deep knowledge of correlation, may ultimately make great contributions here. Perhaps one example could serve to underline the fundamental difference between these two theories. Consider the beryllium atom, and the question of whether this atom has any substantial "p character". The nominal configuration of Be is $1s^2 2s^2$ and the ground state is ^1S, a spherically symmetric state. Now, most quantum chemists would quickly agree that correlating the 2s electrons is important and the way to achieve this is to mix in determinants in which the two 2s electrons are excited into one or the other of the empty p orbitals. This in effect makes the probablility of finding an electron on one side of the nucleus less if there is already an electron of opposite spin on that side. Now, in the KS picture, the (spherical) density can be completely described by using a basis set of s type orbitals only. All of the correlation effects achieved by the configurations involving p orbitals have to be effectuated by the exchange-correlation operator, operating on spherical orbitals. This can be done, in principle, by the exact functional; the extent to which it is achieved by approximate functionals will determine their reliability. Ultimately, the survival of DFT may well depend on our ability to better understand this difference and to use the knowledge from wave function quantum chemistry to guide the development of better functionals.

3. The Local (Spin) Density Approximation and Nonlocal Corrections

The workhorse of applied DFT has been the Local Density Approximation, in which the exchange-correlation energy density is taken from a fully correlated (Quantum Monte Carlo) treatment of the homogeneous electron gas, in just the same way that the kinetic

KS - DFT
$$\left(-\frac{1}{2}\nabla^2 + V(\vec{r}) + \int \frac{\rho(\vec{r}')}{|\vec{r}-\vec{r}'|}d\vec{r}' + V_{XC}(\rho(\vec{r}))\right)\phi_i(\vec{r}) = \varepsilon_i \phi_i(\vec{r})$$

$$\rho = \sum_i n_i \phi_i^2$$

Single configuration

$$V_{XC} \rightarrow \{\phi_i\}$$
$$\nwarrow_\rho \nearrow$$

$\{n_i\}$

OPERATOR

CI
$$E = <\Psi|\hat{H}|\Psi>$$

WAVE FUNCTION
$$\Psi = \sum_j C_j \Phi_j$$

Multiconfigurational

Figure 2. Schematic comparison of Kohn-Sham Density Functional Theory and wave function (Configuration Interaction) theory.

energy density is taken from an electron gas model in the Thomas-Fermi method (of course, in KS theory, the kinetic energy is treated more accurately, through the noninteracting electron gas construct). Nonlocal gradient corrections can also be formulated for the exchange and correlation terms and we will see some of these a little later. But first some unifying concepts can be introduced, if we take time for a brief discussion of density matrices and pair correlation functions (this will be only a sketch, the interested reader may find details in the books of Parr and Yang [20], Davidson [21], McWeeny and Sutcliffe [23] in recent papers of, for example, Becke [24, 25] and Ziegler [26, 27]. Slater's 1951 paper [11] also provides insight.).

The most basic observation to make at the outset is that a many electron Hamiltonian contains only one- and two-particle operators. The wave function depends on the coordinates of all N electrons and is a very complicated function. The energy is given by an integral involving the Hamiltonian and the wave function and $<\Psi \mid \hat{H} \mid \Psi>$ can be simplified by integrating over the "extra" coordinates (and summing over spins). The basic probability distribution is:

$$\Psi(\bar{x}_1, \bar{x}_2 ... \bar{x}_N) \, \Psi^*(\bar{x}_1, \bar{x}_2 ... \bar{x}_N) \tag{30}$$

where x represents all space ($\bar{r}$) and spin(s) coordinates. The first order spinless reduced density matrix is given by:

$$\rho_1(\bar{r}'; \bar{r}) = N \int ... \int \Psi(\bar{r}_1', s_1, \bar{x}_2 ... \bar{x}_N) \Psi^*(r_1, s_1, \bar{x}_2 ... \bar{x}_N) \, ds_1 \, d\bar{x}_2 ... \, d\bar{x}_N \tag{31}$$

where the integration is carried out over all electrons save one and all spins are summed. A matrix form(indices r and r') is maintained to allow the insertion of a Hamiltonian later; then r' will be set equal to $\bar{r}$. The second order spinless reduced density matrix is defined analogously:

$$\rho_2(\bar{r}_1', \bar{r}_2'; \bar{r}_1, \bar{r}_2) = \frac{N(N-1)}{2} \int ... \int \Psi(\bar{r}_1', s_1, \bar{r}_2', s_2, \bar{x}_3 ... \bar{x}_N)$$
$$\Psi^*(\bar{r}_1, s_1, \bar{r}_2, s_2, \bar{x}_3 ... \bar{x}_N) \, ds_1 ds_2 \, d\bar{x}_3 ... d\bar{x}_N \tag{32}$$

where now two electrons are left out of the spacial integration. The diagonal elements of these reduced density matrices are well known. The diagonal elements of the one-matrix

$$\rho_1(\bar{r}_1; \bar{r}_1) = \rho(\bar{r}_1) = N \int ... \int |\Psi|^2 ds_1, d\bar{x}_2 ... d\bar{x}_N \tag{33}$$

is just the density, whereas the 2-matrix yields:

$$\rho_2(\bar{r}_1, \bar{r}_2; \bar{r}_1, \bar{r}_2) = \rho(\bar{r}_1; \bar{r}_2) = \frac{N(N-1)}{2} \int ... \int \left| \Psi^2 \right| ds_1 ds_2 \, d\bar{x}_3 ... d\bar{x}_N \tag{34}$$

a pair correlation function (normalized to the number of pairs). Now the energy can be written in terms of these density matrices:

$$E = E\left[\rho_1(\bar{r}_1';\bar{r}_1),\rho_2(\bar{r}_1;\bar{r}_2)\right]$$

$$= E\left[\rho_2(\bar{r}_1',\bar{r}_2';\bar{r}_1,\bar{r}_2)\right]$$

$$= \int\left[-\frac{1}{2}\nabla_r^2\rho_1(\bar{r}';\bar{r})\right]_{\bar{r}'=\bar{r}}$$

$$+ \int v(\bar{r})\rho(\bar{r})d\bar{r}$$

$$+ \int\int\frac{1}{r_{12}}\rho_2(\bar{r}_1;\bar{r}_2)\,d\bar{r}_1\,d\bar{r}_2 \tag{35}$$

The first term of eqn(35) gives the kinetic energy in terms of the 1-matrix, the second is just the nuclear attraction energy; we will not need these terms further. The final term is a very compact representation of all of the difficulties of quantum chemistry! Knowledge of the second order reduced density matrix would allow us to say exactly "what the electrons are really doing in a molecule [28]". Approximate treatments of electronic structure may be usefully classified and compared by viewing them as involving approximations to the last term in (35). It is useful to remove classical coulomb repulsion from the discussion. This classical energy is given by:

$$J[\rho] = \frac{1}{2}\int\int\frac{\rho(\bar{r}_1)\rho(\bar{r}_2)d\bar{r}_1 d\bar{r}_2}{r_{12}} \tag{36}$$

In this expression the "correlation function" is just the product of two probability functions, the very definition of uncorrelated probabilities. We can write the density matrix for a real, correlated, system with reference to this uncorrelated limit:

$$\rho_2(\bar{r}_1;\bar{r}_2) = \frac{1}{2}\rho(\bar{r}_1)\,\rho(\bar{r}_2)\left[1+h(r_1,r_2)\right] \tag{37}$$

where $h(r_1,r_2)$ is a pair correlation function. This function is normalized to -1.

$$\int\rho(\bar{r}_2)h(\bar{r}_1,\bar{r}_2)\,d\bar{r}_2 = -1 \tag{38}$$

Readers of Slater's 1951 paper or his books will recognize that the pair correlation function is related to the exchange-correlation hole of those works by:

$$\rho_{xc}(\bar{r}_1;\bar{r}_2) = \rho(\bar{r}_2)h(\bar{r}_1,\bar{r}_2) \tag{39}$$

So, in the end we can write the electron-electron repulsion as:

$$V_{ee} = \int\int \frac{1}{r_{12}} \rho_2(\vec{r}_1;\vec{r}_2) d\vec{r}_1 d\vec{r}_2$$

$$= J[\rho] \frac{1}{2} \int\int \frac{1}{r_{12}} \rho(\vec{r}_1)\rho_{xc}(\vec{r}_1;\vec{r}_2) d\vec{r}_1 d\vec{r}_2 \qquad (40)$$

and classify various approximate treatments according to their exchange-correlation holes. The density and exchange-correlation holes can further be decomposed into spin components:

$$\rho_1(\vec{r}_1;\vec{r}_1) = \rho_1^{\alpha\alpha}(\vec{r}_1';\vec{r}_1) + \rho_1^{\beta\beta}(\vec{r}_1';\vec{r}_1)$$
$$\left(\rho = \rho^\alpha + \rho^\beta\right) \qquad (41)$$

so that spin polarized calculations may be performed. The pair correlation functions may also be separated into like-spin ("exchange") and opposite-spin ("correlation") parts:

$$\rho_2(\vec{r}_1',\vec{r}_2';\vec{r}_1,\vec{r}_2) = \rho_2^{\alpha\alpha,\alpha\alpha}(\vec{r}_1',\vec{r}_2';\vec{r}_1,\vec{r}_2) + \rho_2^{\beta\beta,\beta\beta} + \rho_2^{\alpha\beta,\alpha\beta} + \rho_2^{\beta\alpha,\alpha\beta} \qquad (42)$$

The exchange-correlation-hole framework has led to the adoption by some [29-31] of a philosophy of constructing new functionals that imposes constraints derived from properties of the exact exchange and correlation holes. The two simplest constraints on the exchange hole are those of normalisation and the property that the exact exchange-hole exactly cancels the spin density at any given point. The newer nonlocal functionals also satisfy constraints on the asymptotic properties of the exact holes [29, 31]. Such analysis not only provides guidelines for making functionals but also helps to rationalize the success of the existing ones. For example, it is to many surprising that the Local Density Approximation works as well as it does, considering that it is derived from a uniform density gas whereas the density gradients in real atomic or molecular systems are anything but small. In fact, if one looks at the LDA exchange holes for some real systems and compares them with their high accuracy counterpart from *ab initio* calculations, there is every reason for pessimism (see, e.g. ref [32]). However, observable properties, like the energy, involve integration of the hole function along with the density, and they do not depend on all of the details of the hole function, but only on its spherical average. Comparing LDA spherically averaged holes to spherically averaged *ab initio* holes is more reassuring (see ref. [32]). In fact the LDA has provided many valuable results on molecular structure, vibrations, ionization potentials and the like (see refs [26, 33-35] and the chapters by Russo and Ziegler in this book).

We will now briefly review some of the more common functionals that may be met in the literature. Those that follow are of increasing complexity as the level of approximation is improved, but all of them may be expressed in relatively straightforward "programmable" expressions. We write some of them down here to emphasize this fact.

The most commonly used LDA functional involves the Xα (α =2/3) approximation for exchange. The energy density per particle is given by

$$\varepsilon_{X\alpha}\left(\alpha=\tfrac{2}{3}\right)=-\frac{3}{8}\left(\frac{3}{\pi}\right)^{1/3}\rho^{1/3}$$

(43)

which, upon variation in the integral

$$E_X = \int \rho(\bar{r})\varepsilon_X(\bar{r})d\bar{r}$$

(44)

yields the Xα potential:

$$v_{X\alpha}^{\uparrow}\left(\alpha=\tfrac{2}{3}\right)=-2\left(\frac{3}{4\pi}\right)^{\tfrac{1}{3}}\rho_{\uparrow}^{\tfrac{1}{3}}$$

(45)

The next level of sophistication involves the incorporation of correlation, still using the homogeneous electron gas model. Essentially exact results are available from the fit of either Perdew and Zunger [36] or Vosko, Wilk and Nusair [37] to the Ceperley-Alder [38] Quantum Monte Carlo calculations for this model system. The VWN formula for the correlation energy density is given here, in order to make its relative simplicity evident (relative to an "infinite" CI expansion...)

$$\varepsilon_C(r_s)=A\left\{\ln\frac{x^2}{X(x)}+\frac{2b}{Q}\tan^{-1}\frac{Q}{2x+b}\right.$$

$$-\frac{bx_0}{X(x_0)}\left[\ln\frac{(x-x_0)^2}{X(x)}\right.$$

$$+\left.\left.\frac{2(b+2x_0)}{Q}\tan^{-1}\frac{Q}{2x+b}\right]\right\}$$

(46)

In equation (46) $x=r_s^{1/2}, r_s=(3/4\pi\rho)^{1/3}, X(x)=x^2+bx+c, Q=(4c-b^2)^{1/2}$ x_0 is a real root of $(1 + b_1x + b_2x^2 + b_3x^3)$, A is a known constant from the Random Phase Approximation. The parameters x_0, b and c are sufficient to fit the Ceperley-Alder results for all interesting values of $x(r_s, \rho)$ for both the non-spin-polarized and the fully spin-polarized cases. For intermediate spin polarization RPA scaling is used and this also results in a reasonably simple formula to program. The corresponding potential is not fundamentally more difficult.

LDA predictions of structure and vibrational energies are by now well established for a wide variety of systems throughout the periodic table. Although rules of thumb must

TABLE 1. Equilibrium geometries and vibrational frequencies of NiC_2H_4 and PdC_2H_4 obtained by the LCGTO-MCP-LSD method (all distances in Å, angles in degrees, and frequencies in cm^{-1}). (See ref. ref. 39 for details).

		C$_2$H$_4$		NiC$_2$H$_4$		PdC$_2$H$_4$	
		Calc.	Expt.	Calc.	Expt.	Calc.	Expt.
	r_{X-C}(X=Ni,Pd)	(-)	(-)	1.862 (1.972)^c	(-)	2.053 (2.26)	(-)
	r_{C-C}	1.333	1.339	1.425 (1.454)^c	(-)	1.414 (1.36)	(-)
	r_{C-H}	1.107	1.085	1.111	(-)	1.109	(-)
	αHCH	116.3	117.8	114.4	(-)	115.1	(-)
	tilt^a	0.0	0.0	21.8(21)	(-)	20.8(15)	(-)
A_1:	ν_1(C-H stretch)	3053	3026 (3153)^b	2976	2961	3010	2952
	ν_2(C-C stretch)	1654	1623 (1655)	1445	1497	1473	1502
	ν_3(CH$_2$ scissor)	1318	1342 (1370)	1176	1159	1204	1223
	ν_4(CH$_2$ wag)	921	949 (969)	894	901	902	913
	ν_5(X-C stretch)	(-)	(-) (-)	565	(-)	481	(-)
A_2:	ν_6(C-H stretch)	3144	3106 (3234)	3058	(-)	3109	(-)
	ν_7(CH$_2$ rock)	803	826 (843)	760	(-)	792	(-)
	ν_8(CH$_2$ twist)	(-)	(-) (-)	570	(-)	576	(-)
B_1:	ν_9(C-H stretch)	3040	2989 (3147)	2968	(-)	3006	(-)
	ν_{10}(CH$_2$ scissor)	1393	1444 (1473)	1364	(-)	1374	(-)
	ν_{11}(CH$_2$ wag)	926	943 (959)	920	(-)	933	(-)
	ν_{12}(X-C stretch)	(-)	(-) (-)	500	(-)	433	(-)
B_2:	ν_{13}(C-H stretch)	3120	3103 (3232)	3042	(-)	3088	(-)
	ν_{14}(CH$_2$ rock)	1174	1236 (1245)	1140	(-)	1150	(-)
	ν_{15}(CH$_2$ twist)	1016	1023 (1044)	810	(-)	869	(-)

a: Tilt is the angle between the bisector of the HCH bond angle and the C-C bond.
b: Experimental harmonic frequencies are in parentheses.

always be applied with caution, I believe it is fair to summarize present expectations by putting error limits of perhaps 0.02-0.03Å for bond lengths, a degree or two for angles, and 5-10% for vibrational frequencies. For organic systems a rough equivalence appears to be emerging between LDA geometries and frequencies and those calculated at the correlated, MP2, level [26, 33, 34]. For transition-metal systems, Russo's and Ziegler's chapters of this book give several examples to show the reliability of LDA calculations of these properties for even very complex molecules and models. Here, I would like to give just one series of examples (Table 1) to illustrate that structure and vibrations are handled at a very useful level for molecules like ethylene and that introducing a transition metal atom, Ni or Pd, presents no particular problem to these techniques.

The LDA is not, however, omnipotent. Its most significant shortcoming in the present context is in the area of binding energies, a crucial area if one wants to deal with chemical reactivity! Here the LDA tends to overestimate dissociation energies, sometimes very seriously so. This tendency is well documented [26,34, 40-42,Chapters of Russo and Ziegler] so I will only give (Table 2) a few examples of the overbinding for transition metal molecules.

TABLE 2.   Binding energies for transition metal systems (eV) calculated with the local density approximation (LDA) and with non local density gradient corrections (NLDA).

Molecule	LDA	NLDA	exp.	ref
PdH	3.00	2.59	2.43	40
PdC	4.91	3.70	<4.46	40
RhH	3.68	2.93	2.56	40
RhC	7.88	6.27	5.97	40
PdCO	3.08	2.10		40
Ni_5H	3.22	2.23		41
Ni_5O	7.10	5.85		42
Ni_2	3.64	2.88	2.36	41

Fortunately, recent work with nonlocal, gradient corrected functionals appears to go a long way to rectifying the situation. The history of this type of nonlocal functionals has had its ups and downs. Early work by Herman et al. [43] showed on dimensional arguments that in exchange-only DFT the exchange energy has a gradient expansion of the form:

$$E_{X\alpha\beta} = E_{X\alpha} - \beta \int \frac{(\vec{\nabla}\rho)^2}{\rho^{4/3}} \, d\vec{r}$$

(47)

or, in modern notation:

$$E_X^{LDA} = E_X^{LDA} - b \int \frac{(\vec{\nabla}\rho)^2}{\rho^{4/3}} \, d\vec{r}$$

(48)

Although it introduced a second parameter, this so-called $X\alpha\beta$ method was found to give no better, and often worse, results than the $X\alpha$ method which it was intended to improve. It was realized that the gradient-corrected potential was not well behaved in the large gradient limit and the overall mood in the seventies was pessimistic. A key contribution to the practical use of density gradient corrections was made by Langreth and Mehl [46] in the early eighties. They considered the Gradient Expansion Approximation (GEA):

$$E_{XC}[\rho(\vec{r})] = \int A_{XC}[\rho(\vec{r})]d\vec{r} + \int B_{XC}\left[\rho(\vec{r}), (\vec{\nabla}\rho)^2\right] + ...$$

(49)

and performed a full wave-vector analysis of the B_{XC} function. The GEA is only applicable when

$$\frac{|\vec{\nabla}\rho|}{2k_F\rho} \leq 1 \tag{50}$$

and

$$\frac{\nabla^2\rho}{2k_F|\vec{\nabla}\rho|} < 1 \tag{51}$$

where k_F is the local Fermi wave vector

$$k_F = \left(3\pi^2\,\rho\right)^{\frac{1}{3}} \tag{52}$$

The correlation energy was expressed as a sum of dynamic density fluctuations of various wave vector $\mathbf{k}$. They found a strong exponential peak around $k=0$ in B_{XC}. Because the GEA is valid only if inequalities 50 and 51 are satisfied, contributions coming from the small k region are spurious . Langreth and Mehl set to zero the contributions coming from the region where

$$k \leq \frac{f|\vec{\nabla}\rho|}{\rho}, f \approx 0.15 \tag{53}$$

In a sense, the LM functional applies the gradient correction when the gradient is not "pathologically" large and reverts to the LDA in those regions of space where applying the correction would lead to trouble.

Although the LM paper is a landmark, their functional is not in wide use today because further work has led to improvements. For example, the Generalised Gradient Approximation(GGA) is based on a real space analysis of the exchange and correlation holes. The GGA writes the exchange energy in the form:

$$E_X = A_X \int \rho^{\frac{4}{3}} F^{GGA}(s) \, d\vec{r} \tag{54}$$

where s represents the quantity in inequality 50 and the function F^{GGA} embodies the properties of the Fermi hole. Perdew [31] has proposed the explicit form

$$F^{GGA}(s) = \left(1+0.0864s^2/m + bs^4 + cs^6\right)^m$$
$$m = \frac{1}{15} \quad b = 14 \quad c = 0.2 \tag{55}$$

Becke [45] has analysed the asymptotic properties of the Fermi hole and proposed the functional

$$E_X = E_X^{LDA} - \beta \sum_\sigma \int \rho_\sigma^{4/3} \frac{s_\sigma^2}{\left(1 + 6\beta s_\sigma \sinh^{-1} s_\sigma\right)} d\vec{r}$$

(56)

Both of these exchange functionals have given encouraging results and are under active study. They are often complemented with the Perdew [46] functional for correlation:

$$E_C\left[\rho\uparrow, \rho\downarrow\right] = \int \rho \varepsilon_C^{LDA}\left(\rho\uparrow, \rho\downarrow\right) d\vec{r}$$

$$+ \int d^{-1} e^{-\Phi} c(\rho) \frac{\left|\vec{\nabla}\rho\right|^2}{\rho^{4/3}} d\vec{r}$$

ou $\Phi = 1.745 - \tilde{f}\left[c(\infty)/c(\rho)\right] \left|\vec{\nabla}\rho\right| / \rho^{7/6}$

$\tilde{f} = 0.11$

(57)

$$c(\rho) = (0.001667) + \frac{\left(0.002568 + \alpha r_s^2 + \beta r_s^3\right)}{\left(1 + \gamma r_s + \delta r_s^2 + 10^4 \beta r_s^3\right)}$$

$\alpha = 0.023\ 266 \quad \beta = 7.389 \times 10^{-6}$
$\gamma = 8.723 \qquad \delta = 0.472$

d defines the spin polarization and is given by

$$d = 2^{1/3} \left[\left[\frac{1+\xi}{2}\right]^{5/3} + \left[\frac{1-\xi}{2}\right]^{5/3}\right]^{1/2}$$

with $\xi = \left(\rho_\uparrow - \rho_\downarrow\right)/\rho$

(58)

My point in presenting these equations is to show that, although they are a bit more complex than the Xα or VWN expression, they can be written down and programmed quite readily once the spin densities and their gradients have been expressed. In order to perform self-consistent nonlocal calculations it is necessary to evaluate the Kohn-Sham potential which corresponds to the functional derivative of these energy expressions:

$$V_{XC} = \frac{\delta E_{XC}}{\delta \rho}$$

(59)

In much of the literature this has not been done, the reasoning being that the LDA density is likely to be of good acuracy and hence, incorporation of the nonlocal corrections by perturbation theory should suffice. Tests on small molecules, including transition-metal complexes[26, 33] indicate that this is valid for a variety of systems. However, the tests have not been exhaustive and, moreover, a complete theory should include the self-consistent nonlocal density so I think it would be premature to limit attention to the perturbative approach. Calculation of the potential involves the second derivatives of the density, as well as the gradient. The computational expense of calculating the density gradients and eventually the second derivatives will depend on the method chosen to solve the KS equations. We will not discuss this matter in any detail; suffice it to say that methods exist to incorporate the nonlocal energy and potential at an expense which is usually not more than a factor of two or so in computer time. Clearly this is a necessary investment if energetics are to be considered.

4. Techniques for Solving the Kohn-Sham Equations; deMon

Up until now we have defined Density Functional Theory, the Kohn-Sham implementation, and discussed the local and nonlocal approximations for the exchange and correlation components of the Kohn-Sham treatment. The practical solution of these equations has become a thriving field over the last few years. As the successes of DFT have become more numerous and more widely known, a growing circle of computational specialists from both the physics and chemistry communities have turned their attention to these methods. Applications to organic chemistry, for example, are mushrooming and more and more workers with a more traditional *ab initio* background are contributing to technique and program development. Inorganic chemistry and transition metal systems in particular have long been at the center of the chemical use of DFT, going back to the early applications of the Xα method and beyond. This community is used to the utility of the DFT techniques in studies of spectroscopy and descriptions of the basic features of electronic structure. What has changed over the last few years is that now, helped by the infusion of "*ab initio* " mentality and programming standards, the complexity of systems which can be treated has increased dramatically. Most notably, gradient-driven methods for searching potential energy surfaces and for optimising geometries have recently become available for DFT methods. These have done much to bring the DFT techniques closer to the standards of functionality of traditional *ab initio* techniques.

The field has evolved to the point that a varied menu of "options" can be chosen, depending on the application. Some of these DFT choices are summarized in Table 3. No claims are made for the completeness of this table (for example, solid-state options are only mentioned in passing); however, it should give the reader a feeling for the activity in the field, along with some leading references. Selections of these have been implemented in modern program packages, several of which are mature enough to either be on the commercial market or approaching that stage. We will discuss these only briefly, using the program names as a convenient way of classifying the techniques. Perhaps the most important choice to be made is that of a representation for the KS orbitals. Essentially all of the "classic" choices of quantum chemistry are in use. DGAUSS [33] and deMon [47] use Gaussian orbitals, AMOL [50] can use Slater functions, DMOL [51] and NUMOL [55] use a numerical representation. To these can be added the possibility of using plane waves, as in the Car-Parrinello [52] and CORNING [53] codes. Each of these has pros and cons and the choice of one method or the other can depend on the mix of systems and properties which one wishes to study. The Gaussians have the advantage of being

TABLE 3. Some DFT Options (with a few leading references)

Basis	$\hat{H}$
Gaussian (DGAUSS [33], deMon [47]	analytical Hpq
Dunlap [48], Pederson [49])	numerical Hpq
Slater (AMOL(HFS) [50])	numerical solution
Numerical (DMOL [51])	scattered wave
Plane Waves (Car-Parrinello [52], CORNING [53])	Car-Parrinello
LMTO (Gunnarsson et al [54]	
"none" (numerical solution) (NUMOL [55])	
Scattered-Wave (MS-Xα [13,42])	

coulomb	exchange	correlation
exact, N^4	exact, N^4	
fit, N^2M	LDA (Xα, α=2/3)	LDF (e.g. VWN)
muffin tin	NLDA ($\rho, \bar{\nabla}\rho$)	NLDF ($\rho, \bar{\nabla}\rho$)
	- fit to auxiliary basis	- fit to auxiliary basis
	- numerical integration	- numerical integration
	- V_x SCF	- V_c, SCF
	- ε_x pert. theory	- ε_c pert. theory
	muffin tin	muffin tin

spin	relativity
restricted	relativistic model core potential
unrestricted	scalar: mass-velocity, Darwin
	spin-orbit coupling
	4-component

optimiser	functionality
point-by-point	geometries
gradient-driven	vibrations
Car-Parrinello	binding energies
	charge analysis
	dipole moments
	molecular electrostatic potentials
	ionizations
	excitations
	polarisabilities
	IR and Raman intensities

symmetry

- for $\hat{H}$ matrix elements, grids, etc.
- block diagonalize
- periodic boundary conditions

familiar to quantum chemists. Much work has been performed on transition-metal systems using these basis sets. Slater functions provide more compact expansions. The numerical functions are flexible, but analysis can sometimes be a bit less "familiar" (we have really gotten used to Gaussians...). A plane wave basis is independent of nuclear position and this facilitates the calculation of forces, geometry optimisations and molecular dynamics simulations. However, one needs to use a large number of plane waves to describe localised systems and this has so far precluded their use for the transition-metal systems of interest here.

Once the orbital basis set has been chosen, the next step is to evaluate the matrix elements of the Kohn-Sham hamiltonian. For AMOL and DMOL this is done by numerical integration. Very efficient techniques have been developed for this over the last few years [49, 50, 51, 55] so that good accuracy can be obtained with a practical number of integration points. NUMOL attempts a numerical solution of the KS equations, rather than a basis set expansion. This involves using a reference basis set expansion followed by calculations of differences. The plane-wave codes use fast fourier transform techniques to advantage. The gaussian methods evaluate the matrix elements analytically, once the coulomb and exchange-correlation potentials or energies have been fitted to auxiliary gaussian basis sets. The fitting procedure for the exchange and correlation terms involves the same type of numerical grid as in the numerical integration techniques. Indeed, these terms can also be integrated numerically to high accuracy within DGAUSS and deMon.

To get a better feeeling for what is involved, let us write down a few equations. As in the Hartree-Fock method, the KS equations, solved with a basis, require the evaluation of one- and two-electron integrals. The one-electron integrals are **exactly** the same as for a Hartree-Fock calculation

$$\left\langle \chi_p | \chi_q \right\rangle, \left\langle \chi_p \left| -\frac{1}{2}\nabla^2 \right| \chi_q \right\rangle, \left\langle \chi_p \left| \frac{Z}{r_A} \right| \chi_q \right\rangle \tag{60a}$$

namely, overlap, kinetic energy, and nuclear attraction integrals. The coulomb integrals:

$$< \chi_p(1) \left| \frac{\rho(2)}{r_{12}} \right| \chi_q(1) > \Rightarrow < \chi_p(1)\chi_r(2) \left| \frac{1}{r_{12}} \right| \chi_q(1)\chi_s(2) > \tag{60b}$$

are, at the outset, also the same as in Hartree-Fock, but if a fit is done, then one index can be saved. The density is represented as a linear combination of gaussians in the auxiliary coulomb basis:

$$\tilde{\rho} = \sum_i a_i f_i \tag{61}$$

And the coefficients a_i are determined so as to minimise the error in the coulomb energy:

$$\left\langle (\rho - \tilde{\rho}) \left| \frac{1}{r_{12}} \right| (\rho - \tilde{\rho}) \right\rangle \tag{62}$$

As a result the two-electron integrals which must be evaluated depend only on three indices, rather than four

$$\left\langle \chi_p \left| \frac{f_i}{r_{12}} \right| \chi_q \right\rangle$$

(63)

A similar fit is performed for the exchange-correlation terms. For example the exchange and correlation contribution to the KS potential is represented as:

$$\tilde{V}_{XC} = \sum_i b_i g_i$$

(64)

The coefficients are determined by least squares. The resulting integrals, which may be evaluated analytically, are:

$$\left\langle \chi_p \left| g_i \right| \chi_q \right\rangle$$

(65)

Again, three indices are involved. (63) and (65) represent the fundamental reason for the relative rapidity of the DFT techniques. They scale formally as N^2M where N is the number of functions in the orbital basis and M is the number of functions in the auxiliary basis. Since M is typically similar to N, one often speaks of N^3 scaling.

The above integrals over gaussians can be evaluated with modern quantum chemical techniques and one of the advantages of the gaussian approach is that future developments from the "Hartree-Fock world" may be readily adopted.

The last point of methodological comparison that we will discuss here (many more of varying level of detail may be found in the literature) concerns the calculation of forces and higher energy derivatives. All of the modern codes calculate the forces and use this information to aid in geometry searches. The calculation is simplest for the plane wave codes because only the Hellmann-Feynman term needs to be calculated. For numerical basis sets, either a straightforward (and eventually time consuming) numerical differentiation may be performed, or the numerical functions can be piecewise fit to polynomials and then analytic derivation may be used. The technology of gaussian DFT forces [47, 56, 57] is essentially similar to that found in the traditional *ab initio* codes except that the implicit dependence of the energy on the coulomb and exchange-correlation fitting coefficients must either be calculated or ignored (and shown to be small). Improvement of the energy derivative component of these techniques is ongoing in several laboratories. A major step, which should be made in the near future, is to calculate the second derivatives. These will require further integral derivatives and also solution of coupled perturbed equations to account for changes in the orbital coefficients following geometry variation. The careful analysis, massive programming effort, and the subsequent validation required to "certify" a DFT second derivative package is just now in its initial stages. It will require several man-years to complete but, like the first-derivative breakthrough before it, it will have a profound effect on the complexity of systems and processes which may be studied.

The above comparisons have necessarily been sketchy. Indeed, all of these techniques and computer codes are continually undergoing improvement. Comparisons of functionality and of efficiency are difficult; one is dealing with moving targets. One code may perform better for a certain type of property, or for fixed geometry calculations but not for optimisations. Another may be "perfect" for one class of compounds but inapplicable to another. Moreover, the relative performance is also a function of the type of computers available and this can change almost over night. It is perhaps best to avoid putting too fine a point on any particular pro or con. Suffice it to say that all of the programs mentioned above have provided first-class results for systems that would be hard, if not impossible, to treat by methods outside the DFT family. The offspring of at least some of them will undoubtedly continue to evolve and contribute to the toolbox of computational chemistry and physics.

The results in the remainder of this chapter have been obtained using the gaussian program deMon, written in my laboratory, primarily by Alain St-Amant as part of his doctoral work. The main characteristics of deMon are summarized in Table 4.

TABLE 4. Some characteristics of deMon [47]

LCGTO - MCP - DF method

LDF, NLDF - perturbative or fully self-consistent

Obara-Saika integrals

all-electron bases (Godbout, Salahub, Andzelm, Wimmer)

MCP for selected elements (systematic preparation in progress)

analytical gradients

standard (BFGS etc.) optimisations or "hybrid" "CP"

harmonic frequencies (numerical differences of gradients)

properties - dipole moments, Mulliken, Mayer, IR and Raman intensities, IP's, polarisabilities

Under development since Oct. '88

about 80 000 lines of Fortran

portable, Unix, COS, VAX/VMS, NOS/VE, etc.

---

5. Applications of DFT to Metal-Ligand Interactions: two case studies

To illustrate the use of these techniques, along with Ziegler's and Russo's chapters, I will present just briefly, two examples from our own work. The first involves the use of a cluster model to represent an extended surface and chemisorption on it. The second is from the field of cluster beams.

5.1. STRUCTURE AND BONDING OF FORMATE ON NI(110) [58]

We have recently carried out a geometry optimisation for all degrees of freedom of formate adsorbed on two different sites modeling the (100) surface of nickel. The clusters (Figure 3) contained either four atoms, to model adsorption with the formate

334

bound to two surface Ni atoms, or seven atoms for binding symmetrically disposed over a single surface atom. In both cases, the geometry of the metal cluster was taken to be that of bulk nickel.

In Table 5 we show the calculated vibrational frequencies and isotopic shifts for these two models and compare them with their experimental counterparts for bulk Ni (110). The frequencies are calculated in the harmonic approximation, the force constants being obtained as finite differences of the analytic energy gradients. This approximation, and the very small cluster models considered are, in fact, adequate to make a credible decision between these two possible structures. The agreement is very good for the structure involving two nickel atoms and much poorer for the other structure. Moreover, it turns out that the structure involving a single nickel atom is not a local minimum, but rather an extremum of a more complicated nature, as indicated by the imaginary frequency for the frustrated translation along the y axis. Although the cluster is small, and edge effects might possibly change the detailed nature of this mode, our experience with surface vibration calculations is that such small clusters are, indeed, reliable.

The calculated geometry and charge analysis already provide some insight into the nature of the binding. The CH part of the adsorbed species is similar to the free neutral radical, whereas the oxygen end more closely resembles the radical anion.

Even such basic information as this goes a long way to delimiting the possible characteristics of the adsorbate and also puts some limits on the mechanisms for catalytic reactions in which it is involved. Calculations of this accuracy on systems of this complexity (several "real" transition metal atoms, several degrees of freedom to optimize) would be very difficult with traditional correlated *ab initio* methods.

5.2. HYDROGEN UPTAKE ON SMALL NICKEL, RHODIUM, AND PALLADIUM CLUSTERS [59]

This final example represents the use of DFT and deMon in what I like to call "discovery mode". Not much is known about the systems under study, small transition-metal clusters interacting, up to saturation, with hydrogen. One gets the feeling of exploring virgin territory and it is very interesting territory, indeed.

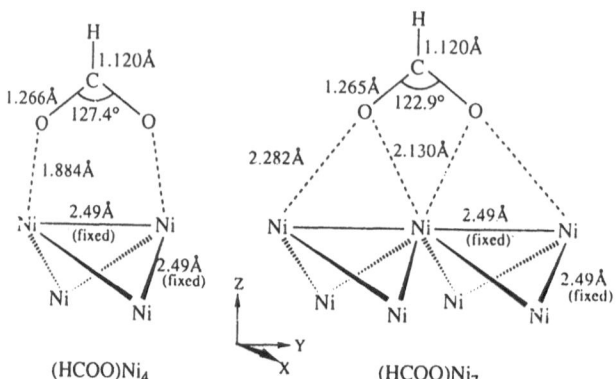

Figure 3. Calculated geometries for (HCOO)Ni$_4$ and (HCOO)Ni$_7$. For details see ref. 58.

TABLE 5. Normal modes and isotopic shifts for (HCOO) Ni_4 and (HCOO) Ni_7 (see ref. 58 for details)

Normal mode[a]	HCOO Calc. Ni_4	Ni_7	Exp.	DCOO Calc. Ni_4	Ni_7	Exp.	isotopic shift Calc. Ni_4	Ni_7	Exp.
a_1									
CH str.	2912	2916	2863	2159	2161	2186	1.39	1.35	1.36
CO str.	1315	13:6	1347	1285	1325	1307	1.02	1.02	1.03
OCO bend	744	648	766	736	641	750	1.01	1.01	1.02
NiO str.(t_z)	368	252	400	363	250	395	1.01	1.01	1.01
ε_2									
τ_z	121	85	-	121	85	-	1.00	1.00	-
b_1									
CH wag.	949	954	-	813	812	839	1.17	1.17	-
τ_y	295	286	-	254	247	-	1.16	1.16	-
t_x	35	41	-	33	40	-	1.06	1.03	-
b_2									
CO str.	1559	1523	1557	1558	1521	1524	1.00	1.00	1.02
HCO bend	1309	1263	-	966	930	928	1.36	1.36	-
NiO str(t_x)	382	117	-	371	114	-	1.03	1.03	-
t_y	108	imag.	-	106	imag.	-	1.02	-	-

a) For example, t_x and τ_x denote hindered translation along the x axis and hindered rotation around the x axis of HCOO, respectively.

The study arose in connection with the cluster beam results of Cox *et al* [60] who found that very small clusters of nickel, rhodium, and palladium could take up amounts of hydrogen which were, at least at first sight, surprising (if not alarming, when one remembers that hydrogen titration is often used in catalysis to evaluate surface areas, assuming that one hydrogen atom is taken up per surface metal atom). The H/M ratio was found to be as high as 8/1 for Rh_2 and Rh_3.

Alain St-Amant has used deMon to optimize a large number of structures for these systems [59]. Calculations were performed both at the local and at the nonlocal level for a variety of starting configurations, starting with the bare monomer, dimer, or trimer and adding hydrogen atoms until these would no longer bind. Full gradient-driven geometry optimisations were performed(in fact, these results could not have been obtained even two or three years earlier, before the advent of the gradient techniques).

I will show only a very small sampling of the results here, to illustrate the fascinating structures that have come out of the calculations and to indicate the great variety of the binding which is possible even in such simple clusters. Figure 4 shows the structures and the nonlocal binding energies. As usual the local approximation yields binding energies(relative to dissociation into a cluster with n-2 hydrogens and a hydrogen molecule, except for n=1) which are much larger than the nonlocal energies. There is hence a tendency to maintain binding somewhat longer in the series, if the LDA is used. All indications are that the nonlocal energies should be more reliable. We emphasize at

336

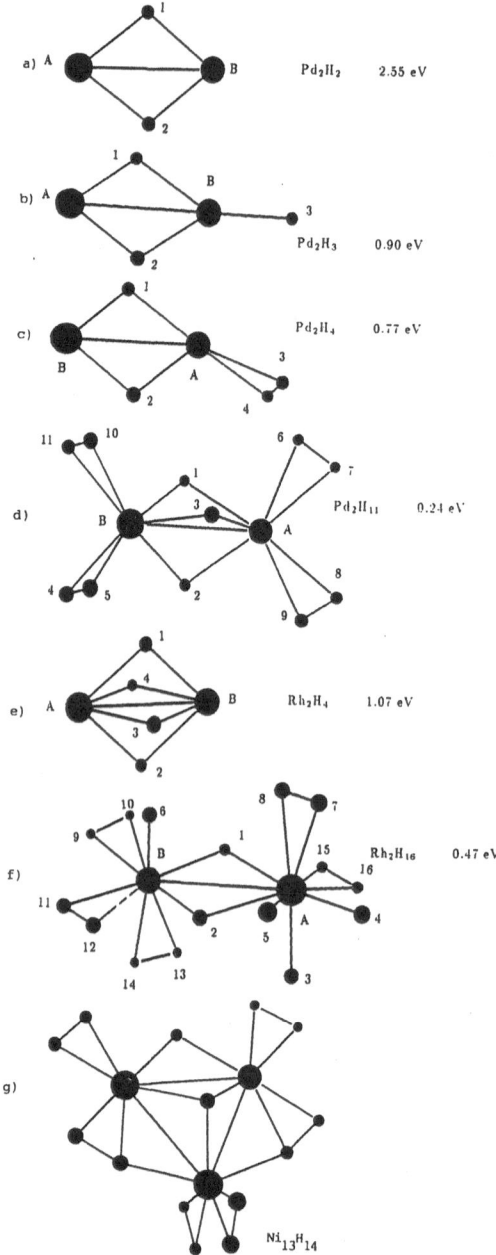

Figure 4. Calculated structures and binding energies for hydrogen interacting with small transition-metal systems (see ref. 59 for details).

the outset that there is no guarantee that the structures shown are absolute, or even local minima (since vibrational analyses have been carried out only in a few cases). They simply represent a minimum or another type of extremum in the catchment region of the starting point. In several cases this has been varied in an attempt to sample more than one catchment region but no claim is made for completeness. But we can say that if the structures shown do not represent the global minimum, they do represent bound structures and it is plausible in most cases that they will be stable or metastable. Clearly, it would be wonderful if experimental structures could be obtained to test these predictions (the thought crosses my mind that only a few years ago pleas for spectroscopic measurements, such as photoelectron spectroscopy for cluster beams, seemed a bit far-fetched...).

Consider first, the structure of Pd_2H (Fig. 4a). The hydrogen is bridge bound, analogously to what would be expected for an extended surface (i.e. the site of highest available coordination is occupied). The binding is "hydridic", Mulliken analysis showing a net negative charge on the H of 0.3 electrons. The second hydrogen also adds to the bridge. For Pd_2H_3, and Pd_2H_4 however, new possibilities arise (Figs 4b and 4c), a terminal H or, even more interestingly, an H_2 unit that migrates to the terminal position. This corresponds to a hydrogen molecule chemisorbed with a weakened H-H bond (R= 0.86Å, B.O. = 0.74). It is only recently that such a type of binding has been discovered in transition-metal cluster compounds and it is fascinating to learn that it may occur so early on in the series of hydrogenated clusters.

Continuing the series, one finds that bridge, terminal and H_2 binding can coexist in some of the clusters. The largest bound cluster of Pd_2 is Pd_2H_{11} (Fig. 4d) which has an esthetically appealing structure with three bridge-bound H's and four H_2 units. It is predicted to be bound by 0.7 eV at the local level but by only 0.2 eV when nonlocal corrections are included.

The rhodium dimer is more voracious for hydrogen (as is the case for the bulk metals). Binding energies are higher and more hydrogen may be taken up. In fact, Rh_2 will bind four hydrogens in the bridge position (Fig. 4e). This can, perhaps, be rationalised, relative to palladium, by realizing that there are fewer d electrons in Rh and this may lead to less "electronic congestion" with the hydridic hydrogens. There is also a greater tendency for rhodium structures to involve terminal monatomic ligands. The largest bound structure is Rh_2H_{16} (Fig 4f), in agreement with the experimental conclusions of Cox *et al* .

For the trimers, similar scenarios develop. For example, Ni_3 adsorbs a single hydrogen atom in the three-fold site, to the surprise of noone. Higher congeners, show both 3-fold and 2-fold bridge bonds, terminal hydrides and H_2 units similar to those just discussed. However, there is even more variety here. Fig 4g shows the structure of Ni_3H_{14}. This is bound at the local level but near the limit of instability at the nonlocal level. Nevertheless, four different types of bonding are illustrated, 3-fold bridge, 2-fold bridge, terminal H_2 and yet another type of H_2 in which the H-H bond is further stretched and weakened and one of the two hydrogen atoms starts to bind to a second nickel atom (B.O. = 0.3). It is more than a little tempting to associate this situation with some intermediate step in the process for dissociation of the hydrogen molecule.

Clearly, these small clusters are capable of a great variety of chemistry and it is exciting to consider the possibility of using calculations such as these, along with experiments, to gain valid insight into their properties and make appropriate comparisons with surfaces and with real world catalyst.

338

6. Concluding Remarks

I hope to have shown in this chapter that the density functional theory is not only elegant and formally appealing, as *"ab initio"* as any of the other *ab initio* methods, but also that it can facilitate real discovery in transition metal chemistry and yield insight at a very high and trustworthy level.

Let me summarize the situation for metal-ligand bonding as it now stands, in my view. The following properties may be studied with a very good confidence level:

- geometries

- vibrational frequencies

- charge distributions, dipole moments

- photoelectron spectra

- bond dissociation energies

- chemisorption energies

- mechanistic insight based on the above properties

It could be emphasized that the binding energy items on this list have only become available(and there is still much testing to do) in the last months, with the advent of the nonlocal corrections.

What does, or should, the future hold? In my view the time is ripe for a full scale attack on chemical reactivity using density functional techniques. The early indications are that activation energies should be handled at least reasonably in many cases by the nonlocal functionals and these should be "pushed until they break". The prospect of completing at least one or two layers of the "onion" described in the introduction, with a transition-metal reaction in the middle, is exhilarating.

ACKNOWLEDGEMENTS

I am far more grateful than can be expressed here to the talented students, postdocs, and colleagues who have contributed to the work described here, and to those who have helped me attain whatever understanding and appreciation of density functional theory I may possess. The Natural Sciences and Engineering Research Council (Canada), the Network of Centres of Excellence in Molecular and Interfacial Dynamics (Canada), and the Fonds pour la formation de chercheurs et l'aide à la recherche FCAR (Québec) have lent ongoing support to this work and I am truly grateful for this. Support from Hitachi, Kodak, the Institut Français du Pétrole, the Petroleum Research Fund administered by the American Chemical Society, and from NATO has also been crucial to the work. Finally, computing resources from NSERC, CEMAID, the Ontario Centre for Large Scale Computation, Cray Canada, and the Services Informatiques de l'Université de Montréal are gratefully acknowledged.

REFERENCES

1. Fine, R. (1991), personal communication.
2. Clementi, E. (1990), MOTECC, Modern Techniques in Computational Chemistry, ESCOM, Leiden.
3. e.g. Field, M.J., Bash, P.A. and Karplus, M. (1990), J. Comp. Chem. **11**, 700.
4. Szabo, A. and Ostlund, N.S. (1982), Modern Quantum Chemistry, MacMillan, New York.
5. Salahub, D.R. and Zerner, M.C. (eds.), (1989), The challenge of d and f Electrons, ACS Symposium Series, 394, American Chemical Society, Washington..
6. Ashcroft, N.W. and Mermin, N.D., (1976), Solid State Physics, Holt Rhinehart and Wilson, New York.
7. Thomas, L.H. (1926), Proc. Cambridge Phil. Soc. **23**, 542.
8. Fermi, E. (1926), Z. Phys. **48**, 542.
9. Von Weiszacker, C.F. (1935), Z. Phys. **96**, 431.
10. Dirac, P.A.M. (1930), Proc. Cambridge Phil. Soc. **26**, 376.
11. Slater, J.C. (1951), Phys. Rev. **81**, 385.
12. Schwarz, K. (1972), Phys. Rev. **B5**, 2466
13. Johnson, K.H. (1973), Adv. Quantum Chem. **7**, 143; Slater, J.C. (1974), The Self-consistent Field for Molecules and Solids, McGraw Hill, New York.
14. Sambe, H. and Felton, R.H. (1975), J. Chem. Phys., **62**, 1122.
15. Dunlap, B.I., Connolly, J.W.D. and Sabin, J.R. (1979), J. Chem. Phys., **71**, 3386, 4993.
16. Ellis, D.E. and Painter, G.S. (1970), Phys. Rev. **B2**, 1887.
17. Andersen, O.K. (1975), Phys. Rev. **B12**, 3060.
18. Hohenberg, P. and Kohn, W. (1964), Phys. Rev. **136**, B864.
19. Kohn, W. and Sham, L.J. (1965), Phys. Rev. **140**, A1133.
20. Parr, R.G. and Yang, W. (1989), Density Functional Theory of Atoms and Molecules, Oxford.
21. Davidson, E.R. (1976), Reduced Density Matrices in Quantum Chemistry, Academic, New-York.
22. Levy, M. (1979), Proc. Nat. Acad. Sci. USA, **76**, 6062; (1982), Phys. Rev. **A26**, 1200.
23. McWeeny, R. and Sutcliffe, B.T. (1976), Methods of Molecular Quantum Mechanics, Academic, London.
24. Becke, A. (1986), J. Chem. Phys. **85**, 7184.
25. Becke, A. (1983), Int. J. Quantum Chem., **23**, 1915.
26. Ziegler, T. (1991), Chem. Rev. **91**, 651.
27. Tschinke, V. and Ziegler, T. (1990), J. Chem. Phys. **93**, 8051.
28. Mullikens, R.S. often quoted.
29. Becke, A.D. (1988), Phys. Rev. **A38**, 3098.
30. Levy, M. and Perdew, J.P. (1985), Phys. Rev. **A32**, 2010.
31. Perdew, J. and Wang, Y. (1986), Phys. Rev. **B12**, 8800.
32. Gunnarsson, O., Jonson, M. and Lundqvist, B.I. (1979), Phys. Rev. **B20**, 3136.
33. Labanowski, J.K. and Andzelm, J.A. (eds.) (1991), Density Functional Methods in Chemistry, Springer-Verlag, New York.
34. Salahub, D.R. (1987), Adv. Chem. Phys., **67**, 447.
35. Jones, R.O. and Gunnarsson, O. (1989), Rev. Mod. Phys. **61**, 689.

36. Perdew, J.P. and Zunger A. (1981), Phys. Rev., **B23**, 5048.
37. Vosko, S.H., Wilk, L. and Nusair, M. (1980), Can. J. Phys. **58**, 1200.
38. Ceperley, D.M. and Alder, B.J. (1980), Phys. Rev. Lett. **45**, 566.
39. Papai, I., St-Amant, A., Ushio, J. and Salahub, D.R. (1990), Intern. J. Quantum Chem., **S24**, 29.
40. Papai, I., Goursot, A., St-Amant, A. and Salahub, D.R. (1992), Theoret. Chim. Acta. (in press).
41. Mlynarski, P. and Salahub, D.R. (1991), J. Chem. Phys., **95**, 6050.
42. Fournier, R. and Salahub, D.R. (1991), Surf. Sci. **245**, 263.
43. Herman, F., van Dyke, J.P. and Ortenburger, I.B. (1969), Phys. Rev. Lett. **22**, 807; (1970), Int. J. Quantum Chem. **S3**, 827.
44. Langreth, D.C. and Mehl, M.J. (1981), Phys. Rev. Lett. **47**, 446; (1983), Phys. Rev. **B28**, 1809; - erratum **B29**, 2310.
45. Becke, A.D. (1988), Phys. Rev. **A38**, 3098.
46. Perdew, J. (1986), Phys. Rev. **B33**, 8822; erratum, **B38**, 7406.
47. St-Amant, A. and Salahub, D.R. (1990), Chem. Phys. Lett., **169**, 387; St-Amant, A. (1992), Thesis, Université de Montréal.
48. Dunlap, B.I. (1990), Phys. Rev. **A41**, 5691.
49. Pederson, M.R. and Jackson, K.A. (1990), Phys. Rev. **B41**. 7453.
50. Boerrigter, P.M., teVelde, G. and Baerends, E.-J. (1988), Int. J. Quantum Chem., **33**, 87: Versluis, L. and Ziegler, T. (1988), J. Chem. Phys., **88**, 322.
51. Delley, B.J. (1990), J. Chem. Phys., **92**, 508.
52. Car, R. and Parrinello, M. (1985), Phys. Rev. Lett. **55**, 2471.
53. Teter, M.P., Payne, M.C. and Allan, D.C. (1989), Phys. Rev., **B40,** 12255.
54. Gunnarsson, O., Harris, J. and Jones, R.O. (1977), Phys. Rev., **B15**, 3027.
55. Becke, A.D. and Dickson, R.M. (1990), J. Chem. Phys., **92**, 3610.
56. Fournier, R., Andzelm, J. and Salahub, D.R. (1989), J. Chem. Phys., **90**, 6371.
57. Fournier, R. (1990), J. Chem. Phys., **92**, 5422.
58. Ushio, J., Papai, I., St-Amant, A. and Salahub, D.R. (1992), Surf. Sci. Lett., **262**, 134.
59. St-Amant, A. (1992), Thesis, Université de Montréal; St-Amant, A. and Salahub, D.R., to be published
60. Fayet, P., Kaldor, A. and Cox, D.M. (1990), J. Chem. Phys. **92**, 254.

DENSITY FUNCTIONAL MODEL CALCULATIONS FOR HOMOGENEOUS AND HETEROGENEOUS CATALYSIS

N. RUSSO
Universita della Calabria
Dipartimento di Chimica
I-87030 Arcavacata di Rende (CS)
Italy

ABSTRACT. Selected molecular structure calculations in the framework of a Density Functional-based scheme Linear Combinations of Gaussian-Type Orbitals-Model Core Potential-Local Spin Density (LCGTO-MCP-LSD) method are reported. Transition-metal carbene $RhCH_2^+$ and $RhCF_2^+$ ions and a carbon dioxide-transition metal complex ($Ni(PH_3)_2CO_2$) have been chosen as model systems for homogeneous catalysis while the interaction of hydrogen with palladium, rhodium and bimetallic PdSn and RhSn clusters is presented as an example of the possibility of DFT quantum chemical calculations to explain significant elementary steps in heterogeneous catalysis mechanisms. The accuracy of these calculations are further examples of the great usefulness of the most recent development of DFT-based methods for the study of the chemical and physical properties of large and complex systems.

1. Introduction

With the recent developments of theory [1-19] and the implementations of related software packages [20-21], the Density Functional Theory (DFT) has emerged as a powerful tool for elucidating the concepts of chemistry in addition to those of solid state and molecular physics. The recent applications (study of electronic, geometrical, and vibrational properties of molecules, clusters, and surfaces; accurate thermodynamical data; reliable potential energy surfaces for chemical reactions with transition metal containing systems; reproduction of ultraviolet, auger and x-ray core spectra and quantum molecular dynamics) bode well for an increasing use of DFT-based methods also in fields that are the current domain of ab-initio Hartree-Fock (HF) and post-HF methods (i.e. organic and biological systems). Because the field of transition metal compounds has been always in a central position in theoretical chemistry, the DFT applications in this area become very significant. In fact, notwithstanding that there is a considerable interest for organometallic molecules and more in general for transition-metal containing systems, the electronic structure and the nature of the bonding of these systems are not as well understood as the properties of organic molecules. When transition-metals are involved, the HF approximation (that in general well describes the ground state of first-row element containing molecules) is not appropriate. Only when post-SCF calculations (Hartree-Fock+ Configuration Interaction) are performed employing a very large number of determinants, the reliability of the results can be improved. The inclusion of correlation

D. R. Salahub and N. Russo (eds.), Metal-Ligand Interactions: from Atoms, to Clusters, to Surfaces, 341–366.
© 1992 *Kluwer Academic Publishers. Printed in the Netherlands.*

drastically increases the computational effort, and also with the availability of supercomputers, the use of fully correlated methods is generally limited to small molecules. For these reasons a great attention has been addressed to search for alternative methods such as those based on Density Functional Theory. The recent implementations of these methods (i.e. the introduction of Model Core Potentials [22], Gaussian-type [22] or Slater-type [23] orbitals, plane- waves [24] and augmented plane-waves [25], analytical formulation of energy gradients [9,13], non-local corrections [11, 12, 26], new density functionals [4], calculation of electronegativity [4] and time-dependent formulation [27]) can allow a systematic study of the chemistry of the transition metal elements.

In this paper we present the results of LCGTO-MCP-LSD (a DFT method) calculations for systems containing transition metals as models for better understanding some processes occurring in homogeneous and heterogeneous catalysis.

The plan of the paper is as follows. In the next section the method used and their more significant implementations are briefly described. The following section will deal with the applications. Two model systems (transition-metal carbene and nickel-carbon dioxide complexes) have been selected as examples of molecules involved in significant homogeneous catalytic processes. For heterogeneous catalysis the results of the study of hydrogen adsorption on palladium, rhodium, and bimetallic PdSn and RhSn catalysts are proposed.

2. Basic Theory and Recent Developments

In this section the basic theory of the LSD method and its recent implementations are outlined. More extensive details are available in other articles of this book and in the recent literature [1-19].

2.1. BASIC THEORY

The LCGTO-LSD method, originally proposed by Sambe and Felton [28] has been considerably extended by Dunlap, Connolly and Sabin [29] and more recently by Andzelm et al. [22] and Salahub [30].

Following the works of Hohenberg and Kohn [31] and Kohn and Sham [32] the exact ground state energy $E^{(N)}$ for a spin-polarized system of N electrons in the field of a nuclear charge Z can be expressed in terms of the spin-up $(n_+(r))$ and spin-down $(n_-(r))$ densities where $f_{i\sigma}$ is the occupation number (0 or 1) for the normalized orbital $\psi_{i\sigma}(r)$.

$$\left(-1/2\nabla^2\right) + V_{coul} + v_{eff}^{\sigma}\left(n_+, n_-, r\right) \psi_{i\sigma}\left(r\right) = \varepsilon_{i\sigma}\psi_{i\sigma}(r) \qquad \left(\sigma= +/-\right) \quad (1)$$

$$n_{\sigma}\left(r\right) = \sum_{r} f_{i\sigma} \mid \psi_{i\sigma}\left(r\right) \mid^2 \tag{2}$$

This means that

$$\sum_{i\sigma} f_{i\sigma} = N \tag{3}$$

The single-particle effective potential is

$$v_{eff}^{\sigma}\left(n_{+},\,n_{-},\,r\right) = \frac{\partial E_{xc}\left[n_{+},\,n_{-}\right]}{\partial n_{\sigma}\left(r\right)} \tag{4}$$

with

$$n(r) = n_{+}(r) + n_{-}(r)$$

$E_{xc}[n_{+},\,n_{-}]$ is the exchange-correlation functional.

Now, the (non-relativistic) ground state energy is:

$$E^{N} = \sum_{i\sigma} f_{i\sigma}\epsilon_{i\sigma} - \int dr \int dr' \frac{n(r)\,n(r')}{|r-r'|} + E_{xc}[n_{+},\,n_{-}] -$$

$$-\sum_{\sigma} \int dr\, n_{\sigma}\left(r\right) v_{xc}^{\sigma}\left[n_{+},\,n_{-},\,r\right] \tag{5}$$

The functional $E_{xc}[n_{+},n_{-}]$ is not known for an inhomogeneous system of interacting electrons and some approximations have been suggested. In the most widely used approximation, the LDA (Local Density Approximation) or its spin-polarized generalization LSD (Local Spin Density), $n(r)$ is supposed to vary slowly and the energy can be expressed as

$$E_{XC}^{LSD}\left[n_{+},n_{-}\right] = \int dr\,n(r)\,\epsilon_{xc}\left(r_{s},\xi\right) \tag{6}$$

where $\epsilon_{xc}\left[r_{s},\,\xi\right]$ is the exchange-correlation energy per particle of a spin-polarized homogeneous electron gas

$$r_{s}\left(r\right) = \left(\frac{3}{4}\pi\,n\,(r)\right)^{\frac{1}{3}}$$

and

$$\xi(r) = \left(n_{+}(r) - n_{-}(r)\right)/n(r)$$

are the usual electron gas parameters.

(7)

The corresponding exchange-correlation contribution to the effective potential is

$$v_{\text{eff}}^{\sigma(\text{LSD})}\left[n_+, n_-; r\right] = \frac{d\left(n(r)\, \varepsilon_{\text{xc}}\left(r_s, \xi\right)\right)}{dn_\sigma(r)}$$

$$= \varepsilon_{\text{xc}}\left(r_s, \xi\right) - \frac{1}{3}\, r_s \left(\frac{\partial}{\partial r_s}\right) \varepsilon_{\text{xc}}\left(r_s, \xi\right) +$$

$$+ \sigma\left(1 - \sigma\xi\right) \left(\frac{\partial}{\partial \xi}\right) \varepsilon_{\text{xc}}\left(r_s, \xi\right)$$

(8)

In the LSD method the Vosko, Wilk and Nusair (VWN) parameterization [33] for the correlation is used

$$r_s \left(\frac{\partial \varepsilon_c}{\partial r_s}\right) = A\left(1 + b_1 x\right) / \left(1 + b_1 x + b_2 x^2 + b_3 x^3\right)$$

where

$$b_1 = \left(bx_0 - c\right) / x_0 c$$
$$b_2 = \left(-b + x_0\right) / x_0 c$$
$$b_3 = -1/x_0 c$$

with x_0, b and c constants.

2.2. MODEL CORE POTENTIALS (MCP)

In many atomic and molecular problems, the electrons of a system can be divided into valence and core electrons. The purpose of pseudopotential or Model Potential (MP) theory of atoms and molecules is to allow an accurate treatment of valence electrons, that are in the chemical idea responsable for the formation of chemical bonds, while the role of core electrons is to provide the potential field in which the valence electrons move (for an exhaustive review see ref.34). Starting from the work of Hellmann [35] , many pseudopotential techniques have been developed in the field of solid-state physics [36]. The applications of pseudopotentials or MP to quantum chemistry are more recent [37-45]. A number of different pseudopotential or MP methods have been proposed within both Hartree-Fock [39] and LSD [22] approaches. An Huzinaga-type Model Core Potential has been developed for the LSD method [43-45]. This scheme of MCP optimization allows correct atomic results to be obtained with truncated valence basis sets when the MCP parameters are properly optimized. Because the $n_s(r)$ is defined in terms of molecular orbitals (see. eq.2) we can separate them into valence (v) and core (c) orbitals. Assuming the orthogonality between v and c orbitals and introducing the projection operator (P^σ) the Kohn-Sham equations can be rewritten as

$$\left(F^{\sigma} - P^{\sigma}F^{\sigma}\right) \psi_i^{\sigma} = \sum_j^{v} \epsilon_{ij}^{\sigma} \psi_j^{\sigma} \qquad (9)$$

$$F^{\sigma}(r) = h(r) + \int \frac{n(r')}{|r-r'|} dr' + v_{xc}[n_+(r), n_-(r)]$$

$$P^{\sigma} = \sum_c |\psi_c^{\sigma}\rangle \langle \psi_c^{\sigma}| \qquad (10)$$

$$h(r) = \frac{1}{2}\nabla^2 - \sum_I \frac{Z_I}{r_I}$$

After symmetrization and diagonalization, since the operator F^{σ} is the same for both v and c orbitals, we can write

$$F^{\sigma} - \sum_c 2\epsilon_c^{\sigma} |\psi_c^{\sigma}\rangle \langle \psi_c^{\sigma}| \psi_v^{\sigma} = \epsilon_v^{\sigma} \psi_v^{\sigma} \qquad (11)$$

now F^{σ} can be separated as

$$F^{\sigma} = F_v^{\sigma} + V_{MCP}^{\sigma} \qquad (12)$$

where the valence part is

$$F_v^{\sigma}(r) = \frac{1}{2}\nabla^2 - \sum_I \frac{Z_I^v}{r_I} + \int \frac{n_v(r')\,dr'}{|r-r'|} + v_{xc}[n_{+v}, n_{-v}] \qquad (13)$$

and the MCP operator is

$$v_{MCP}^{\sigma} = F^{\sigma} - F_v^{\sigma} \qquad (14)$$

If we assume that the core orbitals of the atoms do not overlap each other and that the cross terms in the exchange-correlation potential can be neglected, the V_{MCP}^{σ} can be divided as

$$V_{MCP}^{\sigma} = \sum_I \frac{-Z_c^I}{r_I} + \int \frac{n_c^I(r')\,dr'}{|r-r'|} + v_{xc}[n_{+c}^I, n_{-c}^I] \qquad (15)$$

To simplify the calculations, a further approximation is introduced assuming for both spins a single MCP

$$V_{MCP} = \frac{1}{2}\left(V_{MCP}^+ + V_{MCP}^-\right) \qquad (16)$$

The analytical form of the MCP is

$$V_{MCP}(r) = \sum_k A_k \frac{\exp\left(-\alpha_k r^2\right)}{r}$$

(17)

with the constraint

$$\sum_k A_k = N_c \qquad\qquad N_c = \text{number of core electrons}$$

(18)

The following terms can be added to the reference atomic calculation to take into account the "scalar" relativistic effects (in the relativistic MCP) :
-) the mass velocity correction due to the relativistic increase of the mass of an electron with its velocity is

$$h^1_{mv} = \frac{1}{8}\,\alpha^2\,p^4 = -\frac{1}{8}\,\alpha^2\,\nabla^4$$

(19)

where p is the linear momentum operator.
-) The Darwin correction that is due to the small-scale irregular motion of an electron around its mean position can be expressed as

$$h^1_D = \frac{1}{8}\,\alpha^2\,\nabla^2\,V^0_T$$

(20)

V_T is the non relativistic potential.
The highest contribution to this operator comes from the nuclear potential (V_N)

$$V_N = \sum_\mu Z_\mu\,/\,r_{1\mu}$$

(21)

$$h^1_D \cong \frac{1}{8}\,\alpha^2\,\nabla^2\,V_N = \frac{1}{8}\,\alpha^2\,\sum_\mu -4\pi\,Z_\mu\,\delta\left(r_1 - r_\mu\right)$$

(22)

The determination of the MCP coefficients is performed numerically through an interpolation procedure as described in ref. 22.

2.3. ANALYTIC ENERGY GRADIENTS

The use of Gaussian-type basis functions allows the possibility to evaluate derivatives analytically for both Coulomb and exchange-correlation energies [9, 13]. The energy gradient is written as

$$\frac{\partial E_{LSD}}{\partial x} = F_{HFB} + F_D$$

(23)

where the Hellman-Feynman and orbital basis incompleteness forces F_{HFB} and the F_D terms arising from the incompleteness of the density fit are

$$F_{HFB} = \sum_{pq} P_{pq} \left\{ \frac{\partial h_{pq}}{\partial x} + \sum_r \rho_r \left[\frac{\partial (pq)}{\partial x} \| r \right] + \left[\frac{\partial (pq)}{\partial x} \mu_{xc} (r) \right] \right\} +$$
$$+ \frac{\partial U_n}{\partial x} - \sum_{pq} W_{pq} \frac{\partial [pq]}{\partial x} \tag{24}$$

$$F_D = \sum_r \rho_r \left[\frac{\partial (r)}{\partial x} \| (\rho - \rho') \right] \tag{25}$$

where W_{pq} is, as in the HF gradient implementation, the energy-weighted density matrix and the symbol $\|$ denotes the $1/r_{12}$ operator in the Obara and Saika formulation [46]. The three-index integrals that are essential in the LSD method are calculated as follows

$$I_c = \left[a (1) b (1) \| (c_2) \right]$$
$$I_{xc} = \left[a (1) b (1) (c 1) \right] \tag{26}$$

where I_c and I_{xc} represent the coulomb and overlap integrals respectively.

1.4 Non-Local Corrections

It is well known [30] that the LSD approach overstimates the binding energies. The possibility to take into account the non-local correction terms using the density gradient, largely corrects this error. The total energy (E^{NLSD}) considering the gradient exchange E_x^G and correlation E_c^G energies can be written as

$$E^{NLSD} = E^{LSD} + E_x^G + E_c^G \tag{27}$$

Different expressions for E_x^G and E_c^G are available. The most used are those proposed by Becke [7, 11] and by Perdew [12] for exchange and correlation respectively:

$$E_x^G = h \sum_\sigma \int \rho_\sigma x_\sigma^2 / \left(1 + 6 b_x \sin h^{-1} x_\sigma \right) dr \tag{28}$$

with

$$x_\sigma = \frac{|\nabla \rho|}{\rho \, \sigma^{4/3}}$$

$$E_c^G = \int f(\rho_\sigma, \rho_\alpha) \exp\{[-g(\rho)] \, |\nabla\rho|\} |\nabla\rho|^2 \, dr \qquad (29)$$

In the deMon package the generalized gradient approximation (GGA) [12] for the exchange energy can be used:

$$E_x^{GGA} = 3/4 \left(\frac{3}{\pi}\right)^{1/3} \int dr \, [\rho \, (r)^{4/3} \, F(s)] \qquad (30)$$

with

$$F(S) = (1 + 1.296 \, S^2 + 14S^4 + 0.20S^6)^{1/5}$$

$$s = \frac{\nabla\rho \, (r)}{2\rho \, k_F} \, (e^{-\phi})$$

$$k_F = \left(3 \, \pi^2 \, \rho\right)^{1/3}$$

2.4. BASIS SETS

Adapting the method of Huzinaga et al. [40, 42, 47] sets of orbital and auxiliary basis sets have been developed for the LCGTO-MCP-LSD method [48].
The following expansion pattern is used

$$\left(k_{s1}.......k_{sn} / k_{p1}.......k_{pn} / k_{d1}\right)$$

where k_{li} denotes the number of GTO's in the i-th contraction function of l symmetry. The LCGTO have the form

$$l_i = \sum_{j=1}^{l_i} dl_{i,j} \, gl_{i,j} \, (\alpha l_i, j; \sigma)$$

$$l = s, p, d$$

$$gl_i \text{ denotes a GTO}$$

Auxiliary basis sets are defined to fit the charge density $\tilde{\rho}$ and the exchange-correlation potential $\tilde{U}$ as

$$\left(\tilde{U}\right), \tilde{\rho} = \sum_{n}^{M} a_n \, f_n$$

where the f_n are gaussian functions

2.5. COMPUTATIONAL DETAILS AND MOLECULAR TESTS

The employed orbital (OBS) and auxiliary basis sets (ABS) together with the Model Core Potentials (MCP) are collected in Table 1. The notation for OBS is that proposed by Huzinaga [47]. For ABS and MCP the following notations are used:

$$(N_c; n_v: n_s \, n_p \, n_d) \qquad \text{MCP}$$
$$(k_s, k_{spd}; l_s, l_{spd}) \qquad \text{ABS}$$

where N_C is the number of electrons put into the frozen core; n_v is the number of s-type gaussians used to fit the effective potential of the core electrons; n_s, n_p and n_d are respectively the number of s-, p- and d-type gaussians used to fit the s, p and d orbitals of the frozen core, k_s (l_s) and k_{spd} (l_{spd}) are respectively the number of s-type gaussians in the CD(XC) basis and the number of s-, p-, d-type gaussians constrained to have the same exponent in the CD(XC) basis.

The calculations concerning the model complexes for homogeneous catalysis have been performed employing the deMon package [20], while the study of the hydrogen interaction with Pd, Rh and bimetallic PdSn and RhSn was made with the previously described LCGTO-MP-LSD set of programs [51] without analytical energy gradient and non-local corrections. The spectroscopic constants were calculated by fitting the potential curve with third-degree polynomials and the binding energies by using the interpolated total energies.

The reliability of the method, MCP and basis sets used have been tested on a series of diatomics (see Table 2) for which experimental and/or high level theoretical results are available.

TABLE 1. OBS, ABS and MCP for atoms used in the computations

Atom	OBS	ABS	MCP	Ref.
H	(41/1)	(5,1;5,1)	-	48
C	(5211/411/1)	(5,2;5,2)	-	48
O	(5211/411/1)	(5,2;5,2)	-	48
F	(6311/311/1)	(5,2;5,2)		48
P	(311/211/1)	(2,3;2,3)	(10;4:6,4)	49
Ni	(311/31/311)	(3,4;3,4)	(12;5:7,4)	49
Rh	(2211/2111/121)	(3,4;3,4)	(30;6:9,6,4)	48
Pd	(2211/2111/121)	(3,4;3,4)	(30;6:9,6,4)	50
Sn	(2211/211/1)	(4,4;3,3)	(46;7:9,8,6)	43

TABLE 2. Spectroscopic constants for Rh-C, Pd-H, Sn-H, Ni-H and Ni-C ground states from theory and experiment.

Molecule	Method	Re/Å	$\omega e/cm^{-1}$	De/eV	Ref.
Rh-C	LCGTO-MCP-LSD	1.625	1008	7.70 (6.30)	52, PW
	HF-CI	1.715	821	2.92	53
	exp.	1.614	1050	6.01	54
Pd-H	LCGTO-MCP-LSD	1.540	-	3.00	55
	exp.	1.535	-	2.43	56
Sn-H	LCGTO-MCP-LSD	1.775	1202	2.99 (2.51)	57, PW
	exp.	1.781	1188	~2.73	58
Ni-C	LCGTO-MCP-LSD	1.609	1042	5.78	59
	HF-CI	1.699	-	3.25	60
Ni-H	LCGTO-MCP-LSD	1.456	2234	3.52 (3.06)	61, PW
	HF-CI	1.471	1982	2.79	62
	GVB	1.450	1911	2.76	63
	exp.	1.476	1927	2.61, 3.07	58

As is shown in Table 2, the LCGTO-MCP-LSD results for the spectroscopic constants are in very good agreement with the available experimental data, with the exception of binding energy that are overestimated. The introduction of the non-local corrections removes these errors (see values in parentheses). On the basis of these results and others available in the literature [52, 59] we can conclude that the employed basis sets and model core potential are able to give correct results in molecular calculations.

3. Model Systems for Homogeneous Catalysis

As previously mentioned two different systems have been chosen as models. The former example concerns the structure and the properties of two transition-metal-methylidene complexes while the second treats the Ni-CO_2 complex widely studied because of its use in the activation processes of CO_2 molecule.

3.1. BONDING IN RH-CX$_2^+$(X=H,F) SYSTEMS

Transition-metal complexes with CH_2 ligands are of current interest in modern organometallic chemistry [65]. These compounds have been postulated as intermediates for many catalytic reactions, particularly alkyl decompositions [66], olefin metathesis [67] and Fischer-Tropsch reductive polymerization of CO [65].

Metal-carbene complexes show mainly electrophilic properties (i.e. Lewis base adduct formation via attack at the carbon center [67] and stoechiometric cyclopropanation of olefins [68]) while the metal-alkylidene ones are nucleophilic (i.e. give Wittig-type alkylations [69], Lewis acid adduct formation [70] and olefin metathesis[71]). These differences in the chemical reactivity can be explained with the presence of a different metal-carbon bonding. The "low-valent" metal fragments (i.e. W(CO)$_5$) give carbenes in which a σ-donor/π-acceptor metal bond is formed, while the "high-valent" metal moiety gives alkylidenes in which a covalent double bond is present. The presence of heteroatoms in the methylene moiety stabilizes the carbene complexes. This means that in, the carbene complexe formation the CXY ligand interacts in its singlet (σ^2) state, while in the alkylidene compounds the ligand interacts with the metal fragment in its triplet state (σπ). Previous theoretical studies performed employing the GVB method on different metal-CH$_2^+$ systems confirm these indications. Theoretical studies on metal-methylidene bonds are essentially concentrated on model systems such as CrCH$_2^+$ [72-75], MnCH$_2^+$ [72], FeCH$_2^+$ [76], and RuCH$_2^+$ [73, 77]. Apart from the computational advantage in the study of model systems, we note that recently many of these molecules [78, 79] have been isolated experimentally using Fourier Transform Mass Spectrometry (FTMS), Ion Cyclotron Resonance Mass Spectrometry (ICR) and Ion Beam (IB) techniques. We have studied the RhCH$_2^+$ and RhCF$_2^+$ complexes.

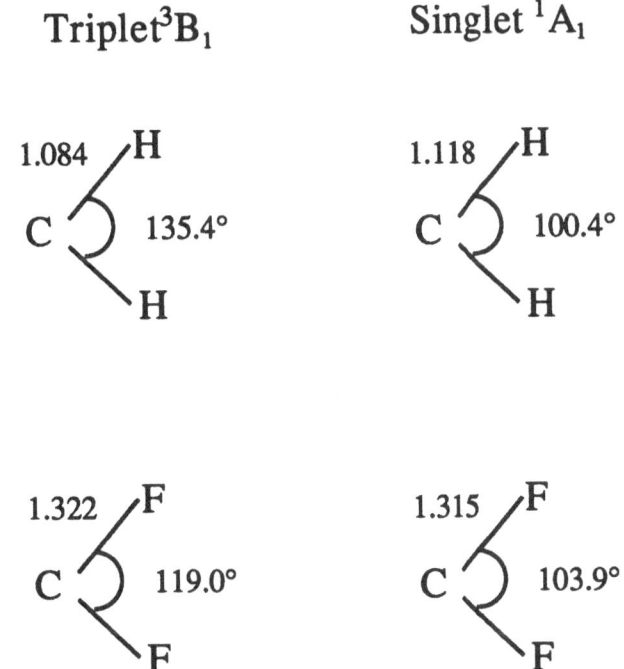

Figure 1. LCGTO-LSD optimum geometries for $^1A_1$ and $^3B_1$ states of CH$_2$ and CF$_2$ molecules (bond lengths are in Å and valence angles in degrees).

The former has been studied experimentally by FTMS [80] and its bond dissociation energy as well as its gas phase chemistry is known.

Before studying the complexes, we report the calculated properties of the two ligands used. The full optimized geometries are reported in figure 1 and the singlet-triplet energy splittings (ΔE S-T), computed as energy difference between singlet and triplet states, are collected in Table 3.

For CH_2 an accurate comparison with the experimental and theoretical geometries is possible. For the $^1A1$ state our calculation gives a bond length of 1.118 Å and a bond angle of 100.4 degrees, while for the $^3B_1$ state we found C-H= 1.084 Å and HCH=135°. The experimental values are C-H=1.110 Å and HCH=102.4° [81] and C-H=1.070 Å and HCH=133.0° [82] or 136.0° [81] for the $^1A_1$ and $^3B_1$ states respectively. The HCH bond angle computed at HF level is 131° for the $^3B_1$ state and 104.7° for the $^1A_1$ one. A better agreement is obtained with the optimization at the higher level of theory. In fact, at the MP2/6-31G* level, the following results are obtained [83]: C-H=1.109 Å and HCH=102.1° for the $^1A_1$ state and C-H=1.077 Å and HCH=131.6° for the $^3B_1$ one. At GVB-POL-CI [84] the values are: C-H=1.113 Å and HCH=101.8° for the $^1A_1$ state and C-H=1.084 Å and HCH=133.2° for the $^3B_1$ state.

The singlet-triplet separation in CH_2 has long been a subject of controversy, but recent experimental studies [87] have established its value to be 9.09 $\pm$ 0.21 Kcal/mol.

Several calculations (see Table 3) agree in predicting a triplet ($^3B_1$) ground state but the value of ΔE_{S-T} is correctly reproduced only if a high level of theory is employed. For this reason the determination of triplet-singlet energy separation in CH_2 is a severe test for quantum mechanical methods. As is shown in Table 3 the HF method fails in the splitting prediction. Our value is in good agreement with both experimental and other high level calculations such as full-CI , GVB and CCCI. When the hydrogen atoms are substituted with a more electronegative atoms an increase of ionic bond character is expected. The C-X bond will utilize C p orbitals, since they have the lowest valence ionization potential, leaving more s character for nonbonding C σ orbital. For these reasons in CF_2 a singlet σ^2 ground state can be predicted. Theoretical and experimental results give the singlet

TABLE 3. Singlet-Triplet splittings (ΔE_{S-T}= E singlet- E triplet) in CH_2 and CF_2 from theory and experiment (all values are in Kcal/mol)

Molecule	CH_2	CF_2	Ref.
HF	26.1	-32.5	84
Full-CI	12.0		85
CI-SD	13.5		86
MP4	16.8		83
GVB	9.1	-47.2	84
CCCI	9.0	-57.5	84
LCGTO-NLSD	9.1	-54.1	PW
Thermochem. Est.	-	-50 $\pm$ 10.8	84
Experimental	9.09 $\pm$ 0.21	-56.6	87, 88

state much more stable than the triplet one. From Table 3 it is evident that, also in this case, the HF method gives an unreliable result for ΔE_{S-T}, while the LCGTO-LSD value is in good agreement with experimental [88] evidence and with values obtained by GVB and CCCI [84] calculations.

Although the $RhCH_2^+$ system has been isolated its geometrical and electronic structure is unknown. Concerning the geometry, different structures (see scheme 1) can be considered as candidates for the absolute minimum. In addition, in order to find the ground state, for each isomer we have calculated both the singlet and triplet states.

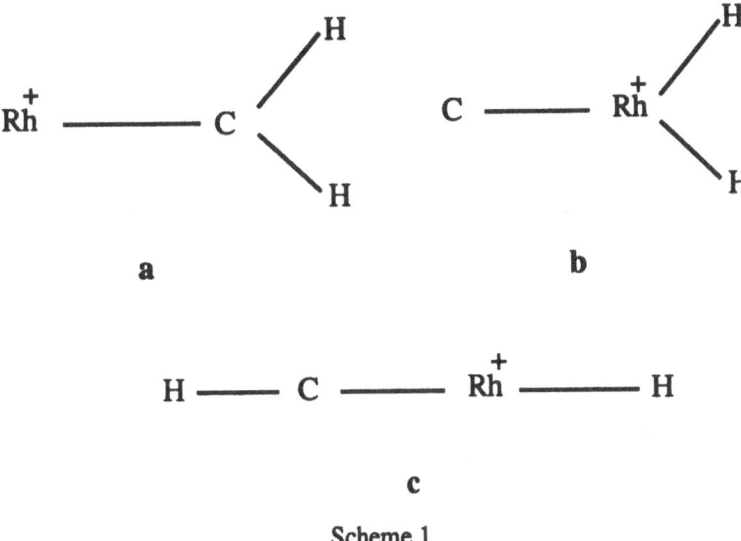

a

b

c

Scheme 1

TABLE 4. LCGTO-MCP-LSD Singlet-Triplet splittings (ΔE_{S-T}) and geometrical parameters in $RhCH_2^+$ and $RhFH_2^+$ molecules in their absolute minima. Bond lengths are in Å, valence angles in degrees and energies in Kcal/mol.

Molecule	Rh- C	X- C -X	ΔE_{S-T}
$RhCH_2$	1.757	124.1	-14.9
$RhFH_2$	1.868	111.7	11.4

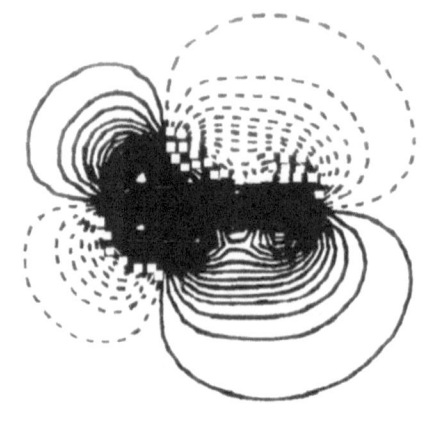

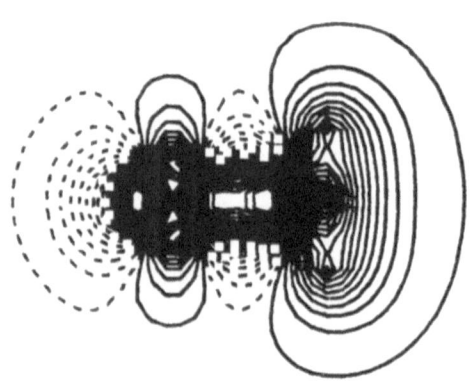

Figure 2. Orbital wave functions for π and σ orbitals of $RhCH_2^+$

Full geometry optimization and vibrational analysis indicate that the structure **a** is the most stable, while **b** is a secondary minimum in the potential energy surface (lies at 46.8 Kcal/mol for $RhCH_2^+$) and the structure **c** is a transition state.

The ground state for the absolute minimum is found to be the $^1A_1$ (C_{2v} symmetry) for $RhCH_2^+$ and $^3A_1$ for $RhCF_2^+$. The singlet-triplet gap together with the optimized geometrical parameters of the ground state are reported in Table 4.

From $RhCH_2^+$ gas phase chemistry [80] it is possible to have some indications about the structure of the complex. In fact the study of the H/D exchange reactions provides evidence against the structures **b** and **c**. The patterns of the dehydrogenation and CH_2 elimination reactions strongly support the structure **a** as the absolute minimum [80]. Concerning the electronic ground state we found that in going from $RhCH_2^+$, to $RhCF_2^+$, it changes from $^1A_1$ to $^3A_1$. If we consider that the ground states of the CH_2 and CF_2 ligands are different (triplet versus singlet) we can regard these results with confidence. The analysis of the bonding in the ground state of $RhCH_2^+$ shows the presence of σ and π bonds between Rh and the CX_2 moiety (see figure 2).

The Rh-C σ bond has an overlap population of 0.79 with 0.94 and 0.27 electrons ascribed to Rh and CH_2 respectively. The bond is essentially due to the overlap between d orbitals of Rh and p orbitals of carbon. The Rh-C π bond has an overlap population of 0.52 with 1.15 and 0.52 electrons associated to rhodium and ligand respectively. Also in this case, the bond formation is largely due to the overlap between Rh 4d and C 2p orbitals. These results are very similar to those obtained for $RuCH_2^+$ employing the GVB method [73, 77]. For the bond formation in the triplet excited state the same mechanism of $RuCH_2^+$ bond is found. The electron density map for the ground state of $RhCH_2^+$ is reported in figure 3.

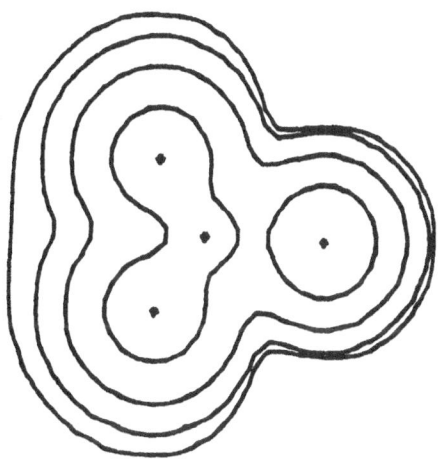

Figure 3. Contour plot of the charge density for $RhCH_2^+$.

In the case of $RhCF_2^+$ the ground state of the ligand is now a $^1A_1$ and those of the complex, a triplet. A comparison between the bond order in the two molecules reveals some significant difference. In the ground states the bond order decreases, in going from $RhCH_2^+$ to $RhCF_2^+$ by 0.53 (1.95 versus 1.42) while a lesser reduction is observed in the case of low-lying excited states. This result is in agreement with the decrease of bond dissociation energy observed experimentally in going from $NiCH_2^+$ to $NiCF_2^+$ (86 versus 47 Kcal/mol) [89]. Our calculated binding energies for $RhCH_2^+$ and $RhCF_2^+$ ground states are 113.1 Kcal/mol and 66.5 Kcal/mol respectively, in agreement with the experimental trend measured for Ni complexes.

3.2. COORDINATION OF CARBON DIOXIDE WITH A NI(0) COMPLEX

A large amount of experimental and theoretical work has been devoted to the organometallic chemistry of carbon dioxide (for a recent review see ref.90).The main reason for this interest is the activation of carbon dioxide, which is the most abundant source of C1 chemistry and the possibility of energy storage. Both these processes are expected to occur through transition metal catalysis. Several transition metal complexes able to coordinate CO_2 or to act as catalysts in CO_2 fixation, have been synthesized and characterized although CO_2 very rarely forms stable adducts with transition metals. Nothwithstanding the great number of theoretical studies performed in these last years (for an excellent review see ref.91 and the appropriate chapter of this book) , an exhaustive description of bonding between transion metals and CO_2 ligand is lacking. The main open question concerns the coordination mode and the factors determining the bond formation. Several coordination modes have been proposed [92,93]. In Figure 4 the η^2 side-on (a), η^1 -O end-on (b) and $\eta^2$1-C (c) modes for the nickel(0)-carbon dioxide model complex are shown.

Until now, the calculations have been performed using the experimental geometry taken from the x-ray study of (carbon dioxide)-bis(tricyclohexylphosphine)nickel ([Ni(CO_2)(PCy$_3$)$_2$]) complex [94] or with a partial geometry optimization (generally the Ni-C distance and OCO bond angle) [92, 93, 95] using symmetry constraints. In all these works the studied coordination modes are considered as minima in the potential energy surface. For the first time, in our study, a full geometry optimization without symmetry constraints has been performed for the three considered coordination modes of figure 4. Results show that only the η^2 side-on (a) and η^1 -O end-on (b) are minima while the η^1 - C (c) is not a stable structure because it collapses into the η^2 side-on during the

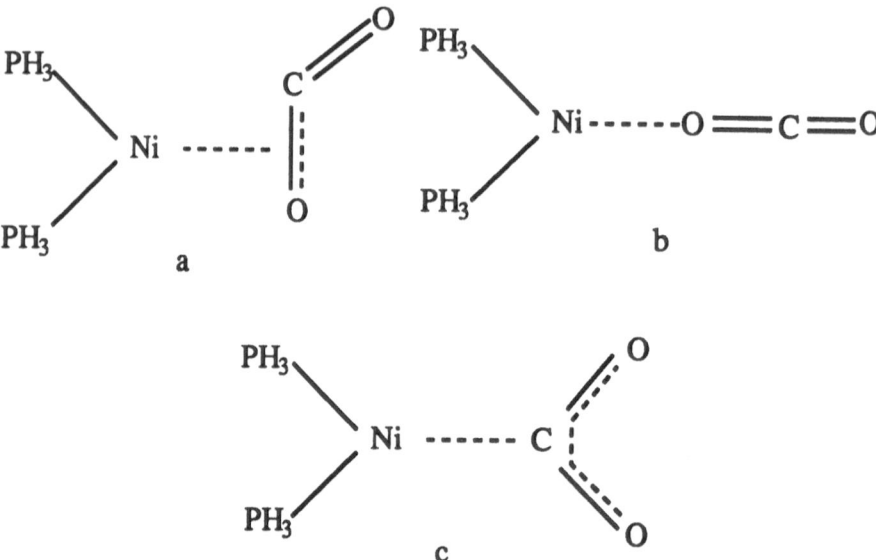

Figure 4. Selected possible coordination modes for CO_2 interaction with Ni(0)(PH$_3$)$_2$ fragment.

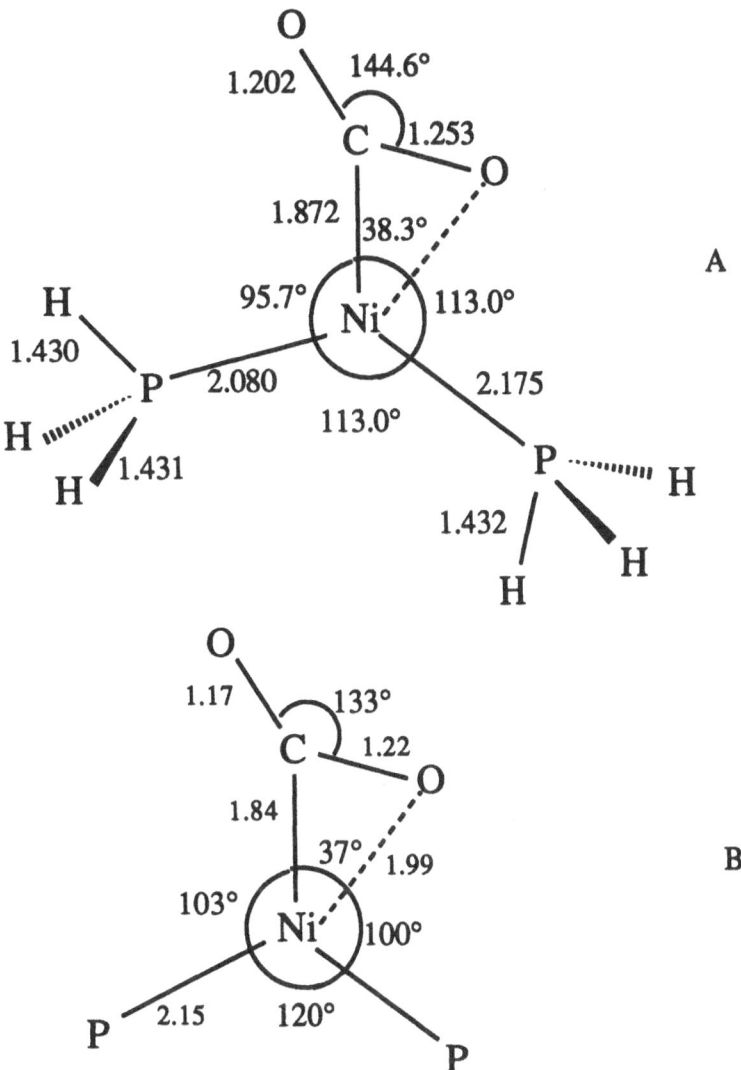

Figure 5. LCGTO-MP-LSD optimized structure for the η^2 side-on CO_2 coordination mode (A); main X-Ray structural parameters (B).

optimization step. The absolute minimum is the η^2 side-on and the η^1 end-on lies at 12.8 Kcal/mol. The calculations have been redone using also the NH_3 ligand instead of PH_3 and also in this case the absolute minimum is the η^2 side-on with η^1 -O end-on as a relative minimum and with the η^1 -C as not a stable structure.

TABLE 5. Relative energies (Kcal/mol) for different coordination modes in L_2NiCO_2 (L=NH$_3$, PH$_3$) complexes from different theoretical computations (est.=estimation; not stable means that this mode is not found as a minimum).

Method	Coord. mode	L	ΔE	Ref.
CAS-SCF	η^2 side-on	NH$_3$	0.	91, 92
	η^1 -O end-on		39.8	
	η^2 -C		46.8	
HF	η^2 side-on	PH$_3$	0.	93
	η^1 -O end-on		27.8	
	η^2 -C		32.2	
HF	η^2 side-on	PH$_3$	0.	95
	η^2 -C		23.3	
SD-CI	η^2 side-on	PH$_3$	0.	
	η^2 -C		3.8	
Full-CI est.	η^2 side-on	PH$_3$	0.	
	η^2 -C		12.1	
LCGTO-MP-LSD	η^2 side-on	PH$_3$	0.	PW
	η^1 -O end-on		13.8	
	η^2 -C		not stable	

The optimized structure of η^2 side-on is reported in figure 5 together with the most significant experimental parameters derived from the X-ray analysis [94].

The agreement is very good especially for the bond lengths. The values of some valence angles are slightly different with respect to the experimental ones. This is mainly due to the fact that the X-ray study has been performed for the ([Ni(CO$_2$)(PCy$_3$)$_2$]) complex in which the P atoms are surrounded by tricyclohexyl groups and a high steric hindrance is present. Results are in good agreement also with a recent SD-CI study [95] in which the C-O and Ni-C bond lengths and OCO valence angle have been optimized.
In Table 5 we have collected the relative stabilities of L_2Ni CO$_2$ (L=NH$_3$, PH$_3$) complexes obtained from different levels of theory. All the computations give the η^2 side-on as the most stable coordination mode as was indicated by the X-ray experiment [94].

As it is shown in Table 5, in all the previous works the η^2 -C coordination mode is considered a significant point in the potential surface, but our full geometry optimization clearly reveals that this geometry is not a minimum.

From this result it is evident that the full geometry minimization is necessary for a reliable description of coordination modes of CO$_2$. Furthermore, our computation confirms that the Dewar-Chatt-Duncanson model is able to correctly describe the bonding (in the η^2 side-on form) between the metal and the carbon dioxide as well the bond in the olefin transition metal complexes and it is in agreement also with previous theoretical works [91-95]. In fact,from the Mulliken population analysis it is clear that in the η^2 side-on topology there is a favorable overlap between the metal dπ and the CO$_2$ π^* orbital.

4. Cluster Model Calculations for Heterogeneous Catalysis

In the past years evidence has been obtained indicating that many binary alloys (e.g. NiAl, PtSn, PtGe, RhSn, PdSn, NiSn, RuCu, NiAl) considerably improve the catalytic processes on monometallic systems. In fact, the bimetallic catalysts are more stable, enhance the selectivity and lower the rate of desorption products. For these reasons bimetallic catalysts are extensively used in industrial processes for catalytic reforming and in fundamental science as model systems for the investigations of elementary mechanisms of catalytic phenomena. Although the bimetallic catalysts have been the subject of many investigations, the elementary mechanism responsible for the improved performance is not well established. Two schools of thought have tried to explain this phenomenon. The so-called "electronic explanation" introduced by Davis et al [96], Burch et al [97] and Figueras and co-workers [98] suggests that the electronic properties of the transition metals are modified by the presence of the second metal (generally a non-transition metal). On the contrary, the geometric reasons [98,99,100] emphasizes that the atoms of the second metal divide up the transition metal surface into small ensembles which are unable to catalyze the undesiderable reactions (e.g. the formation of carbonaceous residues). In the absence of detailed experimental evidence, it is reasonable to postulate that both the hypotheses can be considered for the explanation of the better catalytic efforts of bimetallic systems.At the quantum mechanical level very few works have been devoted to this subject . To our knowledge, besides the very recent works on atomic H on Pd, Rh and RhSn, RhZn and PdSn clusters [57,101], only the interaction of CO with Ni_2 Al and Ni_3 Al clusters [102], performed at the GVB level, and the interaction of carbon monoxide with NiCu clusters [103] using the SCF-LSD-SW method exist. In the former work, the influence of Al atom on the bonding properties of the CO/Ni system is explored while in the latter the effects of chemisorption and alloying on the magnetism of nickel clusters is reported.

One of the most important steps in many catalytic reactions (e.g. in hydrogenation and dehydrogenation reactions) is the interaction of hydrogen between the solid metal catalysts. In addition, hydrogen is extensively used to characterize metal-alloy systems and many interesting aspects of hydrogen-metal interactions (e.g. bulk and surface diffusion, spillover phenomena, complex desorption behaviour) have been recently observed [104-107].The PdSn and RhSn systems studied show well documented catalytic activities. PdSn improves the dehydrogenation reactions (e.g. conversion of cyclohexanone to phenol, cyclohexylamine to aniline, cyclohexane to benzene and 2-propanol to acetone) [108]. The RhSn silica-supported catalyst was found to exhibit a very good selectivity and a high activity in the hydrogenation of ethyl acetate into ethanol [107]. With the aim to investigate the influence of the second metal (Sn) on the hydrogen adsorption and diffusion properties several clusters that simulate both the (111) and (100) surfaces have been considered (the larger are reported in figure 6 and for others see ref. 57). Firstly the H adsorption and diffusion (in the bulk and on surface) have been studied on pure Pd and Rh systems.

It is well known, from previous experimental [105, 109] and theoretical [50] works that atomic hydrogen chemisorbs preferentially on the three-fold and four-fold high symmetry sites of (111) and (100) surfaces of Pd and Rh (see figure 6). For this reason the H adsorption has been studied only on these sites. Results are reported in Table 6 .

The reliability of computations can be verified by looking at the results for the equilibrium distance and equilibrium vibrational frequencies for which experimental or previous theoretical data are available [50, 105, 109]. The experimental ωe for H adsorbed on the palladium (111) surface is 998 cm^{-1} while our value for Pd_4 - H and

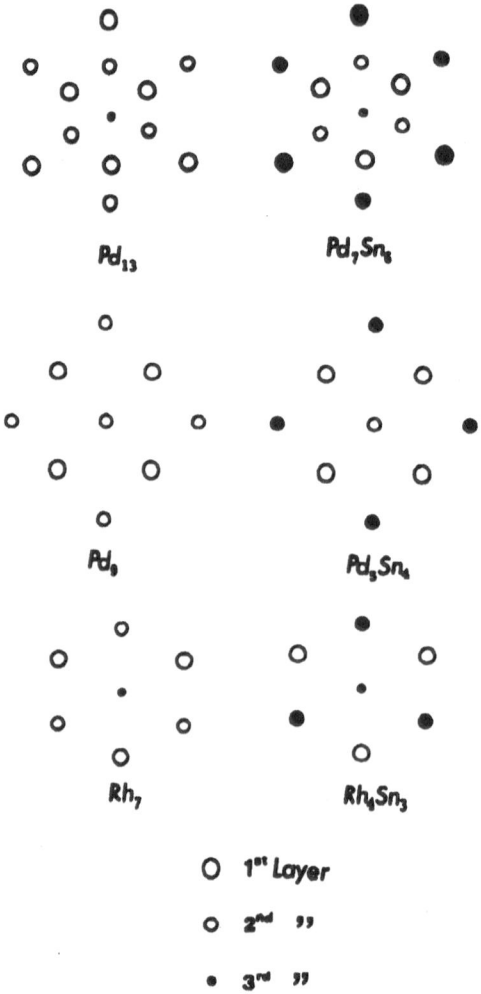

Figure 6. Structures of the studied clusters.

Pd_{13} - H systems is 1166 and 1113 cm^{-1} respectively. A good agreement with HF-CI (Pd_3 -H) and previous LSD (Pd_{10}-H) results is found for the R values. The binding energies (calculated without the non local corrections) decrease toward the experimental value of 2.6 eV for Pd(111) as the cluster size increases (calculation on a Pd_{16} -H cluster give a D_e of 3.3 eV [57]). The same good agreement is found for H on Pd(100) clusters and for Rh containing systems.

From the analysis of Table 5 it is clear that the main effect of the presence of the tin atom on both palladium (111) and (100) surfaces and in the rhodium (100) clusters is the

TABLE 6. LCGTO-MCP-LSD spectroscopic constants for hydrogen chemisorption on clusters simulating the three-fold and four-fold sites of (111) and (100) surfaces of Pd, Rh, PdSn and RhSn clusters respectively. In all cases the Sn atoms lie in the second layer. R (in Å) is the equilibrium H distance from the plane of the first layer cluster. The equilibrium vibrational frequencies (ω_e) are in cm^{-1} and the binding energies (D$_e$) are in eV.

Cluster	Surface	R	ω_e	D$_e$
Pd$_4$ - H	(111)	0.79	1166	3.8
Pd$_3$Sn - H		1.08	988	2.4
Rh$_4$ - H		0.99	1439	4.1
Rh$_3$Sn - H		1.10	1224	2.8
Pd$_{13}$ - H		0.92	1113	3.4
Pd$_7$Sn$_6$ - H		1.01	1173	3.2
Rh$_7$ - H		0.82	993	3.6
Pd$_9$ - H	(100)	0.08	669	3.3
Pd$_5$Sn$_4$ - H		0.13	883	3.0

reduction of binding energies. This effect is less marked if the Sn atom is substituted in the second layer and disappears when the tin atoms are placed in the third layer. The binding energy of Pd$_4$ - H is 3.8 eV. If the Pd atom of the second layer is replaced with Sn the D$_e$ becomes 3.1 eV and in the case of tin substitution in the first layer it decreases to 2.4 eV. The reduction of D$_e$ in the larger Pd$_{13}$ - H and Pd$_7$Sn$_6$ - H cluster is of about 0.2 eV. Similar trend is found in the case of Sn replacement in Rh systems.

The systematic decrease of hydrogen binding energy could have significant consequences for catalysis because the less tightly bound hydrogen is more accessible for certain types of reactions. In addition we note that the reduction of binding energy favours the desorption processes of the reaction products. These results have been employed [57] for an explanation of a recent desorption spectra of H$_2$ from RhSn/SiO$_2$ catalyst [106, 107]. Using the orbital energy diagrams (figure 7) and Mulliken population analysis the decrease of binding energy due to the presence of the non-transition metal can be rationalized.

As it is known, the main effect of H adsorption on Pd$_4$ is the split-off state below the d band at -11.1 eV [105]. This level is bonding with 52% H(s) and 42% Pd(d) character. At -6.4 eV another level with predominant bonding character is due to the H(s) (10%), Pd(s) (40%) and Pd(d) (51%). Finally, from fig. 7 we can see that the first antibonding level is localized at -3.4 eV. With the Sn replacement two split-off states result together with a nonbonding level. The lowest level (the composition is 21% H(s), 12% Pd(d) and 56% Sn(s)) is bonding with respect to both Pd-H and Sn-H , while the second (26% H(s), 32% Pd(d) and 22% Sn(s)) is Pd-H bonding and Sn-H antibonding. Because the antibonding level at -9.0 eV is now occupied the reduction of the binding energy is justified. Similar arguments can be used to rationalize the results for Rh containing clusters. As mentioned before a high-T thermodesorption peak has been observed for the RhSn/SiO$_2$ catalyst [106, 107]. For an explanation of this peak, that is present only for H chemisorption at high temperature, diffusion studies in and on cluster models have been performed. The calculation performed on Pd$_{13}$ - H and Pd$_7$Sn$_6$ - H clusters simulating the (111) surface show that the energy barrier for H diffusion on the bulk increases by about 0.8 eV when six Sn atoms are present . This result can be related to the explanation of the

Pd$_4$ HPd$_4$ H   Pd$_3$Sn HPd$_3$Sn H   Sn$_4$ HSn$_4$ H

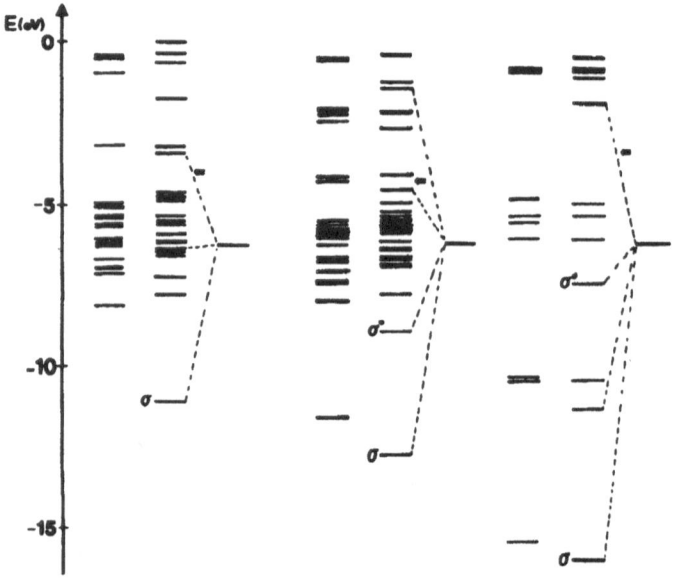

Figure 7. Orbital energies for the interaction of H with Pd$_4$, Pd$_3$Sn and Sn$_4$.

high-T desorption peak. The hydrogen penetrates into the bulk during the cooling from high temperature and is trapped here at lower temperature. The presence of tin requires a higher energy to release the hydrogen. The H diffusion on the surface has been studied employing Pd$_8$ - H, Pd$_6$Sn$_2$ - H, Pd$_{11}$ - H and Rh$_8$Sn$_3$ - H systems. In all cases the presence of tin increases the surface diffusion barrier. Correlations with the high-T desorption peak are possible also with this result because the diffusion paths leading to proximity of the hydrogen to tin are energetically costly [57].

5. Conclusion

The basic theory and the more recent developments of the LCGTO-MCP-LSD method have been briefly reviewed and calculations for some significant systems involved in homogeneous (RuCH$_2^+$, RuCF$_2^+$ and [Ni(CO$_2$)(PH$_3$)$_2$]) and heterogeneous (Rh$_n$ -H, Pd$_n$-H, Rh$_n$Sn$_m$ -H and Pd$_n$Sn$_m$ -H) catalytic processes have been reported. Geometric, spectroscopic and energetic parameters are reproduced with accuracy and are in agreement with available experimental data and other high level theoretical calculations. The results testify to the reliability of the LCGTO-MCP-LSD Gaussian-based Density Functional method in fields that are directly connected with homogeneous and heterogeneous catalysis. A number of new implementations (e.g. other electronic properties, analytical second derivatives and the possibility to localize the transition states) that are in progress in different laboratories can open new challenges for Density Functional studies of large and complex chemical systems including the full determination of the potential energy surfaces of catalytic reactions.

ACKNOWLEDGEMENT

This work was carried out with financial support of MURST, and CNR (Progetto Finalizzato Chimica Fine II, Progetto 90.00554.CT03, e 90.04158.CT12). I would like to thank Dennis Salahub, Alain St-Amant, Annick Goursot and Marirosa Toscano for their interest, collaboration and stimulating discussions.

REFERENCES

1. Trickey, S.B. (ed.) (1990), Adv. Quant. Chem. Vol. 21, Special issue on Density Functional Theory of Many-Fermion Systems.
2. Dreizler, R.M. and Da Providencia, J. (eds.), (1985), Density Functional Methods in Physics, Plenum, New York.
3. Dahl, J.P. and Avery, J. (eds.), (1982) Local Density Approximations in Quantum Chemistry and Solid State Physics, Plenum, New York.
4. Parr, R. and Yang, W. (eds.) (1989) Density-Functional Theory of Atoms and Molecules, Oxford Univeristy Press, New York.
5. Kryatchko, E. and Ludena E. (1990), Energy Density Functional Theory of Many Electron Systems, Kluwer, Dordrecht.
6. March, N. (1991), Electron Density Theory of Atoms and Molecules, Academic Press, New York.
7. Becke, A. D. (1989), Int. J. Quantum Chem. Symp. **23**, 599.
8. Dunlap, B. I. , Andzelm, J. and Mintmire, J. W. (1990), Phys. Rev. **A42**, 6354.
9. Fournier, R., Andzelm, J. and Salahub, D. R. (1989), J. Chem. Phys. **90**, 6371.
10. St-Amant A. and Salahub D. R. (1990), Chem. Phys. Lett. **169**, 387.
11. Becke, A. D. (1988) Phys. Rev. **A38**, 3098; (1988), J. Chem. Phys. **88**, 2547.
12. (a) Perdew, J. P. (1986), Phys. Rev. **B33**, 8822.
    (b) Perdew, J. P. and Wang, Y., (1986), Phys. Rev. **B33**, 8800.
13. Fournier, R. (1990), J. Chem. Phys. **92**, 5422.
14. Andzelm, J and Dunlap, B. I.(1991), Phys. Rev. in press.
15. Jones, R. O. and Gunnarson, O. (1990), Rev. Mod. Physics **61**, 689.
16. Ziegler, T. (1990), Chem. Rev. **91**,651.
17. (1989), J. Chim. Phys. (Paris) **86**, special issue on Density Functional Theory and its Applications.
18. Salahub, D. R. and Zerner, M.C. (eds.) (1989), The Challenge of d and f Orbitals: Theory and Computations, ACS Symposium Series, **394**.
19. Labanowski, J. and Andzelm, J. (eds.) (1991), Density Functional Methods in Chemistry Springer Verlag, New York.
20. St-Amant A. (1991), Ph. D. Thesis, Université de Montréal, Canada.
21. Andzelm, J. DGauss program, Cray Research Inc. Minneapolis, USA.
22. Andzelm, J., Radzio, E. and Salahub, D. R. (1985), J. Chem. Phys. **83**, 4573.
23. Baerends, E. J., Ellis, D. E. and Ros, P. (1973), Chem. Phys. **2**, 41.
24. Teter, M. P., Payne, M. C. and Allan, D. C. (1989), Phys. Rev. **B40**, 1225.

25. Wimmer, E., Krakauer, H., Weinert, M. and Freeman, A. J. (1981), Phys. Rev.**B24**, 864.
26. Mlynarski, P. and Salahub, D. R. (1991), Phys. Rev. **B43**, 1399.
27. Bartolotti, L. J. (1981), Phys. Rev. **A24**, 1682
28. Sambe, H. and Felton, R. H. (1975), J. Chem. Phys. **62**, 1122.
29. Dunlap, B. I., Connolly, J. W. D. and Sabin, J. R. (1979), J. Chem. Phys. **71**, 3396; 4993
30. Salahub, D. R. (1987), Adv. Chem. Phys. **69**, 447.
31. Hohenberg, P. and Kohn, W. (1964), Phys. Rev. **B136**, 864.
32. Kohn, W. and Sham, L. J. (1965), Phys. Rev. **A140**, 1133.
33. Vosko, S. H., Wilk, L. and Nusair, M. (1980), Can. J. Phys. **58**, 1200.
34. Szasz, L. (1985), Pseudopotential Theory of Atoms and Molecules, Wiley, New York.
35. Hellmann, H. (1935), J. Chem. Phys. **3**, 61.
36. Bachelet, G. B., Hamann, D. R. and Schluter, M. (1982), Phys. Rev. **B26**, 4199 and references therein.
37. Cohen, M. L. (1983), Int. J. Quantum Chem. Symp. **17**, 583 and references therein.
39. Durand, Ph. and Barthelat J-C. (1975), Theoret. Chim. Acta **38**, 283.
40. Bonifacic, V. and Huzinaga, S. (1974), J. Chem. Phys. **60**, 2779.
41. Huzinaga, S., Seijo, L., Barandiaran, Z. and Klobukowski, M. (1987), J. Chem. Phys. **86**, 2132.
42. Seijo, L., Barandiaran, Z. and Huzinaga, S.(1989), J. Chem. Phys. **91**, 7011.
43. Andzelm, J, Russo, N. and Salahub, D. R. (1987), J. Chem. Phys. **87**, 6562.
44. Andzelm, J, Russo, N. and Salahub, D. R. (1987), Chem. Phys. Lett. **142**, 169.
45. Musolino, V., Toscano, M. and Russo, N. (1990), J. Comp. Chem. **11**, 924.
46. Obara, S. and Saika, A. (1986), J. Chem. Phys. **84**, 3963.
47. Huzinaga, S.(ed.) (1984), Gaussian Basis Sets for Molecular Calaculations, Elsevier, Amsterdam.
48. (a) Godbout, N., Salahub, D. R., Andzelm, J. and Wimmer, E. Can. J. Chem., in press.
    (b) Data Bank, Département de Chimie, Université de Montréal, Canada.
    (c) Data Bank, Dipartimento di Chimica, Universita della Calabria, Italy.
49. Peluso, A., Goursot, A. and Salahub, D.R. (1990), Inorg. Chem. **29**, 1545.
50. Andzelm, J. and Salahub, D. R. (1986), Int. J. Quantum Chem. **19**, 1091.
51. Salahub, D. R. (1987), Adv. Chem. Phys. **69**, 477.
52. Russo, N., Andzelm, J. and Salahub, D.R. (1987), Chem. Phys. **114**, 331.
53. Shim, I. and Gingerich, K.A. (1985), Surface Sci. **156**, 623.
54. Herzberg, G. (1966), Molecular Spectra and Molecular Structure Vol. 3, Van Nostrand Princeton.
55. Baykara, N. A., Andzelm, J., Salahub, D. R. and Baykara, S. Z. (1986), Int. J. Quantum Chem. **29**, 1025.
56. Talbert, M. A. and Beauchamp, J. L. (1986), J. Phys. Chem. **90**, 5015.
57. Rochefort, A., Andzelm, J., Russo, N. and Salahub, D. R. (1990), J. Am. Chem. Soc. **112**, 8239.
58. Huber, H. and Herzberg, G. (1979), Molecular Spectra and Molecular Structure. Constants of Diatomic Molecules. Van Nostrand, Princeton.
59. Andzelm, J., Goursot, A., Russo, N. and Salahub, D. R. (1990), J. Chem. Phys. **93**, 2919.

60. Panas, I., Schule, J., Brandemark, U., Siegbahn, P. and Wahlgren, U. (1988), J. Phys. Chem. **92**, 3079.
61. Fournier, R. (1989), Ph. D. Thesis, Université de Montréal.
62. Walch, S.P. and Bauschlicher, C. W. (1983), J. Chem. Phys. **78**, 4597.
63. Goddard, W., Walch, S. P., Rappe, A. K. and Upton, T. H. (1977), J. Vac. Sci. Technol. **14**, 416.
64. Russo, N. and Toscano, M. unpublished results.
65. Dotz, K.H., Fischer, H., Hofmann, P., Kreissl, F.R., Schubert, U. and Weiss, K. (1984), Transition Metal Carbene Complexes, Verlag Chemie, Deerfield Beach.
66. Fischer, E. O. (1976), Adv. Organomet. Chem. **14**, 1.
67. Kuo, G. H., Helquist, P.and Kerber, R. C. (1984), Organometallics **3**, 806 and refences therein.
68. Casey, C. P., Niles, W. H.and Tukada, H. (1985), J. Am. Chem. Soc. **107**, 2924 and references therein.
69. Pine, S. H., Zahler, R., Evans, D. A. and Grubbs, R. H. (1980), J. Am. Chem. Soc. **102**, 3270.
70. Schrock, R. R. (1975), J. Am. Chem. Soc. **97**, 6577.
71. Kress, J. and Olsborn, J. A. (1983), J. Am. Chem. Soc. **105**, 6346.
72. Vincent, M. A., Yoshloka, Y. and Schaefer III, H. F. (1982), J. Phys. Chem. **86**, 3905.
73. Carter, E. A. and Goddard III, W. A. (1986), J. Am. Chem. Soc. **108**, 4746.
74. Carter, E. A. and Goddard III, W. A. (1986), J. Phys. Chem. **88**, 4148.
75. Alvarado-Swaisgood, A. E., Allison, J. and Harrison, J. F. (1985), J. Phys. Chem. **89**, 2517.
76. McKee, M. L. (1990), J. Am. Chem. Soc. **112**, 2601.
77. Carter, E. A. and Goddard III, W. A. (1986), J. Am. Chem. Soc. **108**, 2180.
78. Jacobson, D. B. and Freiser, B. S. (1985), J. Am. Chem. Soc. **107**, 2605 and references therein.
79. Elkind, J. L. and Armentrout, P. B. (1987), J. Phys. Chem. **91**, 2037 and references therein.
80. Jacobson, D. B. and Freiser, B. S. (1985), J. Am. Chem. Soc. **107**, 5870.
81. Hellwege, K. H. (1976), Structural Data on Free Polyatomic Molecules, Springer Verlag, Berlin.
82. Jensen, P., Bunker, P. R. and Hoy, A. R. (1982), J. Chem. Phys. **77**, 5370.
83. Luke, B. T., Pople, J. A., Krogh-Jerpersen, M., Apeloig, Y., Chandrasekhar, J. and von Rague' Schleyer, P. (1986), J. Am. Chem. Soc. **108**, 260.
84. Harding, L. B. and Goddard III, W. A. (1978), Chem. Phys. Lett. **55**, 217
84. Carter, E. A. and Goddard III, W. A. (1987), J. Phys. Chem. **91**, 4651.
85. Bauschlicher, C. W. and Taylor, P. R. (1986), J. Chem. Phys. **85**, 5936.
86. Lucchese, R. R. and Schaefer III, H. F. (1977), J. Am. Chem. Soc. **99**, 6765.
87. Bunker, P. R, Jensen, P., Kraemer, W. P. and Beardsworth, R. (1986), J. Chem. Phys. **85**, 3724 and references therein.
88. Koda, S. (1982), Chem. Phys. **66**, 386.
89. Halle, L. F., Armentrout, P. B. and Beauchamp, J. L. (1983), Organometallics **2**, 1829.
90. Aresta, M. and Schloss J. V. (eds.) (1990), Enzymatic and Model Carboxylation and Reduction Reactions for Carbon Dioxide Utilization, Kluwer, Dordrecht.

366

91.   Dedieu, A. and Ingold, F. (1990), in Aresta, M. and Schloss J. V. (eds.), Enzymatic and Model Carboxylation and Reduction Reactions for Carbon Dioxide Utilization, Kluwer, Dordrecht p. 23.
92.   Dedieu, A. and Ingold, F. (1989), Angew. Chem. Int. Ed. Engl. **28**, 1694.
93.   Sakaky, S., Kitaura, K. and Morokuma, K. (1982), Inorg. Chem. **21**, 760.
94.   Aresta, M. and Nobile, C. F. (1975), J. Chem. Soc. Chem. Commun. 639.
95.   Sakaky, S., Koga, N. and Morokuma, K. (1990), Inorg. Chem. **29**, 3110.
96.   Davis, B. H., Westfall, G. A., Watkins, J. and Pezzanite J. J. (1972), J. Catal. **42**, 283.
97.   Burch, R. and Garla, L. C. (1981), J. Catal. **71**, 360.
98.   Coq, B. and Figueras, F. (1984), J. Catal. **85**, 197.
99.   Sinfelt, J. H., Carter, J. L. and Yates, D. J. C. (1972), J. Catal. **24**, 283.
100.   Dautzenberg, F. M. , Helle, J. N., Biloen, P. and Sachtler, W. M. H. (1980), J. Catal. **63**, 119.
101   Andzelm, J., Rochefort, A., Russo, N. and Salahub, D. R. (1990), Surface Sci. **235**, L319.
102.   Tatar, R. C. and Messmer, R. P. (1987), J. Vac. Sci. Technol. **A5**, 675.
103.   Salahub, D. R. and Raatz, F. (1984), Int. J. Quantum. Chem. Symp. **18**, 173.
104.   King, T. S., Wu, X. and Gerstein, B. C. (1986), J. Am. Chem. Soc. **108**, 6056.
105.   Eberhardt, W., Louie, S. G. and Plummer, E. W., (1983), Phys. Rev. **B28**, 465.
106.   Candy, J-P., Ferretti, O. A., Bournonville, J-P. and Mabilon, G. (1985), J. Chem. Soc. Chem. Commun. 1197.
107.   Candy, J-P., Ferretti, O. A., Mabilon, G., Bournonville, J-P., El Mansour, A., Basset, J. M. and Martino, G. (1990), J. Catal. **112**, 201.
108.   Masai, M., Honda, K., Kubota, A., Ohamaka, S., Nishikawa, Y., Nalchara, K., Kishic, K. and Ikeda, S. (1977), J. Catal. **50**, 419.
109.   Christmann, K. (1988), Surf. Sci. Rep. **9**, 1.

A GENERAL ENERGY DECOMPOSITION SCHEME FOR THE STUDY OF METAL-LIGAND INTERACTIONS IN COMPLEXES, CLUSTERS AND SOLIDS.

T. ZIEGLER
University of Calgary
Department of Chemistry,
Calgary, Alberta, Canada
T2N 1N4

ABSTRACT. A decomposition scheme is presented for the metal-ligand interaction energy in complexes, clusters and solids. The scheme is based on density functional theory. It allows for a breakdown of the interaction energy into steric and electronic factors, where the electronic factors can be decomposed further into contributions from ligand-to-metal charge donations and metal-to-ligand back-donations. The scheme allows in addition for the incorporation of relativistic effects .

1. Introduction

The interaction between transition metals and molecular fragments is a common theme in studies on discrete transition metal atoms and complexes as well as clusters and surfaces. The interaction might take place on a transition metal surface, **1**

Hollow position   Top position

1a   **1b**

in a bridge between two or more metals of a polynuclear complex, **2**, or simply on a single metal centre in a mononuclear compound, **3**.

Of interest in such studies is the relative stability of CO chemisorbed at the hollow site, **1a**, and top position, **1b**, the coordinative preference of acetylene parallel, **2a**, or

367

D. R. Salahub and N. Russo (eds.), Metal-Ligand Interactions: from Atoms, to Clusters, to Surfaces, 367–396.
© 1992 *Kluwer Academic Publishers. Printed in the Netherlands.*

in plane	perpendicular	in plane	perpendicular
2a	**2b**	**3a**	**3b**

perpendicular, **2b**, to the metal-metal bond, the preference of olefin to complex in or perpendicular to the coordination plane of $PtCl_3^-$. We shall in the following discuss a method which allows for a quantitative estimate of the interaction energies in systems such as **1-3** as well as a qualitative analysis of the factors responsible for the coordinative preferences in **1-3**.

The method is based on Approximate Density Functional Theory [1] which over the past decade has emerged as a tangible and versatile computational method. It has been employed successfully to obtain thermochemical data [2,3]; molecular structures [4,5]; force fields and frequencies [6]; assignments of NMR [7,8]-, photoelectron [9] -, E.S.R [10] - , and UV- spectra [9]; transition state structures as well as activation barriers [11]; dipole moments [12] and other one-electron properties. Thus, approximate DFT is now applied to many problems previously covered exclusively by *ab initio* Hartree-Fock (HF) and post-HF methods. The recently acquired popularity of approximate DFT stems in large measure from its computational expedience which makes it applicable even to large size molecules at a fraction of the time required for HF or post-HF calculations. More importantly, perhaps, is the fact that expectation values derived from approximate DFT in most cases are better in line with experiment than results obtained from HF calculations. This is in particular the case for systems involving transition metals. An analysis of why approximate DFT affords more reliable results than HF has recently been published by Cook and Karplus [13] as well as Tschinke and Ziegler [14].

2. General Theory

2.1. THE KOHN-SHAM EQUATION

The total energy of an n-electron system can be written [15b] without approximations as

$$E = -\frac{1}{2} \sum_i \int \phi_i(\vec{r}_1) \nabla^2 \phi_i(\vec{r}_1) d\vec{r}_1 + \sum_A \int \frac{Z_A}{|\vec{R}_A - \vec{r}_1|} \rho(\vec{r}_1) d\vec{r}_1$$

$$+\frac{1}{2} \int \frac{\rho(\vec{r}_1)\rho(\vec{r}_2)}{|\vec{r}_1-\vec{r}_2|} \, d\vec{r}_1 d\vec{r}_2 + \sum_{A} \sum_{\neq B} \frac{Z_A Z_B}{|\vec{R}_A - \vec{R}_B|} + E_{XC} \qquad (1)$$

The first term in Eq. (1) represents the kinetic energy of n non-interacting [15] electrons with the same density $\rho(\vec{r}_1) = \sum \phi_i(\vec{r}_1)\phi_i(\vec{r}_1)$ as the actual system of interacting electrons. The second term accounts for the electron-nucleus attraction and the third term for the Coulomb interaction between the two charge distributions $\rho(\vec{r}_1)$ and $\rho(\vec{r}_2)$. The last term contains the exchange-correlation energy, E_{XC}. The exchange-correlation energy can be expressed in terms of the exchange-correlation energy densities [15], $\varepsilon_{xc}^{\gamma\gamma'}(\vec{r}_1)$, as

$$E_{XC} = \sum_{\gamma} \sum_{\gamma'} \int \rho_1^{\gamma}(\vec{r}_1)\varepsilon_{xc}^{\gamma\gamma'}(\vec{r}_1)d\vec{r}_1 \qquad (2)$$

where γ and γ' run over α as well as β spins. The functions $\varepsilon_{xc}^{\gamma\gamma'}(\vec{r}_1)$ contain all information about exchange and correlation between the interacting electrons as well as the influence[15b] of correlation on the kinetic energy. The functional form of the exact exchange-correlation energy densities, $\varepsilon_{xc}^{\gamma\gamma'}(\vec{r}_1)$, is not known. However, good approximations are available. The homogeneous electron gas has been particularly instrumental [16,17] in fostering useful approximate expressions for the exchange-correlation energy density. These expressions have $\varepsilon_{xc}^{\gamma\gamma'}(\vec{r}_1)$ as a simple function of the density and are refered to as local approximations. The simple HFS or Xα method [16] retains only $\varepsilon_{xc}^{\gamma\gamma'}(\vec{r}_1)$ for $\gamma=\gamma'$, whereas the Local Density Approximations (LDA) [17] contains contributions from $\gamma=\gamma'$ as well as $\gamma\neq\gamma'$. Langreth [18a] and Mehl, Becke [18b] and Perdew [18c] have in a series of pioneering papers eliminated many of the shortcomings inherent in the local approximations by introducing correction terms based on electron density gradients. These theories are refered to as nonlocal. Nonlocal corrections are essential for a quantitative estimate of bond energies [18e-k] as well as metal-ligand bond distances [18l,m]. They are also of importance for other properties [18m].

The one electron orbitals, $\{\phi_i(\vec{r}_1); i=1,M\}$, of Eq. (1) are solutions to the set of one-electron Kohn-Sham equations [1]

$$[-\tfrac{1}{2}\nabla^2 + \sum_A \frac{Z_A}{|\vec{R}_A - \vec{r}_1|} + \int \frac{\rho(\vec{r}_2)}{|\vec{r}_1 - \vec{r}_2|}\, d\vec{r}_2 + V_{XC}]\, \phi_i(\vec{r}_1) = h_{KS}\, \phi_i(\vec{r}_1)$$

$$= \varepsilon_i\, \phi_i(\vec{r}_1) \qquad\qquad\qquad\qquad\qquad\qquad\qquad\qquad (3)$$

where the exchange-correlation potential V_{XC} is given as the functional derivative of E_{XC} with respect to the density[1]

$$V_{XC}[\rho] = \frac{\delta E_{XC}[\rho]}{\delta\rho} \qquad\qquad\qquad\qquad\qquad (4)$$

It is relatively simple to derive an expression for the variational potential V_{XC} once a particular form for $\varepsilon_{XC}^{\gamma\gamma}(\vec{r}_1)$ has been selected. With V_{XC} at hand the set of Kohn-Sham equations in Eq.(3) can be solved leading to a set of one electron Kohn-Sham orbitals $\{\phi_i(\vec{r}_1, i=1,M)\}$, from which the total energy E of Eq. (1) as well as other expectation values can be calculated.

2.2. THE GENERALIZED TRANSITION STATE METHOD [19,20]

It is customary to study the metal-ligand bonds in **1-3** by a fragment approach based on the frontier orbitals of the ligand, fragment A, and the transition metal system, fragment B. We shall in the following illustrate how such a fragment analyses can provide a quantitative estimate of the interaction energy between A and B as well as a qualitative analysis of the interaction in terms of steric and electronic terms. The same fragment analysis can also be used to analyze the C-C bond of the simple ethane molecule, **4**, in terms of the interaction between two methyl fragments, or the multiple metal-metal bond in binuclear complexes, **5**, from the interaction between two monomers.

4   **5**

We consider the combined system $\overline{AB}$ as made up of the two fragments A and B. The two fragments A and B are brought together in a rigid fashion without changing the relative positions of the atoms within each fragment. The electron densities of A and B at infinite separations can be written in terms of the occupied and virtual spin-orbitals of A and B as

$$\rho_0^A (\vec{r}_1) + \rho_0^B (\vec{r}_1) = \sum_{i(A)}^{occ(A)} \phi_i^A(\vec{r}_1) \phi_i^A(\vec{r}_1) + \sum_{i(B)}^{occ(B)} \phi_i^B(\vec{r}_1) \phi_i^B(\vec{r}_1) \qquad (5)$$

or simply

$$\rho_0^A (\vec{r}_1) + \rho_0^B (\vec{r}_1) = \sum_j P_{jj}^0 \phi_j^0 (\vec{r}_1) \phi_j^0 (\vec{r}_1) \qquad (6)$$

where j runs over occupied and virtual orbitals of A as well as B. Further $P_{jj}^0 = 1$ if j is occupied whereas $P_{jj}^0 = 0$ in the case where j is a virtual orbital .

The combined basis set of occupied and virtual orbitals on A and B { $\phi_j^0 (\vec{r}_1)$, j=1,M} is able to describe the total system $\overline{AB}$ in any conceivable state, real or fictitious. In combining the two fragments, orbitals on A and B will begin to overlap and interact. Thus, the orbitals of $\overline{AB}$ in state Θ will be a linear combination of the full set { $\phi_j^0 (\vec{r}_1)$, j=1,M} with

$$\psi_i^{\overline{AB}(\Theta)} (\vec{r}_1) = \sum_\mu^M C_{i\mu}^{\overline{AB}(\Theta)} \phi_\mu^0 (\vec{r}_1) \qquad (7)$$

whereas the corresponding density is given by

$$\rho_\Theta^{\overline{AB}} (\vec{r}_1) = \sum_\nu^M \sum_\mu^M P_{\nu\mu}^{\overline{A}\,\overline{B}(\Theta)} \phi_\nu^0 (\vec{r}_1) \phi_\mu^0 (\vec{r}_1) \qquad (8)$$

where $P_{\nu\mu}^{\overline{AB}(\Theta)}$ is the bond order matrix

$$P_{\nu\mu}^{\overline{AB}(\Theta)} = \sum_{i}^{occ} C_{i\mu}^{\overline{A}\,\overline{B}\,(\Theta)} C_{iv}^{\overline{AB}\,(\Theta)} \tag{9}$$

We might also write the total density as

$$\rho_{\Theta}^{\overline{AB}}(\vec{r}_1) = \rho_0^A(\vec{r}_1) + \rho_0^B(\vec{r}_1) + \Delta\rho_{\Theta}^{\overline{AB}}(\vec{r}_1) \tag{10}$$

where

$$\Delta\rho_{\Theta}^{\overline{AB}}(\vec{r}_1) = \sum_{\mu}^{M}\sum_{v}^{M} \Delta P_{\mu v}^{\overline{A}\,\overline{B}(0\to\Theta)} \phi_\mu^0(\vec{r}_1)\,\phi_v^0(\vec{r}_1) \tag{11}$$

is the deformation density due to the change in the electron distribution as $\overline{AB}$ in the state Θ is formed from A and B , and

$$\Delta P_{ij}^{\overline{AB}(0\to\Theta)} = P_{ij}^{\overline{AB}(\Theta)} - P_{ij}^0 \tag{12}$$

It is readily shown [19a] that the energy of formation for $\overline{AB}$ from the fragments A and B is given by

$$\Delta E = E_{\Theta}^{\overline{AB}} - E_0^A[\rho_0^A] - E_0^B[\rho_0^B] =$$

$$+ E_{el} + E_{dlxc} + \sum_{\mu}^{M}\sum_{v}^{M} F_{v\mu}^{(0\to\Theta)} \Delta P_{v\mu}^{(0\to\Theta)} \tag{13}$$

Here

$$E_{el} = -\sum_{gA}^{N(A)} \int \frac{\rho_0^B(\vec{r}_1)Z_{gA}}{|\vec{r}_1 - \vec{R}_{gA}|} d\vec{r}_1 - \sum_{gB}^{N(B)} \int \frac{\rho_0^A(\vec{r}_1)Z_{gB}}{|\vec{r}_1 - \vec{R}_{gB}|} d\vec{r}_1$$

$$+ \quad \int \frac{\rho_0^A(\vec{r}_1)\rho_0^B(\vec{r}_2)}{|\vec{r}_1 - \vec{r}_2|} d\vec{r}_1 d\vec{r}_2 \quad +1/2 \sum_{gA}^{N(A)} \sum_{gB}^{N(B)} \frac{Z_{gA} Z_{gB}}{|\vec{R}_{gA} - \vec{R}_{gB}|} \qquad (14a)$$

is the electrostatic interaction between the fragments A and B at the positions they take up in the combined compound $\underline{AB}$, and

$$E_{dlxc} = E_{xc}[\rho_0^A + \rho_0^B] - E_{xc}[\rho_0^A] - E_{xc}[\rho_0^B] \qquad (14b)$$

is the difference in exchange correlation energy between the charge distribution $\rho_0^A + \rho_0^B$ with A and B at infinite separation, $E_{xc}[\rho_0^A] - E_{xc}[\rho_0^B]$, and A and B at the positions they will take up in $\overline{AB}$, $E_{xc}[\rho_0^A + \rho_0^B]$.

The matrix element $F_{\nu\mu}^{(0\rightarrow\Theta)}$ in the last term of Eq. (13) is given by

$$F_{\nu\mu}^{(0\rightarrow\Theta)} = \int \phi_\mu^0(\vec{r}_1) \, h_{ks}[\rho_0^A + \rho_0^B + 1/2 \, \Delta\rho_\Theta^{\overline{AB}}] \, \phi_\nu^0(\vec{r}_1) d\vec{r}_1 \qquad (15a)$$

where $h_{ks}[\rho_0^A + \rho_0^B + 1/2 \, \Delta\rho_\Theta^{\overline{AB}}]$ is the Kohn-Sham operator

$$h_{ks}[\rho] = -\frac{1}{2}\nabla^2 + \sum_A \frac{Z_A}{|\vec{R}_A - \vec{r}_1|} + \int \frac{\rho(\vec{r}_2)}{|\vec{r}_1 - \vec{r}_2|} d\vec{r}_2 + V_{XC}[\rho] \qquad (15b)$$

defined with respect to a density, $\rho = \rho_0^A + \rho_0^B + 1/2 \, \Delta\rho_\Theta^{\overline{AB}}$, which is intermediate between the superimposed charge distribution of the two fragments, $\rho_0^A + \rho_0^B$, and the final density of AB, $\rho_0^A + \rho_0^B + \Delta\rho_\Theta^{AB}$. We note that $F_{\nu\mu}^{(0\rightarrow\Theta)}$ as well as $\Delta P_{\nu\mu}^{(0\rightarrow\Theta)}$ both are zero if ν and μ belongs to different symmetry representations.

The expression in Eq. (13) can also be used to calculate the difference in energy between two states of $\overline{AB}$, Θ_1 and Θ_2, with the same geometry but different densities. The energy difference is given by

$$\Delta E = E_{\Theta_2}^{\overline{AB}} - E_{\Theta_1}^{\overline{AB}} = \sum_{\mu} \sum_{\mu} F_{\nu\mu}^{(\Theta_1 \to \Theta_2)} \Delta P_{\nu\mu}^{(\Theta_1 \to \Theta_2)} \qquad (16)$$

where $\Delta P_{\nu\mu}^{(\Theta_1 \to \Theta_2)}$ is the change in the P-matrix associated with the density difference

$$P_{\Theta_2}^{\overline{AB}} - P_{\Theta_1}^{\overline{AB}} = \sum_{\mu}^{M} \sum_{\nu}^{M} \Delta P_{\nu\mu}^{(\Theta_1 \to \Theta_2)} \phi_{\mu}^{0}(\vec{r}_1) \phi_{\nu}^{0}(\vec{r}_1) \qquad (17)$$

and $F_{\nu\mu}^{(\Theta_1 \to \Theta_2)}$ is a matrix element with respect to the Kohn-Sham operator defined by a density which is the average of the densities due to Θ_1 and Θ_2. The method outlined here is a generalization of Slater's transition state method [16]. The generalization [19] allows for a calculation of energy differences involving geometry changes, Eq. (13), whereas the original method [16] only considered changes with fixed molecular geometry such as ionizations and excitations, Eq. (16).

3. Energy Decomposition

There are a number of schemes available by which one can analyze the interaction energy between two fragments. Schemes have been proposed by Morokuma and co-workers [21],Whangbo and Wolfe and co-workers [22] Bernardi, Bottoni et al. [23] Stone and Erskine [24], Bagus et al. [25] as well as Fujimoto [26]. We shall here describe a scheme due to Ziegler and Rauk [19b,c] which makes use of the generalized transition state method [19a].

3.1. THE PREPARATION ENERGY CONTRIBUTION

Consider as an example the Re-CH3 bond in CH3Re(CO)5, **6**, generated from the two radical fragments CH3, **7**, and Re(CO)5, **8**. The free radicals **7** and **8** do not have the

| **6** | **7** | **8** | **9** | **10** |

same structure as the CH3, **10**, and Re(CO)5, **9** , frameworks in the combined CH3Re(CO)5 complex, **6**. Thus the CH3 framework, **10**, is trigonal pyramidal whereas the free CH3 radical is planar, **7**. Further the Re(CO)5 framework, **9**, in **6** is square pyramidal whereas the free Re(CO)5 radiacal is trigonal bipyramidal, **8**. The first step towards the formation of CH3Re(CO)5, **6**, involves a deformation or "preparation" of CH3 and Re(CO)5 to the shapes they will adopt in the combined complex, **6**. The energy required for this deformation is referred to as

$$E_{prep} = \sum_{i}^{Frag} E_i(\text{Fragment in free state})$$
$$- \sum_{i}^{Frag} E_i(\text{Fragment with structure as in combined complex}) \qquad (18)$$

The term E_{prep} is in general positive (destabilizing) and amounts to 31 kJ mol^{-1} for CH3Re(CO)5, Table I. The contribution is 25 kJ mol^{-1} from CH3 and 6 kJ mol^{-1} from Re(CO)5.

TABLE 1. Decomposition of Re-CH3 Bond Energy in CH3Re(CO)5

E_{prep}	E_{el}	E_{dlxc}	E_{exrp}	E_{st}^b	$E_{cp}^{A_1}$	E_{cp}^{E}	E_{rel}	ΔE^a
31	-420	-343	1021	288	-437	-26	-25	200

^{a}Total interaction energy is given by $\Delta E = E_{st} + E_{cp}^{A_1} + E_{cp}^{E} + E_{rel}$
^{b}Steric interaction energy is given by $E_{st} = E_{prep} + E_{el} + E_{dlxc} + E_{exrp}$

3.2. STERIC INTERACTION ENERGY AND EXCHANGE REPULSION

The two distorted fragments **9** and **10** are in the second step brought together from infinite separation and allowed to take up the positions they will have in the combined complex **6**. However, for the moment we allow the combined system to be described only by the basis spanned by the occupied orbitals of the two fragments A and B. The combined system is thus described by the wave function where the set of orbitals

$$\Psi^{\Theta_0} = |\Phi_1^A, \Phi_2^A, ..., \Phi_{n_A}^A, \Phi_1^B, \Phi_2^B, ..., \Phi_{n_B}^B| \tag{19}$$

$\{\Phi_i^A; i=1,n_A; \Phi_i^B; i=1,n_B\}$ is obtained from a mutual orthogonalization of the set of occupied orbitals on A and B given by $\{\phi_j^0(\vec{r}_1), j=1,n_A+n_B\}$, Eq. (6). The density corresponding to Ψ^{Θ_0} is simply given by [26b]

$$\rho^{\Theta_0}(\vec{r}_1) = \sum_\mu^{occ} \sum_\nu^{occ} S_{\mu\nu}^{-1} \phi_\mu^0(\vec{r}_1) \phi_\nu^0(\vec{r}_1) \tag{20}$$

where $S_{\mu\nu}^{-1}$ is an element in $\mathbf{S}^{-1}$ which is the inverse to the overlap matrix $\mathbf{S}$ between the occupied orbitals on A and B, $\{\phi_j^0(\vec{r}_1), j=1,n_A+n_B\}$. The contribution to the interaction energy from the second step, ΔE_{st}, is refered to as the steric interaction energy and given by

$$\Delta E_{st} = E_{el} + E_{dlxc} + E_{exrp} \tag{21}$$

according to Eq. (13), where

$$E_{exrp} = \sum_\mu^M \sum_\nu^M F_{\nu\mu}^{(0 \to \Theta_0)} \Delta P_{\nu\mu}^{(0 \to \Theta_0)} \tag{22}$$

The electrostatic term E_{el} is in general negative (stabilizing) if both A and B are neutral. This is so since the electron density on A will penetrate [19a,b] the shielding of the nuclei on B due to the electron density on B, **11b**. A similar penetrating effect applies for the density on B. The electrostatic term E_{el} amounts to -420 kJ mol^{-1} in the case of $CH_3Re(CO)_5$, Table 1. The nuclear-nuclear repulsion will eventually make E_{el} repulsive (positive) for small inter-atomic distances. However, such distances are normally not realized in equilibrium structures.

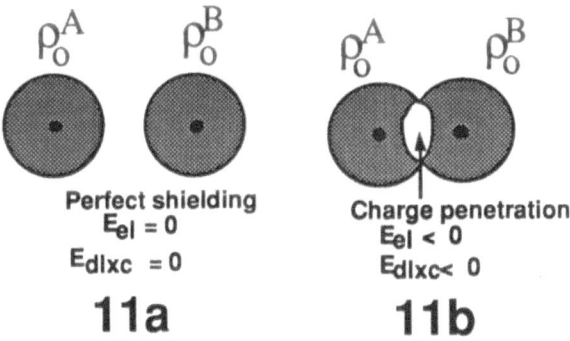

ρ_0^A ρ_0^B ρ_0^A ρ_0^B

Perfect shielding
$E_{el} = 0$
$E_{dlxc} = 0$

11a

Charge penetration
$E_{el} < 0$
$E_{dlxc} < 0$

11b

The exchange correlation energy E_{xc} is in general negative, and regions of high local density contribute in absolute terms more to E_{xc} than regions of low density. The local density around A and B will increase as we bring together the two fragments, **11a**, and superimpose their respective charge distributions ρ_0^A and ρ_0^B, **11b**. We have as a consequence that $E_{xc}[\rho_0^A] + E_{xc}[\rho_0^B] > E_{xc}[\rho_0^A + \rho_0^B]$ and $E_{dlxc} < 0$. Thus, E_{dlxc} of Eq.(21) ,which we have defined in Eq.(15) is in general negative [19b,c] with a value of -343 kJ mol^{-1} in the case of CH3Re(CO)5, Table 1.

The last term in Eq.(21) takes into account the change in density

$$\Delta\rho^{(0\rightarrow\Theta_0)}(\vec{r}_1) = \rho^{\Theta_0}(\vec{r}_1) - \rho_0^A(\vec{r}_1) - \rho_0^B(\vec{r}_1) \qquad (23)$$

due to the mutual orthogonalization of the occupied orbitals on A and B,

$\{\phi_j^0(\vec{r}_1), j=1,n_A+n_B\}$, resulting in the set $\{\Phi_i^A; i=1,n_A; \Phi_i^B; i=1,n_B\}$.

We might write

$$\Delta\rho^{(0\rightarrow\Theta_0)}(\vec{r}_1) = \sum_\mu^M \sum_\nu^M \Delta P_{\nu\mu}^{(0\rightarrow\Theta_0)} \phi_\mu^0(\vec{r}_1) \phi_\nu^0(\vec{r}_1) \qquad (24)$$

where according to Eq. (20)

$$\Delta P_{\nu\mu}^{(0\rightarrow\Theta_0)} = S_{\nu\mu}^{-1} - \delta_{\nu\mu} \qquad (25)$$

We have further that $F_{\nu\mu}^{(0\rightarrow\Theta_0)}$ is a matrix element of the form given in Eq.(14) with the Kohn-Sham operator of Eq. (15) defined by the intermediate density $\rho_0^A(\vec{r}_1)+\rho_0^B(\vec{r}_1)+1/2\,\Delta\rho^{(0\rightarrow\Theta_0)}(\vec{r}_1)$.

3.3. EXCHANGE REPULSION AND FOUR-ELECTRON TWO-ORBITAL INTERACTIONS

The term E_{exrp} of Eq. (22) is directly related to " four-electron two orbital " interactions in qualitative molecular orbital theory (QMOT) [27]. The total energy is given in qualitative theories usually as the sum of the energies for the occupied orbitals.

$$EQMOT = \sum_{i}^{occ} n_i\varepsilon_i \tag{26}$$

Consider the interaction, **12**, between the two orbitals ϕ_1^0 and ϕ_2^0, **12**, where ϕ_1^0 and ϕ_2^0 are separated in energy by $\Delta E = \varepsilon_1 - \varepsilon_2$, with $\Delta E < 0$. This interaction will result in a bonding combination ϕ_+ of energy $\varepsilon_+ = \varepsilon_1 - \Delta E_1$ as well as an anti-bonding combination ϕ_- of energy $\varepsilon_+ = \varepsilon_2 + \Delta E_2$. The orbital ϕ_- will usually be raised [28] more in energy (by ΔE_2) compared to ϕ_2 than ϕ_+ will be lowered (by ΔE_1) compared to ϕ_1, thus $\Delta E_2 > \Delta E_1 > 0$. The total interaction energy in **12** will, in the case where both ϕ_1 and ϕ_2 are occupied, be destabilizing since one pair of electrons after the interaction will be raised more in energy (by ΔE_2) than the other pair will be lowered (by ΔE_1) in energy. In fact, the contribution to E_{QMOT} is given by [27]

$$2\Delta E_2 - 2\,\Delta E_1 = 2(S_{11}^{-1}h_{11} + 2S_{12}^{-1}h_{12} + S_{22}^{-1}h_{22}) \tag{27}$$

The four-electron two-orbital repulsion might be divided up into two interactions, **13a** and **13b**, each involving a pair of electrons with the same spin.

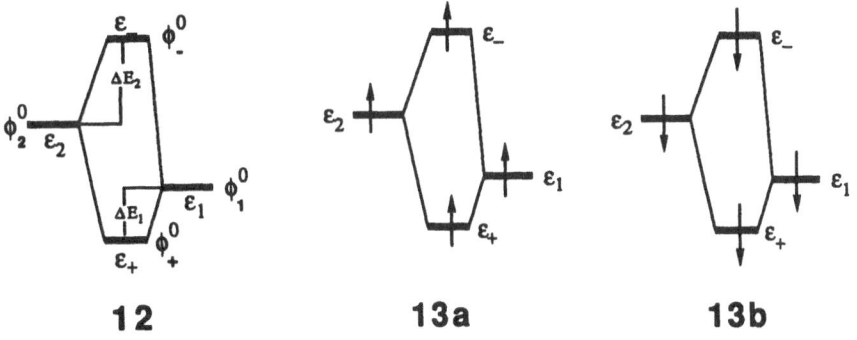

12　　　　　　　**13a**　　　　　　　**13b**

The E_{exrp} term represents the sum over all four-electron two-orbital repulsions with the "effective " Hamiltonian matrix element $h_{\nu\mu}$ replaced by $F_{\nu\mu}^{(0\rightarrow\Theta_0)}$. We can carry out the relation to QMOT one step further by considering the overlaps to be small, in which case [26b].

$$S_{\nu\mu}^{-1} \sim \begin{cases} - S_{\nu\mu} + \displaystyle\sum_{l \neq \nu;\, \mu} S_{\nu l}S_{l\mu} \; ; \; \nu \neq \mu \\[2em] 1 + \displaystyle\sum_{l \neq \mu} S_{\mu l}S_{l\mu} \; ; \; \nu = \mu \end{cases} \tag{28}$$

Let us further replace $F_{\nu\mu}^{(0\rightarrow\Theta_0)}$ with the "effective " Hamiltonian matrix element $h_{\nu\mu}$. This matrix element is given in QMOT[27] by

$$h_{\nu\mu} \sim S_{\nu\mu} [h_{\nu\nu}^0 + h_{\mu\mu}^0] \tag{29}$$

where $h_{\nu\nu}^0$ and $h_{\mu\mu}^0$ are the (negative) orbital energies of the occupied orbitals $\phi_{\mu}^0(\vec{r}_1)$ and $\phi_{\nu}^0(\vec{r}_1)$ on the separate fragments. Substituting Eqs. (28) and (29) into Eq. (22) and retaining only terms to second order in $S_{\nu\mu}$ affords

$$F_{\nu\mu}^{(0\rightarrow\Theta_0)} \Delta P_{\nu\mu}^{(0\rightarrow\Theta_0)} \sim \begin{cases} -S_{\nu\mu}^2 [h_{\nu\nu}^0 + h_{\mu\mu}^0] \; ; \; \nu \neq \mu \\[2em] 0 \qquad\qquad\qquad\; ; \; \nu = \mu \end{cases} \tag{30}$$

The right hand side of Eq. (30) is positive for any pair of orbitals, ϕ_{μ}^0 and ϕ_{ν}^0 , since both $h_{\nu\nu}^0$ and $h_{\mu\mu}^0$ are assumed negative. It is thus clear that the total sum in

Eq.(22) must be positive. The term E_{exrp} amounts to 1021 kJ mol^{-1} in the case of $CH_3Re(CO)_5$, Table 1. We present in Figure 1 a schematic representation of E_{st} and its components E_{exrp}, E_{el} and E_{dlxc}. The term E_{exrp} is often referred to as exchange repulsion [29].

3.4. EXCHANGE REPULSION AND INCREASE IN KINETIC ENERGY

The exchange repulsion term E_{exrp} has been viewed in the previous section in terms of interaction diagrams, 13, involving occupied orbitals. It is possible to obtain a more fundamental understanding of exchange repulsion by considering the individual contributions to E_{exrp} from the kinetic energy

$$E_{exrp}^{kin} = - \sum_{\mu}^{occ} \sum_{\nu}^{occ} \Delta P_{\nu\mu}^{(0\rightarrow\Theta_0)} \int \phi_\mu^0(\vec{r}_1)\nabla^2 \phi_\nu^0(\vec{r}_1)d\vec{r}_1 \quad (31)$$

and the Coulomb interactions, $E_{exrp}^{Coul} = E_{exrp} - E_{exrp}^{kin}$. The kinetic term E_{exrp}^{kin} can be analysed by making use of the approximation in Eq. (28) and retaining terms up to first order in $S_{\nu\mu}$. In this case

$$E_{exrp}^{kin} = 1/2 \sum_{\mu}^{occ} \sum_{\nu}^{occ} S_{\nu\mu} T_{\nu\mu} \quad (32)$$

where

$$T_{\nu\mu} = \int \phi_\mu^0(\vec{r}_1)\nabla^2 \phi_\nu^0(\vec{r}_1)d\vec{r}_1 \quad (33)$$

The contributions to $T_{\nu\mu}$ and $S_{\nu\mu}$ come in the case of small overlaps from the tail regions of $\phi_\mu^0(\vec{r}_1)$ and $\phi_\nu^0(\vec{r}_1)$ where the two functions overlap, 14. The two functions and their Laplacians are in this region of the same sign, 14, with the result that each product term $S_{\nu\mu} T_{\nu\mu}$ in Eq. (32) is positive, and the same must thus hold for the sum, E_{exrp}^{kin}.

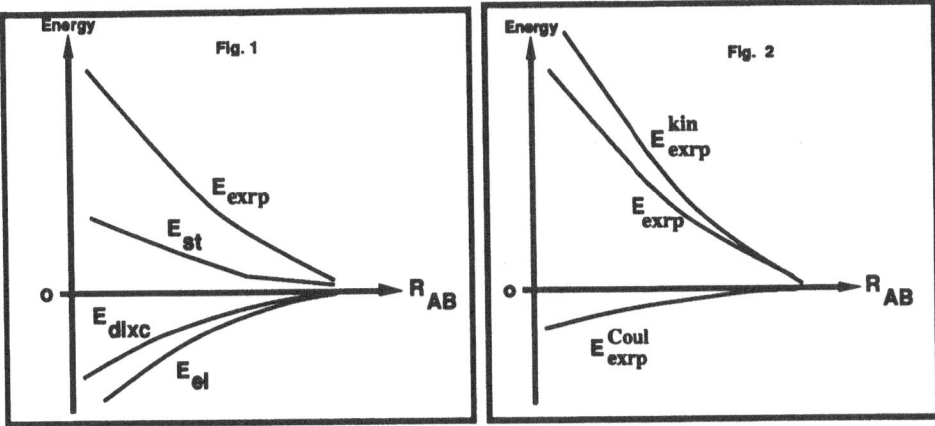

Figure 1. A schematic representation of E_{st} and its components E_{el}, E_{dlxc} and E_{exrp} as a function of the distance R_{AB} between the two fragments

Figure 2. A schematic representation of E_{exrp} and its components $E\backslash S(kin,exrp)$ and $E^{Coul,exrp}$ as a function of the distance R_{AB} between the two fragments A and B.

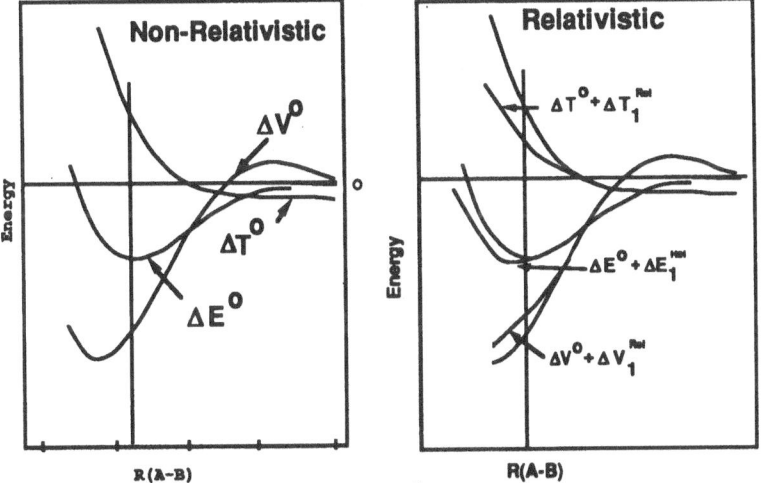

Figure 3. Total non-relativistic energy, ΔE^0, for the diatomic molecule AB, $\overline{}$ relative to the separate atoms A and B. Total energy is decomposed into its kinetic, ΔT^0, and Coulomb components, ΔV^0, as $\Delta E^0 = \Delta T^0 + \Delta V^0$.

Figure 4. Influence of $\Delta T^{Rel, 1}$ and $\Delta V^{Rel, 1}$ on the bond, energy, ΔE, and equilibrium bond distance, R(A-B), of AB, $\overline{}$. Solid lines represent non-relativistic curves of Figure 3.

382

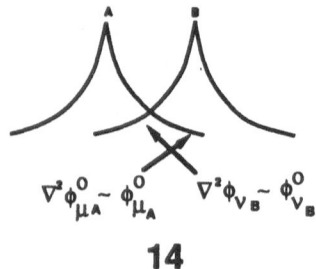

$$\nabla^2\phi^0_{\mu_A} \sim \phi^0_{\mu_A} \qquad \nabla^2\phi^0_{v_B} \sim \phi^0_{v_B}$$

14

The kinetic term, E^{kin}_{exrp}, is the dominating [30] contribution to E_{exrp} and largely responsible for the destabilizing influence of exchange repulsion on the interaction between A and B. A schematic representation of E_{exrp}, E^{kin}_{exrp} and E^{Coul}_{exrp} is given in Figure 2.

It follows from our analysis that interactions between occupied orbitals, **13**, are repulsive and destabilizing due to an increase in the kinetic energy, E^{kin}_{exrp}. This increase is largely responsible for keeping the two fragments A and B apart at their equilibrium distance. This point is underlined further in Figure 3 where we depict the gain in energy, ΔE^0, associated with the formation of a diatomic molecule $\overline{AB}$ from two atoms A and B. The formation energy, ΔE^0, can be written as a sum of its kinetic, ΔT^0, and Coulomb component, ΔV^0, as $\Delta E^0 = \Delta T^0 + \Delta V^0$. The minimum for ΔE^0 at the equilibrium distance is brought about as a consequence of the fact that the increase in ΔT^0 at shorter distances "out-runs" the decrease in ΔV^0, Figure 3. The increase in ΔT^0 is largely due to the interactions between occupied orbitals. These interactions (overlaps) might involve core orbitals on the two centers, core-valence overlaps or valence-valence overlaps.

3.5. STABILIZATION DUE TO CHARGE TRANSFER AND POLARIZATION

We have up to now brought the two fragments together under the restriction that the electrons were confined to the orbitals occupied in the separate fragments. We shall now relax this restriction by allowing the virtual orbitals on A and B to participate in the bonding of $\overline{AB}$. That is, we perform a full SCF-calculation on $\overline{AB}$. The relaxation will result in a polarization of charge within each fragment as well as charge transfer between the fragments. The wave function for $\overline{AB}$ in state Θ is given by

$$\Psi^\Theta = |\psi^{\overline{AB}(\Theta)}_1, \psi^{\overline{AB}(\Theta)}_2, \ldots, \psi^{\overline{A\,B}(\Theta)}_n| \tag{34}$$

where the set of orbitals $\psi_i^{\overline{AB(\Theta)}}$ is defined in Eq. (7). The total density associated with $\psi^{AB(\Theta)}$ was introduced in Eq. (8) as $\rho_{\Theta}^{\overline{AB}}$. We can write it as

$$\rho_{\Theta}^{\overline{AB}} = \rho_0^A(\vec{r}_1) + \rho_0^B(\vec{r}_1) + \Delta\rho^{(0\to\Theta0)}(\vec{r}_1) + \Delta\rho^{(\Theta0\to\Theta)}(\vec{r}_1) \qquad (35)$$

where $\Delta\rho^{(\Theta0\to\Theta)}(\vec{r}_1)$ represents the change in density due to the involvement of the virtual orbitals on A and B, thus

$$\Delta\rho^{(\Theta0\to\Theta)}(\vec{r}_1) = \sum_\mu^{all} \sum_\nu^{all} \Delta P_{\nu\mu}^{(\Theta0\to\Theta)} \phi_\mu^0(\vec{r}_1)\phi_\nu^0(\vec{r}_1) \qquad (36)$$

The change in energy associated with $\Delta\rho^{(\Theta0\to\Theta)}$ is given by

$$E_{cp} = \sum_\mu^{all} \sum_\nu^{all} F_{\nu\mu}^{(\Theta_0\to\Theta)} \Delta P_{\nu\mu}^{(\Theta_0\to\Theta)} \qquad (37)$$

where $F_{\nu\mu}^{(\Theta_0\to\Theta)}$ is a matrix element of the form given in Eq. (14) with the Kohn-Sham operator of Eq. (15) defined by the intermediate density $1/2\rho^{(0\to\Theta0)}$, $+ 1/2\rho^{(\Theta0\to\Theta)}$. The expression in Eq. (37) provides a direct link between any differential change in the density represented by $\Delta P_{\nu\mu}^{(\Theta_0\to\Theta)}$ and the associated energy change given by $F_{\nu\mu}^{(\Theta_0\to\Theta)} \Delta P_{\nu\mu}^{(\Theta_0\to\Theta.)}$. It is further possible from Eq. (37) to block out contributions from different symmetry representations as

$$E_{cp} = \sum_\Gamma E_{cp}^\Gamma \qquad (38)$$

where

$$E_{cp}^\Gamma = \sum_{\mu(\Gamma)}^{all} \sum_{\nu(\Gamma)}^{all} F_{\nu(\Gamma)\mu(\Gamma)}^{(\Theta_0\to\Theta)} \Delta P_{\nu(\Gamma)\mu(\Gamma)}^{(\Theta_0\to\Theta)} \qquad (39)$$

The symmetry blocking is possible since $F_{\nu\mu}^{(\Theta_0 \to \Theta)}$ as well as $\Delta P_{\nu\mu}^{(\Theta_0 \to \Theta)}$ both are zero unless μ and ν belong to the same representation. Both the symmetry blocking of Eq. (38) as well as the direct relation between any differential modification of the density, $\Delta P_{\nu(\Gamma)\mu(\Gamma)}^{(\Theta_0 \to \Theta)}$, and the corresponding energy change, $F_{\nu(\Gamma)\mu(\Gamma)}^{(\Theta_0 \to \Theta)} \Delta P_{\nu(\Gamma)\mu(\Gamma)}^{(\Theta_0 \to \Theta)}$, are useful features of the energy decomposition method.

The methyl complex $CH_3Re(CO)_5$ of conformation **6** has a C_{3v} point group symmetry. The CH_3-Re interaction has two symmetry contributions, E_{cp}^{Γ}, to E_{cp} of Eq.(38). The first comes from the a_1-representation where an occupied σ_{CH_3} orbital of (say) α-spin interacts with an empty σ-orbital on the metal fragment, **15a**, and an empty σ_{CH_3} orbital of β-spin interacts with an occupied σ-orbital on the metal fragment, **15b**. The interactions will give rise to transfer of charge between the fragments and result in the formation of two equivalent and occupied spin-orbitals representing the CH_3-Re σ-bond. The energy term $E_{cp}^{A_1}$ related to the interactions in **15** is a measure of the CH_3-Re σ-bonding interaction and amounts to -437 kJ mol^{-1}, Table 1.

15a   **15b**

The second symmetry interaction is much weaker. It takes place in the e-representations and involves a transfer of charge from occupied metal-based d-orbitals to the empty σ_{C-H}^{*} orbitals on CH_3, **16**. This interaction is much like hyper-conjugation in organic molecules. The interaction energy E_{cp}^{E} due to,**16**, was calculated to be -26 kJ mol^{-1}, Table 1.

There is a relationship between the expression in Eq. (37) for E_{cp} and the theories by Sanderson [31], Pearson [32], as well as Parr and Pearson [33] in which the bond energy of $\overline{AB}$ is related to differences in electronegativity and hardness between the two fragments A and B. Most of these theories relate the stability of $\overline{AB}$ to the gain in energy associated with the transfer of charge from one fragment to the other. The stabilization due to charge transfer and polarization is as already mentioned represented by E_{cp} in our theory. However, the steric term, E_{st}, is of equal importance for the

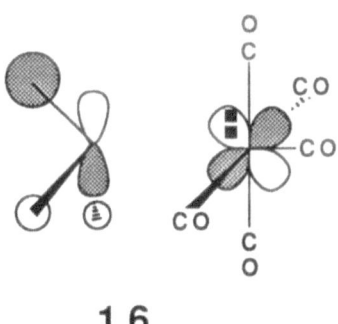

1 6

interaction between A and B, although its influence in most cases is destabilizing. It is not straightforward to see how E_{st}, and in particular the exchange repulsion part, E_{exrp}, enters into the theories based on electronegativity and hardness .

3.6. RELATIVISTIC EFFECTS [33]

The total non-relativistic interaction energy between A and B can be written according to our decomposition scheme as

$$\Delta E = E_{prep} + E_{el} + E_{dlxc} + E_{exrp} + \sum_{\Gamma} E_{cp}^{\Gamma} \qquad (40)$$

Valence electrons in atoms and molecules have a finite (albeit small) probability of being close to the nuclei and they can as a consequence acquire high instantaneous velocities. In fact, the velocities for the valence electrons can approach that of light as they pass in close proximity to heavier nuclei with $Z > 72$. It is for this reason not too surprising that relativistic effects become of importance for the chemical properties of compounds containing 5d-block elements in the third transition series or 5f-block elements in the actinide series. Relativistic effects have in fact a profound effect on the periodic trends within a triad of transition metals [34, 35].

The first order relativistic correction to the non-relativistic Hamiltonian can be written [36] as

$$\hat{H}_1 = \hat{H}_{MV} + \hat{H}_{Darw} + \hat{H}_{SO} \qquad (41)$$

The so-called mass-velocity term $\hat{H}_{MW}$, which represents the first order relativistic correction to the non-relativistic kinetic energy , is given by

$$\hat{H}_{MV} = -\frac{\alpha^2}{8}\sum_i \nabla_i^4 = -\frac{\alpha^2}{8}\sum_i \hat{p}^4(i) \tag{42}$$

whereas the Darwin term, from the Zitterbewegung of the electrons, after neglecting some numerically insignificant two-electron operators [34c], takes on the form

$$\hat{H}_{Darw} = \frac{\alpha^2}{8}\sum_i \nabla_i^2 \, (V_N(\vec{r_i})) \tag{43}$$

where $V_N(\vec{r_i})$ is the electron-nuclear attraction potential. The Darwin term represents the firt order relativistic correction to the Coulomb interaction. The spin-orbit operator $\hat{H}_{SO}$ does not contribute to the total energy of the closed shell molecules considered in the following and need thus not be specified here. In Eqs. (42) and (43) α represents the fine-structure constant [36]. The contribution to the interaction energy, ΔE, from relativistic effects will be referred to as E_{rel}. It is in most cases stabilizing and amounts to -26 kJ mol^{-1} for $CH_3Re(CO)_5$.

The major contribution to E_{rel} comes from $\hat{H}_{MV}$ and represents [34a,c] the first order relativistic correction to the exchange repulsion. We have already mentioned that the destabilizing exchange repulsion is due to the kinetic term E_{exrp}^{kin} . The first order relativistic correction affords

$$E_{exrp}^{kin(rel)} = -\frac{\alpha^2}{8}\sum_\mu^{occ}\sum_\nu^{occ}\Delta P_{\nu\mu}^{(0\to\Theta_0)}\int \phi_\mu^0(\vec{r}_1)\nabla^4 \phi_\nu^0(\vec{r}_1)d\vec{r}_1 \tag{44}$$

The term in Eq. (44) is only of significance if the molecule under investigation involves elements of high nuclear charges. Further, the only important contributions stems from terms in Eq. (44) where $\mu = \nu$, and μ represents a core orbital with a high amplitude near the nucleus. The diagonal $\Delta P_{\mu\mu}^{(0\to\Theta_0)}$ terms can, according to Eqs. (25) and (28), be written as

$$\Delta P_{\mu\mu}^{(0\to\Theta_0)} \sim \sum_{l\neq\mu} S_{\mu l}S_{l\mu} \tag{45}$$

They are positive definite and represent an increase in density near the nuclei as the two fragments A and B are brought together. The corresponding contributions to Eq. (44) can be written as

$$E_{exrp}^{kin(rel)} \sim -\frac{\alpha^2}{8} \sum_{\mu}^{occ} \sum_{l \neq \mu}^{occ} S_{\mu l} S_{l \mu} \int \nabla^2 \phi_{\mu}^0((\vec{r}_1) \nabla^2 \phi_{\mu}^0(\vec{r}_1) d\vec{r}_1 \quad (46)$$

where we have made use of the fact that ∇^2 is a hermitian operator. The total sum in Eq. (46) is negative (stabilizing) since each individual term is negative definite. The first order relativistic correction is thus seen to make four-electron two-orbital interactions less destabilizing by reducing the kinetic energy. This is brought out in Figure 4. where we give the bond energy, $\Delta E = \Delta E_0 + \Delta E_1^{Rel}$, of the diatomic molecule $\overline{AB}$ with the first order relativistic correction ΔE_1^{Rel} included. The bond energy is decomposed into a kinetic part given by $\Delta T = \Delta T_0 + \Delta T_1^{Rel}$ as well as a Coulomb part given by $\Delta V = \Delta V_0 + \Delta V_1^{Rel}$. The relativistic correction to the kinetic energy, ΔT_1^{Rel}, originates from the mass-velocity term $\hat{H}_{MV}$ of Eq. (42) whereas the Coulomb correction, ΔV_1^{Rel}, stems from the Darwin term, $\hat{H}_{Darw}$, of Eq. (43). The kinetic term, ΔT_1^{Rel}, is dominating and largely responsible for the relativistic stabilization of the A-B bond.

The primary contributions to ΔT_1^{Rel} comes from $E_{exrp}^{kin(rel)}$ of Eq. (46). Of particular importance are those terms in which μ represents a 1s core orbital on a 5d- or 4f elements. The Laplacian for such an orbital is substantial and $-\Delta T_1^{Rel}$ as a consequence large. The relativistic reduction, $E_{exrp}^{kin(rel)}$, in the exchange repulsion is further seen to contract the A-B bond distance, Figure 4. A more detailed discussion of the importance of relativistic effects and their origin can be found elsewhere [34a-c].

4. Applications

We shall in the following apply the energy decomposition scheme outlined in Section 3 to the M-CO bond in binary carbonyls. The same scheme can be used to analyze the M-CO interaction in polynuclear complexes, clusters and metal surfaces.

4.1. METAL-CARBONYL LIGAND DISSOCIATION ENERGIES [35c]

Calculated values for the first ligand dissociation energy, ΔH, for $M(CO)_6$ (M= Cr,Mo,W) are compared in Table 2 to experimental data from different sources. Our calculated ordering for the first ligand dissociation energy, given as ΔH_{Mo} < ΔH_W ~ ΔH_{Cr} is in line with data from solvent kinetics [38] as well as Bernstein's [39] results based on a photochemical study, but differs from the ordering ΔH_{Cr} < ΔH_{Mo} < ΔH_W obtained by Lewis et al. [40] in their laser pyrolysis work.

TABLE 2. Calculated values for the first ligand dissociation energy, ΔH, in $Cr(CO)_6$, $Mo(CO)_6$ and $W(CO)_6$

$M(CO)_6$		ΔH (kJ mol^{-1})		
	cal.	exp.[a]	exp.[b]	exp.[c]
$Cr(CO)_6$	147	162	155	154
$Mo(CO)_6$	119	126	142	169
$W(CO)_6$	142	166	159	192

[a] Ref. 37. [b] Ref. 38. [c] Ref. 39.

The first ligand dissociation energy can be decomposed as

$$\Delta H = -E_{prep} - E_{st} - E(a_1) - E(e) - E_{rel} \qquad (47)$$

Here E_{st} is the steric interaction energy between CO and the $M(CO)_5$ fragment whereas $E(a_1)$ represents the contribution to ΔH due to the donation **17a** from σ_{CO} to the LUMO on $M(CO)_5$ and $E(e)$ is the contribution to ΔH due to the back-donation **17b** from the HOMO of $M(CO)_5$ to the π_{CO}^* orbital on CO. The contribution from relativistic effects is given as E_{rel}. We should in principle include a contribution, E_{prep}, from the energy required to relax the geometry of free $M(CO)_5$ to that of the $M(CO)_5$ framework in $M(CO)_6$. We have not included such a relaxation energy. We expect it to be small (< 5 kJ mol^{-1}) since several experimental and theoretical studies indicate that d^6 pentacarbonyls have a square-pyramidal ground state conformation. It can be seen from Table 3, where ΔH is decomposed into its various components, that the steric interaction energy E_{st} and the term $E(a_1)$, representing the donation **17a**, are quite similar for the three hexacarbonyls, whereas $E(e)$, from the back-donation **15b**, is more important for

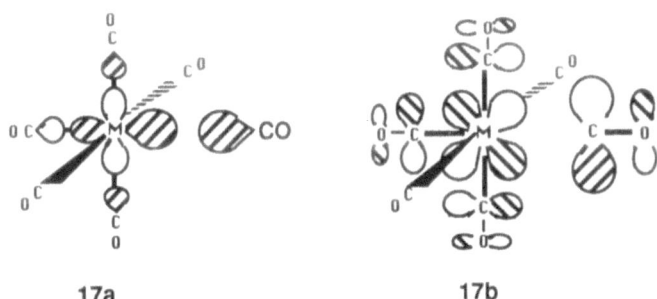

17a   17b

Cr(CO)$_6$ than for either Mo(CO)$_6$ or W(CO)$_6$. The back-donation **17b** is thus in the nonrelativistic limit responsible for ΔH_{Cr} being larger than ΔH_{Mo} or ΔH_W. Relativity will, however, strengthen the W-CO bond to the extent where ΔH_W becomes similar to ΔH_{Cr}, Table 3. It follows from our analysis that donation, E(a$_1$), and back-donation, E(e), are of equal importance for the synergic M-CO bond. A more detailed analysis of the Cr-CO bond in Cr(CO)$_6$ has been given by Baerends and Rozendaal [40].

The calculated values for the first ligand dissociation energy, ΔH, in Ni(CO)$_4$, Pd(CO)$_4$ and Pt(CO)$_4$ are given in Table 4. The ordering within the triad is : $\Delta H_{Ni} \gg \Delta H_{Pt} > \Delta H_{Pd}$. The first ligand dissociation energy is known only for Ni(CO)$_4$. The experimental value of 104 kJ mol^{-1} differ by 2 kJ mol^{-1} from our theoretical estimate.

TABLE 3. Decomposition of the first ligand dissociation energy, ΔH, in Cr(CO)$_6$,Mo(CO)$_6$ and W(CO)$_6$. Energies in kJ mol^{-1}.

M(CO)$_6$	-E$_{st}$	-E(a$_1$)	-E(e)	- E$_{rel}$	ΔH
Cr(CO)$_6$	-193	163	177	-	147
Mo(CO)$_6$	-197	165	148	3	119
W(CO)$_6$	-197	160	154	25	142

TABLE 4. Decomposition of the first ligand dissociation energy, ΔH, in Ni(CO)$_4$,Pd(CO)$_4$ and Pt(CO)$_4$. Energies in kJ mol^{-1}.

M(CO)$_4$	-E$_{prep}$	-E$_{st}$	-E(a$_1$)	-E(e)	-E$_{rel}$	ΔH
Ni(CO)$_4$	-9	-169	145	139	-	106
Pd(CO)$_4$	-11	-225	143	113	7	27
Pt(CO)$_4$	-11	-261	146	113	52	38

The first ligand dissociation energy in $M(CO)_4$ can be decomposed according to Eq. (47). Now E_{prep} represents the energy required to deform $M(CO)_3$ from its trigonal planar equilibrium geometry to the trigonal pyramidal conformation of the $M(CO)_3$ framework in $M(CO)_4$. Further E_{st} is the steric interaction energy between CO and $M(CO)_3$. Finally, the two terms $E(a_1)$ and $E(e)$ are the electronic contributions to ΔH from the a_1 and e representations as $M(CO)_4$ is formed from $M(CO)_3$ and CO under C_{3v} constraints.

The HOMO of $M(CO)_3$, **18a**, is a metal based nd-orbital with out-of-phase contributions from σ_{CO} and in-phase contributions from π_{CO}^*. It will interact with π_{CO}^* of the incoming CO ligand. The contribution from this type of interaction is given by $E(e)$ in Table 4 and represents the metal to ligand back-donation. The LUMO of $M(CO)_3$, $3a_1$ on **18c**, is a metal based (d,s,p)-hybrid orbital with in-phase contributions from π_{CO}^*. At lower energy is an occupied orbital on $M(CO)_3$ also of a_1 symmetry, $2a_1$ of **18b**. The σ_{CO} orbital of the incoming ligand will interact repulsively with $2a_1$ since both orbitals are occupied, **18b**, and this interaction will contribute significantly to E^0. Further, there will be a stabilizing interaction in the a_1 representation, **18c**, between $3a_1$ and σ_{CO} on the incoming ligand in which density is donated from σ_{CO} to $3a_1$. This interaction serves, together with a certain amount of polarization of charge from $2a_1$ on $M(CO)_3$ to $3a_1$ on $M(CO)_3$, to reduce the repulsive interaction in **18b**. The contribution to the donation from ligand to metal fragment in **18c** as well as the polarization from $2a_1$ to $3a_1$ is given in Table 4 as $E(a_1)$.

It can be seen from Table 4 that the nonrelativistic ordering for ΔH in $M(CO)_4$, $\Delta H_{Ni} \gg \Delta H_{Pd} > \Delta H_{Pt}$, is the result of an increase in the steric interaction energy E^0 through the triad, primarily due to an increase in the repulsive interaction **18b**, as the overlap in **18b** is larger for M = Pt and Pd than for M = Ni. The π-interaction in **18a** is further seen to enhance ΔH_{Ni} compared to ΔH_{Pd} and ΔH_{Pt}, Table 4. The relativistic effects will strengthen the Pt-CO bond compared to the Pd-CO linkage by reducing the exchange repulsion in **18b** as discussed previously in Section 3.6. The M-CO bond in $M(CO)_4$ is

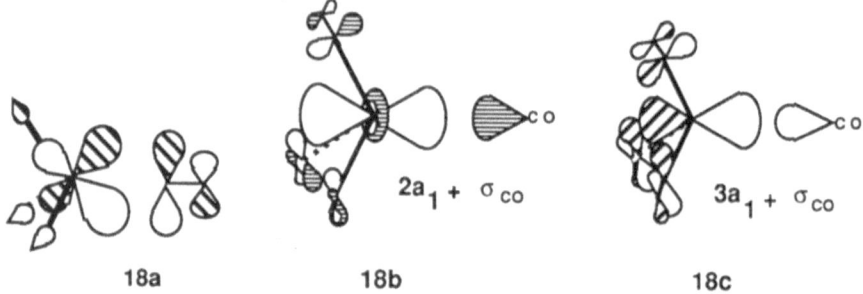

18a	18b	18c
	$2a_1 + \sigma_{CO}$	$3a_1 + \sigma_{CO}$

19a   **19b**

weakened considerably by the two orbital four-electron repulsive interaction in **18b**, in particular for M = Pd and Pt. A similar interaction between a metal-based occupied d-orbital and σCO is not present among the hexacarbonyls. The corresponding M-CO bond is as a consequence much stronger.

The three pentacarbonyls $Fe(CO)_5$, $Ru(CO)_5$, and $Os(CO)_5$ all have a trigonal bipyramidal structure, **19a**. Dissociation of the first CO ligand in the pentacarbonyls can take place from either an axial or an equatorial position. We shall here study the dissociation of CO from the equatorial position. The energy required for the equatorial CO dissociation on the singlet surface can be written as

$$\Delta H = -E_{prep} - E^0 - E(a_1) - E(b_1) - E(b_2) - E_{rel} \qquad (48)$$

The CO-dissociation energies decomposed according to Eq. (48) are given in Table 5. The first term, $-E_{prep}$, of Eq. (48) represents the energy required to deform $M(CO)_4$ from its singlet equilibrium conformation **19b** to the structure of the $M(CO)_4$ framework in **19a**.

There is, as $M(CO)_5$ is formed from CO and $M(CO)_4$ under C_{2v} constraints, a strong repulsive interaction, **20a** , between the incoming occupied σCO-orbital and the occupied metal based $1a_1$-orbital of $M(CO)_4$ which will contribute strongly to the steric interaction energy E^0 of Eq. 48, Table 5. The repulsive interaction in **20a** can be reduced by donation of charge, **20b**, from σCO to the $2a_1$ LUMO of $M(CO)_4$ as well as polarization

TABLE 5. Calculation of equatorial ligand dissociation energies, ΔH, for $Fe(CO)_5$, $Ru(CO)_5$ and $Os(CO)_5$. Energies in kJ/mol^{-1}.

$M(CO)_5$	$-E_{prep}$	$-E^0$	$-E(a_1)$	$-E(b_1)$	$-E(b_2)$	$-E_{rel}$	$-\Delta H$
$Fe(CO)_5$	-10	-213	205	129	73	-	85
$Ru(CO)_5$	-16	-324	235	120	71	6	92
$Os(CO)_5$	-17	-400	263	121	72	59	98

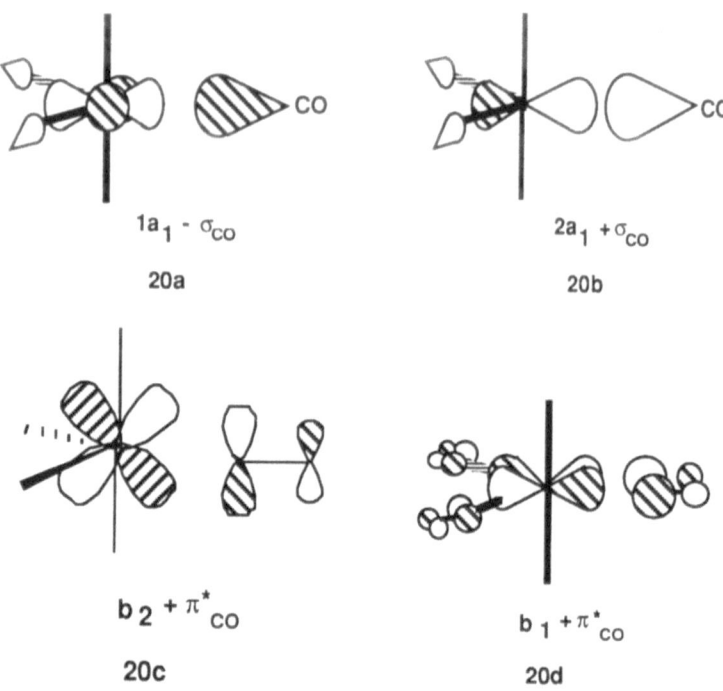

$1a_1 - \sigma_{CO}$

20a

$2a_1 + \sigma_{CO}$

20b

$b_2 + \pi^*_{CO}$

20c

$b_1 + \pi^*_{CO}$

20d

of charge from the occupied $1a_1$ orbital on $M(CO)_4$, **20a**, to the $2a_1$ LUMO, **20b**, of $M(CO)_4$. The transfer of charge from $1a_1$ to $2a_1$ as well as the interaction in **20b** will contribute with $-E(a_1)$ to ΔH , Table 5. Thus $-E(a_1)$ represents the contribution from the ligand to metal σ-donation to the M-CO bond. The π^*_{CO}-orbitals of the incoming CO ligand can interact with the occupied b_1 , **20c**, and b_2 , **20d**, orbitals of $M(CO)_4$. The two π-back-donation interactions **20c** and **20d** will contribute to ΔH with $-E(b_1)$ and $-E(b_2)$, respectively. The b_1 orbital, **20d**, is of higher energy than b_2, **20c**, and is thus better able to interact with the incoming π^*_{CO}-orbital . The contribution $-E(b_1)$ from **20d** is for this reason somewhat larger than the contribution $-E(b_2)$ from **20c**, Table 5.

It is clear from Table 5 that the ligand dissociation energy ΔH is much smaller for $Ru(CO)_5$ and $Os(CO)_5$ than for $Fe(CO)_5$ as the second- and third-row pentacarbonyls have a larger steric interaction energy E^0 than $Fe(CO)_5$, primarily as a result of stronger repulsive interactions for M= Ru and Os than for M=Fe in **20a**. The sum of the two contributions $-E(b_1)$ and $-E(b_2)$ from the π–back-bonding interactions **20c** and **20d** is larger for $Fe(CO)_5$ than for $Ru(CO)_5$ and $Os(CO)_5$, whereas the oppposite is the case for

TABLE 6. Applications of energy decomposition scheme

Fragment A	Fragment B	Ref.
Cu^+, Ag^+, Au^+	C_2H_4	[19b]
$PtCl_3^-$, $Pt(PH_3)_2$		
$Ni(CO)_3$	$CO, CS, N_2, PF_3, CNCH_3$	[19c]
$Zr(BH_4)_3$ and others	BH_4^-	[42]
ML_3 ; (M= Cr,Mo,W; L=H,Cl,OH,NH$_2$)	ML_3	[43,44]
$M(PH_3)_2Cl_2$; (M= Cr,Mo,W,Mn,Tc,Re)	$M(PH_3)_2Cl_2$	[43,45]

TABLE 7. Applications of energy decomposition scheme

Fragment A	Fragment B	Ref.
$M(PH_3)$ (M= Ni,Pd,Pt)		
$M(PH_3)_4^+$ (M= Co,Rh,Ir	C_2H_2, C_2H_4, O_2	[46]
$Ru(CO)_4$	CX, X_2, CX_2, H_2CX $X=O,S,Se,Te$	[47]
$M(CO)_4^{+n}$ (n=0;M=Ni,Pd,Pt), (n=1;M=Co,Rh,Ir)		
$M(CO)_5$ (M=Fe,Ru,Os)	H^+	[48]
MCl_3 (M=Ti,Zr,Hf); $Co(CO)_4$	$OH, OCH_3, SH, NH_2, PH_2,$ CH_3, SiH_3, CN, H	18h
$M(CO)_5$; (M=Cr,Mo,W) $M(CO)_4$; (M=Fe,Ru,Os) $M(CO)_3$; (M=Ni,Pd,Pt)	$CO, AsH_3, PH_3, PF_3, CS$	[35c,19e]
$M(CO)_5$; (M=Mn,Tc,Re) $M(CO)_4$; (M=Co,Rh,Ir) $ThCl_3, HfCl_3$, Cp_2M (M= Sc,Y,La,V,Mn,Tc,Re) $CpM(CO)_n$ (M=Ni,Co,Mn,Fe,Cr)	H, CH_3	[18f,35a,35b]
$Pt(PH_3)_2$	$HP=CH_2$	[49]
Cu_n (n=1,6)	CO	[50]
Al_n	CO	[51]
Ag_n	O_2	[52]

the contribution -E(a_1) due to the σ-donation, **20a**. The relativistic contribution is further instrumental in stabilizing the Os-CO bond compared to the Ru-CO bond, Table 5.

4.2. OTHER APPLICATIONS

The decomposition scheme outlined in Section 3 have been applied extensively . References to most of the applications are given in Tables 6 and 7.

ACKNOWLEDGMENT

This investigation was supported by the Natural Sciences and Engineering Research Council of Canada (NSERC). We also acknowledge access to the Cyber-205 installations at the University of Calgary.

REFERENCES

1. (a) Parr, R.G. and Yang, W. (1989), Density-Functional Theory of Atoms and Molecules, Oxford University Press, New York.
   (b) Kryachko, E.S. and Ludena, E.V. (1990), Density Functional Theory of Many Electron Systems; Kluwer Press, Dordrecht.
   (c) Ziegler, T. (1991), Chem.Rev., **91**, 651.
2. Becke, A.D. (1989), Int. J. Quantum Chem. , **S23**, 599.
3. (a) Ziegler, T., Tschinke, V., Versluis, L., Baerends, E.J. and Ravenek, W. (1988), Polyhedron, **7**, 1625.
4. Versluis, L. and Ziegler, T. (1988), J.Chem. Phys., **88**, 322.
5. Fournier, R., Andzelm, J. and Salahub, D.R. (1989), J. Chem.Phys, **90**, 6371.
6. Fan, L., Versluis, L., Ziegler, T., Baerends, E.J. and Ravenek, W., (1988), Int. J. Quantum Chem., **S22**, 173.
7. (a) Bieger, W., Seifert, G., Eschrig, H. and Grossman, G. (1985), Chem.Phys. Lett., **115**, 275.
   (b) Freier, D.A., Fenske, R.F., Xiao-Zeng, Y. (1985), J.Chem.Phys., **83**, 3526.
   (c) Malkin, V.G. and Zhidomirov, Z. (1988), Zh.Strukt.Khim., **29**, 32.
8. van der Est, A.J., Barker, P.B., Burnell, E.E., de Lange, C.A. and Snijders, J.G. (1985), Mol. Phys., **56**, 1.
9. Case, D.A. (1982) , Annu. Rev. Phys. Chem., **33**, 151.
10. (a) Noodleman, L. and Norman, J.G. (1979), J. Chem. Phys., **70**, 4903.
    (b) Noodleman, L. (1981), J. Chem. Phys., **74**, 5737.
    (c) Noodleman, L., Baerends, E.J. (1984), J. Am. Chem. Soc., **106**, 2316.
    (d) Noodleman, L., Norman, J.G., Osborne, J.H., Aizman, A. and Case, D.A. (1985), J. Am. Chem. Soc.,**107**, 3418.
11. Fan, L. and Ziegler, T. (1990) J. Chem. Phys., **92**, 3645.
12. Trsic, M., Ziegler, T.and Laidlaw, W.G. (1976), Chem. Phys., **15**, 383.
13. Cook, M. and Karplus, M. (1987), J. Phys. Chem., **91**, 31.

14. Tschinke, V. and Ziegler, T. (1990), J. Chem.Phys., **93**, 8051.
15. (a) A clear discussion of this point can be found in Ref. 15b.
    (b) Becke, A.D. (1988), J. Chem.Phys., **88**, 1053.
    (c) Becke, A.D. (1989), ACS Symposium Series, **394**, Washington, p. 165.
16. Slater, J.C. (1972), Adv. Quantum Chem., **6**, 1.
17. (a) Gunnarsson, O., Lundquist, L. (1974), Phys. Rev., **B10**, 1319.
    (b) Gunnarsson, O. and Lundquist. I. (1976), Phys. Rev., **B13**, 4274.
    (c) Gunnarsson, O., Johnson, M. and Lundquist, I. (1979), Phys. Rev., **B20**, 3136.
    (d) von Barth, U. and Hedin, L. (1979), Phys. Rev., **A 20**, 1693.
18. (a) Langreth, D. C. and Mehl, M. J. (1983), Phys. Rev., **B28**, 1809.
    (b) Becke, A.D. (1988), Phys. Rev., **A33**, 2786.
    (c) Perdew, J. P. (1986), Phys. Rev., **B33**, 8822. Also see the erratum (1986), Phys. Rev., **B34**, 7046.
    (d) Tschinke V. and Ziegler, T. (1988), Can. J. Chem., **67**, 460.
    (e) Ziegler, T., Tschinke, V. and Becke, A. D. (1987), J. Am. Chem. Soc. **109**, 1351.
    (f) Ziegler, T., Cheng, W., Baerends, E. J. and Ravenek, W. (1988), Inorg. Chem. **27**, 3458.
    (g) Ziegler, T., Tschinke, V., Fan, L. and Becke, A.D.Becke (1989), J. Am. Chem. Soc. **111**, 9177.
    (h) Ziegler, T., Tschenke, V., Versluis, L., Baerends, E. J. and Ravenek, W. (1988), Polyhedron, **7**, 1625.
    (i) Ziegler, T., Tschinke, V. and Ursenbach, C. (1987), J. Am. Chem. Soc., **109**, 4825.
    (j) Becke, A. D. (1989), ACS Symposium Series, **394,** Washington.
    (k) Becke, A. D. (1989), Int. J. Quantum Chem., **S23**, 599.
    (l) Fan, L. and Ziegler, T. (1991), J. Chem. Phys.,**94**, 6057.
    (m) Fan, L., Ziegler, T. (1991, submitted) , J. Chem. Phys.
19. (a) Ziegler, T., Rauk, A. and Baerends, E.J. (1977), Theoret. Chim. Acta (Berl.), **46, 1**.
    (b) Ziegler, T. and Rauk, A. (1979), Inorg. Chem., **18**, 1558.
    (c) Ziegler, T. and Rauk, A. (1979), Inorg. Chem., **18**, 1755.
    (d) Ziegler, T. (1986) in Veillard, A. (ed.), Quantum Chemistry: The Challenge of Transition Metals and Coordination Chemistry, NATO ASI , **C176**, p. 189, Reidel.
    (e) Baerends, E.J., Rozendaal, A., ibdi, p. 159.
20. Morokuma, K. (1977), Acc. Chem. Res., **10**, 294.
21. Wolfe, S., Mitchell, D.J. and Whangbo, M.-H. (1978), J. Am. Chem. Soc., **100**, 1936.
22. Bernardi, F., Bottoni, A., Mangini, A. and Tonachini, G. (1981), J. Molec. Struct. (THEOCHEM), **86,** 163.
23. Stone, A.J. and Erskine, R.W. (1980), J. Am. Chem. Soc. **102,** 7185.
24. Bagus, P.S., Hermann, K. and Bauschlichter, C.W. (1984), J. Chem. Phys., **80,** 4378.
25. Fujimoto, H. (1987), Acc. Chem. Res., **20**, 448.
26. (a) Szabo, A., Ostlund, N.S. (1982), Modern Quantum Chemistry, MacMillan, New York, p. 142.
    (b) O'Shea, S.F. and Santry, D.P. (1975), Theoret.Chim.Acta (Berl), **37**, 1.

27. (a) Albright, T.A., Burdett, J.K. and Whangbo, M.H. (1985), Orbital Interactions in Chemistry, Wiley, New York.
    (b) Burdett, J.K. (1980), Molecular Shapes , Wiley, New York.
28. (a) Fujimoto, H. and Fukui, K. (1972), Adv. Quantum Chem., **6** , 177.
    (b) Kitaura, K., Morokuma, K. (1976), Int. J. Quantum Chem, **10**, 325.
29. (a) van den Hoek, P.J., Kleyn, A.W. and Baerends, E.J. (1989), Comments At Mol. Phys.,**13** , 93.
    (b) van den Hoek, P.J. (1989), Ph.D. Thesis , Free University Amsterdam.
30. Sanderson, R.T. (1976), 2d edn., Chemical Bonds and Bond Energies, Academic Press, New York.
31. Pearson, R.G. (1988), Inorg. Chem., **27**, 734.
32. Parr, R.G., Pearson, R.G. (1983), J. Am. Chem. Soc. **105,** 7512.
33. Pyykkö, P. (1988), Chem. Rev., **88** , 563.
34. (a) Ziegler, T., Snijders, J.G. and Baerends, E.J. (1989). in ACS Symposium Series, **395**, 322.
    (b) Ziegler, T., Tschinke, V. (1990), in ACS Symposium Series, **428**.
    (c) Ziegler, T., Snijders, J.G. and Baerends, E.J.Baerends (1981), J. Chem. Phys., **74**, 1271.
35. (a) Ziegler, T., Tschinke, V., Baerends, E.J. and Snijders, J.G.Snijders (1991, in press), J. Phys. Chem.
    (b) Ziegler, T., Tschinke, V.Tschinke and Becke, A.D. (1987), J. Am. Chem. Soc., **109**, 1351.
    (c) Ziegler, T., Tschinke, V. and Ursenbach, C. (1987), J. Am. Chem. Soc., **109,** 4825.
    (d) Ziegler, T., Tschinke, V. and Becke, A.D. (1987), Polyhedron, **6,** 685.
36. Snijders, J.G., Baerends, E.J.Baerends, Ros, P. (1979), Molec. Phys., **38**, 1909.
37. Angelici, R.J. (1968) , Chem.Rev., **A 3**, 173.
38. Bernstein, M., Simon, J.D. and Peters, J.D. (1983), Chem. Phys. Lett., **100**, 241.
39. Lewis, K.E., Golden, D.M. and Smith, G.P. (1984), J. Am. Chem.Soc., **106**, 3906.
40. Baerends, E.J., Rozendaal, A. (1986), NATO ASI, **C176**, 159.
41. Hitchcock, A.P., Hao, N.G., Werstiuk, N.H., McGlinchey, M.G.and Ziegler, T. (1982), Inorg. Chem., **21**, 793.
42. Ziegler, T. (1983), J. Am. Chem. Soc., **105**, 7543.
43. Ziegler, T., Tschinke, V. and Becke, A. (1987), Polyhedron, **6**, 685.
44. Ziegler, T. (1984), J. Am. Chem. Soc., **106**, 5901.
45. Ziegler, T. (1985), Inorg. Chem., **24**, 1547.
46. Ziegler, T. (1986), Inorg. Chem., **25**, 2723.
47. Ziegler, T. (1985), Organometallics, **4**, 675.
48. van der Knaap, Th.A., Bickelhaupt, F., Kraaykamp, J.G., van Koten, G., Bernards, J.C.P., Edzes, H.Tr., Veeman, W.S., deBoer, E. and Baerends, E.J. (1984), Organometallics, **3**, 1908.
49. Post, D.and Baerends, E.J. (1983), J. Chem. Phys., **78**, 5663.
50. Post, D. and Baerends, E.J. (1982), Surface Science, **116**, 177.
51. van den Hoek, P.J. and Baerends, E.J. (1989), Surface Science, **211**, L791.

PHYSICOCHEMICAL CHARACTERIZATION OF NOVEL POLYMERIC COPPER COMPLEXES WITH LONG-CHAIN ALIPHATIC DIAMINES

C.M. PALEOS, D. TSIOURVAS AND A. MALLIARIS
NRC "Demokritos", Agia Paraskevi
Athens 15310, Greece

J. ANASTASSOPOULOU and T. THEOPHANIDES
National Technical University of Athens, Zografou Campus
Zografou 15773, Athens, Greece

ABSTRACT. The reaction of copper nitrate with long-chain diamines leads to the formation of amphiphilic complexes of the general formula $[CuL_{2n}]$ $(NO_3)_{2n}$, where L=long chain aliphatic diamine. Also replacement of long-chain diamines with mixtures of amines and diamines produced analogous complexes. The d^9 electronic configuration of Cu(II) was confirmed by magnetic susceptibility measurements on powder samples, whereas the divalent nature of the central metal ion was proved by its electronic spectra. The structure and thermal stability of the copper complexes was investigated by FT-IR spectroscopy and mass spectrometry.

1. Introduction

Following our previous studies [1] on the liquid crystalline behavior of copper complexes bearing long-chain aliphatic amines, we have prepared similar complexes by the reaction of copper nitrate with long-chain diamines, in the hope of forming planar liquid crystalline polymeric structures. In these systems the exhibition of liquid crystalline behavior is associated with amphiphilicity [2,3], which is the driving force behind any form of molecular organization, both in the bulk (thermotropic liquid-crystalline materials) and in solution (micelles, vesicles, etc.).

It must be mentioned that amphiphilic character is not limited exclusively to typical surfactants [4-12], but instead, it is exhibited by all those molecules which bear distinct molecular segments, and are consequently able to segregate and form smectic, or discotic textures. In this context, organometallics such as phthalocyanine metal complexes functionalized with long aliphatic chains [13-15] and metal complexes with long-chain aliphatic carboxylates [16] or amines [1] belong to this class of compounds. In these complexes the metal ion and the complexing functional group constitute the polar segment, whereas the aliphatic chains play the role of the lipophilic moiety. The subject of metallomesogens has been recently extensively reviewed [17].

D. R. Salahub and N. Russo (eds.), Metal-Ligand Interactions: from Atoms, to Clusters, to Surfaces, 397–408.
© 1992 *Kluwer Academic Publishers. Printed in the Netherlands.*

2. Experimental

Preparation of copper complexes with diamines: 0.002 mol of copper nitrate [$Cu(NO_3)_23H_2O$], were dissolved in ethanol and to this solution 0.004 mol of decanediamine (or dodecanediamine) disolved in hot ethanol, were added. Immediate precipitation occurred and the material was filtered and dried over phosphorous pentoxide. Elemental analysis of the produced complexes of copper-decanediamine (Complex I) and copper-dodecanediamine (Complex II) agreed with the formula corresponding to two ligands per copper ion, i.e. (Cu_nL_{2n}). Thus, Analysis Calcd for: $CuC_{20}H_{48}N_6O_6$.11 1/2 H_2O (I): C, 42.95%, H, 9.18%, N, 15.02%. Found: C, 42.60%, H, 8.80%, N, 14.78%. Analysis Calcd for $CuC_{24}H_{56}N_6O_6$ (II): C, 49.00%, H, 9.60%, N, 14.28%. Found: C, 48.70%, H, 9.60%, N, 14.11%.

Preparation of copper complexes with mixtures of amines and diamines:

0.002 mol of copper nitrate were reacted as above with the following mixtures of dodecylamine and 1,12-dodecanediamine: a) 0.004 mol dodecylamine and 0.002 mol dodecanediamine (Complex obtained III). b) 0.006 mol dodecylamine and 0.001 mol dodecanediamine (Complex IV) and c) 0.002 mol dodecylamine and 0.003 mol dodecanediamine (Complex V).

Characterization Studies: Optical microscopy was performed on a hot-stage Reichert polarizing microscope (Thermopan). For the DSC studies a Perkin-Elmer DSC-4 in conjunction with a System-4 programmer at a scanning rate of 10°C/min was used. FT-IR spectra were recorded on a BOMEM MICHELSON-100 spectrometer in the region 4000-400 cm^{-1} with a resolution 4 cm^{-1}. Absorption spectra were recorded on a Varian-Cary 210 spectrophotometer. Mass spectrometric experiments were conducted on a Kratos MS-50 TCTA mass spectrometer with thioglycerol as solvent. Data were acquired by using a Kratos DS-55 data system. Room temperature magnetic susceptibilities were measured using a PAR-155 vibrating sample magnetometer.

3. Results and Discussion.

Theoretically, the reaction of copper (II) with bidentate long chain diamines could lead to the following square planar, idealized, structures (Scheme), all in line with the strong Jahn-Teller distortion of the d^9 configuration of Cu(II) [18].

(a) A polymeric network planar structure formed by Cu(II) and diamines as shown in Scheme, A.

(b) A polymeric ring-like structure with Cu(II) ions attached at the amino groups of four long-chain diamines (Scheme, B).

(c) A monomeric chelate structure formed by the looping of the long chain of an aliphatic diamine attaching both of its ends to the same Cu(II) ion (Scheme, C).

However, it should be pointed out that due to the flexibility of the aliphatic amines, a more complex structure corresponding to a crosslinked polymer, may be envisaged. This structure could correspond to an intermediate of the above mentioned structures. The near insolubility of these complexes in all common solvents, did not permit the determination

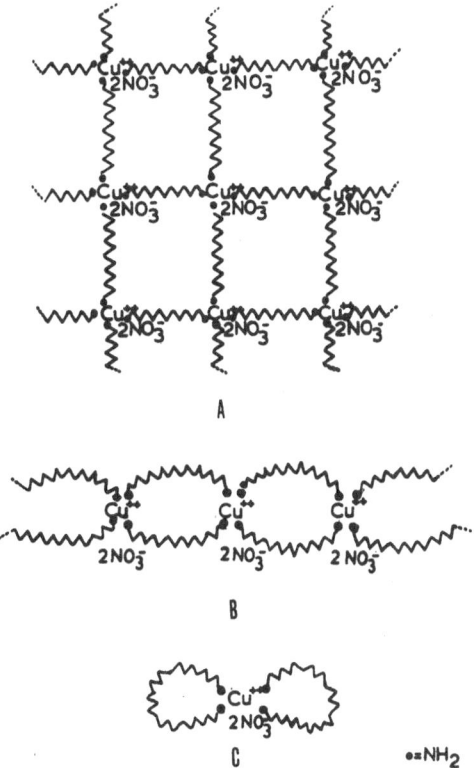

SCHEME

of their molecular weights. Therefore the distinction between the two polymeric structures (Scheme A, B) and the monomeric chelate one (Scheme, C) was not possible by these means. However, as will be discussed below, FT-IR and FAB mass spectra were employed to discriminate monomeric from polymeric structures.

The divalent nature of these copper complexes I and II was established by means of their powder magnetic susceptibilities. Thus, their effective magnetic moments, μ_{eff}, were found equal to ca. 1.8 BM, corresponding to a Cu(II) ion with a d^9 structure and only one unpaired electron. This was also confirmed by their solution absorption spectra (solvent absolute ethanol) which exhibit a broad peak around 600-700 nm [19].

When the materials were examined under crossed polarizers in the optical microscope, both I and II complexes melted in the range 110-112°C to anisotropic melts. As is always the case with amphiphilic liquid crystals, when pressure was exercised on the glass cover, melting was achieved at lower temperatures due to the melting of the aliphatic chains. Both complexes gradually decompose as it becomes evident by the yellowish tint and final darkening of the samples. Note, however, that complex I decomposes completely at about 140°C, whereas complex II decomposes at about 180°C. The differentiation as far as the decomposition temperature is concerned could be attributed to the higher strain existing in complex I compared to II because of the shorter length of its aliphatic chain. By using various mixtures of dodecylamine and dodecanediamine it was attempted to decrease the strain that existed in the complexes and to lower the melting point in order to obtain thermally stable liquid crystals. The results were not the expected ones and the materials decomposed as evidenced by optical microscopy and DSC as will be discussed below. On cooling, and before complete decomposition, textures resembling smectic phases were observed under crossed polarizers, as shown in Figs 1, 2 and 3 for complexes I, II and III, respectively. These textures were attributed to monomeric base units containing one copper ion and the chelated amines. Indeed, mass spectra of thermally treated materials showed the presence of ions containing one copper atom with at least three or four ligands, i.e.$[CuL_3]^+$ and $[CuL_4]^+$. These findings agree with the structures of copper (II) complexes with long chain monoamines, which have been shown to exhibit liquid crystalline behavior [1].

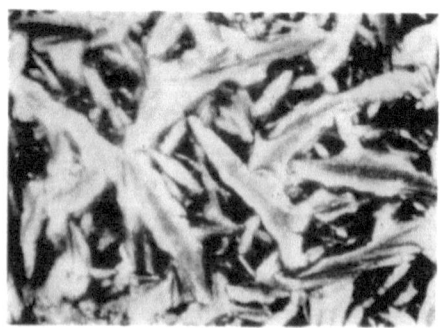

Figure 1. Liquid crystalline texture of complex I observed on cooling.

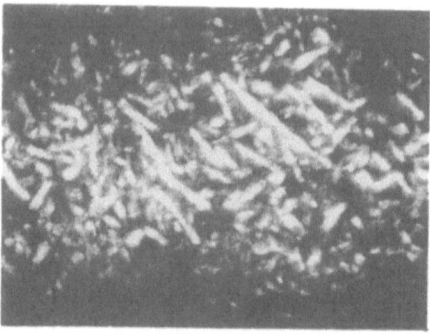

Figure 2. Liquid crystalline texture of complex I observed on cooling.

Figure 3. Liquid crystalline texture of complex III observed on cooling.

The DSC diagrams of these compounds are in line with the gradual decomposition of the complexes as shown in Figs. 4 and 5. Thus, complex II exhibits a weak and broad endothermic peak which is followed by the first exothermic decomposition peak. At even higher temperatures (ca. 180°C) a second exothermic peak appears associated with the complete decomposition of the material. In complex I, on the other hand, the decomposition starts immediately at the onset of the melting of the aliphatic chains, and therefore the endothermic peak of chain melting is not discernible in DSC. Nevertheless, the aliphatic chain melting is clearly observed under the microscope. Decomposition of complex I occurs at about 140°C. The melting behaviour of the complexes III-V is shown in Fig. 6. From these DSC traces it seems that these mixed complexes still decompose in a way corresponding to the molar percentage of mono and diamine. Indeed, the peak corresponding to either the monoamine (peak at lower temperature), or the diamine (peak at higher temperature) increases as the percentage of the corresponding amine in the mixture increases. In contrast, when only long-chain monoamines are used in the synthesis, the resulting monomeric copper complexes are thermally stable [1]. This proves the detrimental role of the diamines in inducing the thermal instability of these materials.

Survey Fourier transform infrared spectra of the ligand 1, 12-diaminododecane and of the complex II, in the solid state, in the region of 4000-400 cm^{-1} are shown in Fig. 7. In the region of νNH_2 absorption (3680-3260 cm^{-1}) are observed considerable changes upon complexation with the metal [20,21]. The δNH_2 bending region (1680-1550 cm^{-1}) shows for complex II an absorption at 1606.5 cm^{-1} assigned to complexed NH_2 bending vibration and a less prominent absorption at 1584 cm^{-1} attributed to the tail free NH_2 groups [20,21]. Spectral changes were also observed upon complexation in the region 1460-1300 cm^{-1}, where the CH_2 bending absorptions are expected to occur, and in the region 1020-720 cm^{-1} of $\nu(C-H)_n$ and $\nu(C-C)$ absorptions.

Comparison of close up spectra of complex II and IV shows a free amine stretching absorption (non-complexed) at 3494.7 cm^{-1} for complex II, which is not observed for complex IV (Fig. 8). This is most likely due to the presence of only 25% diamine in the mixed complex IV which results in a considerable reduction of the amount of free amine groups. In the case of mixed complex IV this absorption is very broad and structureless (Fig. 9). Thus, the presence of the free NH_2 absorptions rules out the monomeric chelate

402

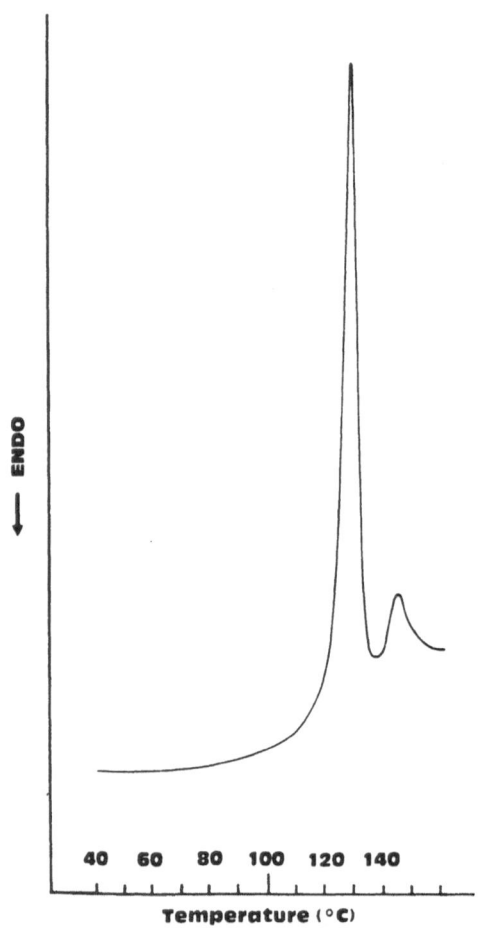

Figure 4. DSC diagram of complex I.

structure (Scheme, C) for the copper complex. This has also been confirmed by mass spectra since fragments of mass $[CuL_2]^+$ were not observed. On the contrary, there were observed fragments of mass $[CuL_3]^+$, $[CuL_4]^+$ and $[Cu_2L_2]^+$, etc. which indicate a polymeric network. However with FT-IR spectra it was not possible to decide between the two polymeric structures, or at least which was the predominant product of the reaction of Cu(II) with long chain diamines.

On heating the samples between 110-140°C, and then allowing them to cool down to room temperature, spectral changes indicative of conformational modifications in the -(CH$_2$) chain were observed. Comparison of the spectra after heating of complex I with those of the free ligands (Fig. 10) showed similarities with the free ligand which indicate

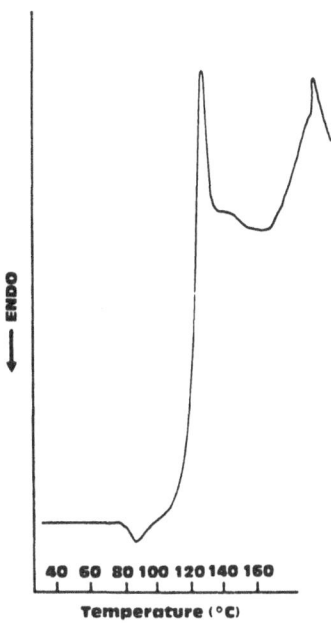

Figure 5. DSC diagram of complex II.

that the first exothermic transition, in both complexes, is due to the breakage of the Cu-N bond and the conformational freedom associated with it.

The main conclusion of the present work is that these novel materials of polymeric Cu(II) complexes with long chain aliphatic diamines as well as mixtures of monoamines and diamines, exhibit smectic textures, only on cooling provided the samples were not heated up to complete decomposition. These textures must be attributed to the monomeric base units containing one copper ion and the amine ligands. Evidently these units are produced by the partial decomposition of the polymeric materials, as the mass spectra indicate. It seems appropriate to assume that the thermal instability of these copper complexes may be attributed to the tensions created in the structure due to the bridging of Cu^{++} with diamines.

404

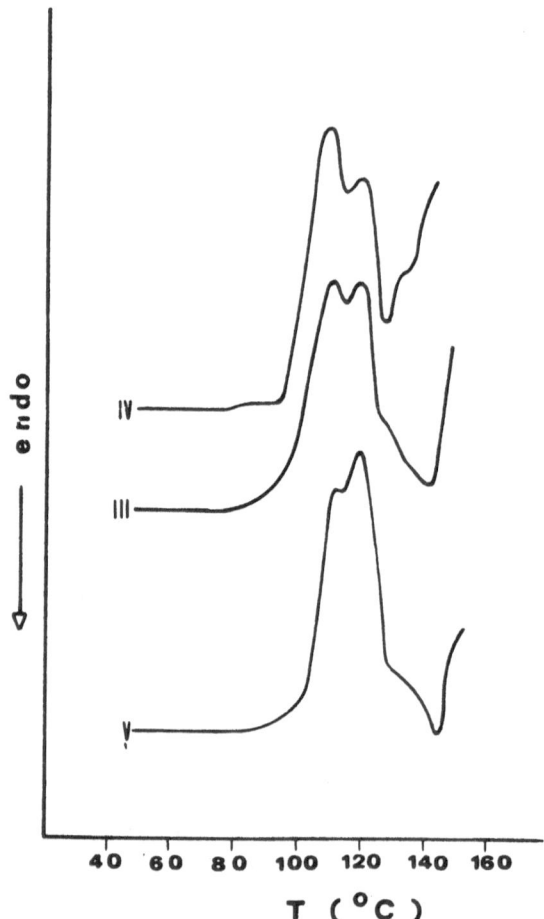

Figure 6. DSC diagram of complex III-V.

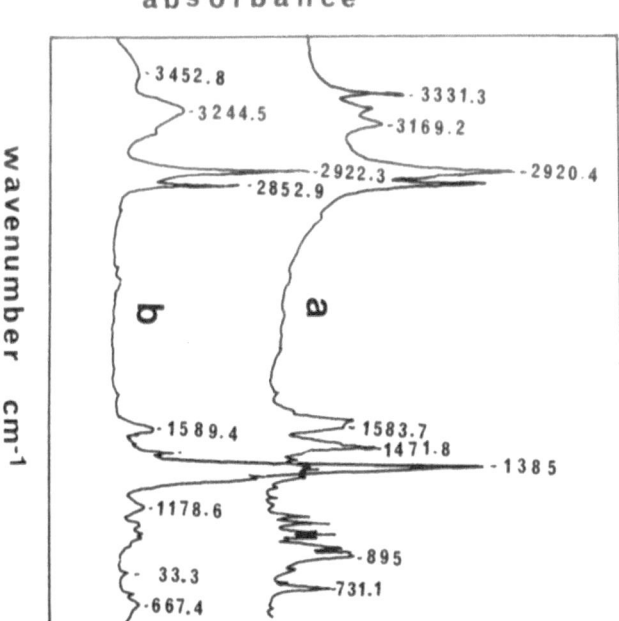

Figure 7. FT-IR spectra of: a) 1,12-dodecanediamine and b) the complex of Cu-1,12 dodecylamine in the region 400-4000 cm⁻¹.

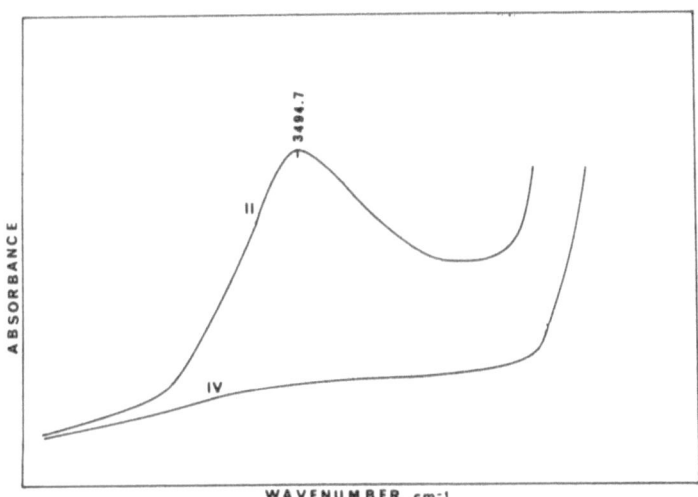

Figure 8. Comparison of close up FT-IR spectra of complex II and mixed complex IV in the region of N-H stretching.

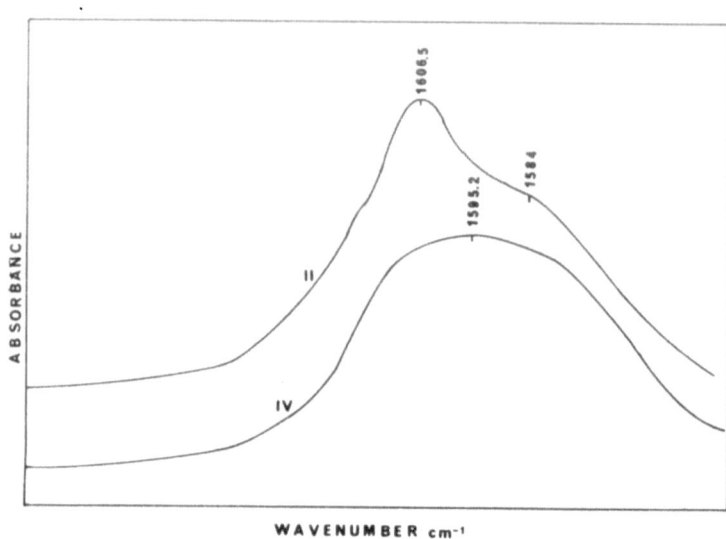

Figure 9. Comparison of close up spectra of complex II and IV in the region of NH$_2$ bending frequency.

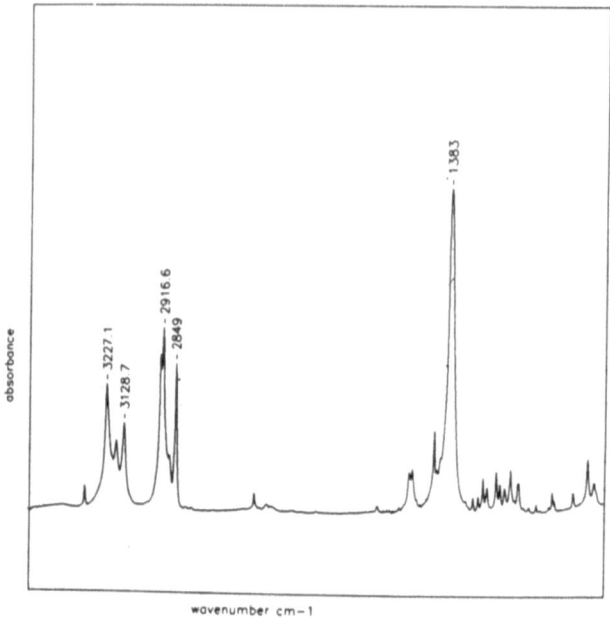

Figure 10. FT-IR spectra of complex I after heat treatment.

REFERENCES

1. Paleos, C.M., Margomenou-Leonidopoulou, G., Anastassopoulou, J. and Papaconstantinou, E. (1988), Mol. Cryst. Liq. Cryst., **161**, 373.
2. Kelker, H. and Hatz, R. (1980), Verlag Chemie (ed.), Weinheim, and references cited therein.
3. Skoulios, A., Guillon, D. (1988), Mol. Cryst. Liq. Cryst., **165**, 317.
4. Bruce, D.W., Dunmur, D.A., Lalinde, E., Maitlis, P.M. and Styring, P. (1986), Nature, **323**, 791.
5. Iwamoto, K., Ohnuki, K., Sawada, K. and Seno, M. (1981), Mol. Cryst. Liq. Cryst., **73**, 95.
6. Paleos, C.M., Margomenou-Leonidopoulou, G. and Malliaris, A. (1985), Chimica Chronica, New Series, **14**, 89.
7. Malliaris, A., Christias, C., Margomenou-Leonidopoulou, G. and Paleos, C.M. (1982), Mol. Cryst. Liq. Cryst., **82**, 161.
8. Margomenou-Leonidopoulou, G., Malliaris, A. and Paleos, C.M. (1985), Thermochimica Acta, **85**, 157.
9. Paleos, C.M., Margomenou-Leonidopoulou, G., Babilis, D. and Christias, C. (1987), Mol. Cryst. Liq. Cryst., **146**, 121.
10. Skoulios, A. and Luzzati, V. (1959), Nature, **183**, 1310.
11. Duruz, J.J., Ubbelohde, A.R. (1972), Proc. R. Soc. Lond., **330**, 1.
12. Malliaris, A., Paleos, C.M. and Dais, P. (1987), J. Phys. Chem., **91**, 1149.
13. Pierocki, C., Boulou, J.C. and Simon, J. (1987), Mol. Cryst. Liq. Cryst., **149**, 115.
14. Guillon, D., Weber, P., Skoulios, A., Piechocki, P. and Simon, J. (1985), Mol. Cryst. Liq. Cryst., **130**, 223.
15. Andre, J.J., Bernard, M., Pierocki, C. and Simon, J. (1986), J. Phys. Chem., **90**, 1327.
16. Amorim Da Costa, A.M., Burrows, H.D., Geraldes, C.F.G.C., Teixeira-Dias, J.J.C., Bazuin, C.G., Guillon, D., Skoulios, A., Blackmore, E., Tiddy, G.J.T. and Turner, D.L. (1986), Liquid Crystals, **1**, 215.
17. Giroud-Godquin, A.M. and Maitlis, P.M. (1991), Angew Chem. Int. Ed Engl., **30**, 375.
18. Cotton, F.A. and Wilkinson, G. (1967), Advanced Inorganic Chemistry, Interscience Publishers, New York, p. 898.
19. Hathaway, B.J. and Tomlinson, A.A.G.(1970), Coordination Chem. Rev., **5**, 1.
20. Levin, I. (1984), Adv. Infrared and Raman Spectra., **11**, 1.
21. Anastassopoulou, J., Paleos, C.M., Theophanides, T., Behnam, V. and Bertrand, M. (1989) in Bal, W and Jezierski, A. (eds.), Proceedings of "Second Symposium on Inorganic Biochemistry and Molecular Biophysics", Wroclaw, p. 13.

INDEX

410

412

421

424

426